Protecting the Ozone Layer

Protecting the Ozone Layer

Science and Strategy

Edward A. Parson

OXFORD
UNIVERSITY PRESS
2003

OXFORD
UNIVERSITY PRESS

Oxford   New York
Auckland   Bangkok   Buenos Aires   Cape Town   Chennai
Dar es Salaam   Delhi   Hong Kong   Istanbul   Karachi   Kolkata
Kuala Lumpur   Madrid   Melbourne   Mexico City   Mumbai
Nairobi   São Paulo   Shanghai   Taipei   Tokyo   Toronto

Library of Congress Cataloging-in-Publication Data
Parson, Edward.
Protecting the ozone layer : science and strategy / by Edward A. Parson.
p. cm.
Includes bibliographical references and index.
ISBN 0-19-515549-1
1. Ozone layer depletion—Prevention—History—20th century.   I. Title.
QC879.7.P37 2003
363.738'757—dc21   2002066256

1 3 5 7 9 8 6 4 2

Printed in the United States of America
on recycled, acid-free paper

To the memory and inspiration of Abram Chayes

Preface

Over the past few decades, the possibility of global environmental limits to human civilization has grown increasingly evident. Our encounter with these limits is occurring slowly but can be seen in the growing number of planetary processes that human activities are disrupting. Changes under way in the atmosphere, oceans, land surface, and distribution and rate of extinction of living species all show a clear and increasing human footprint. Although the reality and potential seriousness of these changes cannot seriously be disputed, large uncertainties remain about how they will develop and what threat they pose to human welfare. Faced with such uncertainty, people disagree—sincerely or opportunistically—over how much precaution to take in responding to the changes.

In addition, managing global environmental stresses poses special difficulties because of their scale and their long-term effects. Global-scale issues must be managed internationally, where institutional capacity is weak and each nation would rather let others bear the burden. Long-term issues rarely appear urgent, so more immediate priorities dominate policy agendas. We are tempted to let our descendants—whom we imagine will be richer and know more—take responsibility. In the face of these difficulties, attempts to manage global environmental risks have thus far been strikingly ineffective, with one notable exception.

This book is about that exception: protection of the stratospheric ozone layer. It challenges the orthodoxy of how it was achieved and argues that we can learn from it and apply to other environmental issues much more than has been recognized. The threat to ozone presents the same features—uncertainty, scale, and slow dynamics—that make all global environmental issues so hard to manage. But its management has been a striking success. With near-universal participation of nations and energetic support from industry, the ozone regime has reduced worldwide use of ozone-depleting chemicals by 95 percent, and use is still falling. Critically, this was achieved not by the initial chemical controls but by their rapid subsequent adaptation in parallel with a flood of innovations to reduce use of the chemicals. In this respect, the ozone regime is the first realization of the widely proposed goal of *adaptive management*—policies and institutions that promote learning about the systems being managed and that adapt in response to what is learned—for any global environmental issue.

The importance of this achievement has not been sufficiently recognized. Many writers have examined the ozone regime, especially in its early years. Initially, observers enthusiastically claimed that it represented a transformation of international politics. Subsequently, however, a revised consensus developed that the ozone issue was so benign that its successful management was more or less preordained. If it is true that ozone was a uniquely easy case, then its successful management can offer little in the way of general insights or lessons for other, more difficult issues.

I began this project concerned that the ozone issue was not adequately documented to support either of these broad claims. I sought to provide an improved record that covered the entire duration of the issue—early frustrations as well as later successes—and that documented scientific and technological developments in enough detail to reveal their interactions with policy and negotiations. Told with this greater scope and detail, the ozone story offers important new insights into regime formation, negotiation strategy, and how scientific knowledge can help shape policy outcomes. It also shows how interactions between treaty revisions, corporate strategy, and technology assessment promoted the rapid innovation that was central to the success of the regime.

Most important, the ozone history shows that seemingly intractable global environmental problems can be successfully managed and offers practical lessons in how to do so. Although each issue is in some respects unique, the success of the ozone regime came in part from processes of general applicability. This means its specific lessons about regime formation, negotiation strategy, and scientific and technical assessment may apply to other issues where conditions are sufficiently similar. Global climate change presents an obvious example. Despite negotiation of two international treaties since 1992, attempts at concrete action on climate change remain stalled. Specific elements of the way adaptive management was achieved on ozone, suitably adapted, hold the best prospect of easing this deadlock. They may also offer guidance on how to manage the additional, linked global-change issues that will appear with increasing frequency and intensity over the coming decades.

Acknowledgments

In the several years it has taken to complete this project, I have accumulated so many debts that my fallible memory will likely omit some, for which I apologize. My first debt is to my interviewees, some of whom spent many hours instructing and correcting me. Of these, Victor Buxton, Alex Chisholm, Robert Hornung, Alan Miller, and Peter Thacher provided complete access to their ozone files, and Daniel Albritton shared his slides from more than ten years of scientific briefings. In addition, six other scholars—Elizabeth de Sombre, Michael Glantz, Peter Haas, Joanne Kauffman, Marc Levy, and Karen Litfin—generously shared interviews and archival material they had gathered.

The project benefited from research assistance by Pierre Blime, Claire Broido, William Dietrich, Sofia Doloutskaia, Michele Ferenz, Gökçe Gazelle, Sydney Rosen, Paul Steinberg, Jenny Stephens, and Caitlin Willoughby; from peerless reference support and referrals by Tom Parris, at the time Harvard's Environmental Resources Librarian; and from editorial and administrative assistance by Susan Hassol, Kate Regnier, Rebecca Storo, and Michelle von Euw. Valuable comments and criticisms on part or all of the project were provided by Stephen Andersen, Jim Anderson, David Ballon, Rosina Bierbaum, Harvey Brooks, Abram Chayes, Antonia Handler Chayes, Bill Clark, Barbara Connolly, David Cutler, Andy Dessler, Neil Donahue, Bob Frosch, Owen Greene, Peter Haas, Jim Hammitt, Jill Horwitz, David Keith, Robert Keohane, Marc Levy, Michael McElroy, Mack McFarland, Ron Mitchell, Michael Oppenheimer, Granger Morgan, Stephen Schneider, Gene Skolnikoff, Rob Socolow, Oran Young, Richard Zeckhauser, and three anonymous referees, as well as seminar participants at Harvard's Center for International Affairs and Atmospheric Research Project; the International Institute for Applied Systems Analysis; workshops for Harvard's Global Environmental Assessment project at the College of the Atlantic and at Airlie House, Virginia; and at the Department of Engineering and Public Policy, Carnegie Mellon University. I received financial support for the project from the Rockefeller Brothers' Fund, the Harvard Committee on the Environment under the V. Kann Rasmussen gift, the U.S. Department of Energy under Award no. DE-FG-02-ER-61941, and the Kennedy School Dean's Research Fund.

I gratefully acknowledge all this help I have received, but three contributions merit special mention. Although its topic lay some distance from his main interests, David Cutler read the entire manuscript in early draft and offered cogent criticisms that went to the heart of its weaknesses. Kristi Olson provided a close, astute review of the complete manuscript. Most of all, my wife, Jill Horwitz, provided support in many forms and large quantities. As intellectual partner, she has been my most determined supporter and most challenging critic throughout the project. As life partner, she has borne an unequal burden in keeping life and home together, and together with generous contributions from David and Susan Horwitz, in caring for our son Matthew. For his part, Matthew reminds me daily of the responsibility we hold to the future. Thank you all.

Contents

Abbreviations

ACS	American Chemical Society
AFEAS	Alternative Fluorocarbon Environmental Acceptability Study
ALE	Atmospheric Lifetimes Experiment
ANPR	advance notice of proposed rulemaking
ASHRAE	American Society of Heating, Refrigeration, and Air Conditioning Engineers
BAS	British Antarctic Survey
BCM	bromochloromethane
BUV	backscatter ultraviolet
CAA	Clean Air Act
CARCE	Committee on Alternatives for the Reduction of Chlorofluorocarbon Emissions
CBM	chlorobromomethane
CCOL	Coordinating Committee on the Ozone Layer
CEFIC	European Federation of Chemical Industries
CEQ	Council on Environmental Quality
CFC	chlorofluorocarbon
CIAP	Climatic Impacts Assessment Program
CISC	Committee on Impacts of Stratospheric Change
CMA	Chemical Manufacturers Association
COAS	Council on Atmospheric Sciences
CPSC	Consumer Products Safety Commission
CT	carbon tetrachloride
DOE	Department of Energy
DPC	Domestic Policy Council
DU	Dobson unit
EC	European Community
EDF	Environmental Defense Fund
EFTC	European Fluorocarbon Technical Committee

EPA	Environmental Protection Agency
FAA	Federal Aviation Administration
FDA	Food and Drug Administration
FOE	Friends of the Earth
FPP	Fluorocarbon Program Panel
FSPI	food service packaging industry
GEF	Global Environment Facility
GWP	global-warming potential
HBFC	hydrobromofluorocarbon
HCFC	hydrochlorofluorocarbon
HFC	hydrofluorocarbon
HOx	hydrogen oxides (HO and HO_2)
HSIA	Halogenated Solvents Industry Alliance
ICI	Imperial Chemicals Industries
ICOLP	Industry Cooperative for Ozone Layer Protection
ICSU	International Council of Scientific Unions
IMOS	Federal Task Force on Inadvertent Modification of the Stratosphere
IPCC	Inter-governmental Panel on Climate Change
km	kilometer
Kte	kilotonne
MC	methyl chloroform
MDI	metered-dose inhalant
MeBr	methyl bromide
mm	millimeter
Mte	megatonne
NAS	National Academy of Sciences
NASA	National Aeronautics and Space Administration
NFPA	National Fire Protection Association
NGO	nongovernment organization
nm	nanometer
NMSC	nonmelanoma skin cancer
NOAA	National Oceanic and Atmospheric Administration
NOx	nitrogen oxides (NO and NO_2)
NPB	n-propyl bromide
NRDC	National Resources Defense Council
NSF	National Science Foundation
ODP	ozone depletion potential
OECD	Organization for Economic Cooperation and Development
OEWG	Open-ended Working Group
OMB	Office of Management and Budget
OTP	Ozone Trends Panel
PAFT	Program for Alternative Fluorocarbon Toxicity Testing
PFC	perfluorocarbon
ppb	parts per billion
ppm	parts per million
ppt	parts per trillion
PSC	polar stratospheric cloud
SAGE	Stratospheric Aerosols and Gases Experiment
SBUV	solar backscatter ultraviolet
SNAP	Significant New Alternatives Program

SST	supersonic transport
STRAC	Stratospheric Research Advisory Committee
TEAP	Technology and Economics Assessment Panel
TOC	Technical Options Committee
TOMS	total ozone mapping spectrometer
TSCA	Toxic Substances Control Act
USDA	U.S. Department of Agriculture
UKDOE	U.K. Department of the Environment
UNDP	UN Development Programme
UNEP	UN Environment Programme
UNIDO	UN Industrial Development Organization
UV	ultraviolet
WMO	World Meteorological Organization
WRI	World Resources Institute

Protecting the Ozone Layer

1

Stratospheric Ozone and Its Protection

Introduction and Background

1.1 Protecting the Ozone Layer

Since 1985, an extraordinary reorganization of a major industrial sector has taken place to protect the global environment. Some several dozen highly useful and versatile industrial chemicals, which were produced in volumes of millions of tons and used to produce goods and services and operate equipment worth hundreds of billions of dollars, have been virtually eliminated. They have been replaced by a host of new chemicals, other technologies, and changes in manufacturing processes that not only are less environmentally harmful but also are in aggregate safer and cheaper, and perform better. The chemicals that have been so sharply reduced are a family of simple molecules derived from the smallest hydrocarbons, methane and ethane, in which atoms of halogens—fluorine, chlorine, and bromine—replace all the hydrogen atoms. The most important of these chemicals are the chlorofluoro-carbons, or CFCs. They were abandoned in order to protect against depletion of the ozone layer.

The ozone layer is located about 12 to 25 kilometers (8 to 15 miles) above the earth's surface, in the stratosphere. It protects life on earth by screening out most of the damaging, high-energy ultraviolet (UV) radiation in sunlight. Beginning in the mid- to late 1960s, scientists started suspecting that pollutants from several kinds of human activities risked disrupting the ozone layer, increasing the intensity of UV radiation reaching the earth's surface. Very little was known about the likely effects of such an increase in radiation, but many possible harms were identified for ecosystems and human health, some of them of the utmost seriousness.

When two scientists in 1974 identified the CFCs as the most serious threat to ozone, a few countries, including the United States, decided to restrict the largest use of the chemicals, as propellants for aerosol spray cans, over the next few years. But it was immediately obvious that the problem was of global scale, so worldwide reductions of the chemicals, not national ones, were needed to solve it. Attempts to develop international cooperation to manage the problem began at nearly the same time as these national policy debates, but the first decade of these attempts ended in repeated failure. This long-standing deadlock ended suddenly in late 1986, when new international negotiations rapidly yielded the 1987 Montreal Protocol with its agreement to cut CFCs by half, the first concrete international measure to control ozone-depleting chemicals. Five subsequent revisions of the Protocol have repeatedly made these chemical controls stricter, advanced their dates, and broadened them to include additional ozone-depleting chemicals. The international measures now in place to protect the ozone layer do not comprise just the Montreal Protocol, but include a collection of linked agreements, institutions, procedures, resources, expectations, and flows of information that make up the ozone regime.

The ozone regime is the most conspicuous success yet achieved in protecting any aspect of the global environment. Its success is evident in several ways. Worldwide production and use of ozone-depleting chemicals has fallen 95 percent from its peak in the late 1980s and continues to decline, and this reduction has been achieved at modest cost. The preliminary signs of recovery of the stratosphere are discernible. The regime enjoys nearly universal global participation, and has directed substantial financial and technological assistance to support developing countries in making the required transition away from ozone-depleting chemicals. The regime made a number of significant innovations in institutions, policy, and negotiations, and many of its features have been either imitated or widely proposed for imitation as the best way to approach other global environmental problems. While these all represent significant dimensions of success, the regime's most important achievements lie in two characteristics that are widely recognized as essential for effective management of the global environment, but have yet been realized nowhere else. First, the typical polarization of environmental issues, with activists demonizing industry as rapacious despoilers and industry grudgingly accepting restrictions while fighting every step of the way, has been almost entirely avoided. The regime has succeeded in engaging the energy, creativity, and enthusiastic support of private firms in finding and implementing ways to reduce ozone loss, truly turning industry from part of the problem into a key partner in the solution. Second, the ozone regime is the only international environmental regime to successfully implement the principle of adaptive management: it has incorporated procedures and institutions to monitor and assess changes in scientific knowledge, technological possibilities, and economic conditions, and to repeatedly adapt its central provisions in response to these changes.

1.2 Goals, Context, and Contributions of the Book

This book tells how the present ozone regime was created, how it evolved, and how it came to be so successful after many years of prior failed attempts to develop international cooperation to manage the issue. It provides a thorough historical account of the development of the issue from the 1960s to the late 1990s. In addition, it seeks to identify the explanatory factors that account for the regime's formation, its major features, its adaptation over time, and its effectiveness. In so doing, it proposes both theoretical arguments of potentially general applicability and practical lessons that the ozone regime may hold for the management of other global issues.

This is not the first history of the ozone issue. Because the success of the international ozone regime is widely recognized, its history has attracted widespread attention. Existing treatments of the issue have included several accounts by participants at various stages in the regime's development, of which the most important is Richard Benedick's detailed record of international negotiations and U.S. domestic politics from 1985 to 1990.[1] Several accounts of specific periods of the issue's development have also been published by those who followed the issue as journalists or popular science writers.[2] In addition, several previous scholarly treatments have examined specific aspects of the ozone regime and used it to develop theoretical arguments about the determinants of international cooperation, the effectiveness of international institutions and regimes, and the relationship between scientific knowledge and policy action.[3]

Relative to these prior accounts, this book makes two major empirical contributions. First, it considers the complete lifetime of stratospheric ozone as a policy issue, from the first domestic controversies and initial attempts to develop international action of the late 1960s and 1970s, through the mature functioning of the international regime in the late 1990s. Second, it engages in sufficient substantive detail each of the three major domains of the issue's history: politics, science, and technology. At one level, the ozone issue was a political problem, which was managed through extended multilateral international negotiations and the coordination of these negotiations with domestic politics in dozens of countries and with transnational issue networks representing the interests of industry, environmentalists, and others. But the issue came onto policy agendas because scientists had identified a potential basis for concern, and arguments about whether scientific evidence of risk was strong enough to proceed with potentially costly regulatory controls were consistently among the most prominent features of policy debates, both domestically and internationally. For its part, the state of scientific knowledge resisted being reduced to such a one-dimensional scale of concern but rather was complex, multidimensional, contested, uncertain, and changing, with degrees of scientific consensus on specific points increasing and sometimes decreasing, and the identity of the most important or controversial points continually shifting over time.

The severity and character of the ozone issue as a political problem also depended acutely on the perceived availability and cost of alternatives to the offending chemicals. Like scientific knowledge of risk, technological knowledge of alternatives was uncertain, contested, and changing, but the significance of this knowledge also depended crucially on who held it and how widely it was shared. Arguments about the availability and cost of alternatives, and control over knowledge about them, were crucial to policy debates and outcomes. Understanding the formation and evolution of the ozone regime depends crucially on understanding the interactions between these domains of politics, science, and technology. This book is the first account of the issue to engage all three domains with enough substantive detail to illuminate these interactions. The book does not consider domestic implementation of international ozone agreements, except where these strongly influenced the development of the international regime.

The ozone regime makes for good storytelling, but its success and the means by which it was achieved are also of great importance, for both practical and theoretical reasons. As a practical matter, history amply attests that it is difficult to achieve international cooperation, even when there are compelling collective interests to be advanced. The environment is only one domain in which such cooperation has proven difficult. Similarly ineffective international cooperation is evident on issues as diverse as trade, development, economic policy, security, public health, and combating drug trade and terrorism. The sad state of international efforts on other global environmental issues—most notably climate change, but also protection of biodiversity; disruption of biogeochemical cycles of nitrogen, sulfur, and phosphorus; control of persistent organic pollutants; and the broader environment-development agenda left neglected since the early 1990s—underscores the unique accomplishments of the ozone regime. Since ozone depletion is one of the most serious global environment risks recognized to date, and the only one to be managed with anything approaching success, the regime compels attention. If there are specific aspects of its management that persuasively account for its successes, these may hold lessons for other global issues. If its management has exploited opportunities that arose from specific characteristics of the issue, or achieved success by mitigating the obstructive effects of other specific characteristics, it might hold lessons applicable to other issues that share these specific characteristics.

The ozone issue is also important theoretically, for it offers important new insights into four long-standing scholarly questions. Each of these questions is rooted principally in a separate line of scholarship, but they are all also of substantial broader importance and generality. First, how do international regimes form and evolve? An extensive body of work in international relations has examined the factors accounting for the initial formation of issue-specific international regimes, the stages of their formation, the processes by which they form, and the determinants of their subsequent evolution and effectiveness.[4] The history of the ozone issue, in which a two-year period of rapid regime formation followed ten years during which the issue was on international agendas but deadlocked, provides an uncommonly sharp empirical record to elaborate and test hypotheses on the conditions and processes of regime formation. In addition, the regime's subsequent evolution provides an exceptional example of rapid regime adaptation and increasing effectiveness in solving the targeted problem. The details of this evolution illustrate specific adaptive processes, not previously identified, through which the regime incorporated and helped generate new scientific and technical information.

Second, what role do negotiations play in determining international policy outcomes? Negotiations between and within organizations are ubiquitous in international policy making, but researchers disagree over how much influence negotiations actually exercise over important policy outcomes. The study of negotiations stresses the interactions between external and internal factors in accounting for outcomes: external factors such as parties' power and preferences determine their options in the absence of negotiated agreement, and interact with internal factors such as bargaining tactics and processes of argument, persuasion, and learning in shaping negotiated agreements.[5] The major lines of international relations theory have placed predominant importance on external factors such as material power (particularly when interests are largely opposed), the structure of the international system, and the institutional context for decision-making (particularly in the context of collective-action problems), which operate by fixing parties' alternatives to negotiated agreement.[6] Recent empirical studies, however, have increasingly found important influences from processes of bargaining, argument, information exchange, and learning that take place within negotiations, especially for issues involving multiple actors, mixtures of opposed and shared interests, uncertainty about the consequences of decisions, long time horizons, and a focus on institution-building rather than immediate action.[7]

These characteristics are typical of global environmental issues, and the detailed record of ozone negotiations allows the influence of internal processes of bargaining, argument, learning, and process control to be traced quite precisely. In particular, the record highlights two aspects of negotiations: the tactical interplay between international negotiations and domestic policy; and the role of arguments based on scientific or technological claims in shaping the feasible bargaining range and supporting or weakening specific bargaining positions within it.[8]

Third, how does—and can—scientific knowledge influence policy? Although simple rationalistic claims that scientific knowledge either can or should fully determine policy choices have been widely rejected, the actual extent and mechanisms of influence of scientific claims and knowledge on policy remain only weakly understood. Recent studies have stressed negotiations between political and expert institutions over where to draw the boundary between their domains of authority, and the role of various intermediary bodies or aggregate social structures of scientists and officials in sustaining this boundary, conveying scientific knowledge credibly, and articulating its policy implications.[9]

The ozone issue has been striking throughout its history for the prominence of scientific arguments in policy debates. In grappling with whether, when, and how to take regulatory action under uncertainty, policy actors have consistently framed their arguments as disagreements over the state of scientific knowledge or over whether available scientific evidence is sufficient to justify action. While both advocates and opponents of action have predictably engaged in the selective appropriation of scientific claims favorable to their case, the issue also has shown more subtle interplay between scientific knowledge and policy debate, of several forms. Most strikingly, the ozone issue highlighted the crucial policy influence of official scientific assessments, as distinct from scientific results themselves, and the mechanisms and conditions for the exercise of such influence.

Finally, how does environmentally significant technological change interact with policy? The rate and character of technological change are among the strongest determinants of the environmental impact of human activities and the effectiveness and cost of measures to reduce impacts. Extensive bodies of both theoretical and empirical research have studied the relative contributions to technological change of autonomous processes, changes in input prices, and specific types of policy.[10] An area of especially sharp controversy has been by how much environmental regulations can induce innovations that reduce the cost of meeting the regulations or, in the extreme, even increase profitability.[11] The rapid development and adoption of technologies to reduce ozone-depleting chemicals was essential to the successful adaptation of the ozone regime. This development was driven by interactions between the regime's regulatory targets and processes for technology assessment, and the strategic responses of industry.

1.3 The Major Arguments in Brief

The bulk of the book is devoted to the empirical account of the ozone issue's history, but the expanded empirical account provides new leverage to develop theoretical arguments of potential importance and generality, in two ways. First, the longer history of the issue developed here, from the long initial period of deadlock through later periods of rapid regime formation and subsequent evolution, provides a richly instructive record of variation in actors, strategies, knowledge, institutional settings, and outcomes that has not previously been available. This expanded record and increased variation allow sharper and better-founded explanations of the origin of the regime than have previously been possible. Second, the expanded treatment of scientific and technological developments in addition to politics and negotiations allows detailed tracing of the processes by which these domains interacted in shaping important policy outcomes. The most important of these processes consist of interactions between knowledge-based and strategic factors, mediated by the set of rhetorical claims available to policy actors to advance and support arguments for their preferred policy positions. The book's three principal theoretical arguments are summarized here, threaded throughout the empirical account, and developed in detail in the final chapter.

The first argument concerns the factors accounting for the rapid formation of the ozone regime between 1986 and 1988 after ten years of deadlock. These factors are best understood within a negotiations framework that distinguishes the establishment of a feasible bargaining range from the determination of a specific outcome within that range. The most important factor accounting for the transition was the availability of an authoritative assessment of relevant atmospheric science, which caused several strong opponents of chemical controls to revise their positions and thereby established a broad new policy consensus supporting modest international controls. A second assessment two years later had a similarly strong effect in consolidating the end of the period of rapid regime formation, and shifting the policy consensus toward complete elimination of the chemicals. The effect of these assessments was not independent of the underlying scientific knowledge that they synthesized for policy makers, but was analytically distinct from it, in that the crucial changes accounting for regime formation were in the assessments, not in the science.

Additional factors of apparent significance in formation of the regime include the gradual accumulation of institutional support for international negotiations, which progressively improved the context for bargaining, and the availability of a negotiating proposal meeting certain conditions for viability, although the evidence for the influence of these is weaker than for the assessments. In addition, while international leadership, by either a single hegemonic nation or a coalition, were not important causes of the regime's formation, a narrower form of leadership by a small faction of activist U.S. officials strongly shaped the specific terms of the initial agreement. Aggressive bargaining tactics by this group, together with a plausible congressional threat of unilateral U.S. action if negotiations failed, were primarily responsible for the agreement to cut the chemicals by half rather than freezing them or making small cuts. In explaining the major terms of the 1987 agreement, the assessment established a policy consensus whose effect was to delimit a feasible bargaining range, while actors' negotiating tactics determined the specific outcome realized within the range.

In contrast to other accounts, I argue that the regime's formation cannot be explained by a breakthrough in development of CFC alternatives and consequent shifts of industry interests. Nor is it explained by any decisive change in scientific knowledge, consensus, or concern about ozone depletion. In accounting for the formation of the ozone regime, even the importance of the 1985 reporting of the Antarctic ozone hole has been substantially overestimated.

The second principal argument concerns the strategy of conducting scientific assessment that allowed the two crucial assessments of 1986 and 1988 to achieve such decisive influence, in contrast with many prior assessments that did not, and the detailed processes by which they exercised their influence. I argue that an effective scientific assessment must be authoritative, in the sense that it makes certain scientific propositions and their policy implications salient enough that policy actors cannot ignore them, credible enough that they cannot arbitrarily deny them, and legitimate enough that they cannot dismiss them as partisan without engaging their substance. Assessments achieve this authoritative status by engaging the participation of a critical mass of the most respected experts on the issue, thereby aligning this group on a single statement of scientific knowledge and its implications, and ensuring that no policy actor can mount a competing assessment in the hope of exploiting differences in conclusions, interpretation, or nuance of language. An assessment that succeeds at this strategy can influence policy outcomes in two ways: by authoritatively resolving specific scientific questions that policy actors have previously identified (however arbitrarily) as decisive determinants of action; or by removing the putative scientific justification for certain policy proposals, thereby making certain positions indefensible and delimiting the range of feasible negotiating outcomes. In each case, the crucial contribution of the authoritative assessment is to make key scientific statements and their implications common knowledge among policy actors, in the game-theoretic sense that all parties know them, all know that all know them, and so on.[12]

The third principal argument concerns the effects of technological options to reduce emissions of the offending chemicals, and the distribution of knowledge about the availability, cost, and other characteristics of such options among policy actors. Although not implicated in the formation of the regime, these factors were crucial determinants of both the long preregime deadlock and the rapid, sustained regime adaptation subsequently. Before regime formation, authoritative technical knowledge about alternatives was narrowly controlled within firms, principally by the CFC manufacturers, who had no interest in sharing it with environmental activists or regulators who wished to use it to compel the firms to change their technologies. Attempts to conduct assessments of technical options were unable to breach this narrow control of knowledge, so it was widely believed that significant cuts in ozone-depleting chemicals would be extremely difficult and costly, and likely dangerous as well. Although activists believed this view to be mistaken, they could never make the case persuasively in policy debates. Absent scientific evidence of a grave, imminent environmental risk, significant controls on the chemicals could never gain serious consideration.

All this changed after formation of the regime. No one knew whether the 50 percent cut adopted in 1987 could be achieved at acceptable cost in the countries that had already eliminated CFC aerosol propellants, the only easy use to cut. With this target already agreed upon and calls spreading for even stricter ones, firms using controlled chemicals had strong interests in reducing their use as much and as fast as possible. Technology assessment processes established under the ozone regime let industry experts collaborate to identify and develop alternatives, solving an urgent business problem that their firms jointly faced. But the same process of developing and evaluating technical options served the regime by providing technical advice of quality never previously available about further feasible reduction opportunities and by continually advancing the margin of reductions that were feasible. Interactions between industry's efforts to identify ways to reduce use and the successive adaptation of regulatory controls under the Protocol, mediated by technology assessment processes, drove a rapid process of technological innovation and diffusion that allowed the regime to move toward complete elimination of the offending chemicals with rapidity, relative ease, and many ancillary benefits that not even the most optimistic could have predicted in 1987. The essence of these interactions lay in the regime's technology assessment processes, which were designed so that the generation and sharing of private technical information jointly provided public benefits to the regime and private benefits to the participating experts and their firms. This was innovation of great novelty, and of both theoretical and practical importance.[13]

In addition to these theoretical arguments, the book articulates several practical lessons that the ozone regime offers for management of other global issues sharing appropriate characteristics. Some of these lessons are closely related to the theoretical arguments discussed above, in that they are conclusions about causal influences on regime formation and development, reframed as strategic guidance to actors seeking to promote these processes. Others concern more prosaic points that are persuasively established in the record and of substantial practical importance, but are of less general theoretical interest. The principal lessons are as follows:

1. For a regime to be able to adapt under changing knowledge and capabilities, its basic architecture must embed provisions to learn about such changes and to modify policies and measures in response to them.

2. Initial quantitative regulatory targets can be crucial in forming an adaptive regime, by setting subsequent adaptive processes in motion, but the precise level of the early targets matters little. They must meet certain strategic conditions, but in no sense need they reflect a correct or optimal response to the issue in benefit-cost terms. Establishing an adaptive regime may require circumventing such arguments over the proper level of quantitative targets.

3. Feedbacks concerned with technological changes may be the most crucial drivers of regime adaptation, so it is at least as important to design a regime to adapt to advancing technological capabilities as to advancing scientific knowledge. The strategic design of the ozone regime's technology assessment processes could readily be applied to other issues, provided that two conditions are met. First, the process must couple the provision of public benefits to the regime with private benefits to participating experts and their firms, by helping firms solve technical problems they face and by providing diffuse commercial and professional opportunities to participants that are linked with successful regime adaptation. The private benefits must be large enough to motivate serious participation, but not so large or so divisive that participants see advantage in seeking to bias the proceedings or withhold information. Second, the process must be protected from bias and capture by involving multiple experts with high stature and closely overlapping expertise but distinct material interests, so ill-founded or partisan claims will be effectively critiqued and deterred.

4. The authoritative-monopoly strategy of conducting scientific assessments discussed above can allow consensual scientific claims to delimit, or in some cases resolve, policy debates.

5. Activists are unlikely to succeed in gaining international agreements by first acting unilaterally, then pressing for others to take the same actions, for both strategic and rhetorical reasons. Further actions must be held in reserve to offer in return for others' initiatives. A prominent exception to this lesson occurs when one consequence of the initial movement is to make it easier for others to follow, or harder for them not to.

6. After an adaptive regime is formed, subsequent scientific and technological developments are more likely to favor progressive increases in the stringency of controls, even if the regime's adaptive measures were honestly designed to admit tightening, weakening, or other changes. I discussed the technological reason above.

The scientific reason arises from the long-term character and slow irreversibility of many environmental problems. If evidence of environmental risk is noisy, and opponents of regulatory action succeed in establishing a high evidentiary standard for initial action, then the first signal that meets this standard is likely to be followed by many progressively stronger signals, even if action is taken to reduce the risk. Since these signals will predictably generate pressure for strong, repeated increases in the stringency of action, those who fear the costs of action should not demand such a high standard for initial action, but rather should support earlier, smaller measures in response to weaker evidence of risk, in order to avoid being trapped in this ratchet.

A final conclusion of the book is a negative one, which corrects a widespread misunderstanding. Many commentators have asserted that the successful ozone regime was largely predetermined by exceptionally benign structural conditions. Given such conditions, some theorists describe ozone depletion as a "most likely" case, for which international cooperation was so likely that only the *nonappearance* of cooperation would have been of theoretical interest. Others argue that the issue's benign conditions cast suspicion on any lessons we might draw from it, since they are unlikely to apply to less benign issues. Two sets of conditions are proposed to make the issue so benign. The first is the achievement of scientific consensus on the nature and severity of the ozone-depletion risk and the causal role of CFCs, reinforced by the specter of the ozone hole. The second is a breakthrough in the availability of CFC alternatives: either a secret breakthrough by a dominant corporate actor led them to support policy change because they expected to benefit, or a widespread recognition that alternatives were feasible made many actors more willing to adopt controls.[14] This book rejects both these accounts, arguing that they reflect ex post rationalization of a too limited empirical view of the issue. The initial decade-long deadlock, which persisted despite repeated attempts to establish international cooperation, rejects the broad claim that the issue was always benign. The account of specific factors changed and unchanged at the time of regime formation rejects the narrower claim that the issue became benign over the crucial period. It is because the ozone regime cannot be explained by uniquely benign structural conditions that the theoretical and practical lessons developed from it are important.

However enriched and expanded from prior accounts, this book is a primarily inductive exercise based on the historical examination of a single case. It builds its conclusions and theoretical arguments on variation over time within that case, and on the detailed tracing of processes of influence. Consequently, these arguments will require testing and elaboration through similarly detailed studies of other issues. It is my hope that the arguments advanced here will stimulate such a research program to advance understanding of the interactions of science, domestic and international politics, and technology and corporate strategy—interactions that are crucial for understanding and improving the management of global environmental change, but which to date have been relatively neglected by research.

1.4 Plan of the Book

The book's empirical material is presented in chapters 3 through 8, with some explanatory and interpretive argument woven throughout, while the final chapter develops in detail the general arguments and lessons sketched in this chapter. To attend to the interactions among multiple parallel histories, involving different sets of actors and grounded in different disciplines, the organization of the empirical material is partly chronological and partly thematic. Chapter 3 details the linked histories of science, policy, and strategy over the first decade of policy concern about the ozone layer (1970 to 1980). After briefly reviewing the controversies of the early 1970s over potential ozone depletion from supersonic aircraft and their significance for subsequent debates, it discusses the domestic controversy over CFCs and ozone depletion in the United States, the parallel but more muted discussions in other countries, and the first unsuccessful attempts to develop international action on the issue.

The major lines of scientific debate, research progress, and controversy over stratospheric ozone from the mid-1970s to the mid-1980s are examined in chapter 4. The discussion begins with the initial qualitative confirmation of the main points of the chlorine-ozone depletion claim in 1976, which coincided with the resolution of the U.S. policy debate over aerosols. It proceeds through 1985, with the seeming contradiction between slowly growing confidence that the main processes operating in the stratosphere were coming to be understood, and two shocking new claims of observed ozone losses that sharply called this confidence into question. In addition to reviewing substantive debates in stratospheric science, the chapter examines the multiple attempts over this period to synthesize and communicate the state of scientific understanding in assessments to inform policy making, culminating in 1985 with the first assessment to significantly influence international policy debates.

Chapter 5 concentrates on domestic and international policy making. It tracks the 1982 resumption of international negotiations to protect the ozone layer after two further unsuccessful attempts. It then follows the progression of these negotiations, and their interactions with domestic policy and corporate strategy, over five years: three years of stagnation leading to the 1985 Vienna Convention, followed by two years of rapid progress culminating in the 1987 Montreal Protocol—the first international agreement on concrete measures to reduce human contribution to ozone depletion.

Chapters 6 and 7 examine major changes, in scientific understanding and technology respectively, that followed the adoption of the Protocol and completed the period of initial regime formation. Chapter 6 returns to the two disturbing claims made in 1985—extreme seasonal ozone loss in Antarctica and large ozone loss worldwide—and traces their investigation over the following three years, their initial resolution in the year after the Protocol, and the consequences of their resolution in spreading calls to eliminate ozone-depleting chemicals. The chapter also discusses the early development of the Protocol's expert assessment panels, the centerpiece of the regime's structure to adapt to changing knowledge and capabilities.

Whereas previous chapters touched briefly on the availability, development, and knowledge of technologies to reduce ozone-depleting chemicals, and their connection to strategic decisions by the major CFC-related industries, chapter 7 focuses exclusively on these issues. The chapter examines the reactions of major producers and users of CFCs to the challenge posed by the negotiation and adoption of the 1987 Protocol. It discusses how major CFC producers revived previously abandoned efforts to commercialize less ozone-damaging chemical alternatives to CFCs. It also examines how CFC users, many of whom faced more serious risks from CFC restrictions than the producers, responded with intense efforts to reduce their dependence on all ozone-depleting chemicals—efforts that reduced most CFC uses much faster than had seemed possible, and directed large shares of former CFC markets away from similar chemicals entirely.

Chapter 8 reintegrates the discussions of science, technology, and policy to examine how the ozone regime has adapted since its formation. It discusses the linked processes of negotiation, assessment, and technological innovation that allowed the Protocol to adapt repeatedly to new scientific knowledge and new technological capabilities, strengthening and extending its control measures four times. The chapter closes with a brief assessment of the status of the regime at the end of the 1990s, including a review of several significant implementation problems. Finally, chapter 9 provides the detailed development of the arguments and lessons sketched here.

Before starting the empirical account, however, chapter 2 makes a brief detour. The book provides a balanced account of policy, science, and technology and their interactions in shaping the development of the ozone issue. But the scientific and technological strands of this story require more background than the policy strand, because they have received the least emphasis in existing accounts and because it was from their deeper historical roots that the first concerns about human disruption of the stratosphere emerged. Chapter 2 provides this background. It presents a brief, historically organized account of highlights in the early science of the stratosphere, the development and use of CFCs, and the initial appearance of environmental concerns about potential human disruption of the stratosphere.

2

Early Stratospheric Science, Chlorofluorocarbons, and the Emergence of Environmental Concern

2.1 Early Stratospheric Science

The stratosphere and its layer of elevated ozone have been matters of scientific interest, investigation, and international cooperation since the mid-nineteenth century. Early understanding emerged from three areas of scientific study: the properties of chemical compounds; the properties of light and its interactions with matter; and the composition and physical properties of the atmosphere.[1] Schonbein discovered ozone in 1839, when he noticed its pungent smell near electrical sparks and lightning, and in the oxygen released from electrolysis of water. He named the substance "ozone" and developed a simple way to measure its concentration by using a chemical-impregnated paper that changed color when ozone was present. The presence of an easily measured, chemically reactive substance in air sparked widespread interest, and within a few years ozone was being measured in hundreds of locations.

Studies of ultraviolet (UV) radiation and its absorption also contributed to early understanding of ozone. The spectrum of sunlight reaching the earth showed a sharp cutoff in the UV region of wavelengths shorter than about 290 nanometers (nm),[2] suggesting that some atmospheric substance might be absorbing radiation with shorter wavelengths. Hartley measured ozone's absorption and found that even tiny quantities absorbed strongly in the 200–300 nm region of the UV, with the edge of the strong absorption region corresponding closely to the sharp cutoff in sunlight. On that basis, he argued that the cutoff might be caused by atmospheric ozone absorbing shorter wavelengths. Since measurements at the earth's surface found far too little ozone to account for the absorption, Hartley proposed in 1880 that the ozone must be high in the atmosphere, a suggestion that two subsequent findings supported. The cutoff in sunlight was found to be just as sharp when looking up from mountaintops or balloons as from the earth's surface. Moreover, using an artificial source of UV light showed that surface air readily transmitted wavelengths down to 250 nm, well below the cutoff in incoming sunlight.

Fabry and Buisson developed the first instrument that could measure ozone remotely. By observing the sun or bright sky at wavelengths where ozone absorbed strongly, their instrument measured total ozone in a column of air from the earth's surface to the top of the atmosphere. In measurements over Marseilles in 1920 they estimated that this total ozone was equivalent to a 3-millimeter (mm) layer of pure ozone at sea level, and that most of the ozone was in a layer about 25 to 30 kilometers (km) above the surface. It is now known that ozone's concentration rises from less than 1 part per million (ppm) in clean surface air to a maximum of about 8 ppm at around 35 km altitude. In view of its profound importance for atmospheric radiation, temperature, and circulation, the amount of ozone present—even at its peak in the stratosphere—is tiny.

The earliest flights of high-altitude weather balloons, around 1900, also contributed to understanding of the stratospheric ozone layer. These flights showed that the well-known trend of temperature declining with altitude stopped at around 10 km and reversed above that point. The newly discovered atmospheric region, in which warmer air lay above denser, cooler air, was called the stratosphere ("layered sphere"). This temperature inversion makes the stratosphere quiet, because it is not subject to the vigorous convection and vertical mixing that characterizes the lower atmosphere, or troposphere ("turning sphere").[3]

Through further developing Fabry and Buisson's technique, Dobson developed the first practical instrument to measure total ozone in 1924. The Dobson instrument compared the relative intensity of light reaching the instrument at two wavelengths, one a maximum and one a minimum of ozone's absorption. Dobson and his colleagues established a small network of stations around the world that made regular ozone observations, from which they quickly discovered that ozone varies strongly with local weather conditions, and by latitude and season. Total ozone over one location can vary by 25 percent over days or weeks due to shifting weather systems, and by as much as 10 percent in its year-to-year average. In addition, there is more ozone at high latitudes than near the equator, and a pronounced seasonal oscillation at higher latitudes, with the highest ozone concentrations in winter and spring.[4]

After ozone's chemical structure was verified in 1899 to be O_3, it was widely believed that ozone was formed in the atmosphere when molecular oxygen (O_2) was split apart (or photolyzed) by absorbing short-wavelength UV radiation. The two oxygen atoms released from this photolysis then each combined with an oxygen molecule to produce a molecule of ozone. But if ozone was produced by UV radiation, which is most intense near the equator and in summer, then finding the most ozone near the poles in winter and spring posed a puzzle. The puzzle increased in 1929 when Götz developed the "Umkehr" (inversion) technique, which gave a crude measure of ozone's vertical distribution by looking straight up with a Dobson meter near sunrise or sunset.[5] Early measurements using this technique, and also from high-flying balloons, found that most ozone was in the lower stratosphere, around 25 km, rather than higher. In terms of altitude, latitude, and season, most ozone was precisely where it should not be.

In 1930, Chapman proposed a chemical mechanism of ozone creation and destruction that resolved the puzzle. In addition to the pair of photochemical reactions that produced ozone, he proposed a second pair that destroyed it. In this second pair, ozone absorbed UV radiation and photolyzed to release an oxygen atom, which then combined with another ozone molecule to form two molecules of diatomic oxygen.[6] With ozone creation and destruction both requiring UV radiation,

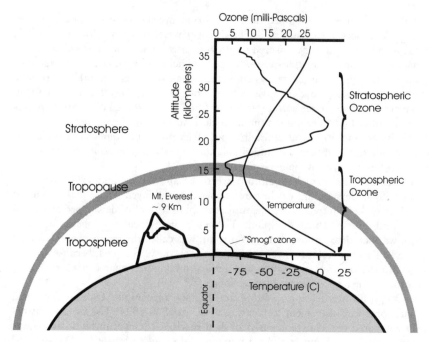

Figure 2.1 The Stratosphere and the Ozone Layer. The figure shows typical vertical profiles of ozone and temperature at tropical latitudes. Temperature decreases with increasing height up to the tropopause, located at an altitude of about 15 kilometers in the tropics and 10 kilometers at high latitudes. Above that point temperature increases and ozone concentration increases sharply, reaching its maximum in the lower and middle stratosphere at altitudes between 20 and 30 kilometers. Source: adapted from UNEP 1999b.

both would proceed fastest in the upper stratosphere and in the tropics, so the largest quantity of ozone would not necessarily be found there. Rather, ozone's distribution would depend both on the processes creating and destroying it, and on atmospheric motions. Chapman's theory also provided the first explanation for the temperature inversion of the stratosphere. Absorption by molecular oxygen effectively screens the shortest-wavelength (highest-energy) UV radiation so it does not penetrate below altitudes of about 80 km; but ozone, which is less tightly bound than O_2, absorbs lower-energy UV with wavelengths up to about 310 nm (and more weakly, even lower-energy visible radiation). The stratosphere is warmer at higher altitudes because it is heated from above by ozone's absorption of UV.

Further explanation of ozone's distribution was provided by Brewer and Dobson, who proposed a stratospheric circulation scheme by which ozone formed in the upper stratosphere in the tropics is transported downward and toward the poles, accumulating in the lower stratosphere at high latitudes.[7] Combined with this model of stratospheric circulation, Chapman's chemistry appeared to give a complete description of the ozone layer, and provided the basis for a vigorous program of predominantly meteorological research on ozone that continued until the 1960s.

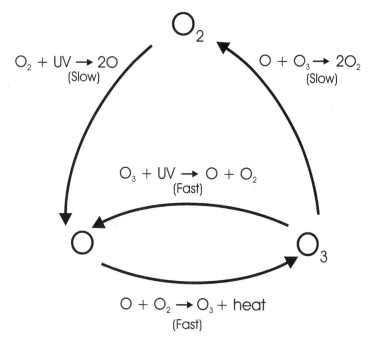

$$O_2$$

$$O_2 + UV \rightarrow 2O$$
(Slow)

$$O + O_3 \rightarrow 2O_2$$
(Slow)

$$O_3 + UV \rightarrow O + O_2$$
(Fast)

$$O \qquad O_3$$

$$O + O_2 \rightarrow O_3 + heat$$
(Fast)

Figure 2.2 Formation and Destruction of Ozone from Molecular Oxygen. Ozone is formed when molecular oxygen is split by UV radiation (left arrow) and the resultant oxygen atoms each combine with an oxygen molecule (bottom arrow). Ozone and atomic oxygen, collectively called "odd oxygen," interchange rapidly by absorbing UV and reacting with molecular oxygen (bottom two arrows). Odd oxygen is destroyed when an ozone molecule and an oxygen atom combine to form two oxygen molecules (right arrow).

Meteorologists were interested in ozone because of its strong local and short-term variation. Since ozone could be observed from the ground, and since most of it was in the lower stratosphere, where it was not created or destroyed chemically, it could be used as a "tracer" to infer movements of stratospheric air that could not be observed directly. Meteorological interest in ozone grew after World War II, with expanded international cooperation under the newly established World Meteorological Organization (WMO) and the International Council of Scientific Unions (ICSU).[8] The network of stations observing ozone was also expanded, particularly for the 1957 International Geophysical Year, including several stations doing the first regular ozone monitoring from Antarctica. This heightened interest initially reflected the widespread belief, found to be erroneous by the mid-1960s, that dynamic couplings between the stratosphere and troposphere were important to understanding and predicting the weather.[9]

In the 1950s, however, improved atmospheric and chemical measurements of several kinds began calling the quantitative accuracy of Chapman's theory into question. These included new laboratory measurements of highly reactive and short-lived chemical species to determine the rates of the relevant reactions, observations of the spectrum of sunlight high in the atmosphere before it is attenuated by atmospheric absorption, and sufficiently accurate and widespread measurements of ozone to estimate its global average concentration. Improved postwar measurements, including measurements of solar radiation from flights of captured German V-2 rockets that found substantially stronger UV than expected, gradually showed that Chapman's chemistry predicted too much ozone—two or three times as much as was actually present. The observed amount of ozone could represent an equilibrium between processes of ozone creation and destruction only if there were other destruction processes, or sinks, in addition to the mechanism identified by Chapman.[10] Since the main constituents of the atmosphere, N_2 and O_2, were known not to react with ozone, the additional sink must be something present at trace concentrations.

Through the 1960s and early 1970s, researchers looked for possible ozone sinks to correct the observed imbalance. This search took place during a period when human activities were beginning to obtrude on the upper atmosphere, with rocket launches, atmospheric nuclear tests, subsonic civil jet aviation in the upper troposphere, and both military aviation and proposed civil supersonic aircraft in the stratosphere. The same period also saw the first emergence of widespread environmental concerns, and the first scientific realizations that human activities were capable of perturbing natural systems at global scale.[11] In the case of stratospheric ozone, the effect of this threefold historical coincidence was a rapid coupling of scientific speculation and policy concern. Each time a new potential ozone sink was identified in the natural stratosphere, it was quickly realized that some current or proposed human activity might augment the sink, and so could disrupt the ozone layer.

Over several years, three classes of potential ozone-destroying processes were identified, involving hydrogen, nitrogen, and chlorine. Each process has the same general form: a pair of chemical reactions in which a catalyst removes an oxygen atom from ozone (leaving molecular oxygen), then gives it to either an oxygen atom or an ozone molecule (forming one or two more molecules of oxygen). The net effect of each process is to produce oxygen molecules from ozone and oxygen atoms, with the catalyst restored at the end.

The first proposed sink involved hydrogen radicals derived from breakdown of stratospheric water vapor, collectively denoted HOx.[12] The HOx cycle was first proposed to operate only above the stratosphere, where high-energy UV radiation could form HO directly through photodissociation of water vapor. It was later suggested that HOx could also be formed in the stratosphere, by reaction of water with the energetic oxygen atoms released by photodissociation of ozone, and that the resultant HOx cycle might be the missing natural sink for ozone. The rates for the two reactions in the HOx cycle had not been measured, but Hunt hypothesized plausible rates that would allow this cycle to yield the observed quantity of ozone. Hampson then noted that if these rates were correct, they implied that the additional water vapor that proposed supersonic airliners would inject into the stratosphere could significantly deplete the ozone layer. Initial calculations suggested that a fleet of 500 supersonic airliners would increase stratospheric water vapor about 10 percent from its natural level of 3 ppm, reducing ozone by about 4 percent on average with larger losses near flight corridors.[13] This hypothesized loss of ozone due to HOx was gaining widespread, although tentative, acceptance just as the controversy over U.S. supersonic aircraft reached its peak in 1970 and 1971. The first measurement of one of the HOx reactions found it much too slow, however, causing attention to shift away from the HOx cycle soon after it was proposed.[14]

The second proposed ozone sink was a catalytic cycle involving oxides of nitrogen, NO and NO_2, collectively denoted NOx. After NO_2 was detected in the lower stratosphere in 1968, Crutzen (1970) proposed a NOx catalytic cycle and suggested it could explain the observed quantity of ozone if total stratospheric NOx was about 10 parts per billion (ppb). No NOx measurements had been made in the stratosphere, but this suggestion was roughly consistent with available measurements higher and lower in the atmosphere.[15] Although the source of stratospheric NOx was initially unclear, three papers in 1971 separately proposed that it could be the nitrous oxide (N_2O) released as a few percent of the product of bacterial breakdown of organic material.[16] Because N_2O is chemically inert and insoluble in water, it can reach the stratosphere. Once there, most of it photolyzes to yield N_2 and O, but about 1 percent reacts with energetic oxygen atoms to yield NO. Although this stratospheric NO arises from small side paths at two stages in the global nitrogen cycle, it is the dominant natural source of stratospheric NOx.

NO also can be produced by high-temperature combustion in air, such as occurs inside a jet engine, so stratospheric aircraft would inject some NO directly into the stratosphere. At a 1971 scientific meeting concerned with impacts of stratospheric flight, Johnston independently identified the NOx cycle and calculated that the NOx in a large aircraft fleet's exhaust would destroy much more ozone than the HOx in the same exhaust. Where the preliminary calculations of HOx-catalyzed ozone loss from a 500-aircraft fleet were about 4 percent, Johnston (1971) calculated NOx-catalyzed losses from this fleet as 3 to 23 percent globally, approaching 50 percent near flight corridors.

The third proposed ozone sink was a catalytic cycle involving chlorine.[17] In contrast to the hydrogen and nitrogen cycles, the chlorine-ozone debate was driven by concern about artificial sources from the start, initially the exhaust of the space shuttle then under development. Each shuttle launch was projected to inject about 10 tons of HCl into the stratosphere. Several groups began studying the shuttle's effect on ozone after the July 1972 publication of its draft environmental impact statement, which discussed various exhaust constituents at length but did not mention chlorine.[18] Early work on chlorine and ozone was contentious, marked by sharp rivalries between research teams, suggestions that the National Aeronautics and Space Administration (NASA) was trying to suppress research results, and widespread confusion over whether researchers were discussing the shuttle or the tiny natural sources of stratospheric chlorine known at the time.[19] In January 1974, a NASA-sponsored scientific meeting reviewed this work and concluded that 50 shuttle launches per year could deplete ozone by 1 to 2 percent (with wide uncertainties), an amount that they called "small but significant." With researchers' attention fixed on flight, however, they failed to notice a much larger source of stratospheric chlorine that had already been identified.

2.2 The Chlorofluorocarbons and Their Atmospheric Impacts

The chlorofluorocarbons (CFCs) are a family of industrial chemicals derived from simple hydrocarbons, principally methane and ethane, by replacing all their hydrogen atoms with fluorine or chlorine. The CFCs with just one carbon atom were

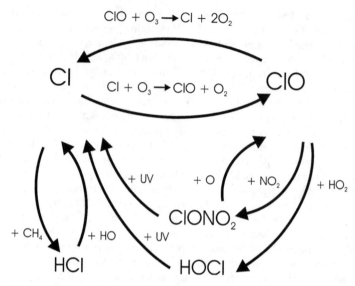

Figure 2.3 Catalytic Destruction of Ozone. CFC molecules carry chlorine to the stratosphere, where they are split by UV radiation to release atomic chlorine. This atomic chlorine destroys ozone catalytically by taking an oxygen atom from one ozone molecule and giving it to another; the net effect is to produce three molecules of oxygen from two molecules of ozone (top pair of arrows), regenerating the chlorine atom at the end of the process. The rate of this ozone-loss cycle is limited by other processes that temporarily convert Cl or ClO into more stable reservoir species (lower arrows).

cheap and easy to produce because their synthesis process—developed in the 1890s—used cheap raw materials, involved few steps, did not require high temperatures or pressures, and gave high yields of pure finished product.

The first commercial application of CFCs, developed in 1928, was as working fluids for refrigerators. The major refrigerants of the time—ammonia, methyl chloride, and sulfur dioxide—were all either dangerously toxic or inflammable.[20] Two scientists in General Motors's Frigidaire division were asked to identify a refrigerant that performed well but was safer, and quickly identified one of the CFCs, CF_2Cl_2 (dichlorodifluoromethane) or CFC-12.[21] General Motors and DuPont formed a joint venture, the Kinetic Chemicals Corporation, to commercialize new refrigeration systems based on this chemical. The new venture sold the first home refrigerators using CFC-12 in 1933 and began producing several other related CFCs (CFCs 11, 113, and 114, and HCFC-22) for other cooling applications—industrial and commercial refrigeration, freezing, and air conditioning—through the 1930s. By the mid-1940s, the new chemicals dominated U.S. refrigeration markets.[22]

The new chemicals made good refrigerants because of their thermodynamic properties: they had high vapor pressure, low heat of vaporization, and boiled near atmospheric pressure at a range of temperatures suitable for various cooling appli-

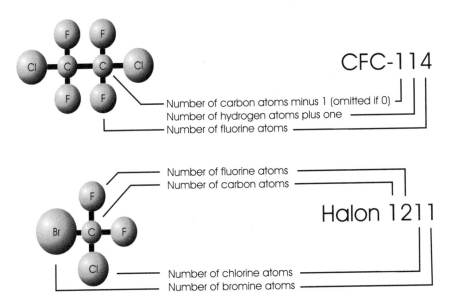

Figure 2.4 Names of CFCs and Halons. CFCs and halons are identified by numbering systems in which each digit is related to the number of one type of atom in the molecule. Source: UNEP Division of Technology, Industry, and Economics.

cations. In the 1950s, their uses expanded to other cooling applications, including the large chillers that made it possible to air-condition large commercial buildings and indoor spaces such as shopping malls. The same properties made the chemicals suitable for several nonrefrigeration applications as well, most of which were developed during World War II and grew to major commercial markets in the following decades.

The largest of these new applications, which by the 1970s became the largest share of total CFC production, was as propellants in aerosol spray cans. The concept of using a pressurized, low-boiling liquid to propel a product through a nozzle as a fine spray was first developed in Norway before the war, then implemented on a massive scale (50 million units) during the war to produce spray insecticides for U.S. troops in the Pacific. These wartime "bug bombs" were propelled by CFC-12, whose high vapor pressure required a heavy can and valve unsuitable for mass-market household products. But mixing CFC-12 with lower-pressure CFC-11 allowed lightweight valves and cans to be used, allowing rapid growth in the 1950s and 1960s as a wide range of toiletries, cleaning products, paints, and insecticides were repackaged as aerosols. By the early 1970s, 200,000 tonnes (metric tons) of CFCs were used in aerosols annually in the United States and the typical U.S. household contained 40 to 50 aerosol cans, half of them propelled by CFCs.

It was also during the war that the Dow Chemical Company first used CFC-12 to blow polystyrene into a rigid foam. Blowing the material into a froth while liquid created a structure of extremely fine bubbles which solidified to form a material that was rigid, lightweight, and an excellent thermal insulator. Several other forms of CFC-blown foams were introduced in the 1950s. Flexible, open-celled foams, particularly polyurethane, were used as cushioning in furniture, automobiles, and mattresses. Rigid insulating polyurethane foam blown with CFC-11 had double the insulation value of fiberglass, and quickly surpassed it as the dominant insulating material in buildings and appliances.

The latest growth use for CFCs, especially the two-carbon CFC-113, was as solvents. Introduced in 1964 as a solvent for dry-cleaning clothes, CFC-113 had only moderate success in this market. But it was ideally suited for cleaning delicate plastics and synthetic materials, and for hand-cleaning of precision metal parts: it was an effective solvent with low toxicity and low contribution to air pollution, and was particularly mild on substrates. Consequently, its use grew rapidly, beginning in the 1970s, in the electronics, computer, and aerospace industries.

A second family of chemicals closely related to CFCs, the halons or bromofluorocarbons, was developed in the 1950s by the U.S. Army to fight fires inside tanks and armored vehicles. Halons suppress fires rapidly at very low concentrations without conducting electricity, leaving any residue, or harming people, in part by scavenging the free radicals that propagate flames. Two halons were commercialized in the 1960s and grew rapidly thereafter. Halon 1301 (CF_3Br), which had extremely low toxicity, was used for flooding enclosed spaces such as computer rooms, vaults, oil-drilling platforms, and telephone exchanges. Halon 1211 (CF_2BrCl), which was slightly more toxic, began widespread use in handheld fire extinguishers in the 1970s.

2.3 Early Environmental Concerns about Chlorofluorocarbons

With the repeated development of major new applications, world CFC production grew nearly 20 percent per year for 40 years, reaching 1 million tonnes (Mte) in

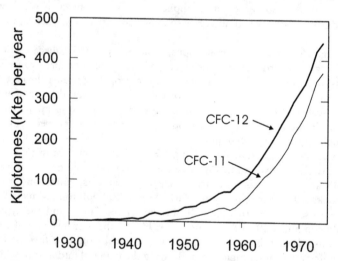

Figure 2.5 World CFC Production, 1930–1974. After four decades of rapid growth, world production of CFCs approached 1 million metric tons in the early 1970s. Source: Alternative Fluorocarbons Environmental Acceptability Study data.

the early 1970s. In 1968, James Lovelock reasoned that because CFCs and similar halogenated gases were so inert, they should make good tracers of atmospheric movements. He began using an extremely sensitive instrument of his own invention to measure these trace gases in the air of southwest England and Ireland.[23] In the summer of 1970, he noted that concentrations of CFC-11 and sulfur hexafluoride (SF$_6$) were higher in air blowing from Europe than from the Atlantic, supporting his conjecture that they could be used to trace large-scale atmospheric movements. He published the observations in *Nature* and reported them at a conference in the summer of 1971, where in a conversation involving Lovelock, Lester Machta of the U.S. National Oceanic and Atmospheric Administration (NOAA), and Ray McCarthy of DuPont, McCarthy was asked to estimate the total historical production of CFC-11. McCarthy's estimate was roughly equal to the quantity that Lovelock's measurements implied was still present in the atmosphere, suggesting that CFC-11 persisted in the atmosphere for many decades.[24]

This conversation had three consequences of later importance for the ozone issue. First, Lovelock and colleagues continued their measurements in 1971–1972 on a ship voyage from the United Kingdom to Antarctica and back, where they found an average of 40 parts per trillion (ppt) of CFC-11 in the southern hemisphere and 50–70 ppt in the northern. Second, Machta reported the strikingly long apparent atmospheric lifetime of CFC-11 to F. S. Rowland at a scientific meeting in January 1972. Third, McCarthy and his colleagues decided to research the environmental fate of CFCs, since large quantities were being released and no one knew what happened to them. They were particularly concerned that CFCs might contribute to smog formation. DuPont began funding environmental research on CFCs and organized a conference on the subject, at which 15 CFC producers agreed to establish a joint research program on the environmental fate of the chemicals.[25] The program was directed by a panel of one senior scientist from each firm, the Fluorocarbon Program Panel (FPP), and established under the auspices of the Chemical Manufacturers' Association (CMA).[26] For its first round of research projects in 1973, the panel took over three projects already initiated by DuPont: further measurements of CFCs' atmospheric concentrations; measurements of CFCs' UV absorption spectra, and studies of CFCs' chemical reactivity in the lower atmosphere.[27]

Intrigued by CFCs' apparent long atmospheric lifetimes, Rowland began studying their fate in fall 1973 with Mario Molina, then a postdoctoral researcher in Rowland's laboratory.[28] After investigating many potential CFC sinks in the lower atmosphere, Molina and Rowland concluded that none was likely to be important and, consequently, that CFCs would slowly mix upward through the stratosphere until breaking up under the strong UV radiation above 25 km, releasing chlorine atoms. They then identified the catalytic chain by which the chlorine atoms released would destroy ozone. Assuming continued world CFC production of about 1 million tonnes per year, they calculated that equilibrium between production of CFCs and their catalytic destruction in the stratosphere would be reached after about a century, and that this steady state would involve large ozone losses. Several other researchers had already recognized pieces of the puzzle—indeed, competition among research groups had grown quite heated—but Molina and Rowland's work first put the pieces together and first identified an important stratospheric chlorine source and threat to ozone.

The three major chemical pathways for ozone depletion—hydrogen, nitrogen, and chlorine—were proposed separately and gained public attention sequentially, but a realistic understanding of the stratosphere and its responses to perturbations from multiple human activities required considering the three pathways together. Study of the three catalytic cycles increasingly identified new processes coupling the cycles, however. As I discuss in chapter 4, the complexity of these couplings and the associated measurement challenges accounted for much of the difficulty of gaining quantitative understanding of ozone loss through the 1970s and early 1980s. While the proposed hydrogen and nitrogen catalytic cycles were neither refuted nor fully understood, their prominence declined through events that appeared to reduce the risk they posed, and on account of the evidently more serious CFC threat. The hydrogen threat appeared to recede when new measurements showed a key reaction to be too slow, then was overtaken by the more serious nitrogen concern pressed by Johnston in 1971. The nitrogen threat in turn receded from public attention as it became clear that very few commercial stratospheric aircraft would fly, although important scientific controversies continued. In contrast, the chlorine-CFC threat arose from a source that was both large and rapidly growing, suggesting larger and longer-lived stratospheric effects than either of the other processes. The CFC threat quickly gained and held high public attention and produced sharp scientific and policy controversy.

The link between even a large reduction of stratospheric ozone and policy concern, however, was not self-evident: that ozone might decline is not sufficient to show why, or whether, we should care. Rather, each time a threat to ozone became prominent, it was coupled with some proposed impact of ozone loss whose human concern was more direct and obvious. For the first several years of debate, the impacts of greatest concern shifted over time, independent of the chemical processes and human activities proposed as the origin of the threat.

The earliest concerns about ozone loss identified it as a problem of climate disruption, either catastrophic or incremental. In 1964, John Hampson proposed that the atmosphere might have alternative circulation modes, one corresponding to the present climate and another to Ice Age conditions. He argued that ozone loss could flip the atmosphere into another mode, bringing potentially catastrophic climatic change. Even absent such catastrophic climate effects, strong enough radiative or dynamic coupling between the stratosphere and troposphere could allow ozone loss to cause significant changes in atmospheric circulation and surface climate.

A second catastrophic view of the effect of ozone loss depended on the resultant increase in UV radiation reaching the earth's surface. Since Chapman first proposed his theory of ozone creation and destruction through absorption of solar radiation, it had been widely believed that the resultant shielding of the earth's surface from high-energy UV was essential for the survival of life.[29] Berkner and Marshall elaborated on this view in the 1960s, hypothesizing that in the earth's ancient shift to an oxygen atmosphere, forming an ozone layer thick enough to screen surface UV was a necessary condition for complex multicellular life to develop on land or in the surface ocean. Several authors subsequently speculated about how extreme losses of ozone due to ancient astronomical events might have caused the massive extinctions in the fossil record.[30]

These suggestions that ozone loss threatened the survival of life on earth attracted harsh criticism, however, since a hypothesis that total loss of ozone would be catastrophic implied little about the effects of partial, possibly small losses.[31] In 1971, J. E. McDonald provided the crucial link between incremental ozone loss, surface UV, and a specific, immediate and documented harmful effect: skin cancer. Asked to investigate an unrelated environmental harm that had been proposed from stratospheric aircraft—that the hazy trails of ice crystals from their exhaust would disrupt the climate by reducing the intensity of visible sunlight reaching the surface—McDonald rejected the effect as insignificant. But he went on to estimate the effects of ozone depletion due to the HOx cycle (the most prominent ozone-loss mechanism of the time) on the incidence of skin cancer, which he suggested might be the most serious effect of ozone loss. Using the known increase in UV intensity at lower latitudes and epidemiological data on skin cancer, he estimated that each 1 percent ozone loss would cause a 6 percent increase in incidence of skin cancer. In the United States, with 120,000 new cases of skin cancers per year at the time, this would mean 7,000 extra cancer cases per year for each 1 percent ozone loss.[32] McDonald presented his cancer estimates in congressional testimony and at a scientific meeting on stratospheric flight in March 1971, and Johnston used McDonald's cancer estimates in his subsequent attempts to publicize the threat of nitrogen-induced ozone depletion.

The estimated multiplier, or elasticity, linking ozone depletion to cancer incidence has varied strongly with empirical work since McDonald's, mostly over values substantially lower than his estimate of 6. Many other health and ecosystem effects of increased UV have also been identified, but skin cancer has remained the best-known and most politically salient effect of ozone loss over the thirty years since McDonald's work.

This review of the science of ozone and the stratosphere, the chlorofluorocarbons and their major uses, and the hypothesized effects of ozone loss completes the background necessary to examine policy debates over ozone loss. Chapter 3 begins with the controversy over supersonic stratospheric aircraft in the early 1970s, and the first policy debates over CFCs and ozone depletion, domestically and internationally, that persisted through the end of the 1970s.

3

Setting the Stage

National Action and Early International Efforts, 1970–1980

The rise and fall of the first period of public concern and political controversy over depletion of the ozone layer was neatly bounded by the decade of the 1970s. As the decade opened, the first claims were being made that supersonic commercial aircraft could destroy ozone. Before this controversy was fully resolved, it was overtaken by the distinct concern that ozone would be destroyed by chlorine, principally from CFCs. Controversy over CFCs raged in the United States for two years, ending with a regulatory decision to ban their use as aerosol propellants. The global nature of the ozone issue was widely recognized from the outset, and several attempts were made to achieve coordinated international CFC control. These all ended in failure by 1980, however, as did a U.S. domestic effort to control total CFC production.

3.1 The Supersonic Transport Controversy and Its Legacy

Through the 1960s, three national programs sought to develop airliners to fly supersonically in the stratosphere, called supersonic transports (SST). The British/French Concorde and the strikingly similar Soviet TU-144 would fly at about 17 km altitude, near the bottom of the stratosphere, at twice the speed of sound (Mach 2). Boeing was developing an SST that would fly higher (at 20 km) and faster (Mach 3), on a schedule several years behind the others. In the United Kingdom and, especially, in the United States, these aircraft met sustained public opposition, principally motivated by concerns about noise and cost.[1] The projects also posed the new risk of polluting the stratosphere, where the exhaust of hundreds of aircraft would represent an unprecedented increase in human disturbance.

Critics raised three separate concerns about stratospheric pollution from SSTs, one physical and one chemical effect of water vapor, and a chemical effect of nitrogen oxides. The first effect, raised in 1963 but soon dismissed, was that ice crystals in the aircraft's exhaust would form a long-lived hazy veil, blocking sunlight and altering the atmospheric heat balance and climate. The second was that water in exhaust could destroy ozone through the HOx catalytic chain discussed in chapter 2. Arising from several scientists' work, this concern became prominent in the final days of the U.S. SST debate due to J. E. McDonald's presentations in March 1971—to a congressional committee and, a few days later, to a scientific advisory meeting on the SST in Boulder, Colorado—claiming that ozone loss, by allowing more intense ultraviolet (UV) radiation to reach the earth's surface, would increase rates of skin cancer.[2] The third concern was that nitrogen oxides (NOx) in exhaust, although emitted in much smaller quantities than water vapor, could augment the more efficient NOx catalytic chain for ozone loss.[3] At the Boulder meeting, Harold Johnston calculated that NOx emissions from a fleet of 500 SSTs would destroy much more ozone than the few percent calculated for the HOx cycle: 10 to 90 percent in his preliminary calculations, later refined to 3–23 percent. Such a large depletion, when combined with McDonald's estimates of the skin-cancer effect of enhanced UV, suggested that SSTs could cause tens of thousands of new cases of skin cancer per year in the United States alone. Johnston, believing that scientists at the Boulder meeting did not take the NOx effect seriously enough, pressed his concerns through political and media channels and gained national attention.[4]

Ozone depletion was a powerful late addition to arguments over the U.S. SST, but the outcome was already determined. The aircraft's supporters, in increasingly desperate attempts to save their lost cause, denounced those raising the risk as unpatriotic or crazy. Its opponents invoked the ozone risk but did not need it to prevail.[5] Indeed, the HOx-depletion risk that SST opponents were advancing just before the crucial congressional votes in March 1971 was already on its way toward being discredited, while the apparently more serious NOx risk was known at the time by only a few scientists. In their attempts to reinstate funding, however, SST supporters kept the issue before the public through the spring of 1971, thereby providing a continuing forum for anti-SST arguments. The NOx risk, which Johnston publicized forcefully through this period, consequently gained much more attention than it would have without attempts to revive the program.[6] The link of ozone loss to cancer caught and held public attention, even though the underlying mechanisms of ozone loss (HOx versus NOx) were controversial and shifting.

Although the last-minute controversy over ozone depletion had little effect on the U.S. SST decision, it did reveal major gaps in scientific knowledge of the stratosphere. To verify and extend the initial depletion calculations of Johnston and Crutzen would require addressing several key questions about NOx in the stratosphere on which little was known: how much is present naturally; its chemical fate, including the existence of inert sinks; and how fast injected pollution is diluted by air movements. Because the Concorde was nearing commercial production, these questions remained important despite the cancellation of the U.S. program.

To investigate these questions, the U.S. Department of Transportation sponsored an ambitious assessment of stratospheric pollution from SSTs, the Climatic Impacts Assessment Program (CIAP). Though charged with "determining the regulatory constraints necessary to safeguard that future stratospheric flights . . . do not make adverse environmental effects,"[7] most of CIAP's efforts went to supporting the research necessary to do such an assessment. From 1971 to 1975, CIAP spent $23 million and involved more than 1,000 scientists from universities, government, and industry, including many from aerospace firms. Six CIAP panels conducted parallel research programs on the stratosphere, its response to pollutants, and resultant biological and socioeconomic impacts, followed by an assessment of the impacts of specific scenarios of stratospheric flight. CIAP's assessment sought to synthesize results and policy implications across all these disparate domains, making it the first integrated environmental assessment. The program's published output ran to roughly 9,000 pages.

CIAP's contribution to atmospheric science, and to the subsequent development of the ozone-depletion issue, was enormous. The program supported crucial research including the first measurements of stratospheric NOx, attracted many new researchers to atmospheric sciences, and helped bridge many artificial disciplinary barriers that hindered development of an integrated understanding of the atmosphere.[8] But as a crash program with an enormous mandate and a three-year deadline, CIAP faced hard challenges and sharp trade-offs. The largest funding went for atmospheric measurements and modeling, with a much smaller share (about $1 million) for laboratory measurements.[9] Model studies had shorter lead times than laboratory or atmospheric measurements, but depended on them for data. Consequently, with few new measurements available by the end of the program,[10] CIAP's model projections were largely based on pre-CIAP chemical kinetics and observations.

The model studies sought first to represent the natural stratosphere, then to simulate the effect of separately injecting water vapor, sulfate aerosols, and NOx. The models excluded one important set of reactions, the "smog" reactions through which NOx *produces* ozone when methane or other hydrocarbons are present. These reactions are the dominant effect of NOx injected at lower altitudes, while ozone destruction dominates at higher altitudes.[11] Although there were good reasons for the exclusion of the ozone-formation reactions—their kinetics were poorly known, and they greatly increased computation times—it caused large overestimates of ozone loss from lower-flying aircraft such as the Concorde. Even subsonic aircraft flying at 9–12 km were calculated to reduce ozone, when in fact they were more likely to increase it.

With the HOx cycle appearing to be less important than the NOx cycle by this time, CIAP's assessment principally undertook a critical review and extension of the NOx depletion estimates initially proposed by Johnston and Crutzen. Although this job was much more complex than the relatively simple initial calculations, it largely confirmed the earlier claims: NOx emissions from a fleet of 500 high-altitude SSTs would reduce global ozone by 10–20 percent.[12] Despite serious efforts to include biological and socioeconomic impacts in the assessment, these panels faced debilitating obstacles, most acutely that their analyses depended on results from the atmospheric panels that were available only at the last minute, or not at all. The only quantitative impact estimate presented was for skin cancer, where the panel proposed reducing the ozone-cancer multiplier from McDonald's estimate of 6, to 2.[13]

The reputation of the entire CIAP project was harmed by a bitter controversy over its executive summary and the summary's presentation at a press conference in January 1975. To meet a congressional deadline, the CIAP program manager and staff had revised the summary hastily and without consulting other participating scientists.[14] The summary and its presentation largely ignored the ozone-depletion effects of a large fleet of high-altitude SSTs, emphasizing instead the *climatic* effects (i.e., change in surface temperature) of the tiny projected near-term fleet of 30 Concordes, which were called "much smaller than minimally detectable."[15] Consequently, the presentation appeared to suggest that CIAP's results had refuted, rather than largely confirmed, the initial ozone-depletion concerns raised by Johnston, McDonald, Crutzen, and others. A wire service report of the press conference made this misinterpretation explicit and was widely repeated, in some cases with scathing attacks on the scientists who had raised the alarm.[16] Moreover, the summary's obscure presentation of such basic points as how many aircraft were assumed to fly in the stratosphere, at what altitude, for how long, and emitting how much NOx made it difficult for even a careful reader to understand the conclusions. A storm of controversy followed, including congressional hearings on whether the summary willfully misrepresented CIAP's findings.[17] It was widely speculated that the summary's authors were under political pressure to ease approval of the Concorde, which was then seeking U.S. landing rights. This suggestion was never confirmed, however, and reviews of the conflict concluded that the misleading summary more likely arose from technical optimism, weak writing, and haste than from an intent to deceive.[18]

Scientific assessments of the impacts of stratospheric flight were also conducted by a U.S. National Academy of Sciences (NAS) committee, and by government advisory committees in the United Kingdom and France.[19] The NAS report, presented as an independent review of CIAP, appeared at the peak of the executive summary controversy. It updated several results and presented new model results that doubled CIAP's ozone-loss projections. It also shared many of CIAP's weaknesses, such as the omission of tropospheric smog chemistry that erroneously implied that even subsonic aircraft would deplete ozone.[20] Its most important innovation was an attempt to put uncertainty bounds on its main conclusions by using subjective probability distributions.

The British and French assessments were of much smaller scale than CIAP's: conducted by government-appointed committees of one to two dozen scientists, chaired and staffed by officials.[21] The three studies drew on the same body of scientific knowledge, and exchanged interim results through joint conferences. The British and French studies, however, examined only lower-flying aircraft such as the Concorde and subsonics, and criticized CIAP for exaggerating these aircrafts' effect by omitting NOx-smog chemistry. They also failed in their attempts to include this chemistry in their models, however, so their ozone-loss calculations had the same bias.[22] The two European studies also differed sharply from CIAP's in their interpretations and policy conclusions, forcefully drawing attention to limitations of available scientific knowledge. The British study argued with particular force that two- or even three-dimensional atmospheric models were essential for credible ozone-loss projections, even though modeling weaknesses compelled them, like CIAP, to rely entirely on one-dimensional models for their depletion estimates. With few stratospheric measurements available, both the British and French reports attached great importance to one post-CIAP measurement that found more nitric acid (HNO_3) in the stratosphere than expected, suggesting that NOx would be tied up in this inactive form and destroy less ozone.[23]

Both European studies focused narrowly on atmospheric sciences, and the U.K. report stated that atmospheric uncertainties rendered any attempt to calculate effects meaningless. Both concluded that scientific evidence provided no basis for restricting stratospheric flight, and criticized CIAP and the NAS study for overestimating ozone loss, attempting more precise quantification of uncertainty than was appropriate, and endorsing regulations that scientific evidence did not warrant. While these criticisms have some merit, the intensity of the attacks suggests that the European studies suffered from the opposite bias—or at least that they presumed that decisive evidence of harm was needed to justify any regulatory action. This stance may have reflected long-established characteristics of British and French scientific and official cultures, or a more direct alignment of these assessments with the political mission of defending the Concorde.[24]

Controversy over supersonic aviation persisted through 1977, as British Airways and Air France sought access for the Concorde at airports in New York and Washington. Although both Federal Aviation Administration (FAA) and Environmental Protection Agency (EPA) officials were initially sympathetic to admitting the Concorde—the FAA even granted tentative approval in March 1975, subject to public hearings[25]—local opponents succeeded in obstructing access to New York's Kennedy Airport until exhausting their legal options in the fall of 1977. European bitterness over the treatment of the Concorde was intense, with substantial consequences for subsequent attempts to protect the ozone layer. The predominant European view of the controversy contained three major errors, however: that the Concorde could have been economically viable, had U.S. protectionist forces not killed it; that ozone depletion was the major pretext for U.S. opposition; and that these concerns were soon shown to be unfounded.[26] The last of these claims was misleading and overstated, the first two simply wrong.

Even immediate U.S. acceptance of the Concorde would not have made it profitable. Like the abandoned American SST, the Concorde was technologically problematic and economically catastrophic.[27] That any Concordes entered commercial service was more testimony to the British and French governments' willingness to subsidize and ability to coerce their aerospace firms and airlines, than to the aircraft's commercial attractiveness. Far from opposing the Concorde, U.S. regulators initially supported it, retreating only in the face of strident public opposition. Moreover, while U.S. opposition was broadly environmental, it principally concerned noise and general ideological opposition to technology, not ozone depletion.[28] Although both the Concorde's supporters and its opponents took the predictable stance on the ozone threat, the ozone issue was entirely peripheral. As for early ozone concerns being refuted, *all* early SST assessments exaggerated ozone loss from lower-flying SSTs like the Concorde, because they could not include the smog chemistry that operated at lower altitudes. But the early ozone-loss calculations and resultant controversy were not about the Concorde, but the higher-flying Boeing SST. Although new results through 1979 reduced calculated loss from these aircraft, even reversing its sign for a few months, later results returned their estimated effect to a substantial depletion.

Although the SST-ozone controversy had little effect on the aircrafts' fate, it had great influence on later scientific and policy debates over ozone. The most direct result was that the research undertaken to study SST effects brought a huge increase in U.S. capacity for stratospheric science, making rapid deployment of effort on chlorine-catalyzed ozone loss possible after the initial statement of the chlorine risk in 1974. In addition, the conflicting views that emerged in the SST debate over how to base public decisions on uncertain scientific knowledge pervaded the ozone policy debate until the 1990s. The suspicion that arose between policy and scientific elites in the United States and Europe had an especially durable influence on later debates. This conflict arose in part from scientific rivalry and different norms about the use of scientific evidence in policy, but was exacerbated by European resentment over "alarmist" American opposition to the Concorde. Since the scientific issues and participants in debates were very similar, it was natural that the same disagreements carried over to the chlorine-ozone issue. The parallels were particularly salient to

British and French officials, who were fighting for Concorde access to the United States in the same months as the American CFC controversy reached its peak.

3.2 The American "Ozone War," 1974–1977

Although conflict over the Concorde continued until 1977, the principal U.S. debate over ozone depletion shifted to CFCs by 1975. As I discussed in chapter 2, four research groups separately identified part or all of the chlorine-ozone chemistry, three of them in examining atmospheric impacts of the space shuttle. Molina and Rowland, the fourth, identified the same chlorine chemistry and also identified the photolysis of CFCs as the major source of stratospheric chlorine. They did their work in late 1973, initially without knowledge of the shuttle controversy and with little connection to the stratospheric research being concluded under CIAP. Their results were mentioned briefly in the CIAP and UK reports,[29] but not included in either study's modeling. Still, several groups that had developed stratospheric models were able quickly to add chlorine chemistry to evaluate Molina and Rowland's work and extend it with quantitative calculations.

Molina and Rowland's argument rested on four basic points. First, the extreme chemical inertness of the CFCs in the lower atmosphere allows them to mix upward through the stratosphere. Second, CFCs absorb the high-energy UV radiation present above about 25 km in the middle stratosphere, breaking apart and releasing free chlorine (Cl) atoms. Third, these free Cl atoms destroy ozone by the catalytic cycle discussed in chapter 2. Finally, continuation of then-current levels of world CFC production, about 800 Kte per year, would bring substantial loss of ozone once stratospheric chlorine reached equilibrium after about a century. They did not calculate ozone loss quantitatively, but only concluded that it was likely to be big enough to be important. Nor did they address the effects of increased UV radiation or potential policy actions, although they began discussing these matters in later publications, congressional testimony, and interviews by the fall. Rather, the paper's fundamental point was the long-term nature of the problem: because CFCs accumulate in the troposphere and are only slowly destroyed in the stratosphere, emissions carry a commitment to future consequences that can only very slowly be reversed. Even if CFC emissions were to cease immediately, ozone loss would roughly double over one or two decades before beginning a 50- to 100-year recovery. This long-term commitment sharply distinguished the effect of CFCs from that of the SST: because SST exhaust would be injected directly into the stratosphere rather than accumulating in the troposphere, stratospheric recovery after curtailment of SST emissions would be much faster.

Recognizing the grave significance of their results, Molina and Rowland waited for the paper to appear before discussing it publicly. During the six-month wait, a few rumors and unauthorized reports circulated, encouraging DuPont officials to shift the emphasis of their CFC environmental research to include preliminary studies of the stratosphere. Still, there was little initial reaction to publication of the *Nature* article.[30] The silence ended abruptly in September 1974, when Rowland and Molina presented their work to an American Chemical Society (ACS) conference and held a press conference discussing its implications.[31]

This conference marked the beginning of a period of intense scientific debate and confused policy conflict in the United States, described by one account as the "ozone war," whose first phase culminated two years later with a regulatory decision to ban CFCs as aerosol spray-can propellants.[32] The issue rapidly gained attention through the fall. National press coverage began with two *New York Times* articles.[33] The National Academy of Sciences (NAS) convened an ad hoc panel on October 26, which recommended that the Academy conduct a full-scale study. On November 20 the Natural Resources Defense Council (NRDC), an environmental advocacy group, petitioned the Consumer Products Safety Commission (CPSC) to ban CFC aerosol spray propellants as hazardous products.[34] By December, the first congressional hearings on CFCs and ozone were held, and two bills had been introduced to restrict CFC aerosols.[35]

Early Industry Response

In the early 1970s, U.S. aerosols used about 200 Kte of CFCs annually. Although other propellants, such as hydrocarbons, ethers, and nitrous oxide, cost less, CFCs were the propellant of choice for the major toiletry uses—hairspray, deodorant, and antiperspirant—as well as for many smaller uses, because they were noncombustible, had no odor, and gave a particularly uniform fine spray. A diverse collection of large and small firms was potentially affected by aerosol restrictions, including CFC producers, marketers of products in aerosol packaging, aerosol fillers, and manufacturers of associated equipment. These firms were highly heterogeneous in their dependence on CFCs and in their scientific sophistication. The chemical manufacturers, the largest firms involved, had the most at stake in absolute, but not in relative, terms. DuPont, the largest producer, had just opened the world's largest CFC plant in Texas in a bid to secure the low-cost position in the market. DuPont was also the producer with the strongest scientific research capacity and had led the establishment of the industry research program on environmental effects of CFCs after Lovelock's measurements showed them to be long-lived. Although producers had established this research program in 1972, the claim that CFCs would harm the *stratosphere* caught them by surprise. Their research had examined CFCs in the lower atmosphere, where they might contribute to air pollution or harm plants, animals, or materials. Since such harms can arise only through the chemicals' reacting, every finding of nonreactivity was thought to vindicate them. In contrast, it is CFCs' very inertness, and their resultant long atmospheric lives, that enables them to deliver chlorine to the stratosphere and accounts for the long time needed to reverse ozone loss. For the first time, chemical inertness rather than reactivity was the source of environmental harm.

Even the scientifically sophisticated chemical manufacturers lacked specific expertise in this new area, and initially were unable to respond to the substance of the claims. The main vehicle for stratospheric research over the prior three years had been CIAP, which had extensive participation from aerospace firms but none from chemical firms. The CFC industry's research program added a stratospheric component in September 1974, but it was not possible for chemical-industry scientists to develop expertise in this area overnight.[36] Caught off guard, industry's initial response was a jumble of generic legalistic arguments, specific scientific claims that appear foolish even in view of the knowledge of the time, and personal attacks on those raising the concern.[37] Developing a coordinated and sophisticated response to the issue took the threatened industries two years, by which time they had lost the aerosol battle.

While all participants in the early debate agreed that more research was needed, they differed over how much should be done, and by whom, before considering

CFC regulation, and what specific findings would warrant either restricting CFCs or setting the issue aside. Industry representatives used one early dynamic modeling calculation to argue that the harm of each year's delay was small, so consideration of potentially unnecessary regulation should await completion of their newly established three-year stratospheric research program.[38]

In 1974, in a statement that proved decisive a decade later, DuPont's McCarthy announced that "if creditable scientific data developed *in this experimental program* show that any chlorofluorocarbons cannot be used without a threat to health, DuPont would stop production of these compounds" (emphasis added).[39] DuPont reiterated the commitment in corporate publications and full-page newspaper advertisements in the spring and summer of 1975, over the signature of Chairman Irving Shapiro.[40] As was widely noted, the pledge retained the discretion to decide what would count as "creditable evidence," and suggested that industry would rely exclusively on results from their own research in forming this judgment.[41]

Early Scientific Results

The controversy over CFCs and ozone provoked a flood of scientific work over the next two years. This work generated much confusion and controversy but in aggregate supported Molina and Rowland's initial claims, while modifying some details and revealing many associated complexities. Several groups quickly replicated and extended Molina and Rowland's work with quantitative ozone-loss calculations using models previously developed to study the effects of NOx from SSTs or HCl from the space shuttle. Molina and Rowland in turn drew on other groups' models to develop their own quantitative estimates. These early calculations gave steady-state global ozone loss from constant CFC emissions ranging from 6.5 to 18 percent, while continued CFC growth gave much larger losses.[42]

Although these early calculations used the best laboratory data available, they were based on few measurements of CFCs and related chemicals in the troposphere, and none in the stratosphere. Obtaining these atmospheric measurements was clearly the most urgent research priority. A November 1974 meeting of scientists from government, universities, and industry reached strong agreement on research needs, identifying four priorities, of which three were atmospheric measurements: (1) measure CFC trends in the troposphere and stratosphere; (2) detect and measure the active species in the chlorine catalytic chain, ClO and Cl, in the stratosphere; (3) measure HCl in the stratosphere, the less reactive Cl reservoir that limits the catalytic chain; and (4) look for mechanisms other than HCl formation that might remove active chlorine from the stratosphere. These four tasks directly addressed the four central points of Molina and Rowland's argument. Measuring CFC trends in the troposphere would verify that CFCs persist long enough for a large fraction of emissions to reach the middle stratosphere. Detecting Cl and ClO would verify that CFCs break apart in the stratosphere and release chlorine and that the chlorine reacts with ozone. Moreover, measuring Cl, ClO, HCl, and other chlorine compounds would help determine the efficiency of the catalytic cycle and improve ozone-loss calculations.

Although with some differences in emphasis, these priorities shaped both the industry research program and the newly established Upper Atmospheric Research Program (UARP) in NASA, which assumed leadership of U.S. stratospheric research after CIAP. It was widely believed that this research program would either confirm or refute the depletion claims, and that a decision whether to restrict CFCs would follow in an obvious and uncontroversial way. But while the research that followed quickly brought strong *qualitative* confirmation of Molina and Rowland's claim, verification of the *quantitative* magnitude and importance of chlorine-catalyzed ozone depletion remained controversial. Indeed, precise quantitative explanation of observed ozone losses and predictions of future losses remained elusive through the 1990s.

Through 1974 and 1975, many measurements found CFCs present and rapidly growing throughout the troposphere and lower stratosphere, with concentrations doubling in four to seven years.[43] These measurements showed that CFC-11's atmospheric life was at least 30 years, but instrument problems and large uncertainties in estimating world emissions prevented further refinement of lifetime estimates. Beginning in late 1974, measurements from high-flying aircraft, balloons, and rocket drops provided the first CFC profiles up through the upper stratosphere. These showed constant concentrations from the ground to the lower stratosphere declining at higher altitudes, a pattern consistent with CFCs' being photolyzed by energetic UV radiation in the middle and upper stratosphere.[44]

Several groups used laboratory work and atmospheric observations to attack the question of whether CFCs photolyze in the stratosphere to release atomic chlorine. Molina and Rowland's initial measurements of CFCs' UV absorption were confirmed and extended by several others, who also verified—as all expected—that the process released chlorine atoms. That CFCs decreased with height in the upper stratosphere also suggested they were being photolyzed there, as did observations of stratospheric hydrogen chloride (HCl), an expected product of the chlorine atoms released from CFCs. Several measurements showed that HCl was undetectable at the tropopause but increased in the stratosphere to a maximum around 25 km.[45] The low values in the troposphere ruled out the possibility that HCl was being transported to the stratosphere, so it had to be formed there from other species, such as CFCs or methyl chloride that delivered the chlorine. The detection of stratospheric hydrogen fluoride (HF) in the summer of 1975 provided further support, since it was thought to arise exclusively from photodissociation of CFCs.[46]

The decrease of CFC and increase of HCl suggested that the chlorine catalytic chain was taking place in the stratosphere, as did laboratory measurements of the reaction of Cl with O_3. But it was widely agreed that decisive evidence of the cycle would be a direct observation of the catalytic intermediary, ClO, in the stratosphere, since the only known source of ClO was the reaction of Cl with O_3.[47] Anderson provided this demonstration with measurements from balloons that first detected ClO in the stratosphere in the summer of 1976.

Other early research yielded revisions to a few key reaction rates, reducing estimated ozone losses by about 20 percent between late 1974 and late 1975.[48] A larger change came in late 1975, when Rowland and colleagues identified a previously overlooked reaction forming chlorine nitrate ($ClONO_2$) that would reduce ozone losses. This compound, formed by reaction of ClO with NO_2, temporarily ties up both N and Cl in unreactive form, reducing ozone losses from both the NOx and the ClOx cycles. Rowland initially advised the NAS panel, based on an old published result, that chlorine nitrate photolyzed so fast that its effect on ozone would be insignificant. But when he and colleagues rechecked chlorine nitrate's lifetime, they found it long enough to significantly reduce ozone loss from both chlorine and nitrogen.[49] Their announcement in early 1976 generated a flurry of hastily publicized claims about the effect of the new result on ozone loss, ranging from small changes to near elimination of losses, many of which were subsequently found to be wrong.[50] Confusion and conflicting claims were so widespread that the NAS panel decided to delay its report in order to resolve the matter.

Response by the Executive Branch

The CFC-ozone issue was sufficiently novel that it was unclear which federal agencies, if any, had the authority to regulate CFCs. At the December 1974 hearings, several members of Congress expressed concern that the authority might not exist and that the agencies had shown little initiative in clarifying it. Responding to this criticism, the heads of the National Science Foundation and the Council on Environmental Quality established a Federal Interagency Task Force on Inadvertent Modification of the Stratosphere (IMOS), with representation from a dozen agencies, to identify the extent and location of federal authority to regulate CFCs, to consider alternative approaches to regulation; and to provide an internal assessment of scientific evidence.[51]

Formed in January 1975, IMOS held a contentious public hearing on February 27,[52] then worked privately through the spring of 1975. Its report, released on June 12, 1975, included a summary of atmospheric science and research priorities, a review of UV impacts, and a preliminary review of CFC uses and potential alternatives. All of these were moving targets, especially atmospheric science. As a vehicle for clarifying regulatory mandates and reaching an agreed regulatory position across executive agencies, IMOS was notably successful. It concluded that authority to regulate CFCs existed only for aerosol uses, and was distributed among the EPA, Food and Drug Administration (FDA), and CPSC for different aerosol products. IMOS recommended against new legislation specific to CFCs, arguing that this would establish a cumbersome, chemical-specific precedent; rather, it recommended passage of the Toxic Substances Control Act (TSCA), which would provide authority to regulate CFCs in all uses.

The most striking aspect of the IMOS report was its statement of an aggressive default position regarding regulation. Stating that CFC releases were "a legitimate cause for concern," the report concluded that "unless new scientific evidence is found to remove the cause for concern, it would seem necessary to restrict uses of (CFCs) 11 and 12 to replacement of fluids in existing refrigeration and air-conditioning equipment and to closed recycled systems or other uses not involving release to the atmosphere." That is, absent new information to refute present concerns, *all* atmospheric CFC emissions should be eliminated. Moreover, IMOS put the onus for reversing this stance onto the NAS panel and committee that had just begun their work. Noting that the Academy was conducting an "in-depth scientific study" of the issue, IMOS recommended that if the Academy bodies "confirm the current task force assessment," federal regulatory agencies should begin rulemaking procedures to restrict CFC uses, which "could reasonably be effective by January 1978."[53]

Predictably, neither industry nor the NAS was pleased with this delegation. Industry objected sharply that IMOS had prejudged the scientific issues the Academy was to consider, by establishing a presumption of regulation.[54] A senior DuPont official wrote to the president of the Academy to ask that the panel resist this inappropriate delegation of regulatory authority by "interpreting the significance of the report to provide guidance to the nonscientist policymakers."[55] Members of the panel, who had thought their mandate was a simple research summary similar to that already included in the IMOS report, were not pleased at having their job greatly expanded. Nor were the members of the Academy committee pleased at having implicit authority for a regulatory decision thrust on them.[56]

Political Fights during Academy Deliberations

After the IMOS report was released, regulatory activity idled while awaiting the Academy reports. Fights for public opinion and over specific proposals raged, however, in municipal, state, and federal bodies, private markets, and the press. Cities and states took the first policy initiatives. The city council of Ann Arbor, Michigan, home of a major stratospheric research group, enacted a symbolic "voluntary" ban on CFC aerosols in the fall of 1974. By June 1975, Oregon had banned the sale of CFC aerosols, effective in March 1977; the New York State legislature had passed a labeling requirement; and bills to restrict CFC aerosols had been introduced in twelve other states and the U.S. Congress.[57]

Federal agencies faced many calls to restrict CFCs but deferred decisions until release of the NAS reports, expressing concern that policies enacted before then would not withstand legal challenge. Following NRDC's first petition to the CPSC in November 1974, the CPSC responded in March 1975 that they were waiting for IMOS to clarify jurisdiction, and then, in July—following a close divided vote—that they were waiting for the Academy reports before making a regulatory decision.[58] The governments of Michigan, Oregon, and New York joined NRDC and the Environmental Defense Fund (EDF) in a September petition to the FDA to restrict CFCs in foods, drugs, and cosmetics, which was also denied pending the Academy report. Ten states joined EDF and NRDC in a second petition to the CPSC in December, citing new scientific results.[59]

Congressional hearings on CFCs and ozone continued through 1975 and 1976, in connection with the Concorde landing-rights controversy and the development of two major pieces of legislation: TSCA and amendments to the Clean Air Act (CAA). In September 1975 hearings, several senators pressed regulatory officials to consider CFC restrictions, with Senator Packwood calling for an immediate aerosol ban.[60] In a position subsequently endorsed by the *New York Times,* one scientist called for immediate restriction of the least essential uses of CFCs, even without decisive scientific conclusions.[61]

Public awareness of and concern about ozone was high: the issue even appeared in a February 1975 episode of the popular television comedy *All in the Family,* when one character brandished a can of hair spray and announced that it was going to destroy the world.[62] Driven by widespread concern and environmental and consumer group boycotts, sales of consumer aerosol products plummeted in 1975. After two decades of 25 percent annual growth, sales of aerosol cans dropped 25 percent in 1975.[63] The conjunction of adverse political and market factors posed difficult strategy and coordination problems to the CFC industries, which they did not manage well. Some groups tried to deny or discredit the scientific claims against CFCs. Though ostensibly intended to disseminate scientific information favorable to industry's positions, this campaign was led by public affairs and marketing officials, and a few industry associations of dubious scientific capability, particularly the Council on Atmospheric Sciences (COAS).[64] This campaign made a series of high-profile blunders. It sponsored a July 1975 U.S. speaking tour by the British atmospheric physicist Richard Scorer, who had gained prominence for attacking ozone-depletion concerns as hysterical and ill-founded but was not conversant with the details of the scientific debate and fared badly in both scientific debates and his press coverage.[65] COAS attracted further derision with an ill-conceived attempt to show that volcanic chlorine would overwhelm CFCs' stratospheric contribution by monitoring an eruption in Alaska in late 1975, and with a May 1976 claim that the newly discovered effect of chlorine nitrate would reduce depletion "nearly to zero"—an extreme interpretation at the time, soon shown to be wrong.

Although the aerosol conflict raised serious questions on which responsible people disagreed, this campaign embarrassed those firms and individuals who cared about their reputations for corporate or scientific responsibility. Some senior executives quietly disavowed the conduct of their marketing and public relations departments,[66] although this two-tracked approach surely held some advantages and might, in some cases, have been intentional. A DuPont letter circulated to Academy members in September 1976 with a highly tendentious interpretation of the committee report released that month received the sharpest censure, an angry published response in which NAS president Philip Handler characterized the action as "unworthy of a great institution that has long been a major contributor to chemical research and to sound technological progress."[67]

Moreover, as the fight continued, large differences of interest within industry emerged. Sectors and firms differed strongly in how much they depended on CFCs and how easily they could replace them. For reasons of cost and performance, different aerosol products used different propellants: alcohol-based products such as hair spray and deodorant used CFCs, while shaving creams and most cleaning products used hydrocarbons and foods used nitrous oxide or carbon dioxide. Non-CFC aerosols were losing sales along with CFC products, because consumers could not tell the difference. Moreover, all CFC aerosol products could use other propellants or delivery systems, although with some cost increase or loss of performance, product quality, or safety. In 1975, many companies began marketing nonaerosol forms of their products (e.g., hand-pump or squeeze containers), or developing new spray systems or propellants. Attempts to hold a united industry stance increasingly conflicted with the interest of those firms least dependent on CFCs in publicly renouncing them. A crucial break occurred in June 1975, when the Johnson's Wax company announced it would eliminate CFCs from its products—which required reformulating only three products of its large line.[68] Other marketers whose aerosols did not use CFCs, or who could easily eliminate them by reformulating products, began advertising their differences—thereby implicitly endorsing the charges against CFCs, despite charges of treachery from their peers.[69]

Perhaps because most participants in the debate tried to ground their policy arguments in scientific claims, movements in the public and political debate followed the direction of announcements of new scientific findings. New findings swung consistently against CFCs through 1975, turned in their favor for a few months in early 1976, then reversed again by the late summer of 1976 as further findings appeared to increase the likely risk.

The NAS Study

After the October 1974 ad hoc panel called for a full study, the NAS study got off to a slow start. The panel on Atmospheric Chemistry was appointed in March 1975, and its 13 members—including one each from Canada, Britain, and Germany—were announced at an American Chemical Society meeting in April; its report was scheduled for April 1, 1976. Following standard Academy practice, the panel membership sought to include relevant expertise but avoided activists strongly committed to a particular view. This panel was to consider atmospheric chemistry and modeling, while the Committee on Impacts of Stratospheric Change would consider impacts of the resultant changes and policy implications.

The new panel and committee initially had modest goals for their reports, but found they came under strong pressure from the outset, and their charge greatly increased as parties began relying on them as the decisive arbiters of the issue. Regulatory agencies resisted calls for immediate action by saying they were awaiting the Academy reports to act. Congressional CFC bills proposed to make regulatory decisions directly contingent on findings by the Academy.[70] IMOS stated that regulation should proceed unless the Academy reports contradicted their presumption of serious concern, and DuPont's letter to the Academy urged them to resist this delegation. Several national governments also decided to await the Academy reports before deciding on CFC regulations, in part because they presumed U.S. action would follow it,[71] while the EPA began organizing an international meeting to consider coordinated CFC controls shortly after the planned release of the Academy studies.

In addition to this external pressure, the panel had the extremely difficult job of assessing a contentious field in which significant new claims were appearing weekly. To the consternation of the sponsoring agencies, the NAS announced that the chlorine nitrate controversy would delay the reports a few months, although chlorine nitrate was only the most serious of several mid-course disruptions setting back the panel's work.[72] After several weeks of intensive work the main sources of discrepancies between modeled effects of chlorine nitrate were identified—principally different means of simplifying how sunlight varies over the day. The panel and committee worked through the summer, while officials waited tensely and the EPA delayed its planned international regulatory meeting to the spring of 1977.

Released in September 1976, the panel report reviewed knowledge of CFC releases, their fate in the troposphere, and their transport and chemistry in the stratosphere. The central conclusion was an estimate based on continued 1973 releases that the "most probable" steady-state ozone loss would be 6–7.5 percent, the upper figure calculated by a one-dimensional stratospheric model and the lower one reflecting a judgmental 20 percent reduction to incorporate the maximum plausible effect of unidentified processes that might destroy CFCs in the troposphere.[73] Adding their judgments of quantitative uncertainties in CFC release rates, stratospheric transport, and key reaction rates gave a combined uncertainty range of "at least" 2–20 percent.[74] The addition of chlorine nitrate to model calculations had reduced projected loss by one-third to one-half, and made the uncertainty bounds wider. In a limited attempt to consider emissions growth, the panel noted that calculated steady-state depletion varied linearly with emissions, so for example doubled constant emissions would give 12–15 percent loss. The only discussion of emission paths other than steady-state was a qualitative statement that if exponential growth resumed, losses would be larger. In a brief consideration of a few other stratospheric pollutants—space shuttle exhaust, N_2O from fertilizer, and methyl bromide—the panel concluded that these should be watched, but posed less immediate risks than CFCs.

The committee, given the even harder job of interpreting the panel's report and other knowledge, offered a set of carefully worded recommendations that were widely, perhaps unfairly, viewed as temporizing. Ambiguities and subtleties in the text allowed advocates on both sides to claim the report had supported their position.[75] Like the panel, the committee highlighted steady-state depletion from continued present CFC production, rather than growth scenarios; in addition, they presented model results showing that the effect of a few years' delay in restricting CFCs would be small.

The committee's central conclusion was that, on the one hand, "selective regulation of [CFC] uses and releases is almost certain to be necessary at some time and to some degree of completeness"; but on the other hand, "neither the needed timing nor the needed severity can reasonably be specified today." Regulation should proceed—not immediately, but "as soon as the inadequacies in the bases of present calculations are significantly reduced, for which no more than two years need be allowed, and provided that ultimate ozone reductions of more than a few percent then remain a major possibility."[76] Clearly uncomfortable with the implicit regulatory authority that had been thrust on them, the committee stressed the political nature of regulatory decisions, and the government's exclusive authority to make them. The report's tone was didactic, seeking to convey general principles for rational policy choice under uncertainty rather than recommending a particular choice. For example, decisions must be made under uncertainty but can be adapted over time; when a decision carries uncertain costs and benefits, delaying it to learn more might be worthwhile; uses should be restricted incrementally, beginning with those that are least costly to restrict; and having better information with which to make decisions in the future requires supporting research on the questions most relevant to the decisions.

The U.S. Aerosol Ban: End of the First "Ozone War"

Both the panel and committee reports were released at a conference on the stratosphere in Logan, Utah, September 15–17, 1976. At this meeting, several important new results were presented that were not available to the NAS panel, including Anderson's report of the first detection of Cl and ClO in the stratosphere and Weiss's report of a ten-year trend of increasing atmospheric N_2O. But the most important event of the meeting was the articulation by senior officials of an aggressive new regulatory position. In a forceful speech anticipating the precautionary principle, Council on Environmental Quality chair Russell Peterson argued that the criminal defendant's presumption of innocence was not appropriate for regulatory decisions under uncertainty.[77] Both Peterson and CPSC commissioner David Pittle criticized the Academy reports for "deciding how much uncertainty was enough for policy, . . . assuming the role of the policy maker."[78] Canadian and Norwegian officials attending the meeting both announced their intention to restrict CFCs.

Whatever the committee had intended, officials immediately began moving toward regulations.[79] United States regulatory agencies had already begun assessing possible responses and commissioned a consultant's study of control options and costs immediately after the Academy's reports.[80] On September 21, IMOS recommended that federal agencies begin rulemaking, noting that the Academy reports were silent on regulatory lags and when to start the process of regulation.[81] On October 12, the FDA announced the start of rulemaking on behalf of the three regulatory agencies; EPA administrator Russell Train made the same announcement at a NATO meeting in Brussels, asking other nations to take similar action and inviting them to a regulatory meeting in Washington in the spring of 1977. DuPont called the decision astonishing and contrary to the NAS's recommendation and vowed to fight, but to no avail.[82]

Late in October 1976, a plan for implementing regulations was announced, in two phases: Phase 1 would restrict nonessential aerosol uses of CFCs 11 and 12 by April 1977, while Phase 2 would restrict additional uses of CFCs and other ozone-depleting chemicals by June 1978.[83] A new inter-agency group worked through the winter to draft the Phase 1 regulations, which were discussed at a public hearing in December 1976, then announced jointly by the three agencies on May 11, 1977.[84] With little further opposition from industry, final regulations were announced unchanged on March 15, 1978.[85] The regulations banned use of CFCs 11 and 12 in nonessential aerosol sprays, preempted most state regulation, and defined criteria for essential-use exemptions.[86] Many applications for essential-use exemptions were filed, but very few—principally for prescription drugs and contraceptives—were granted.[87]

Even by the time draft regulations were announced, U.S. aerosol use of CFCs had fallen by nearly three-quarters, reflecting consumer switching and rapid development of alternatives by aerosol marketers and packaging firms.[88] Several further innovations followed the draft regulations, including a new non-CFC aerosol propellant system announced one day later by the Precision Valve Company, which had been one of the fiercest opponents of the ban.[89] The CFC makers played no role in this innovation, as their products were already premium propellants used only where their favorable properties improved product quality enough to justify their cost, and there was clearly little room in the fast-moving aerosol market for still costlier CFC alternatives of uncertain availability. With such rapid market movement away from CFCs, the ban came into force with little or no disruption.

Before the final aerosol regulations were enacted, the 1977 Clean Air Act amendments changed the terms for future controls of ozone-depleting substances, establishing a coherent statutory basis for regulation and conferring a remarkably strong precautionary obligation on the EPA administrator. The act required the administrator to commission biennial ozone studies from the Academy and to regulate any substance that "may reasonably be anticipated to affect the stratosphere, especially ozone in the stratosphere, and such effect may reasonably be anticipated to endanger public health or welfare."[90] Regulation was thus required under "reasonable anticipation of harm"; and while the administrator was required to consider feasibility and cost of complying, explicit comparison of benefits and costs was not required.[91]

Analysis and Conclusions: U.S. Ozone War

However bitterly fought the first U.S. ozone war was, its resolution was remarkably fast, simple, and seemingly rational, eliminating half the U.S. contribution to ozone risk at low cost. The cogency of the initial scientific claim and the support it gained from subsequent results facilitated this strong resolution. In addition, many policy actors and scientists were primed with knowledge and concern about ozone by prior debates over the SST and, less publicly, the space shuttle—conditions that sharply distinguished U.S. policy debates from those in Europe.

The unique relationship between the IMOS and NAS reports also contributed to this strong policy outcome. IMOS effectively stated a presumption in favor of regulating, and imposed on the NAS committee the burden of saying otherwise. Many other actors also deferred policy decisions pending the Academy's conclusions, effectively delegating to them the decision whether or not controls were warranted. Stuck in this position, the committee made explicit policy recommendations that exceeded its expertise or mandate, which were decisive in promoting the U.S.

aerosol ban. Although the committee's overreaching was an understandable response to the expansive role thrust on them, it established a harmful model for scientific assessments and weakened the credibility of subsequent Academy reports on the issue. This effect was particularly acute internationally, where the only governments who put strong credence in Academy reports were those already disposed to follow the U.S. lead on CFC controls. Those who were initially skeptical viewed the Academy reports simply as scientific supporting documents for the U.S. government position.

A second powerful contributor to the initial policy outcome was the immediate, nearly exclusive focus of the policy debate on aerosols. Advocates fixed on an aerosol ban as the initial goal from the outset and did not waver until it was enacted. All participants understood that aerosols were only half of U.S. CFC use and that the other half would also need attention if the problem was serious enough—indeed, U.S. regulators simultaneously announced their intention to regulate aerosol and nonaerosol uses—but everyone treated aerosols as the first target, for good reasons. Aerosols were the single largest CFC use, growing fast and aggressively marketed; their use was unavoidably emissive; they were known to be readily replaceable in most uses; and the inessential, even frivolous nature of most uses—a convenient packaging mode for toiletries—made a powerful, widely employed media image. These factors made aerosols both an appropriate target for pursuing large, immediate cuts, and a vulnerable one. Moreover, while overall authority to control CFCs was unclear, CPSC clearly had authority to ban CFC aerosols if they deemed them a "hazardous product"—and was, moreover, obliged to answer petitions within 120 days. The NRDC initially targeted aerosols through CPSC, and other activists restricted their initial attacks to aerosols, for these immediate, pragmatic reasons.[92]

A final factor promoting the quick U.S. regulatory outcome was that the major opponents, the chemical and aerosol industries, were so ineffective. United States industry lost the aerosol battle on every front—market, scientific, and political. The sharp popular retreat from aerosols in consumer-product markets required product marketers to defend their sales by loudly abandoning CFCs, a movement that CFC producers were helpless to resist. In the scientific debate, restrained industry voices advocating cautious skepticism, limited delay, and vigorous research were discredited by their more inflammatory allies, tactically outmaneuvered by IMOS's success at shifting the terms of debate to establish a presumption in favor of controls, and ultimately overwhelmed by widespread support for a more precautionary stance. Industry groups did not make the best argument against aerosol controls—that single-use controls are inefficient and ultimately futile—lest they be interpreted as supporting broader controls. In the political arena, they faced a groundswell of state and local opposition threatening a patchwork of inconsistent rules, relative to which federal controls were strongly preferable. Finally, the chemical industry could not risk sustaining full-scale opposition to CFC aerosol controls, which threatened a small fraction of their business, when TSCA under development posed a broader and more serious threat. Industry's loss in the aerosol war was so decisive that the May 1977 announcement of the aerosol ban only formalized the defeat.

Even as the aerosol ban came into force, however, it was clear that unless emissions from other uses, of other chemicals, and in other countries were also reduced, global emissions would overtake their earlier peak by 1985.[93] However reasonable the U.S. aerosol ban was as a first step, this decision and the means of reaching it obstructed later attempts to limit these broader contributions to ozone depletion in several ways. The decision established the precedent of an ad hoc policy response, reflecting the accidental conjunction of strong public pressure and a readily available option, with no strategic view of the problem. Similarly, it established a precedent of controlling CFCs by categorical product bans rather than quantitative controls on aggregate emissions or production. Despite the obvious unfeasibility of extending product bans use by use, many actors continued to promote this approach because it had succeeded once, thus long obstructing a shift in the policy agenda toward more comprehensive and rational approaches. More specifically, two immediate reactions to the ban, which were predictable but widely overlooked, limited its effectiveness. First, ongoing CFC growth in other uses accelerated, as manufacturers pursued other markets for their excess capacity. Second, public and political attention to the issue declined sharply as soon as the initial aerosol regulations were enacted. Most seriously, given the structure of CFC uses, the U.S. aerosol ban created obstacles to further movement both internationally and domestically. Because an even larger share of CFC production went to aerosols abroad than in the United States, an aerosol ban would impose larger harms on foreign producers. They and their governments predictably opposed calls to replicate the U.S. ban, citing their higher costs as well as their greater skepticism about the risk. The United States, having made the easy cuts, would face sharply increasing costs and resistance to moving further but would be unable to persuade others to match them.

From NOx to Nuclear Winter

After CFCs overtook SSTs as a policy issue, debates over stratospheric effects of NOx shifted in character and triggered a separate controversy over the environmental effects of nuclear war. The atmospheric nuclear tests of the 1950s provided a unique historical experiment to test the effect of large stratospheric NOx injections. A 1973 paper suggested the NO produced by tests around 1960 should have caused a 4–5 percent ozone reduction but that this could not be seen in the ozone record, sparking a long debate over the existence and possible causes of ozone trends.[94] A *Nature* paper the next year made the obvious extrapolation that if a few nuclear blasts would destroy 5 percent of ozone, a full-scale nuclear war could cause extreme ozone losses.[95] The director of the U.S. Arms Control and Disarmament Agency began promoting this idea in speeches and publications and commissioned a small NAS study that affirmed, implausibly, that the greatest harm to noncombatant nations in a nuclear war would be loss of the ozone layer. The Pentagon replied that even if a nuclear exchange destroyed 50–75 percent of temperate-latitude ozone, the remaining ozone would still be similar to present levels in the tropics, where life survives; and that the higher-yield Soviet warheads would be responsible for most of the loss.[96] The debate elicited widespread alarm and derision, and CFC producers were caught in the crossfire when the *New York Times* denounced their stance against CFC-ozone claims and the more egregious Pentagon stance in the same editorial.[97]

Although calculated ozone losses from NOx declined by 1980, the journal *Ambio* still asked Paul Crutzen to estimate them for a special 1982 issue on the environmental effects of nuclear war. Crutzen and a colleague first examined whether ozone increases in the upper troposphere might be a larger effect than losses in the stratosphere. But in calculating changes in this photochemistry, they noted that the sunlight driving it would be reduced by smoke and dust from the fires ignited by nuclear blasts. Their estimates of this reduction showed an extreme darkening of the earth's surface lasting many weeks, which would likely preclude crop growth in much of the northern hemisphere—the phenomenon that came to be called nuclear winter.[98]

3.3 Aerosol Debates outside the United States and Early Efforts at International Cooperation

As the world's largest producer and consumer of CFCs and the largest sponsor of stratospheric research in the 1970s, the United States experienced the most heated early controversy over CFCs and led its initial resolution. Other industrialized nations experienced debates that roughly paralleled events and arguments in United States, but which came to diverse ends reflecting distinct institutional and political settings.

Several other nations had extensive stratospheric research programs, built in part around the network of ozone monitoring stations developed for the International Geophysical Year of 1957. The United Kingdom was the early leader in the field, building on pioneering work by Hartley, Dobson, Chapman, and others, until the rapid advance in U.S. research capacity in the 1970s spurred by CIAP. Some research communities concerned with ozone had enjoyed close international cooperation for decades, particularly in atmospheric dynamics. Because ozone was thought to be useful for tracing atmospheric movements and thus for weather forecasting, international programs for standardized monitoring, research, and data sharing began in 1930, and greatly expanded with the postwar founding of the World Meteorological Organization and the IGY. In part because of these scientific networks, information moved rapidly among nations when ozone became a policy issue in the early 1970s. In both the SST and CFCs controversies, new results or arguments from American debates typically appeared within days or weeks in the press and policy statements in Canada, the United Kingdom, Germany, the Netherlands, and the Nordic countries. Although some actors called for CFC controls in all these countries, the most serious and organized—though unsuccessful—campaign for controls was in the United Kingdom. Here, individual Members of Parliament raised the question in the fall of 1974, and by July 1975 calls for aerosol controls were made by the London *Times*, several articles in the *New Scientist*, and the Labour Party opposition.[99]

But these debates were among elites. Despite extensive reporting of the U.S. debate and calls for aerosol controls by a few citizens' groups in Canada, Germany, and the United Kingdom, these other countries saw little public arousal. The only exception was the Netherlands, where a coalition of environmental groups and a socialist political party led a consumer campaign and boycott against CFCs that achieved a 17 percent reduction in aerosol sales in 1975 and a further 19 percent reduction in 1976. As in the United States, Dutch aerosol marketers responded by switching to non-CFC propellants and alternative packaging, but because the Netherlands was a major CFC exporter, production declined by much less than consumption. Moreover, despite these citizen-led accomplishments and various calls for formal regulations, Dutch policy decisions were strongly constrained by the need to coordinate within the European Community (EC).[100]

From the outset, the argument for managing CFCs internationally appeared compelling. Because a dozen countries produced CFCs, and because both CFCs and ozone were mixed globally, even the largest producers could not solve the problem alone. Moreover, enacting domestic controls even as a first step posed legal and bureaucratic challenges, because CFCs fit awkwardly into existing regulatory authority in most countries. Air pollution laws typically targeted ground-level air quality, while hazardous substance and product laws concerned direct and immediate hazards. CFCs, which represented a long-term cumulative hazard from polluting the upper atmosphere, fell through the cracks.[101]

Many saw this logic and pushed for cooperative international action through several bodies. Throughout the U.S. domestic aerosols debate, officials repeatedly called for other countries to enact CFC aerosol controls, or for an international mechanism to control them.[102] Canadian scientists and officials, who had participated in U.S. ozone debates since 1971, were the strongest early proponents of international action. They lobbied other officials bilaterally and led efforts in the United Nations Environment Programme (UNEP) and the World Meteorological Organization (WMO) that resulted in both organizations calling for international coordination of research and policy. While pursuing these international initiatives, the same officials also skillfully used the resultant international statements to advance domestic action in Canada. These efforts bore fruit in November 1976, when an advisory panel report that drew on international statements from UNEP and WMO and the 1976 NAS panel concluded that the risk to ozone justified policy action, and the environment minister announced within days that he planned to restrict aerosols.[103]

Despite the intrinsic advantages of international control, and attempts by Canadian, American, and Nordic officials to persuade other major CFC-producing nations to participate, the first several years' pursuit of international cooperation on the ozone layer were an unmitigated failure. Although the specific conditions of each failure were unique, in aggregate the advocates simply failed to mount enough pressure on the major CFC-producing nations to overcome their vigorous, organized resistance. The forms of resistance varied among the major producer nations. Japanese officials, for example, simply ignored the issue throughout the 1970s, rebuffing all bilateral and multilateral initiatives. A Japanese Environment Agency report questioned the U.S. EPA's concern about ozone, citing the popular belief that Asians were not at risk from UV-induced skin cancer.[104] The Soviet Union supported extensive stratospheric research, but ozone-depletion concerns were quickly declared not sufficiently compelling to justify action.[105] Officials in France and Italy left the issue in the hands of their national CFC producers. In Germany the government was divided, with vigorous activism by a few officials in the Environmental Research Agency (UBA) set against a broader climate of opposition to further regulation.[106] European resistance was led by U.K. and EC Commission officials, who vigorously opposed any consideration of CFC controls and argued that scientific understanding of the stratosphere did not justify them. In addition to rejecting calls for CFC controls, these officials—and other European officials in turn—also rejected American scientific assessments, whether produced by the U.S. government or by the Academy.[107] The United Kingdom, alone among the opponents in facing some domestic pressure for CFC controls, was also the only opponent to sponsor its own scientific assessment of the issue, which appeared in April 1976 at the height of the controversies over U.S. Concorde access and chlorine nitrate.[108] Unlike France and Italy, the United Kingdom kept governmental control of the issue, but still showed substantial solicitude toward, and accorded substantial influence to, its major national producer, Imperial Chemicals Industries (ICI).

Several organizations were plausible vehicles for considering CFC-ozone policy at the international level, including the World Meteorological Organization (WMO), the Organization for Economic Cooperation and Development (OECD), and the United Nations Environment Programme (UNEP)—although none of these was a perfect fit. Other international bodies also considered the issue—so many that by late 1975, the United States needed an interagency committee to coordinate positions across international bodies discussing CFCs and ozone.[109]

Most international bodies simply provided a forum for national officials to discuss the issue. More concrete initiatives took place under WMO, OECD, UNEP, and an ad hoc international process led by activist officials from a few nations, particularly the United States. Of these, WMO showed no interest in making policy contributions, limiting its ambition to expanding its existing ozone monitoring network. Beginning in 1975, WMO issued periodic short statements of concern on the ozone layer, drafted by ad hoc groups of eminent scientists and then approved by WMO's Executive Council, but these restricted their conclusions to recommending expanded ozone research and monitoring under WMO leadership.[110] In this aspiration, WMO had to fight a widespread perception that it was dominated by operational meteorologists of limited scientific ability, a perception that was especially prevalent in those countries—including the United States—where weather forecasting and atmospheric research were housed in separate institutions.[111] When WMO's proposed program received only very limited resources, it was scaled back to improving the calibration and operation of the existing ozone stations—an unglamorous contribution, but one of crucial importance for developing the reliable long-term measurements necessary to identify and analyze ozone trends.

The OECD included all major CFC-producing nations except the Soviet Union. Although primarily a consultative organization, the OECD had the authority to negotiate agreements among its member states, as well as specialized subbodies—the Environment Committee (at ministerial level) and the Chemicals Group under it—whose mandates could have included the CFC-ozone issue.[112] After a May 1975 discussion of CFCs in the Chemicals Group, the United States and Canada submitted notes to the Environment Committee in June, asking the OECD to take action on the issue.[113] The committee responded by asking the two governments to prepare a preliminary report on CFCs. Submitted in November, the report took a moderately strong line—subsequent research had confirmed the ozone-depletion hypothesis, but uncertainties remained and more research was needed—and summarized international CFC production data newly supplied by OECD member states.[114] On the basis of this report, the Environment Committee in March 1976 adopted a work plan charging secretariat staff to investigate CFC substitutes, policy options, and economic impacts of CFC controls, but leaving scientific issues to other bodies. This initiative accomplished little, however, and by late 1976 U.S. officials were pressing to redirect pursuit of international regulatory cooperation to a new ad hoc process that they initiated. The OECD staff report, published in January 1978, relied on consultants' studies previously commissioned by the Dutch government and the U.S. EPA. Like them, it examined only an aerosol ban, concluding unsurprisingly that its cost would be large if enacted rapidly (e.g., in six months), but modest if phased in slowly. With publication of this report, OECD activity on CFCs effectively stopped for three years.[115]

In principle, UNEP showed the most promise as an institution for managing the ozone issue internationally. UNEP had several serious practical limitations, however, including a small and uncertain budget, a mandate that excluded direct programmatic work, and an inconvenient location in Nairobi, Kenya. Still, several early activist officials proposed that the breadth and novelty of the ozone issue, while making the issue an awkward fit for existing organizations, made it appropriate for UNEP. Although UNEP was weak, it was not narrow or entrenched, so it might be able to address the atmospheric science, UV effects, technology, and policy and regulatory issues with the integration that the issue required. UNEP's first executive director, Maurice Strong, proposed stratospheric ozone as an area of UNEP activity in 1973, and UNEP staff were eager to make a mark by showing leadership on the issue.[116]

In its first ozone initiative, UNEP helped fund the June 1975 meeting of scientists to draft WMO's first statement on ozone.[117] UNEP's new executive director, Mostafa Tolba, then used the WMO statement to support his call in the 1975 Governing Council meeting to give high priority to ozone. The council approved the proposal to address ozone under UNEP's environmental assessment mandate and also accepted a U.S. proposal that UNEP convene an international scientific meeting on ozone but rejected a proposal by Canada and Sweden that UNEP also convene negotiations for a treaty to protect the ozone layer.[118] Many national officials doubted UNEP's ability to successfully coordinate either scientific information or policy. In addition to impediments of its budget, mandate, and location, UNEP suffered severe staffing problems at the time and was riven by north-south conflict over how to balance environment and development issues.[119]

UNEP's scientific meeting was held in Washington, D.C., in March 1977. Scientists and officials from 32 countries discussed papers that reviewed areas of relevant scientific knowledge and described existing research programs. The meeting adopted a "World Plan of Action" on ozone, comprising a long list of recommendations for ozone-related research, monitoring, and information-sharing. It also recommended that UNEP establish and chair an international scientific committee to coordinate research and periodically summarize the state of knowledge. Especially in view of the meeting's quasi-official character, its recommendations were impressively detailed and specific, and some turned out to be important: one of them, for example, requested that the pending removal of the United Kingdom's ozone meter at Halley Bay, Antarctica (the instrument that several years later detected the Antarctic ozone hole), be canceled, and the instrument reactivated.[120]

Beyond the scientific agenda, however, participants' goals for the meeting differed so sharply that it ended in deadlock. Some UNEP staff and national officials (particularly from Norway and Canada) wanted UNEP to convene international policy discussions, at least to consider hypothetically what to do if further research and observations revealed a major risk. Europeans, now led by Germany as well as the United Kingdom, argued the risk was not well enough established scientifically to justify any policy discussions, however. In response, the proponents narrowed their ambition for UNEP to international coordination of scientific research. For their part, American officials wanted international policy discussions, but not under UNEP. They had grown frustrated with UNEP's administrative limitations and cumbersome inter-governmental organization during preparations for the Washington meeting, which were slow and precarious despite the loan of several staff by the American and Canadian governments. Instead, U.S. officials sought to pursue international regulatory coordination through an ad hoc process separate from any international organization. The head of the U.S. delegation invited participants to attend the first ad hoc meeting, to be held in Washington two weeks later, but interest was so low that postponing the regulatory meeting was briefly considered. With the scientific plan completed and no progress being made on policy or regulatory issues, the meeting ended two days early.[121]

The same divisions persisted at UNEP's Governing Council meeting in May 1977. The council endorsed the Plan of Action drafted in Washington and agreed to establish the proposed international scientific committee, the Coordinating Committee on the Ozone Layer (CCOL), but divided over whether it was appropriate even to discuss international control of threats to the ozone.[122] UNEP staff and the most activist national officials had great hopes for CCOL, however. They believed that CCOL's coordination of research programs and periodic summaries of progress would support a shared, integrated, international understanding of the issue, and that such an understanding would promote development of appropriate international regulatory controls. They gravely underestimated the magnitude of the task. For several years the CCOL was the only international body reviewing ozone science and conducting assessments, but it was weak and ineffective, as I discuss in chapter 4.

Consequently, there was no international organization willing and able to coordinate international policy on the ozone issue. With the organizations having relevant mandates either uninterested, blocked by forceful opposition of some members, or regarded by important national actors as incompetent, the job fell to the ad hoc process promoted by U.S. officials. This process began with the Washington regulatory meeting of April 1977, held two weeks after UNEP's scientific meeting. As a face-saving measure, the earlier UNEP meeting was described as having provided scientific input to the regulatory meeting, although many of its participants had much larger ambitions that failed. Chaired by the three U.S. agencies regulating aerosols, the meeting was an informal closed gathering of officials from 13 countries (including all major CFC producers) and 5 international organizations. American and Canadian officials discussed their proposed aerosol bans, and the EPA distributed a consultant's study of the cost of aerosol controls. Other governments favored awaiting more scientific information before regulating, noting that regulatory lags are longest in the United States, so they could afford to wait and still act quickly if necessary. Nevertheless, all participating governments except Japan signed a non-binding resolution supporting reductions in nonessential aerosols.[123]

A few further national regulatory initiatives had occurred by the time of the Washington meetings, and the regulatory meeting set more in motion. Canada's initial voluntary cuts of 50 percent in aerosols took effect in late 1977.[124] In Germany, environmental officials had initially pursued an aerosol ban, but their case was weakened by two assessment results: the 1976 NAS committee's statement that controls could wait two years for more research; and a Germany study of control costs that considered only an immediate ban on all aerosols, which predictably projected large losses.[125] Opposed by other ministries and lacking support for binding controls from these studies, the officials instead pursued voluntary controls negotiated with industry. In early 1977, the minister of interior and the aerosol industry agreed to voluntarily reduce CFCs 11 and 12 in aerosols voluntarily, by 30 percent from 1976 levels by 1979.[126] Switzerland mirrored the German aerosol cut, while Denmark had achieved a 24 percent reduction by 1978 with no regulatory measures.[127] In the Netherlands, the government relied entirely on the NAS report for scientific assessment and commissioned a consultant to assess the economic effects of banning aerosols. In contrast to the German study, this one concluded the impacts would be "tolerable."[128] The Environment Ministry sought an aerosol ban, but government-industry consultations of early 1977 settled on more limited measures: warning labels on CFC aerosols, and a resolution to proceed with aerosol controls if other major producer nations did so and the second NAS assessment, expected in 1979, reported no new contrary scientific information.[129]

After the Washington meetings and several months of informal consultations, the EC Commission in August 1977 proposed a recommendation that member states freeze their production capacity for CFCs 11 and 12. With this commission initiative and their own voluntary aerosol cut, German officials deferred plans for a ban or compulsory labeling, and began working with the commission to develop European controls compatible with trade harmonization. In May 1978, the Council of Ministers acted on the commission's proposal, adopting a recommendation for a production-capacity freeze and measures to promote development of substitutes, rejecting calls for stronger measures from the Netherlands (compulsory labeling) and the European Parliament (a ban on nonessential aerosols).[130]

Sweden and Norway, whose officials had been sympathetic to CFC controls from the start, also moved to action after the 1977 Washington meetings. Like the Netherlands relying heavily on NAS reports for scientific assessments, both nations enacted aerosol bans rapidly and with little controversy. At the government's request in late 1976, the Swedish Academy of Sciences reviewed the NAS report and the EPA consultant's economic analysis, and recommended controlling aerosols.[131] After industry representatives reported that they would not be substantially affected by a CFC aerosol ban,[132] the government announced in December 1977 that it planned to ban manufacture and import of CFC aerosols, except for medical products, in June 1979. Norway announced an equivalent ban in December 1978, to take effect in July 1981. These bans posed very little domestic hardship for either country. Neither of them manufactured CFCs, while only Sweden had a small aerosol-filling sector employing about 500 people. Both bans were enacted over objections from Finland, which produced most of the aerosols sold in the Nordic market.

Despite internal divisions, the German government followed the U.S. lead and in 1978 hosted the annual CCOL meeting and a second ad hoc regulatory meeting one week later, this time with all major CFC-producing nations attending except Japan and the Soviet Union.[133] The force behind the regulatory meeting, as for much early German action on the issue, was Schmidt-Bleek of the environmental research agency Umweltbundesamt (UBA), who also commissioned the American environmental group NRDC to survey CFC policies and statutory authority in 12 countries.[134] EPA deputy administrator Barbara Blum, leading the U.S. delegation, again called for multilateral aerosol cuts, supported by Canada, Norway, Sweden, Denmark, and the Netherlands. Sweden and Norway both announced their planned aerosol bans. Germany took a weaker position despite its role as host, reflecting the delicate agreement reached among ministries to support voluntary reductions but not regulation. The United Kingdom, France, and Italy continued to argue that available scientific evidence did not support any form of reductions, and forced the EC to block a resolution calling for specific quantitative reductions. Instead, the meeting adopted a weaker resolution calling for work toward unspecified reductions in global CFC emissions. Industry representatives were not invited to the meeting, and the British firm ICI protested publicly that CFC control decisions were being discussed by politicians with no technical understanding of the issues.[135] The threat that ICI perceived was not real, however: the meeting achieved no concrete agreement and no convergence of views on regulation and had no follow-up. Proponents of this process identified this meeting as the end of hope for its success.[136]

Blockage by major European CFC producers remained crucial at this stage, as the primary activity on the issue shifted to the EC immediately after Munich. In the first meeting of the German presidency (the first half of 1979), the German environment minister forcefully called for the council to enact stronger European action and asked the commission to submit a proposal during his presidency on which a decision to reduce CFC aerosols could be based.[137] As in the ad hoc process, the strongest opposition came from the United Kingdom, where the weak domestic support that had previously existed for controls had faded, and several powerful forces were now aligned against controls. The Thatcher government, elected in April 1979, was aggressively antiregulatory. ICI, the largest CFC producer in the United Kingdom, was also Britain's largest industrial firm and the CFC producer whose markets depended most strongly on aerosols. The scientists and officials working on the issue were the same ones earlier involved in the fight over the Concorde, many of whom retained a distrust of American environmental claims based on the earlier controversy. Even the *New Scientist*, which prior to 1978 had published several evenhanded discussions of the CFC issue and associated scientific controversies, adopted a hostile stance toward claims of risk after publishing a 1978 interview with the aggressive British skeptic James Lovelock.[138] Supported by France and Italy, U.K. opposition prevented concrete progress in the EC as it had in the ad hoc process, with U.K. officials now supporting their opposition with reference to a second British scientific assessment completed in 1979.

As the German environment minister had requested, the commission brought forward a proposal for EC action in May 1979, but it was weak and cosmetic. It proposed to reaffirm and elaborate the production capacity cap already adopted, add a requirement to reduce CFC aerosol use by 30 percent by 1982 (a reduction that was already achieved by market shifts), report production and consumption data to the commission, and reassess the issue periodically, beginning in 1980. The Dutch, Danes, and European Parliament sought stronger measures, without success. The commission refused even to recommend a Europe-wide extension of the Dutch labeling requirement. The proposal was adopted as a council decision in March 1980. After this European action, the possibility of stronger national action in the more concerned nations was even more restricted. One Dutch environment minister sought 50 percent CFC aerosol cuts, without success. After these defeats, the ozone issue disappeared from high-level national policy agendas in EC nations until 1986.[139]

The last meeting of the ad hoc process took place in Oslo in April 1980, this time without an accompanying CCOL meeting. Since all attempts to urge controls on the major European producers had failed, this meeting was planned as an informal session for activists to discuss how to put more effective pressure on the EC, in particular whether to propose that CFCs be included in the newly signed Convention on Long-Range Transboundary Air Pollution. Not all activists agreed with the tactic of excluding major European states, however, particularly after Denmark proposed to raise the meeting's prominence by inviting environment ministers. In the end one representative of the EC Commission attended, who summarized the recent council decision and noted that both the EC and the OECD were to reconsider CFCs in 1980. Officials from the U.S. EPA announced their plan to cap overall CFC production and discussed possible substitutes and forms of controls. Canadian officials weakly committed to "investigating ways to achieve the same objective." Although other delegates praised these steps and the principle of aggregate emission reductions, none made further commitments and no concrete agreements were reached.[140] The report of the meeting noted that a few countries' aerosol reductions could be offset by growth in nonaerosol uses of other countries and suggested that protecting the ozone layer might require an international treaty. One U.K. official denounced the meeting as illegitimate due to the exclusion of the United Kingdom and France, and argued that calling for CFC controls carried no political cost for Denmark or the Netherlands—Denmark because it had no aerosol producers, and the Netherlands because it used very little CFC propellant.[141]

Shortly after the Oslo meeting, EPA officials sought to persuade French officials bilaterally to accept the Oslo recommendations, although France was not invited to the meeting. The reception was hostile, partly because the French suspected the United States of promoting the Oslo meeting to circumvent existing institutions and thus have more control over shaping the agenda—a tactic the United States had used in 1978 over European objections to shape the OECD agenda for regulating toxic chemicals. Bilateral consultations with several other nations, including Japan, also made no progress.[142] The ad hoc process was dead.

The Significance of Early International Efforts

These early international efforts accomplished little. In addition to the U.S. aerosol ban, Canada, Sweden, and Norway (none of them a major CFC producer) enacted similarly strong aerosol bans by 1980, while Germany, the Netherlands, and the EC enacted various weaker measures. Although there were significant constituencies for stronger measures in Germany and the Netherlands, as well as in Denmark, these were thwarted by complexities of coordinating policy within the EC and by significant domestic opposition in Germany. Other major producer nations did nothing, with stances ranging from disinterest to hostility. The nations that undertook some action all accepted the NAS's scientific assessment of the issue with little or no national review, while the firmest opponent of action, the United Kingdom, was the only nation outside the United States that had similar scientific capacity on the issue and conducted its own scientific assessments.

Beyond these national measures, a very limited international framework was established to coordinate scientific research and monitoring under UNEP's CCOL. Though the CCOL accomplished nothing in the 1970s, many continued to hope it would catalyze a scientific consensus that would eventually prompt international policy cooperation. Moreover, the dual scientific and regulatory meetings in Washington and the subsequent ad hoc process provided some stimulus to both the real and the symbolic actions of other nations. Swedish and Norwegian officials had been sympathetic to action since the first emergence of the issue, but were moved to action by these international discussions. German activist officials used the international meetings to advance policy proposals and deliberations domestically, although they were constrained to stay in step with the EC and blocked within it by the firm opposition of the United Kingdom and France. The EC Commission responded to pressures both from the international process and internally from Germany and Netherlands to acknowledge the problem and enact initial measures, although these were so modest as to be largely symbolic. The constitutional status of environmental issues in the EC was ambiguous, and the commission had a strong interest in establishing its own authority by establishing the EC as the appropriate level for action. But with three of the four largest member-states opposing action, this institutional shift to the EC level accorded little influence to the activists, even while limiting their discretion to act nationally.

In view of these small advances, the attempt to develop substantive international cooperation to protect the ozone layer through this period must be judged a complete failure. No government supported action through any international process that had not already indicated willingness to act unilaterally. The activists failed to muster the resources needed to either persuade or pressure the major opponents—first the United Kingdom, with at least passive support from France, Italy, Japan, and the Soviet Union—to accept action. The reasons for this failure are instructive both for understanding the subsequent success in gaining international cooperation on ozone, and for general questions of the conditions and determinants of international action.

A possible, but ultimately unsatisfactory, explanation is that the attempts failed because they should have failed. That is, the opponents were correct that the risk was not well enough established to justify international action. Several pieces of evidence undercut this claim. The evidence of risk was persuasive enough between 1975 and 1978 to move action in several countries despite no one's having a material interest in the controls, and appeared still stronger in 1979 and early 1980. Opponents' claims that they resisted hysteria and weighed available scientific evidence more fairly are weakened by the tactics of argument they employed, in that they selectively highlighted favorable results and mixed scientific with political arguments at least as aggressively as did the most vigorous activists. The related argument that aerosol bans were such ill-conceived controls that it was rational to reject them independent of the perceived risk, while superficially more attractive, is also unsatisfactory. An aerosol ban had powerful advantages as a pragmatic first step, and its flaws were not widely recognized at the time. The efficiency advantages of more flexible regulations, widely recognized by the late 1980s, were at the time little known outside the economics profession. The only proponents of more rational broad-based CFC production at the time were European officials, and the specific terms of their proposed production-capacity cap were so transparently intended to exert no actual effect that their general arguments, however reasonable, were ignored.

Moreover, these arguments do not explain why the advocates, right or wrong, failed to move their opponents. Evaluating the effectiveness of the international approaches pursued requires examining the resources each one provided to resolve differences of opinion or preference, to persuade or pressure the reluctant, to conclude agreements, and to coordinate actions to mutual benefit. Both the UNEP-led CCOL process and the U.S.-led ad hoc process had weaknesses that rendered them unable to accomplish any of these tasks.

The CCOL sought, by reviewing research and conducting assessments, to narrow disagreements over the content and significance of scientific knowledge, thereby helping to narrow policy disagreements by founding them on a common body of agreed scientific knowledge. The CCOL's weaknesses are discussed in detail in chapter 4, but in brief it failed because it lacked the resources or stature to articulate a broad scientific consensus, and because it had no policy body as its client. However strong and persuasive the CCOL reports might have been (and they were not), no body with policy-making authority on the issue was under any obligation to receive, consider, or respond to them.

The ad hoc process suffered from distinct but equally systematic and decisive obstacles. It was doomed by its lack of institutionalization, continuity, and connections. United States officials, especially in the regulatory agencies, attempted through this process to persuade other major CFC-manufacturing nations to follow them in an aerosol ban. They created this process because no existing international institution appeared suitable for the task. They mistrusted UNEP for its weak administrative capacity and proclivity for ideological disputes, while believing that the WMO lacked a suitable mandate and sufficient interest in the job. The reasons for rejecting the OECD are less clear. That the Soviet Union was not a member was clearly a problem, but other reasons that were given, such as inadequacies of staff or scientific expertise, are not plausible. The most likely reason was that U.S. officials wanted to move rapidly, expecting that they could quickly persuade others to follow through informal discussions, but that staff deference and diplomatic process in the OECD would too readily allow opponents to subvert or block initiatives. Activist officials from other nations shared both the regulatory objective and the reservations about UNEP but still favored working through existing multilateral organizations, either the OECD or UNEP, and went along with the ad hoc process only to maintain unity with the United States. This was particularly true of Canadian officials, who strongly preferred working through existing multilateral organizations to counterbalance U.S. dominance.[143]

In pursuing the ad hoc process, the proponents revealed two strong assumptions about how easily and quickly agreement on international action could be achieved. First, international scientific assessment was not necessary to develop the substantive basis for international agreement: that is, either policy agreement could be achieved without assessment, or the existing U.S. assessments would meet the need. Second, the American policy choice, an aerosol ban, was also appropriate for other nations. More broadly, much of the U.S. approach suggested an expansive willingness to extend U.S. domestic regulations internationally: in discussions of the initial U.S. aerosol ban, activists even proposed that the global dimension of the problem could be addressed simply by applying the U.S. aerosol ban extraterritorially to U.S. firms' operations and holdings abroad.[144]

Under some conditions, such ad hoc processes can be effective vehicles for international deliberations, with their flexibility and informality conferring significant advantages. Participation can be tuned to include precisely those required for an effective agreement—in this case, the major CFC producers—while informal procedures limit opportunities for procedural blockage. Where there is a real prospect of mutual persuasion or joint problem-solving to identify joint actions that all prefer, informal proceedings can help to realize such benefits. But the setting for this ad hoc process was not favorable. The European opponents had specific, well-founded reasons for opposing the proposed aerosol ban. Because aerosols were a larger share of European markets, a ban would impose disproportionate costs on their producers—a disparity that grew through the 1970s, because the consumer-driven aerosol decline was much smaller in Europe than in the United States.[145] The opponents knew that they preferred the status quo to any joint decision involving real reductions on their part, and had no interest in being persuaded otherwise. More generally, the U.S. approach of attempting to extend their domestic regulations internationally, and choosing an ad hoc process rather than an established institution to do so, incensed many European officials. At the time of this process, European chemical producers were objecting to provisions of TSCA for being discriminatory in trade, even as the United States was promoting an approach to international regulation of toxics very similar to TSCA.[146] European officials were particularly outraged at U.S. promotion of an ad hoc meeting on international coordination of toxic chemicals policy, which they saw as an attempt to seize control of the agenda and circumvent discussions on the same subject under way in the OECD.[147]

Faced with such opposition, the ad hoc process commanded virtually no resources to change opponents' views, coerce them, or otherwise advance agreement on joint action. Without a formal institutional base for negotiations, there was no secretariat, chair, or bureau with a stake in the success of the process beyond their substantive policy preferences. Meetings were short and infrequent, had no preparatory sessions, and had no prepared agenda or background papers unless the host government took the initiative to prepare or commission them. Essential background work was done hastily or not at all. These conditions offered little chance for the serious, detailed discussions necessary to make progress toward resolving, or even elaborating differences.

Moreover, just as CCOL's effectiveness was impaired by having no official client for its assessments, the ad hoc discussions were impaired by having no officially provided scientific assessment that they were obliged to consider. In this respect, the flexibility and informality of the ad hoc process served the opponents' interests, since no participant could be compelled to address authoritative scientific statements. The activists' frustration was palpable when British, French, and Italian representatives rejected in Munich the same scientific points that their representatives (in some cases the same people) had accepted one week earlier at the CCOL meeting in Bonn.[148] While the option of blocking concrete decisions through procedural maneuvers was unavailable, major players wishing to avoid coercion or embarrassment could—like Japan and the Soviet Union after the initial 1977 meeting—simply not attend. With no serious prospect of the activists moving further themselves to the detriment of the opponents, such nonparticipation carried neither cost nor risk. Facing resolutely hostile opponents of action, and with no resources to advance discussions, the ad hoc process accomplished nothing. Except for Germany, whose stance was divided and mercurial, the only countries willing to coordinate aerosol controls through the ad hoc process were those that were already inclined to adopt such controls and did so unilaterally. Other than the possible early inspiration of Sweden, this process effected no advance in international action.

3.4 Industry's Initial Research into Chlorofluorocarbon Alternatives

From the initial resolution of the first ozone controversy in the fall of 1976, U.S. regulators' intention to control both aerosol and nonaerosol uses of CFCs was clear. Facing the threat of broader CFC restrictions, the major U.S. producers began modest programs to evaluate other chemicals as possible substitutes for CFCs. Although not facing this immediate threat, other producers also began researching substitutes as defense against longer-term regulatory threats.

The desired chemicals would match CFCs' performance and desirable properties, but not destroy ozone. A small set of plausible candidates is evident from basic chemical principles, and the major firms all began investigating the same set. These chemicals, like CFCs, are halogenated derivatives of methane or ethane, but differ from them in one of two ways. Either they are not fully halogenated (i.e., they include hydrogen), making them more reactive in the troposphere so only a small fraction can reach the middle stratosphere; or they include only fluorine, which forms the unreactive species HF so quickly and stably in the stratosphere that it makes no contribution to ozone depletion. The first group, substituted alkanes that include hydrogen, chlorine and fluorine, are called HCFCs (hydrochlorofluorocarbons); the second group, with no chlorine, are called HFCs (hydrofluorocarbons). Considering only one- and two-carbon molecules, 14 HCFCs and HFCs are possible. Most of these had already been synthesized in the laboratory, and firms began investigating them for suitable thermodynamic properties, synthesis routes, and toxicity, while industry's Fluorocarbon Panel sponsored research on their atmospheric chemistry beginning in 1976.[149]

The most readily available alternative was HCFC-22 (CHF_2Cl), which was already produced at large scale for use in home air conditioners. Unlike other HCFCs, 22 can be synthesized by the same simple process as CFCs 11 and 12, and 11/12 plants can be retrofitted to produce it, so it is cheap and its production can be expanded rapidly. For these reasons, HCFC-22 was briefly considered as a substitute aerosol propellant, although its high vapor pressure would require mixing it with something else in order to use lightweight cans and valves. It was not pursued, however, when suspicious early toxicity results suggested it might be inappropriate for personal-care products.[150] In principle HCFC-22 could be used for additional refrigeration applications and for foam blowing, although its high vapor pressure and low coefficient of performance would require major equipment redesign and would significantly degrade the energy efficiency of refrigeration equipment. Despite its disadvantages, DuPont officials argued in 1978 that HCFC-22's proven high-volume production process made it the *only* serious near-term alternative for non-aerosol uses.[151]

More attractive substitutes would have thermodynamic properties as close as possible to those of CFCs 11 and 12, allowing them to serve present uses with minimal changes to equipment or operations. The most important property is a chemical's vapor pressure over the relevant range of operating temperatures. Preliminary studies quickly showed that one alternative, HFC-134a, matched CFC-12's vapor pressure closely over a wide temperature range, while two others, HCFCs 123 and 141b, matched that of CFC-11 reasonably well.[152] Firms also investigated several other alternatives that matched the properties of 11 and 12 less well, but could potentially substitute in some applications with equipment redesign.

Firms initially kept the size of their alternatives research programs secret, whether for commercial confidentiality or out of fear of embarrassing comparisons with their expenditures on research and public-relations to attack the charges against CFCs.[153] DuPont made the only two public announcements, reporting a cumulative research effort of $10 million in early 1979, and of $15 million in mid-1980.[154] Patent applications and publications from this period show that ICI, DuPont, and Daikin Kogyo all developed synthesis processes and catalysts to produce HFC-134a; that several firms including DuPont and Allied developed new refrigerant and foam applications of HCFC-22, including blends; and that DuPont developed a process to use HCFC-123 instead of CFC-11 to blow polystyrene foam. DuPont researchers also published discussions of new aerosol, solvent, foam, and refrigeration uses for 10 CFCs and HCFCs.[155]

For all proposed alternatives, the most serious obstacle was cost. Although various lab-scale synthesis routes had been known since the 1960s, all posed serious challenges for commercial-scale production. They required multistep processes or high temperatures, gave low product yields, had short catalyst lifetimes, or generated toxic by-products. Primarily for these reasons, both DuPont and ICI concluded after preliminary investigation that even the most promising substitutes would cost two to five times more than CFCs. In June 1980, DuPont reported ceasing work on 8 of the 14 initial candidates, for various reasons. Work continued on six others considered more promising, but this too ceased within a year.[156]

3.5 The Attempt to Enact Comprehensive U.S. Chlorofluorocarbon Controls, 1977–1980

When U.S. regulators announced their intention to control CFCs in October 1976, they stated target regulatory schedules for both aerosols (Phase 1) and other uses (Phase 2). They reaffirmed their intention to control nonaerosol uses when announcing aerosol regulations in March 1978, and the interagency work group commissioned an economic assessment of comprehensive CFC controls from the Rand Corporation later that spring. Nonaerosol regulations raised new political difficulties, however. At an October 1977 public hearing to gather information on nonaerosol uses, more than 100 witnesses testified that CFCs were essential to their businesses. Recognizing that nonaerosol uses might be much harder to control than aerosols, the work group began considering regulations to target emissions rather than uses. At a second public hearing in February 1979, they announced that draft Phase 2 regulations would be delayed at least until late 1979, after the expected release of new assessments being conducted by the Rand Corporation and the National Academy of Sciences (NAS).

The Rand study surveyed all nonaerosol CFC uses and assessed the economic impact of both conventional regulatory controls and novel approaches exploiting economic incentives, such as tradable permits. Based on detailed sector-by-sector analysis, the report estimated a cost schedule for reductions of CFCs 11, 12, and 113, measured in new units called "permit pounds" that weighted each chemical by its chlorine content to approximate its contribution to ozone depletion. The estimated reduction schedule was highly inelastic: a tax of $1.00 per pound, which would nearly triple the price of CFCs, was estimated to reduce consumption by only 20 percent, while reductions beyond 25 percent were judged technically infeasible at any price. Without restrictions, the study projected that U.S. CFC use would grow 6 percent annually through 1990.[157]

The new NAS ozone assessment combined separate reports of three bodies. The panel on Stratospheric Chemistry and Transport provided a summary of atmospheric science similar to that of the 1976 panel. The Committee on Impacts of Stratospheric Change (CISC) synthesized the panel report and a discussion of health and ecological effects. Finally, the newly established Committee on Alternatives for the Reduction of Chlorofluorocarbon Emissions (CARCE) assessed CFC uses, technical alternatives, and policy options to reduce them. CARCE had a broad mix of members including academics and industry, labor, environmental, and consumer groups, and included panels on industrial technology and the socioeconomic impacts of regulations.

The panel summarized advances in scientific knowledge since 1976, on the basis of which they substantially increased their projections of ozone loss: a "best estimate" of 18.6 percent steady-state depletion from constant emissions, which decreased to 17.7 percent when temperature and water-vapor feedbacks were included, and dropped further to 16.5 percent, in a manner the panel admitted was "somewhat arbitrary," to accommodate claimed but undemonstrated CFC sinks. The corresponding uncertainty range was estimated as 5 to 28 percent, with most of the uncertainty arising from uncertain reaction rates. For the first time, the panel also calculated future depletion under increasing or decreasing emissions: a 25 or 50 percent cut in steady-state emissions reduced depletion to 13.2 or 9.4 percent, while 20 years of 7 percent emission growth, constant thereafter, increased steady-state depletion to 57 percent, of which 25 percent would be realized by the year 2025. These calculations with emissions growth were presented with strong qualifications, since such large ozone loss would alter the stratosphere's thermal structure and dynamics beyond the model's validity. Still, the panel judged that depletion from this scenario would "very likely" exceed 30 percent.[158]

The two committee reports were published together with jointly prepared findings and summary. CISC summarized the panel report, giving somewhat greater emphasis to the emissions growth scenario, and estimated the effects of 16–30 percent ozone depletion: 44–100 percent increase in surface UV, several hundred thousand extra U.S. cases of nonmelanoma skin cancer per year, and (with less confidence) several thousand extra cases of melanoma. Summarizing the preliminary state of research on ecosystem effects, they concluded it was not possible to tell with any confidence whether the ecological effects of 16 percent ozone loss would be serious or not.[159]

CARCE was charged with assessing CFC uses and alternatives, their feasibility and costs, and policy responses, a task for which much of the required information had never previously been compiled. The committee drew on an early draft of the Rand study, as well as CMA data, the OECD's study on CFC use in developing countries, DuPont's submissions to the EPA, and direct consultations with about 40 firms, industry associations, consultants, and research centers. Their principal conclusion was that CFC emissions would continue to grow in the United States and worldwide. Although they identified viable alternatives for some nonaerosol uses, all had significant problems. Reporting substantial controversy among technical experts over what magnitude of reductions was feasible, they drew a conclusion that was highly pessimistic, although less so than the Rand study: the maximum further reduction in U.S. emissions that was technically feasible *at any price* was 50 percent, representing a 15–20 percent reduction in world emissions. Consequently, aerosol cuts abroad would achieve more and cost less than any further controls in the United States. With further controls on existing CFC uses so difficult, they suggested that CFC regulation might have to be confined to banning new uses.[160] Although the committee also concluded that the EPA should *consider* further CFC controls, the overall tone of the report was of hopelessness—unsurprisingly, since the committee, like the Rand authors, was largely forced to rely on asking industry experts how much they could reduce their use and emissions of CFCs.

Despite the pessimism of both assessments, the EPA attempted to use them to support its pursuit of further domestic action, announcing its intention to propose further CFC controls at the Oslo ad hoc meeting on April 15, 1980. The EPA's announcement noted that nonaerosol uses grew 9 percent annually between 1975 and 1979, and proposed capping and reducing U.S. use of six CFCs and one halon, in the context of an eventual worldwide reduction of 50–70 percent.[161] DuPont denounced the announcement, terming the Oslo meeting "scientifically and politically unconstructive" and calling for an international blue-ribbon panel of scientists, industry leaders, and government officials to propose scientific and regulatory programs to study the issue further.[162]

After several delays, the EPA issued a formal statement of its intention in an

"Advance Notice of Proposed Rulemaking" (ANPR) in October 1980. Tentative, weakly argued, and containing significant errors, the proposal reflected the now widespread pessimism about technical feasibility of nonaerosol reductions and the efficacy of further U.S. unilateral action, as well as EPA organizational and staffing problems.[163] It proposed only the precautionary measure of holding U.S. CFC production at current levels, citing three supporting arguments: increased estimates of steady-state depletion; continued growth in nonaerosol uses; and the difficulty of detecting a global ozone loss, even if one were actually occurring. The ANPR elaborated on the two broad approaches to controls that had been first presented in the Rand report: further bans on specific CFC uses; and a novel system of permits to produce or use CFCs, which could be transferred among producers or exchanged among chemicals, with each chemical weighted by its chlorine content. This highly innovative approach was the first proposal for a full tradable environmental permit system, and the first attempt at joint control of multiple chemicals weighting each by the harm it imposes.[164]

The EPA faced major obstacles in taking these regulatory proposals forward. Although other uses were growing, early public and political attention had so powerfully emphasized aerosols that attention declined as soon as aerosol controls were enacted. In addition, much of the inter-agency coordinating structure that had supported development of aerosol controls, including IMOS, became inactive after these were issued. Although the EPA's now clear authority to regulate ozone-depleting substances eliminated the jurisdictional confusion that had made IMOS necessary, IMOS had also provided an effective forum for interagency consultation over regulations affecting many of their interests. Confusion over formal jurisdiction had been eliminated, but the practical need for broad coordination and consensus-building, and high-level sign-off, was even greater for the more challenging Phase 2 regulations, even as the structure for providing such consultation lapsed.[165]

Most significantly, the CFC industries mounted a campaign against the new regulations of a scale and unity they had never achieved during the aerosol debate. DuPont distributed a briefing book to their CFC customers, urging them to contact the EPA with their concerns, in August 1979—more than a year before the EPA actually issued the ANPR—and followed with a series of publications over the next two years rebutting the NAS and Rand studies that the EPA was using to support its regulatory proposal. Although new scientific results had, on balance, made CFCs appear more damaging, the DuPont material forcefully attacked the scientific foundation for CFC regulations. DuPont continued to emphasize the theoretical character of ozone depletion; noted that UV radiation has beneficial as well as harmful effects; and argued that any further CFC controls should await clear detection of global-scale ozone loss. They exploited seeming differences among scientific assessments by quoting summary language from two British assessments to criticize the NAS report.[166] During 1980, DuPont also highlighted new results that reduced estimated depletion and that appeared too late to be available to the NAS panel. The CMA Fluorocarbon Panel also circulated several reports with selective summaries of recent research results to advance the anti-regulatory case.[167]

DuPont also supported their case against broader CFC controls by reporting on their alternatives research program in highly unfavorable terms, arguing that the promising compounds had no commercial synthesis processes, questionable environmental acceptability, or other problems. Exploiting prominent concerns about energy shortages, DuPont also sponsored a study of how banning CFCs would increase energy losses, because available substitutes (principally HCFC-22) would degrade performance of refrigeration equipment and insulating foams.[168] Still, their report also stated that alternatives could be commercialized in five to ten years including toxicity testing and application development, clearly implying that they foresaw no insurmountable technical obstacles.[169] This point was largely missed, however, as most observers took the message as additional evidence of the difficulty of reducing CFC use.

At the initiative of DuPont and two user-industry associations, U.S. industry also made an important organizational innovation in the summer of 1980 by forming the Alliance for Responsible CFC Policy. The Alliance included CFC producers (who were few in number and mostly large) and CFC users from all sectors (who were many and diverse, including many very small firms), organized around their common interest in continued availability of the chemicals. The strategy of building an Alliance of all CFC sectors was intended to prevent further targeting of vulnerable usage sectors for controls, as had happened to aerosols. Moreover, the Alliance between many widely distributed small businesses, and a few large firms with powerful scientific and legal resources, made a politically potent combination. The small members provided a wide grassroots network to mobilize constituent appeals to Congress, claiming that they needed CFCs to stay in business,[170] while the large members could sponsor sophisticated legal and technical interventions. Moreover, the coordination of responses by producers and users sought to avoid repeating the aerosols debacle, in which uncoordinated, sometimes contradictory, and erroneous statements from multiple industry groups and scientifically ignorant public-affairs officials had seriously embarrassed the industry. DuPont, which had been most embarrassed during the aerosol conflict, solidified its position as the scientific voice of CFC industries by filling the Alliance's science adviser position.

The Alliance mounted a show of political force in late 1980, generating more than 2,000 comments from its members criticizing the ANPR and developing substantial congressional hostility to further regulations.[171] Although EPA officials had expected a sympathetic industry response to the novel market-based regulatory approach they proposed, even this aspect of the proposal drew uniform hostility.[172] In the entire public response to the ANPR, only four favorable comments were recorded.[173]

Faced with such effective opposition, as well as the clear difficulties of controlling nonaerosol uses, the EPA retreated. Citing four factors—"complex economic and administrative considerations . . . revealed by comments," new scientific developments that lowered estimated ozone depletion, slowing of world CFC growth rates, and the need for a comprehensive policy review by the incoming administration—the EPA announced in the spring of 1981 that it would postpone any new CFC regulation until at least December 1981.[174]

With the inauguration of the Reagan administration and the appointment of Anne Gorsuch as EPA administrator, the issue fell into neglect. Uncertainty regarding the new administration's direction in environmental policy persisted through the summer of 1981,[175] and EPA officials avoided stating their intentions, although Gorsuch's comments at her confirmation hearings—including a statement that ozone depletion was a "highly controversial" theory—gave some hint of what was to come. In August, an EPA toxics official testified that the outstanding ANPR did not constitute a statement that EPA intended to regulate CFCs further, "either now or in the future."[176] It became clear through late 1981 that the new EPA management was actively hostile to enacting new regulations or, for that matter, enforcing existing law. Riven with internal conflict, the EPA continued to meet its minimal statutory obligations on CFCs—commissioning biennial Academy assessments and sending reports to Congress—but serious consideration of further action on CFCs was inconceivable.

3.6 The Early Ozone Debates: Explanation and Significance

By late 1980, the attempt to secure international agreement on aerosol controls and the U.S. attempt to restrict growth of broader CFC uses had both failed. Though other factors operated, a confused view in the United States of the strategic relationship between domestic and international action contributed to both failures. The two failures were connected because advocates failed to effectively link domestic and international action. The international effort failed because U.S. officials sought international replication of existing U.S. domestic controls. However sincerely these officials may have wanted global action, their initiative was viewed as opportunistic and not serious, because it would impose costs on others but none on its proponents. The activists missed the opportunity to link international controls to the prospect of further U.S. action. In addition, the ad hoc process had no institutional foundation to apply sustained pressure on opponents to admit the risk, or for others to press the activists to advance more reasonable proposals. Sustaining such pressure through periods of varying domestic political conditions is the essence of securing international agreements.

In turn, the failure of the international process doomed the subsequent U.S. attempt to enact comprehensive CFC controls. All participants in this debate noted the global character of the problem and the obvious consequence that unilateral U.S. action would not solve it, but they drew opposite conclusions for U.S. action. The ANPR argued that unilateral U.S. leadership would move others to act by demonstrating feasibility and signaling commitment. Industry argued the reverse, noting (reasonably) that other major producers had refused to follow the initial U.S. aerosol ban, and arguing (less reasonably) that since all governments had access to the same scientific evidence, others' refusal must have indicated that they judged the evidence inadequate to warrant a regulatory response.[177]

Both these arguments are suspect as bases for U.S. action or inaction in this instance. Absent effective domestic-international linkage, U.S. leadership served more to obstruct than to promote international action, since the more the United States moved ahead, the more the international agreements it sought would impose disproportionate costs on others. Moreover, as a practical matter, U.S. domestic leadership could have little international effect in 1980, since all international processes considering the issue had collapsed and no new forum was in view. Not until 1986 did the activists effectively link domestic to international action and signal their seriousness, by proposing approaches that took the larger burden on themselves. On the other hand, industry's claim that other nations failed to follow the U.S. lead because they doubted the scientific evidence was disingenuous, in view of the vigor of industry's own efforts to ensure that the international process failed.

Other than prior failure of the international process, what accounted for the failure of the U.S. attempt? Although the proposed tradable-permit system was a lightning rod for criticism, it is unlikely that this bore much responsibility, as the theoretical advantages of such approaches were becoming familiar and the problems with the specific proposal were readily correctable. Nor can either the international or the domestic failures be attributed to weak U.S. political resolve, or to the election of Reagan as president. The international process had clearly failed by 1978, and it was only the determination of Carter administration officials that pushed the process forward to a third international meeting, however futile, in early 1980. Even this postscript to the international process took place months before the Reagan campaign had emerged as a serious threat. The weakening of the U.S. domestic initiative also substantially predated Reagan's electoral threat. Although the ANPR appeared only in October 1980, the activists began losing their resolve nearly a year before, discouraged first by the prevailing pessimism about technical feasibility of reductions and later by the political effectiveness of the Alliance, and also responding to the suggestion of reduced risk from new scientific results that appeared through 1980.[178] In addition to their more effective political organization, industry groups had by this time grown more scientifically sophisticated, and were able to advance an effective critique of the 1979 NAS study that EPA used to support its regulatory proposals as they had not been able to do in 1976. They also skillfully exploited differences in interpretation and policy conclusions between the NAS and U.K. assessments, and were able to use the need for international action as an argument against, rather than in favor of, U.S. leadership.

It is also possible that the sequencing of controls (that is, the decision of activists and regulators to pursue aerosols first) may have contributed to the failure to enact broader controls. Although industry spokespeople attacked U.S. regulators as hypocritical bullies for attacking the vulnerable aerosol sector—bullies for picking on the weak, hypocrites for denouncing aerosols while benefiting from CFCs in other nonessential uses such as air conditioning—controlling aerosols first had both a strong tactical and substantive rationale. Attempting to restrict all uses in the first step would likely have encouraged all affected industries, including those with no evident alternatives to CFCs, to mobilize a coalition able to block any measures. But taking the initial expedient step on aerosols, rather than waiting until coherent principled measures could be adopted, risked inciting the mobilization of opponents and the distraction of activists after the first step was accomplished, just as occurred. Unless such a first step also puts in place forces that facilitate subsequent steps, no further step may be feasible. Because the specific means used to reduce aerosol CFCs were largely irrelevant to reducing use in other sectors, the aerosol bans put no such forces in place.

Following the failure of both the ad hoc process and the U.S. attempt at comprehensive controls, the issue fell into a period of dormancy. Although the incoming Reagan administration in the United States played no role in setting the stagnation in place, it willingly sustained it. Under the new administration, a much more compelling rationale was needed to propose far-reaching regulations, which was not available through the early 1980s. In fact, many took the 1980 decline in steady-state depletion estimates to indicate that the issue had become less serious. Between these factors and the loss of public attention after implementation of the aerosol bans, the issue dropped almost entirely off the policy agenda in the United States, which for several years had been the most vigorous promoter of the issue. Even most of the environmental groups that had been active during the aerosol wars stopped paying attention.[179]

In the ozone controversies of the 1970s, scientific argument was always prominent. Parties consistently attempted to advance their preferred policy positions with support of scientific arguments, but scientific argument was outweighed by other factors in determining the outcomes of the policy debates, such as the superior political organization of U.S. industry, the widespread view that large nonaerosol reductions might be technically infeasible, and activists' inability to apply effective influence in international forums. These early debates did, however, define the basic lines of argument over the policy significance of scientific knowledge, consensus, and controversy that became increasingly prominent through the 1980s and 1990s. For example, these early debates defined the two questions that for 10 years remained the accepted standard for whether scientific evidence provided a warrant for policy action: How much depletion is projected in steady-state, and can a significant global depletion yet be observed? As policy interest in the stratosphere declined after 1979, so did public attention to scientific work in the field. Investigation continued, however, as did assessments—many of them continuing to discharge obligations established during the initial few years of intense political attention. Despite the gradual accumulation of knowledge and confidence, these assessments began making significant contributions to policy debates only after 1985. Chapter 4 reviews the development of scientific knowledge and controversy, and the parallel development of assessments of that evolving scientific knowledge, between 1976 and 1985.

4

The Search for Knowledge-Based Resolution

Science and Scientific Assessment, 1976–1985

4.1 Stratospheric Science, 1976–1985

From Qualitative Confirmation to Quantitative Argument

Chapter 3 discussed how Molina and Rowland's initial claim that CFCs could deplete ozone was qualitatively confirmed within two years, through persuasive demonstration that each step of their proposed mechanism occurred. These results did not close the debate, however, but shifted it to the quantitative magnitude, and by implication the importance, of CFC-induced ozone loss.[1] Persuasively demonstrating the quantitative importance of the depletion risk was a much greater challenge than merely demonstrating its existence. It required showing not only that each step of the depletion process occurred, but also that it occurred fast enough and was not offset or overwhelmed by other processes. This required progress in several linked areas of research, each posing distinct scientific challenges.

A first research area concerned the budget of chlorine (and later bromine) in the troposphere and stratosphere—how much is introduced by human activities and natural sources, and how much is removed by natural processes—that determines how much CFC emissions will perturb stratospheric chlorine. A second area concerned how a specified perturbation of chlorine would change the overall chemistry of the stratosphere, including ozone. Progress in both areas, particularly the second, depended on three distinct types of research: laboratory studies of chemical kinetics to identify the relevant chemical reactions and measure their rates; model calculations to integrate the results of many chemical processes and atmospheric movements; and measurements of the quantities of various species actually present in the stratosphere. A final set of atmospheric questions concerned the statistical analysis of ozone measurements to detect global trends. These questions figured only marginally in early policy debates but became increasingly pivotal after 1979 as statistical methods advanced, new satellite data sets became available, and the first claims were advanced that global ozone losses could be observed. Projections of future stratospheric change also depended strongly on the nonatmospheric question of whether and how much future emissions would grow, which was largely ignored until about 1983 but then became controversial as CFC production resumed growing after several years of stagnation.

Two other research areas are in principle of central importance to policy debates over ozone depletion: the effects of ozone loss and resultant UV increases on health, resources, and the environment; and the feasibility, costs, and effects of measures to reduce ozone loss. Even if depletion could be projected perfectly, judging its importance and what should be done about it would require highly contestable evaluations of the importance of the resultant impacts and the cost and difficulty of measures to reduce it. Although rarely engaged explicitly, these arguments underlay many conflicts over quantitative depletion projections.

This chapter reviews progress and controversy in these research areas between 1976 and 1985, to inquire whether and how scientific knowledge, consensus, and new results interacted with policy debates and negotiations. At the end of this period, a broad consensus was forming that the stratosphere and its response to human perturbations were on the threshold of being well understood—a confidence that was shattered as soon as it developed by two shocking observations of ozone depletion reported in 1985, in Antarctica and worldwide. The chapter also reviews the repeated attempts by several institutions over the same period to summarize and synthesize this evolving scientific knowledge in assessments to inform policy making. The discussion of assessments does not stress their substantive content, which largely reproduces the development of scientific knowledge outlined in the previous sections. Rather, it emphasizes variation among assessments in their organization and participation, the methods and processes they used to review and synthesize current knowledge, and how their emphasis and conclusions compared with current knowledge. The closing section reviews the influence of primary scientific results and assessments over this period on subsequent international policy debates. It argues that only one assessment exercised any such influence, and examines the characteristics and external conditions that allowed it do so.

Projecting Changes in Stratospheric Chlorine

Whether CFCs have an important effect on the stratosphere depends in part on how much additional chlorine they deliver relative to naturally occurring chlorine. Determining this requires sufficiently precise estimates of the three major components of the atmospheric chlorine budget: human emissions of relevant chemicals; natural emissions; and natural destruction processes, or "sinks," that remove the chemicals at or near the earth's surface and so reduce the quantity that can reach the stratosphere. The existence and magnitude of both natural chlorine sources and tropospheric sinks were controversial in the 1970s, but largely resolved by 1980. In contrast, estimating industrial emissions was initially thought straightforward, but became increasingly controversial in the 1980s.

Estimates of U.S. production of CFCs and other halocarbons were published beginning in 1958. Similar data were not available internationally because other producer-country governments did not, or in some cases could not, collect it. The 20 largest CFC producers began voluntarily reporting total world production of CFCs 11 and 12 in 1975, however, with estimates of how much was used in each major sector (e.g., aerosols, refrigeration). Industry also estimated annual world emissions using a model of how long CFCs remained inside different products, ranging from immediate release from aerosol and solvent uses to years of residence inside some closed-cell foams and sealed refrigeration systems.[2] The largest source of the estimated 5 percent uncertainty in global emissions estimates was the need to estimate unreported production in eastern Europe and the Soviet Union.

Natural sources of stratospheric chlorine are important because it is reasonable to presume that the more is there naturally, the less a specified increase from human activity is likely to matter.[3] But natural sources, like anthropogenic sources, can reach the stratosphere in quantity only if they have long tropospheric lifetimes. Most natural chlorine emissions, such as salt from ocean spray and HCl from volcanoes, are in chemical forms that quickly dissolve in rainfall, react chemically, or are deposited on surfaces. They may contribute much chlorine to the lower atmosphere, but almost none of it reaches the stratosphere. The initial studies of CFCs and ozone assumed that natural chlorine sources were all too short-lived to reach the stratosphere. Three contrary claims were soon made, of which one was found to be correct.

It was first proposed that volcanic eruptions could deliver large quantities of HCl directly to the stratosphere, but it was quickly shown that rapid rainout would make this at most a small source, even from explosive eruptions.[4] A more serious claim that carbon tetrachloride (CCl_4) could be a natural source of long-lived atmospheric chlorine was suggested by Lovelock's initial observations, which found CCl_4 as well as CFC-11 in remote locations. Since initial estimates of historical release were 40 times smaller than the average concentration he observed, Lovelock argued this CCl_4 must be primarily natural in origin. Estimates of CCl_4 emissions are highly uncertain, however; because most is used as an intermediate in CFC manufacture, its emissions are the small difference of two large numbers. A large upward revision of estimated historical emissions in 1976 suggested they could account for all the CCl_4 observed in the atmosphere.[5] Substantial uncertainties in even current CCl_4 emissions persisted through the early 1980s, and Lovelock continued to argue for a primarily natural source, but a growing body of atmospheric measurements suggested that the behavior of CCl_4 was best explained by a large industrial source, no natural source, and no sink except the stratosphere.

An important natural source was discovered, however, in methyl chloride (CH_3Cl), which is formed near the ocean's surface from methyl iodide released by seaweed. Measurements of its atmospheric concentration initially showed large discrepancies, which converged by 1980 to suggest that methyl chloride contributed 0.6 to 0.7 ppb of natural chlorine to the stratosphere.[6] By this time it was also widely agreed that methyl chloride was the only significant natural source of stratospheric chlorine, so quantifying its contribution closed the debate over natural sources of ozone depletion. This debate had in any case tended to obscure the basic fact that even a large natural source would not grow over time, so a growing anthropogenic source would eventually dominate any natural source. This was the case with CFCs, whose contribution to stratospheric chlorine surpassed that of natural methyl chloride by 1980 and continued to grow.

A significant tropospheric sink for CFCs, however, would represent a more enduring reduction of their threat to ozone. Vertical mixing in the stratosphere is so slow that even a very slow process destroying CFCs in the troposphere would substantially reduce the quantity that goes high enough to destroy ozone. For example, with CFC-12's atmospheric lifetime estimated as 100 years if it was destroyed only in the stratosphere, a tropospheric sink destroying only 1 percent per year would cut CFC-12's contribution to ozone loss by half; a 2 percent per year sink, by two-thirds.[7] The CFC industry research program began looking for tropospheric sinks in 1972, since they might indicate pathways by which CFCs could be toxic or could contribute to smog formation. Early CFC measurements quickly ruled out a tropospheric sink of more than a few percent per year, which in the context of this early research (i.e., before stratospheric ozone depletion from CFCs was proposed) suggested the chemicals were benign. This early research also suggested that even a 1 percent per year sink was unlikely, but measurements were not precise enough to verify this until the early 1980s.

There are two ways to look for sinks, both difficult: directly, by looking at each proposed place CFCs might react; and indirectly, by examining the relationship between emissions and atmospheric concentrations over time. About a dozen potential sinks were proposed in the early CFC debate. The 1976 NAS panel examined them directly, rejecting nearly all and concluding that the effect of the remaining few was at most one-fifth the loss rate in the stratosphere.[8] By 1979 these few sinks were dismissed, but a new proposal was made that CFCs could be dissociated by heat or light while adsorbed on the surface of desert sand grains. Although this sink was periodically touted for several years as a potential flaw in ozone-depletion projections, it was clear from the outset that it was too small to be of quantitative importance.[9] Whatever the fate of particular sinks, however, seeking sinks directly had the more fundamental limitation that it could test only those that had been proposed, not demonstrate whether others existed.

Consequently, a comprehensive assessment of sinks would require observing them indirectly, by comparing current CFC concentrations against historical emissions or changes in concentrations against current emissions. Unfortunately, neither emissions nor concentrations were well enough known in the 1970s to for these methods to give precise sink estimates. Even current emissions were known only with 5–10 percent error, while concentration measurements were complicated by large apparent variations with season and location, and by the vulnerability of the standard instrument, the electron-capture detector, to large systematic errors. In 1977, both published estimates of CFC-11 concentrations and different researchers' measurements of standard samples in a controlled test showed discrepancies of 20 to 40 percent.[10]

Given these limitations, the first few years of CFC monitoring provided only the crudest bounds on their atmospheric lifetimes. To improve these estimates, a network of intercalibrated stations was established beginning in 1978 to monitor CFCs and other trace gases in five remote, widely separated locations. This project, the Atmospheric Lifetimes Experiment (ALE), sought to observe concentration trends so precisely that atmospheric lifetimes could be estimated after only a few years of operation, despite continuing uncertainty in emissions.[11] The first ALE data, released in 1983, showed lifetimes of 83 years for CFC-11 and 169 years for CFC-12, which were modestly reduced to 75 and 110 years in 1985—consistent with the only sink for CFCs being photolysis in the stratosphere.[12] Although the prospect of major CFC sinks had become highly unlikely before ALE data were published, these results definitively closed the debate by demonstrating that there were no significant sinks. By the mid-1980s, questions of atmospheric budgets and lifetimes had to be revisited for additional chemicals identified as potential ozone depleters, and ALE was replaced in 1986 by an expanded network measuring more gases.

Projecting Changes in Ozone

Given an estimate of how much chlorine industrial chemicals will add to the stratosphere, the second step of inference is to determine how this change will alter the composition of the stratosphere, including ozone. This is a complex question, which depends on many chemical and photochemical processes in the stratosphere, on physical movements of the air, and on interactions between these processes. Understanding these effects required linked advances in three areas. Laboratory measurements were needed to isolate and measure the relevant chemical and protochemical reactions, identifying which species react, how fast they react at relevant temperatures and pressures, and what products they yield. Because the stratosphere contains many trace species that interact through multiple chemical processes and are physically transported by winds, calculating how much of each species should be present also required model calculations to integrate representations of all the important chemical and radiative processes, and of air movement and mixing. Finally, measurements of trace species concentrations in the stratosphere with sufficient time and space resolution were needed, to test model calculations and to constrain the values of reaction rates that were not well enough known. When there are persistent discrepancies between what models say should be in a particular region of the stratosphere and what repeated measurements find to be there, these discrepancies identify priorities for further model and laboratory work. In aggregate, progress in these three areas produced new questions and discrepancies about as fast as it resolved old ones, leading to the seemingly paradoxical result that much more was understood about the stratosphere in 1985 than in 1976, but the number and severity of identified discrepancies between observations and model calculations were roughly unchanged. The following sections briefly review progress in each of these domains in turn.

As late as 1970, direct measurements of many important stratospheric reactions were not possible because laboratory methods were inadequate to deal with the rapid rates and short-lived unstable species involved. Instead, many reaction rates were inferred indirectly, by analogy to known rates of related but more manageable processes or by jointly measuring the results of several processes, but these estimates were highly uncertain. Through the 1970s, large advances in kinetics techniques increasingly allowed direct measurement of both known reactions and newly hypothesized ones, yielding vastly improved knowledge of stratospheric photochemistry by the mid-1980s. Each such advance changed understanding of both the natural stratosphere and the changes that would result from a specified addition of pollutants. The aggregate effect of these advances was to make projected ozone depletion progressively larger between 1975 and 1979, then progressively smaller through 1983.

The reactions to be considered included the basic processes of ozone formation and destruction identified by Chapman in the 1930s, and the reactions that initiate, sustain, link, and terminate the catalytic ozone-destruction cycles involving hydrogen, nitrogen, and chlorine. The first significant changes came from new measurements of the main reactions in the chlorine catalytic cycle. Through these reactions, atomic chlorine reacts with ozone as it cycles back and forth to chlorine monoxide, is temporarily tied up in inactive HCl by reaction with methane, and is released to active form again by reaction with HO. New measurements of two of these reactions in 1975 reduced calculated ozone depletion by about one-third from previous values.[13] Although several suggested processes to tie up chlorine in inactive form were soon dismissed, chlorine nitrate was found in 1976 to be an important reservoir for both chlorine and nitrogen.[14] Chlorine nitrate formation reduced ozone loss from both chlorine and nitrogen, and made the effect of each species depend on how much of the other was present: higher chlorine weakened ozone loss from nitrogen, and vice versa.

Beginning in 1977, a series of important new measurements was made of reactions of the hydrogen species HO and HO_2, which were widely recognized as important uncertainties. The hydroxyl radical HO plays a crucial role in the chemical balance of the stratosphere—both directly, through the hydrogen catalytic cycles, and indirectly, by modulating the strength of the nitrogen and chlorine cycles. HO affects the nitrogen cycle by reacting with active NO_2 to form the stable reservoir HNO_3, so higher HO reduces ozone loss from nitrogen. HO affects the chlorine cycle in the opposite direction, attacking the stable reservoir HCl to release active Cl, so higher HO increases ozone loss from chlorine. The sensitivity of ozone to both CFCs and NOx consequently depends strongly on how much HO is present.[15] No adequate measurements of HO in the lower stratosphere were available until the 1990s, however. Rather, HO concentration was inferred indirectly from measurements of other species that react to produce or consume HO and from the rates of these reactions. Between 1975 and 1979, most new measurements increased the inferred HO concentration, making ozone appear more sensitive to chlorine and less sensitive to nitrogen. Between 1979 and 1983, most changes went the other way, reducing inferred HO and hence making ozone less sensitive to chlorine and more sensitive to nitrogen.[16] When the estimated HO concentration peaked in 1979, the calculated effect of SSTs briefly changed sign to an ozone increase. Although the estimated effect returned to an ozone loss only a few months later, this reversal was used for several years by skeptics to suggest that ozone-loss concerns were likely to be generally exaggerated.[17]

Of these new measurements, the first reported and most important was a new rate for the reaction of NO with HO_2, reported in early 1977. Using a new laser magnetic resonance technique, Howard and colleagues remeasured several rates involving HO and HO_2 directly and found this reaction to be 40 times faster than the prior estimate.[18] By sharply increasing the inferred HO concentration, the new rate increased calculated ozone loss from chlorine and decreased that from nitrogen. In addition, since this fast rate allowed NO to enhance the chlorine cycle by producing HO, it also introduced a positive interaction between the nitrogen and chlorine cycles, through which nitrogen and chlorine together would destroy more ozone than the sum of their separate effects. In total, the new rate doubled the ozone loss calculated from CFCs and greatly reduced that from SSTs. Since several other rates involving HOx had been estimated jointly with this one, the new measurement also called the others into question. For the reaction of HO with HO_2, which was especially difficult to control, repeated remeasurements in 1976 and 1977 gave progressively slower rates. Since this reaction *consumes* HO, a slower rate *increases* the inferred HO concentration and makes ozone more sensitive to chlorine and less sensitive to nitrogen. With both these rates revised, it became clear that model calculations would better match atmospheric observations if the rate of a third reaction, that of HO_2 with ozone, was also substantially faster than its old estimate. A new measurement found this rate to be even faster than proposed, further increasing ozone's estimated sensitivity to chlorine.[19]

In addition to new rates for known reactions, the late 1970s saw increasing identification and measurement of previously neglected chemical pathways. The most important were a series of newly identified reactions coupling the major cat-alytic cycles. These reactions formed reservoir species that tied up active radicals in more stable forms for periods from a few minutes to a few days, until broken apart by photolysis or attack by other radicals. Consequently, the first-order effect of each of these new reactions was to reduce ozone's sensitivity to the species involved. But depending on details of how the new reservoirs were formed and destroyed, these couplings could also provide new pathways for ozone destruction or produc-tion, greatly increasing the complexity of ozone-loss calculations. Chlorine nitrate was the first of these chain-coupling species to be identified. It was initially assumed that once it was formed, chlorine nitrate would eventually photolyze to reproduce the chlorine and nitrogen species that formed it (ClO and NO_2), so its only effect would be to temporarily hold chlorine and nitrogen in inactive form, reducing the contribution of each to ozone loss. But 1979 measurements showed that it mainly photolyzed in a different way (to Cl and NO_3), forming a new ozone-depleting cycle. In aggregate, chlorine nitrate still reduced ozone loss due to chlorine, but by substantially less than originally thought.[20]

By 1979, reactions analogous to chlorine nitrate formation were identified that coupled the hydrogen and chlorine cycles, and the hydrogen and nitrogen cycles. The hydrogen and chlorine cycles were coupled by the formation of hypochlorous acid (HOCl). Since this process also tied up some active chlorine in unreactive form, its first-order effect was to reduce ozone loss due to chlorine. As with chlorine nitrate, however, uncertainty over the fate of hypochlorous acid made its overall effect on ozone uncertain for several years. By 1981 it was recognized that HOCl would photolyze fast, with a significant fraction following a pathway that created another ozone-loss cycle. Considering these effects together, this coupling had a very small effect on calculated ozone loss.[21] Similarly, the hydrogen and nitrogen cycles were coupled by the formation of peroxynitric acid (HO_2NO_2). Although the direct effects of this reaction were on the hydrogen and nitrogen cycles, by reducing the calculated concentration of HO it also indirectly reduced ozone's sensitivity to chlorine. Industry representatives seized on early results showing this reaction to be fast, arguing that models needed to provide a full treatment of peroxynitric acid to be credible.[22] But as with hypochlorous acid, the aggregate effect of peroxynitric acid on ozone loss was found by 1981 to be small.[23]

Chlorine is not the only halogen that can destroy ozone. As early as 1975, bromine was known to have an ozone-destruction cycle stronger than chlorine's, but its quantitative importance was initially dismissed because the atmospheric con-centration of bromine was so low, roughly 10 ppt versus 2 ppb of chlorine. Bromine began to appear more important by 1979, however, because the kinetic changes that increased ozone's sensitivity to chlorine did the same for bromine, and because a new catalytic cycle was identified in which BrO reacts with ClO to regenerate bromine and chlorine atoms, without the need for atomic oxygen. The effect of these changes was that even 20 ppt of bromine, only twice the 1979 level, could augment ozone destruction due to chlorine by 5 to 20 percent.[24]

By 1980, confidence was growing that the most important stratospheric chem-istry was coming to be understood, particularly the reactions of the chlorine and nitrogen cycles.[25] Reactions involving the hydrogen cycle were not so well estab-lished, however, and several researchers noted that remaining discrepancies between models and atmospheric observations would be improved if the HO concentration in the lower stratosphere—which was still inferred, not observed—was substantially lower than models were calculating.[26] Several new measurements beginning in summer 1980 found slower rates of HO formation and faster rates of HO destruction, confirming this speculation and reversing the trend of the prior three years.[27] Because the new results reduced the inferred concentration of HO, they made ozone less sensitive to chlorine and more sensitive to nitrogen. Calculations of steady-state ozone loss from CFCs dropped from 20 percent in 1979 to 5–9 percent in 1981, with nearly all depletion once again occurring in the upper stratosphere.[28] A few further kinetic changes continued to reduce ozone's sensitivity to chlorine through 1983, then reversed the trend in 1984. Calculations of steady-state ozone loss incorporating these changes dropped to 2–5 percent in 1983, then increased to 5–9 percent in 1984. In 1984, NASA staff judged the recent changes to be so minor that they did not, as usual, hold a large workshop to prepare their required report to Congress, instead drafting a staff assessment based on prior publications and meetings.[29] By 1985, the kinetics of gas-phase stratospheric reactions appeared to be a largely settled field.

In contrast, the possibility of heterogeneous-phase chemistry in the stratosphere—reactions that occur on the surfaces of liquids or solids—was a continuing concern. Molina and Rowland had noted the potential importance of heterogeneous processes in their 1974 paper, suggesting that they represented a possible, though unlikely, decomposition route for CFCs. Beginning in 1984, Sato and Rowland argued that heterogeneous processes involving chlorine, particularly the reactions of chlorine nitrate with HCl and with H_2O that regenerate active chlorine, could be important in the stratosphere. Noting that these two reactions are very slow in gas phase but proceed rapidly on surfaces of even the most inert laboratory materials, they speculated that the reactions could also occur on surfaces of dust, sulfate aerosols, or ice crystals in the stratosphere. Rowland also suggested that these reactions, if they occurred on the sulfate aerosols injected into the lower stratosphere from the 1982 eruption of El Chichon, might account for the apparent ozone decreases in the lower stratosphere in 1983 and 1984, where gas-phase models said that ozone should be increasing.[30]

Modeling the Stratosphere

While laboratory kinetics experiments refine understanding of one chemical process at a time, calculating the aggregate state of the stratosphere or the change caused by a pollutant requires considering the effects of dozens to hundreds of chemical processes simultaneously, and their interactions with dynamic air movements. A quantitative understanding of the combined effects of all these processes requires use of computational models, which represent each chemical and dynamic process numerically, and calculate their combined effects in a simplified, simulated atmosphere.

The relative importance of chemistry and dynamics varies with height. In the upper stratosphere, above about 35 km altitude, radiation is so intense and photochemical processes so rapid that species react much faster than they move. Consequently, the atmosphere at this height is essentially in photochemical equilibrium, allowing its chemical state to be calculated without considering transport. By assuming photochemical equilibrium, and by considering a limited subset of the most important chemical reactions, the earliest ozone-depletion calculations, such as those of Johnston (1971), could be done (laboriously!) on hand calculators. At lower altitudes, radiation is less intense and chemistry slower, so transport becomes progressively more important. Understanding the state of the middle and lower stratosphere requires simultaneous calculations of chemistry and transport, which are possible only using computer models.

The detail and complexity of model calculations are limited by computing speed, which requires compromises and simplifications in model-building. The most fundamental simplifications are of dimensionality, reducing the continuous three-dimensional atmosphere to an averaged representation in either one or two dimensions. One-dimensional (1-D) models represent a single column of globally averaged air extending from the tropopause (or sometimes the earth's surface) to the top of the atmosphere. All horizontal movement is excluded, as are the effects of latitude and season. The only air movement represented is vertical mixing, which is usually summarized in a single parameter, the "eddy diffusion coefficient," describing how fast constituents mix vertically at each height. The model calculates how the atmosphere's chemical composition and other properties vary with altitude, with each calculated value taken to represent a global average at that altitude. Two-dimensional (2-D) models perform equivalent calculations for a vertical slice of atmosphere extending from pole to pole, like a paper-thin slice of an apple. In these models, calculated values represent averages over a ring around the earth at each latitude and altitude (a "zonal average"). Unlike 1-D models, 2-D models can represent air movements in both the vertical and north-south directions, and so can calculate how the composition of the atmosphere varies with latitude and season.

The advantages of two- and even three-dimensional models were recognized from the beginning of ozone policy debates, and 1-D models were continually attacked for their simplistic treatment of transport. It is problematic to interpret the output of a 1-D model in terms of any real atmospheric observations. The eddy-diffusion parameter, a computational fiction representing a globally averaged rate of vertical mixing, was particularly troublesome. Its vertical profile was estimated by observing the vertical distribution of some inert or chemically well-understood trace species. Different trace species gave widely different profiles of the mixing rate, however, which in turn gave differences up to a factor of 2 in calculated ozone loss.[31] Moreover, representing the atmosphere by a single column introduced a variety of errors and uncertainties, which could not be fully captured even by using a wide range of alternative profiles for the vertical mixing rate.[32]

Despite these known limitations of 1-D models, and despite vigorous pursuit of 2-D models beginning as early as 1973, 1-D models remained the principal tools for stratospheric calculations through the mid-1980s.[33] Until then, the additional computational burden of even a 2-D model required using highly simplified chemistry, a disadvantage that persisted for years because rapidly increasing chemical complexity was a moving target for 2-D modelers.[34] Most modelers judged it more important to include a full, current set of chemical processes than to represent two-dimensional movement. While some policy actors repeatedly argued that only 2-D models could give believable policy-relevant conclusions, they continually overestimated how soon 2-D models would attain sufficient chemical realism to be used in this way.[35]

Still, 1-D models saw important improvements through this period. By 1979, they had substantially improved treatments of transport, the averaging of photochemical processes over the daily cycle, and the transmission and scattering of radiation through the atmosphere.[36] A particularly important class of improvements, first recognized as necessary in 1976, were temperature feedbacks. Because both CFCs and ozone absorb infrared radiation, they warm the atmosphere in their vicinity. Changes in the vertical distribution of CFCs and ozone consequently change the atmosphere's temperature distribution. Since the rates of the reactions that determine the ozone balance vary with temperature, such a change introduces feedbacks into ozone-depletion calculations, which 1-D models began to include in the early 1980s. An especially important temperature feedback was that CFCs could warm the tropical tropopause, allowing more water vapor to reach the stratosphere. One early calculation suggested that this effect could increase CFCs' ozone depletion by 4 percent.[37]

Throughout the period considered, trends in models' calculated ozone depletion differed strongly between the upper and lower stratosphere. In the photochemically controlled upper stratosphere, where neither chemical reservoirs nor transport is important, the calculated effects of chlorine injection changed little from 1974 to 1985: all calculations projected large ozone losses. In the more complex middle and lower stratosphere, where chemical reservoirs and transport are both important, calculated changes in ozone varied sharply over time. Projections for the lower stratosphere ranged from large losses in the late 1970s to increases that nearly offset higher-altitude losses in the early 1980s. These changes in projections for the lower stratosphere accounted for nearly all variation over this period in calculated total ozone losses.

Although early model calculations of ozone loss examined only the effect of increases in chlorine, human activities were known to be perturbing the stratosphere in several ways. Aircraft operating in the upper troposphere and lower stratosphere, as well as emissions of nitrous oxide (N_2O) from fertilizer, were increasing stratospheric NOx,[38] while increasing emissions of methane (CH_4), partly natural and partly anthropogenic, increased stratospheric HOx.[39] In addition, human emissions

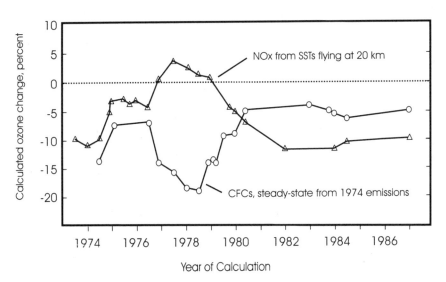

Figure 4.1 Projected Steady-State Ozone Depletion. The future ozone loss projected to result from constant CFC emissions, and from the NOx in SST exhaust, varied over time with new measurements of reaction rates and improvement of models. Projected losses from CFCs approached 20 percent in 1979, then declined to a few percent in 1982. Sources: Wuebbles 1983; NASA/WMO 1986.

of CO_2 and other greenhouse gases were warming the surface troposphere and cooling the stratosphere, changing both the rates of stratospheric reactions and the amount of water vapor entering the stratosphere across the tropical tropopause.

Realistic calculations of projected ozone depletion required considering all these human perturbations to the stratosphere together, but computational and modeling limits restricted early models to examining one change at a time. The first studies to consider two changes together appeared in 1978. They found, as expected, that increasing CO_2 reduced ozone loss from CFCs,[40] while CFCs and NOx showed nonlinear interactions: their combined effect on ozone was smaller than the sum of the separate effects, in some cases smaller than the effect of CFCs alone.[41] By 1981, the first model runs appeared that simultaneously considered CFCs and other halocarbons, NOx from aircraft, N_2O from fertilizer, and CO_2. These showed very small total ozone depletion, typically with a large loss in the upper stratosphere nearly offset by a large increase in the lower stratosphere.[42] Adding increases in methane further strengthened the result that upper-stratosphere losses were offset by lower-stratosphere gains, yielding near-zero total depletion even under small increases in CFCs.

As these multiple-perturbation studies increasingly revealed the importance of changes in the lower stratosphere, and consequently of adequate representation of atmospheric transport, 1-D models appeared to have reached the limits of their usefulness. Although simple 2-D models had been available since 1975, the first steady-state depletion calculation from a 2-D model incorporating the same chemistry as contemporary 1-D models became available in 1981.[43] The results were similar to those obtained with earlier, chemically simpler 2-D models: projected global-average ozone losses were similar to those in 1-D models, with the largest losses at high latitudes, particularly in winter and spring.[44]

A series of model runs published in 1983 and 1984 raised the possibility that ozone depletion might not vary linearly with chlorine inputs, as had been widely believed. One analysis noted that some models that showed ozone loss up to 7–10 percent when stratospheric chlorine reached 8–10 ppb, showed no loss—or even an increase in ozone—as chlorine was initially increased up to about 3 ppb.[45] This raised the disturbing possibility that ozone might remain constant or increase for several decades of increasing chlorine inputs, then begin a sharp decline. A still more disturbing result, published in 1984, suggested that depletion might become nonlinear at high chlorine levels.[46] This phenomenon, which came to be called the "chlorine catastrophe," attracted great attention. The result arose through a titration effect in the lower stratosphere when the concentration of active chlorine exceeded that of NOx. With nearly all NOx tied up as chlorine nitrate, further chlorine inputs increased ClO and ozone loss much more rapidly than total chlorine. The 1-D model that exhibited this effect calculated less NOx than most models, so its nonlinear depletion appeared when Cl reached about 15 ppb, a level that plausible CFC growth rates could attain within a few decades. When other models introduced large chlorine growth, they found similar nonlinearities, although models with higher NOx required higher chlorine to trigger the effect.

When early studies with 2-D models showed less nonlinearity with high chlorine, it was widely concluded that the nonlinearity was an artifact of 1-D modeling, mitigated in 2-D models by their ability to represent north-south transport in the lower stratosphere. Most of the 2-D models studied had high NOx, however, and those with the lowest NOx showed nonlinearity comparable to the original study, suggesting that the phenomenon was not primarily a 1-D modeling artifact, but could also appear in 2-D models given the required NOx/Cl relationship. Although the causes and plausibility of the nonlinearity were never fully resolved, the mechanism anticipated the extreme depletion later identified in the Antarctic, as discussed in chapter 6. Without the novel mechanisms that drive the Antarctic depletion, the high levels of chlorine needed to trigger nonlinearity made the issue appear to be of little practical importance, although the authors noted that the heterogeneous reactions being suggested by Rowland and colleagues (discussed below) would greatly lower the chlorine threshold at which nonlinearity would appear.[47]

A major 1985 model comparison exercise summarized the rapidly developing state of ozone-loss projections. Six 1-D models projected steady-state ozone depletion ranging from 4.9 to 7 percent, or 6.1–9.4 percent when radiative feedbacks were included.[48] A few early 2-D model results showed global-average depletion similar to 1-D models or slightly larger, with total-column losses two to four times larger near the poles than near the equator—the sum of a large decrease in the upper stratosphere at all latitudes, and an increase in the lower stratosphere near the equator that turned to a decrease poleward of about 40 degrees. Since increases in the lower stratosphere occurred only in the tropics, they would not—as had been inferred from 1-D model results—offset upper-stratosphere losses at mid and high latitudes.[49] Doubling atmospheric concentrations of CH_4 and CO_2 and increasing N_2O by 20 percent reduced ozone loss in six 1-D models nearly to zero with chlorine at 8 ppb, and to a few percent with chlorine at 15 ppb, in each case with large ozone losses in the upper stratosphere offset by increases in the lower stratosphere. The effect of multiple perturbations was not additive, and continued to show nonlinearities at high chlorine. For example, one calculation with CO_2, CH_4, and N_2O concentrations growing at present rates gave very small ozone loss if CFC emissions grew by less than 1.5 percent per year, but large depletion (10 percent after 70 years and rapidly increasing) with 3 percent annual CFC growth.[50] These calculations continued to show substantial sensitivity to kinetic uncertainty and large discrepancies between models even with all inputs fixed, particularly for species other than ozone. For one model that projected steady-state depletion of 7.7 percent, a Monte Carlo simulation gave a range from 1.9 to 13.5 percent (one standard deviation), with a long tail extending in the direction of large depletion.[51]

The first time-dependent 2-D model run with multiple perturbations, completed in 1985 and widely publicized at international meetings in 1986, underscored the emerging result that even small CFC growth could give substantial ozone loss at mid to high latitudes, since the lower-stratosphere increases that offset upper-stratosphere losses in 1-D models were concentrated near the equator. With continued growth of N_2O and CH_4 and constant CFC emissions, this model projected very small loss globally, but 8 percent loss at 60 degrees North by 2030; with 3 percent CFC growth, losses by 2030 were 6 percent globally, but 16 percent at 60 degrees North.[52] With one exception, no attempt was made through 1985 to consider heterogeneous reactions in stratospheric models. The exception was a study that separately introduced two hypothesized heterogeneous reactions into a 1-D model with multiple perturbations. In each case, the newly introduced reaction changed the result from nearly no ozone loss to a large loss.[53] The 1985 assessment merely noted heterogeneous reactions as being of potential, but unproven, importance.[54]

Atmospheric Measurements

Stratospheric models produce two kinds of calculations: simulations of the present stratosphere, and projections of the changes that would follow a specified injection of pollutants. These two kinds of calculations are closely linked. Each change in chemistry changes both the calculated state of the present stratosphere and the calculated effects of pollutants. Models are validated, and their projections of future changes gain confidence, by checking how well they reproduce the distribution of species observed in the present stratosphere. In principle, models could also be validated experimentally, by subjecting the atmosphere to a known perturbation and comparing the observed response against that calculated by a model. While it would clearly be irresponsible (even if it were possible) to make large intentional disruptions of the atmosphere simply to test whether we have modeled it correctly, it would be equally so to give no credence to model projections until we see how well they predict future ozone loss after decades more of CFC emissions. This latter approach, although periodically advocated as the appropriate standard of proof on which to base policy on model projections, would renounce any attempt to anticipate and forestall the environmental risk. Rejecting both these experimental approaches leaves only the possibility of weaker model validation through sufficiently accurate replication of the present stratosphere.

There were, however, two large-scale historical "experiments" with the atmosphere that offered some prospect of a stronger test of stratospheric theory: an extreme burst of solar radiation (a "solar proton event") in August 1972 and the atmospheric nuclear tests of the 1950s and early 1960s. The solar event generated a large increase in NO in the upper stratosphere near the poles, which should have increased nitrogen-catalyzed ozone loss—a conjecture that could be tested because, by good luck, satellite instruments measuring both solar radiation and total ozone were in operation at the time. A calculation using the measured increase in radiation to drive an early 2-D model estimated 16 percent ozone loss in the Arctic upper stratosphere, while ozone measurements found a 20 percent decrease in that region, which appeared shortly after the event and lasted several weeks.[55] Despite some discrepancies between satellite and ground-based ozone measurements, and between the observed and calculated patterns of ozone loss by latitude and over time, this reasonable agreement between the calculated and measured effect of the event looked like strong confirmation of current photochemical theory, at least in the upper, photochemically controlled region of the stratosphere.[56]

Atmospheric nuclear tests provided a more ambiguous, and ultimately less satisfactory test of theory. Initial calculations suggested that the blasts should have injected enough NO into the stratosphere to reduce ozone by a few percent, but the ozone record appeared not to show the loss and model refinements to calculate expected depletion as a function of time and latitude only slightly improved the agreement.[57] Suggestions that the 10 percent global ozone increase through the 1960s might represent atmospheric recovery from the tests were widely attacked because the timing was implausible: the ozone minimum occurred too early to be caused by the tests, and the subsequent increase continued too long and went too high.[58] A proposal that mid-1960s ozone data lacked a peak that should have been present given an increasingly suspected influence of the 11-year solar cycle, could not be resolved.[59] When late-1970s changes in kinetics data reduced the calculated effect of NOx, the question of whether these tests produced an observable ozone signal became of lower priority, and was not ultimately resolved until 1988.[60] In sum, the response of ozone to nuclear tests and to the 1972 solar event both attracted attention for a few years, but failed to yield the crisp confirmation of stratospheric theory that had been hoped. Moreover, since both these events involved injections of NOx, their role in testing projections of chlorine-induced depletion was at best indirect. Consequently, model testing through the 1970s and 1980s consisted principally of improving measurements of key species in the present stratosphere.

While observations of atmospheric ozone began more than a century ago, stratospheric measurements of key nitrogen, chlorine, and hydrogen species began only in the 1970s. There are two broad approaches to these measurements: remote measurement, in which an instrument looks up from the ground or down from a satellite; and in situ measurement, in which an instrument is carried through the stratosphere on a balloon, airplane, or rocket. Remote measurements, like those from the Dobson ozone instrument, rely on the target species absorbing or emitting radiation at an identifying wavelength, so it can be seen from a distance and distinguished from other species. These are not possible for all species, and when they are, they usually reveal only the total quantity of the target species in the instrument's line of sight, although clever instrumental techniques can sometimes allow separation of a few broad vertical bands. Precise vertical profiles, and measurements of how multiple species vary together, require in situ measurements.

Atmospheric measurements through the early 1980s verified that the major catalytic cycles occurred by the proposed pathways, and never seriously called into question the early qualitative confirmation provided by the initial observations of CFC losses, HCl, and ClO.[61] There were always substantial discrepancies between models and observations throughout the period, but particular discrepancies came and went with new observations and changes in key reaction rates. For example, the 1977 increase in the rate for HO_2 + NO led all models to overpredict ozone in the lower stratosphere, until the subsequent increase in the rate for HO_2 + O_3 corrected the problem.[62] The total body of stratospheric measurements available through this period remained too thin, however, to either clearly confirm or refute the claim that CFCs would bring large future depletion of ozone. After 1982, new measurements and better coverage, particularly from new satellite instruments, resolved some long-standing questions, although important gaps and discrepancies between calculations and observations remained in 1985.

Measurements of stratospheric nitrogen species, particularly NO and NO_2 (collectively called NOx), were the top observational priorities of the early 1970s because they were thought to be crucial in determining the effect of NOx injections from supersonic aircraft. By the end of the decade, many in situ measurements of NO were obtained. These showed good agreement in daytime vertical profiles, a strong diurnal cycle (NO increases rapidly at sunrise and decreases slowly through the night), and substantial seasonal variation.[63] Many in situ measurements of nitric acid (HNO_3) were also available after 1974, which agreed well and were relatively uncomplicated by fast daily chemistry. NO_2 was harder to measure, and only remote measurements at sunrise and sunset were available in the 1970s. While these agreed well among themselves, the strong daily cycle involving NO, NO_2, and possibly other nitrogen species made them hard to interpret. Spatial coverage of all these measurements was limited, and it was clear that strong variability both diurnally and spatially—such as the "Noxon cliff," the sharp drop in total-column NO_2 north of 50 degrees latitude in winter[64]—required better spatial coverage and simultaneous measurements of multiple species.

Atmospheric measurements of chlorine species began in 1974, with early measurements of altitude profiles of CFCs and HCl, and the first detection of stratospheric ClO in 1976. Further chlorine measurements showed several troubling anomalies, however, which persisted into the 1980s. For HCl, the problem was discrepancies between different measurement techniques, which persisted through 1981. While all measurements showed the expected qualitative behavior—concentration increasing from the tropopause to a maximum in the mid- to upper stratosphere—different techniques found large differences in the location and size of the maximum, and in vertical structure. Since HCl constitutes the largest share of stratospheric chlorine, accurate measurement of its profile was essential. The latitudinal and seasonal variation of HCl (and HF, which is often measured at the same time) was clarified only after intercomparison projects in 1982 and 1983, in which several balloon-based instruments measuring HCl and other species were launched together to identify interinstrument differences without the confounding effect of variation by location or season.[65]

Measurements of ClO showed several anomalies: large apparent variation with latitude and season, which illustrated the limitations of one-dimensional models; much more fine-scale vertical variation than expected; and, most disturbingly, two extremely high early measurements, the highest reaching 8 ppb. This was more chlorine than should have been present in all forms, and eight times more than the highest ClO calculated by any model.[66] These high ClO concentrations were never explained, and were particularly troubling because they occurred together with normal ozone levels. Although they were also never replicated, they were for years advanced as grounds for skepticism about the entire chlorine-ozone hypothesis, suggesting that important chemistry—perhaps an unidentified additional source of stratospheric chlorine, or an unidentified mechanism by which chlorine produced ozone—had been overlooked.[67] As the key intermediate in chlorine-catalyzed ozone loss, ClO was crucial, and the struggle over interpretation of its measurement was intense. The earliest debates suggested that finding ClO at all would be adequate verification of ozone depletion, but the debate shifted, after ClO was detected, to whether it was present in the predicted quantity.[68] Although many subsequent measurements converged on roughly consistent, lower values,[69] industry representatives argued that the anomalous early measurements still called for a program to measure the vertical profile of total stratospheric chlorine. The industry research program began funding this measurement in early 1977. The first results, reported in 1980, found about 3 ppb of total chlorine near 20 km, roughly consistent with both model predictions and the sum of the increasingly consistent body of measurements of ClO and HCl (excluding the high early values).[70] For other potentially important chlorine species, few or no measurements were available through 1980—a single measurement of chlorine nitrate that provided only an upper bound of 1 ppb (about double the amount calculated by models), and no measurements of ClO_2 or HOCl.[71]

Stratospheric measurements continued to accumulate through the early 1980s, with modest increases in time and space coverage and increasing attempts to measure multiple species simultaneously. Important gaps and discrepancies remained, however. The most important gap was the lack of any measurement of the crucial hydroxyl radical HO in the lower stratosphere. There were HO measurements available above 30 km by 1975 and ground-based measurements of the total column by 1980.[72] These constrained the range of possible values in the lower stratosphere, but not enough to resolve the range of HO concentrations calculated by models at this altitude, which spanned a factor of 3. The difference between these "high-HO" and "low-HO" states was of fundamental importance, determining whether ozone in the lower stratosphere was controlled by HOx or by NOx, and whether chlorine had any significant effect on ozone at these altitudes.[73] Direct measurements of HO in this crucial region were obtained only in the 1990s.[74]

In addition, measurements of even the most important species remained sparse in spatial and temporal coverage; few simultaneous measurements of multiple reactive species were available; and no adequate measurements were available, even in 1985, of the major reservoir species such as chlorine nitrate, hypochlorous acid, and peroxynitric acid.[75] An ambitious new in situ observation technique to measure extremely fine-scale vertical profiles of multiple species, proposed in 1979, suffered several years of setbacks. The "reel-down" technique used a cable to reel an instrument platform 20 kilometers down and up below a balloon floating in the upper stratosphere.[76] Multiple instruments on the platform would provide simultaneous profiles of O, O_3, ClO, and HO_2, as often as 10 times per day. Due to equipment problems, the first test flight was delayed until September 1982; the first measurement was obtained in September 1984.[77]

In 1983 and 1984, satellite data began to fill some of the important gaps and clarify some discrepancies. Data from instruments on the Nimbus-7 satellite, released in early 1984, provided the first global distributions of N_2O, CH_4, H_2O, HNO_3, and NO_2. These new data sets greatly improved the ability to validate models, especially 1-D models, and enabled the first real progress in understanding behavior of nitrogen at high latitudes. The new global NO_2 data showed that the Noxon cliff, the sharp decline in NOx at high latitudes in winter, corresponded to the edge of the polar vortex, stimulating a new explanation based on conversion of NOx to NO_3 and N_2O_5 inside the vortex for which further satellite observations of N_2O_5 and NO_3 provided tentative support.[78]

Despite progress in both models and measurements, important discrepancies between them remained throughout this period. In 1979, the most important discrepancies were that models predicted too much ClO in the lower stratosphere; too much NOx in the upper stratosphere, and too high a ratio of HNO_3 to NO_2 in the lower stratosphere.[79] By 1982, the most important discrepancies shifted to models predicting too little ClO in the upper stratosphere (fueling continued speculation that some unknown process may convert HCl to ClO above 35 km), and underpredicting the dissociation of CFCs as they rise through the lower stratosphere.[80] These were resolved by a 1983 correction to the absorption spectrum of O_2, which indicated that UV would penetrate farther through the stratosphere and dissociate CFCs at lower altitude than previously thought. By 1985, the major discrepancies shifted once again. Most seriously, models now predicted 30 to 50 percent too little ozone in the upper stratosphere, and about 20 percent too much near the ozone maximum between 20 and 30 km. Models also predicted too much HO in the total profile, and—despite major advances in NOx chemistry—still did not adequately represent the behavior of HNO_3 at high latitudes in winter.

Emissions Scenarios

Nearly all uncertainties that were prominent in policy debates between 1975 and 1985 were questions of atmospheric science, concerned with how much ozone loss would result from specified CFC emissions. The largest source of uncertainty in projections of future ozone loss, however, was how much CFC emissions would grow. Under almost any plausible chemical and dynamic assumptions, large continued growth in CFC emissions would eventually bring large ozone depletion. Questions about how much emissions were likely to grow, however, were strangely absent from the policy debate until the mid-1980s. In part, this neglect was a consequence of the requirements of developing and comparing stratospheric models. Understanding differences between models required standardizing them on common input assumptions, including future emissions. From the start of the ozone debate, modelers followed the approach of Molina and Rowland's initial paper and expressed their results in terms of steady-state ozone loss from constant emissions. Steady-state is entirely reasonable as a way to standardize models for comparison because it is simple, it focuses attention on long-term consequences, and it can serve as a proxy for other scenarios as long as depletion is linear (in which case depletion from any other specified emission trend will vary in parallel with steady-state depletion). Although these virtues do not imply that constant emissions is the best guess, or even a reasonable guess, for actual future CFC emissions, steady-state model projections were frequently mistaken for projections of actual future depletion. The prominence in policy debates of the wide swings in steady-state depletion resulting from new kinetic results tended to obscure the highly conservative market-growth assumptions on which these calculations were based and also the fact that future ozone loss under market growth could greatly exceed steady-state loss. Despite the crucial importance of market-growth assumptions for future ozone loss, these questions were never seriously considered in early policy debates.

Granted, any attempt to base present control decisions on hypothetical future trends in the absence of control embeds deep difficulties of self-reference. Taking it as given that emissions are harmful, the faster they grow without controls, the stronger will be the case for imposing controls. Consequently, the more likely it is that controls will be imposed in the future, the less the urgency to do so now. In the more concrete—and self-serving—terms advanced by the industry's Fluorocarbon Panel, the threat of potential regulation will deter investment and market development even if no regulation is enacted—thus making it unnecessary to enact regulation.[81]

Controversy over CFC growth first peaked in the United States in 1979 and 1980, as the EPA was developing its proposal for nonaerosol regulations. The flat CFC markets of the 1970s had combined two opposing effects, large reductions in aerosol uses and substantial growth in nonaerosol uses. The EPA projected that nonaerosol growth would drive world production to surpass its 1974 peak by the mid-1980s, while the 1979 NAS assessment included one scenario with 7 percent annual growth through the 1980s.[82] DuPont and the Fluorocarbon Panel forcefully attacked these projections, charging the Rand analysts with naiveté regarding the practical determinants of investment decisions and asserting that various limits, including limited resources of fluorspar (the mineral from which fluorine for CFCs is extracted) rendered such growth rates infeasible.[83] The attack appeared to deter analysts from further consideration of growth scenarios. No attempt was made to repeat a Rand-style analysis of CFC markets for more than five years; the 1979 NAS panel gave much more prominence to steady-state calculations than to the CFC growth scenario, although Rand had identified the growth scenario as most likely; and no scientific assessment gave prominent treatment to emissions growth scenarios until 1985.[84] The recession of the early 1980s also helped to suspend controversy over future growth, lending seeming support to industry claims that CFC production would not grow again. After increasing steadily through the 1970s, U.S. nonaerosol use fell slightly in the early 1980s, although nonaerosol use worldwide still grew slowly. In contrast to 1979 projections that CFCs would grow 3 to 7 percent annually through the 1980s, world production was flat until 1983.[85]

The related question of whether current CFC emissions were being accurately reported also became controversial when the Fluorocarbon Panel's published estimates came under attack in 1981. The panel pooled production reports from 19 participating firms, and added staff estimates of nonreported production in Argentina, India, the Soviet bloc, and China to generate a world total. Their July 1981 estimate showed world production down 20 percent from 1974 to 1980. But in April 1982, Rowland argued that atmospheric measurements of CFC-12 showed emissions must be substantially higher than the Panel's estimates.[86]

The major weakness in the panel's estimates was production in the Soviet bloc and China. For this region there was only one publicly available data source, which showed Soviet production for 1968–1975. In constructing their global estimates, the panel had arbitrarily assumed 3 percent annual growth in Soviet production before and after this seven-year period, increased this estimate by 15 percent to account for production in eastern Europe, and assumed Chinese production to be zero.[87] The panel initially responded to Rowland's criticism by adopting revised estimates of sectoral emissions from Rand and EPA, and by changing their assumed growth rate for Soviet production to 18 percent per year, the average rate reported over the seven-year period for which data were available.[88] This revision increased estimated 1980 global production from 696 to 743 Kte, 18 percent below the 1974 peak rather than 20 percent, followed by 2 percent growth in 1981.[89] These revisions reduced, but did not eliminate, the discrepancy between production estimates and concentration measurements.[90]

In any case, adjusting global estimates by arbitrary changes in the assumed growth of unknown Soviet production was at best a stopgap solution. Continuing to assume 18 percent annual production growth through the 1980s would give the Soviet Union half of world CFC production by 1990.[91] Moreover, atmospheric measurements continued to suggest discrepancies with the panel's revised emission estimates. Faced with these difficulties, the panel announced in June 1984 that they would stop making any estimates for eastern Europe, the Soviet Union, or China, but would publish only reported production.[92] There was consequently no longer a single standard source of production estimates to use as an input to model calculations. This new, narrower production total fell 7 percent in 1982, then grew 8 percent in 1983.[93]

As the recession eased in 1983 and 1984, CFC production began to grow and questions of potential future growth became controversial once again, principally due to projections circulated by the EPA. In their early work on greenhouse gases and climate change, EPA officials used a set of CFC growth scenarios developed at the OECD in 1981–1982.[94] Convinced that CFC growth was an important and neglected question, they commissioned three independent studies to use different methods to project future CFC growth. All three projected similar growth, from 2 to 6 percent annually over the next few decades.[95] Industry groups responded to these projections in two ways. The 1983 research report of the Fluorocarbon Panel obscured the importance of growth by using a short time horizon, illustrating differences between steady-state and growth scenarios only through 2010.[96] Others attacked the analyses, arguing that fluorspar shortages or regulatory uncertainty would restrain expansion, or generally attacking the exercise as speculative and meaningless.[97] One industry study argued that the high growth of the 1970s would never return, even in developing countries, and estimated future world growth of

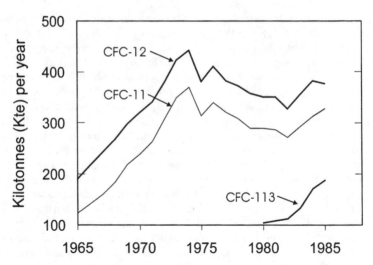

Figure 4.2 World CFC Production through 1985. After dropping sharply in the mid-1970s, world CFC production resumed growing in 1982. Source: AFEAS data.

1.5 percent per year.[98] But by 1985, industry had lost this battle, for global production had clearly reached its bottom and resumed growing. Fluorocarbon Panel estimates released in late 1985 showed 8 percent production growth in both 1983 and 1984, with reported production of CFCs 11 and 12 reaching 694 Kte in 1984.[99] In retrospect, it is remarkable that CFC growth remained controversial for so long. CFCs had grown rapidly for decades as they expanded into new uses, until stopped by aerosol bans and recession. Continued 10–15 percent growth was implausible, with established markets becoming saturated and no major new applications evident; but absent strict controls, sustained zero growth was equally implausible.

Debates about growth projections were initially confined to CFCs 11 and 12, but other chemicals—other CFCs, as well as methyl chloroform, carbon tetrachloride, and brominated chemicals such as halons and methyl bromide—contributed to the same ozone-depletion chemistry to some degree. It was recognized early in the debate that these chemicals might be important, but they were not high priorities for controls, either because very little was produced (other CFCs and halons) or because natural sources or tropospheric sinks significantly reduced their effect on ozone (methyl chloroform, methyl bromide, and HCFC-22).[100] In the early 1980s, however, indications of rapid growth made several of these chemicals appear more important. Methyl chloroform emissions appeared to be growing rapidly, from atmospheric concentrations increasing 6 percent per year and from its known rapid substitution for more strictly controlled solvents.[101] CFC-113 was the fastest growing CFC, principally due to its use in electronics: its production doubled from 1976 to 1979, and reached 15 percent of all CFCs by 1984, while its first atmospheric measurements, released in 1984, showed its concentration increasing more than 15 percent per year.[102] Atmospheric measurements of HCFC-22 found more than production estimates implied, although its large-scale use as a chemical intermediary made estimating emissions particularly uncertain. Estimated releases of carbon tet-rachloride and halons also were highly uncertain.[103]

Although it appeared from 1981 that other halogenated chemicals could aug-ment ozone loss from CFCs 11 and 12 by one-third, no international production data were available for any of them. UNEP began asking governments for data on other chemicals in 1982 but few responded, some citing concerns of commercial confidentiality and others unable to obtain the data.[104] Under some conditions, such concerns may be reasonable: when, as often, only one firm produces a chemical in a country, national statistics reveal firm production. But the Fluorocarbon Panel reporting process had already addressed this problem for CFCs 11 and 12, by having a third party aggregate data before release. The availability of this obvious means to protect whatever secrecy firms in fact possess about their production, and the Fluorocarbon Panel's refusal to broaden its reporting to other ozone-depleting chemicals, both suggest that confidentiality concerns were being used tactically, to divert regulatory attention from chemicals other than CFCs 11 and 12.[105]

As model studies increasingly examined the effects of trends in multiple gases such as CO_2, CH_4, and N_2O as well as halocarbons, projections of these gases were increasingly important, because the result that multiple perturbations gave very small ozone losses in 1-D models was quite sensitive to the relative growth rates of different gases. Emission projections for these other gases were more difficult than for halocarbons, because they have multiple natural and anthropogenic sources and even their current budgets were not well understood. For example, early studies typically assumed CO_2 concentrations would double preindustrial levels in 50 years, reducing CFC-induced ozone loss by about half, but such rapid CO_2 growth came to be viewed as implausible by 1982.[106]

Halogenated chemicals differ in their quantitative contributions to ozone deple-tion. As more chemicals were recognized as potential depleters, finding a way to compare their quantitative effects became increasingly important. Determining the priority for controlling various chemicals, or controlling multiple chemicals jointly, requires some metric to quantify their relative harm. EPA officials had proposed the first such metric in their proposed 1980 regulation, by weighting each chemical by its chlorine content. Although this general approach represented an important breakthrough, the particular metric was flawed because factors other than chlorine content also affect a chemical's contribution to depletion—such as its susceptibility to tropospheric attack and the altitude at which it photodissociates in the strato-sphere. A valid metric of relative harm requires integrating these factors in an at-mospheric model, first done with the Livermore model in 1980. Called the chemi-cal's "ozone depletion potential" (ODP), the new metric measured its contribution per kilogram to ozone depletion, standardized relative to CFC-11.[107]

Observed Ozone Trends

When concerns about ozone loss first surfaced in the early 1970s, there was an extensive, if flawed, record of ground-based observations of total ozone dating back 20 to 50 years. Within months, two related controversies over this record appeared, which persisted until the late 1980s: how quickly the observing network could detect a decline in global ozone if one were occurring, and what trend in global ozone, if any, the record showed.

Early analyses of ozone data found that northern hemisphere ozone decreased in the late 1950s, then increased 5–6 percent through the 1960s. No corresponding trend was evident in the southern hemisphere.[108] The increase of the 1960s varied strongly with latitude, so the calculated average trend for the hemisphere depended on how individual stations were weighted. A small decline followed in the 1970s.[109] The cause and significance of these trends was not understood. Some industry representatives argued that the 1960s increase showed CFC-ozone depletion could not be a serious concern, while also noting that ozone variations were largely unexplained, so CFCs should not be blamed when ozone showed a decline.[110]

In the early ozone record, there appeared to be a cycle of about 11 years associated with the solar cycle.[111] By the late 1970s, this cycle was widely accepted on theoretical grounds, although its size, its precise timing, and how confidently it could be seen in the record remained controversial.[112] As this cycle became accepted, it was increasingly argued that new methods of statistical time-series analysis could control for natural cycles in ozone, allowing any new trend to be quickly detected and attributed to CFCs. The first application of these methods to ozone examined one station's data from 1932 to 1974, and found no significant trend after removing cycles of roughly 2 and 11 years: the increase of the 1960s and the decrease of the 1970s were both captured by natural cycles.[113] Applying the same techniques to a hypothetical global network suggested that if a 1.5 percent global ozone decrease occurred, it should be detectable with 95 percent confidence after only six years.[114] From 1975 to 1986, industry representatives used analyses of this kind to argue that any ozone loss could be detected in time to take action, so the test of whether CFCs in fact deplete ozone—and the basis for CFC controls—should be the clear detection of a reduction in global ozone. Other analysts argued that the smallest detectable loss was larger, perhaps as large as 10 percent, based on different assumptions about instrument noise, biases, and drifts, as well as the possibility of natural trends or slower cycles than those identified.[115] This range of estimates of the minimum detectable loss narrowed only a little through the early 1980s, to about 1–6 percent.[116] The WMO's third official statement on ozone expressed frustration with the debate in an uncharacteristically strong criticism of claims of precision from ozone trend analysis.[117]

Any attempt to make action depend on prompt detection of ozone loss faced two difficulties: ozone variability, and weaknesses of the observing network. Ozone is highly variable in both space and time. Spatial variability, even on a continental scale, introduces uncertainty into attempts to calculate the global average at any time; time variability, including annual and longer cycles, obscures attempts to detect a long-term trend. In addition, the ozone observing network suffered multiple practical problems. Spatial coverage was sparse and biased.[118] Nearly all stations used the Dobson instrument, designed in the 1920s to observe short-term fluctuations, which required substantial skill in calibration and operation, and was ill-suited for detecting long-term trends. In addition, stations were of highly variable quality in operations and record-keeping, few had good long records, and one-third of all stations did not report their data regularly.[119]

To argue that taking action should await a clearly observed ozone loss is to take an extreme stance on the standard of proof required to justify regulation, renouncing any attempt to forestall the risk of ozone loss before it occurs. The stance is especially extreme given the long delay of stratospheric response: even when models calculated future depletion of 20–30 percent, they typically calculated that the depletion already realized should be only a percent or two, a level hard to distinguish from random fluctuations. Still, this stance gained surprisingly widespread support, despite wide differences in the claimed minimum detectable trend.[120]

Despite its known weaknesses, the Dobson network was all that was available, and several researchers looked for long-term trends in its data. Three independent analyses were undertaken in the early 1980s under Fluorocarbon Panel sponsorship. Using Dobson data through 1979, the three found that the small apparent decline through the 1970s could be fully accounted for by long-term cycles, leaving a small, statistically insignificant global increase from 1970 to 1979.[121]

Adding data through 1984 to these analyses gave no significant trends, while adding 1985 gave the first appearance of small but significant negative trends. [122] These changes principally reflected a 5–8 percent decrease in 1983, but the significance of this sharp one-year drop was unclear since it made the statistical detection of a trend depend strongly on how long a series of observations was used. Moreover, three natural events of 1982 and 1983 combined to depress ozone, weakening any attempt to attribute the decline to human pollution: a strong El Niño, the April 1982 eruption of El Chichón, and declining solar radiation following the 1981 solar maximum.[123] On the other hand, a newly identified mechanism by which pollution from aircraft or other sources could create ozone in the upper troposphere suggested that increases there could be masking decreases in the stratosphere.[124] Model calculations at this time of how much ozone loss should have occurred were close to zero, however, weakening the power of a no-trend observation as a test of theory and making projected future depletion depend even more strongly on future emissions.

Although total ozone data showed little trend, predicted losses were consistently larger in the upper stratosphere, around 40 km, than in the total column. Consequently, ozone data at this altitude offered the best chance to observe an early depletion signal. Data on the vertical profile of ozone were even weaker than total ozone data, however. Beginning in the 1960s, two types of ground-based measurements were available from a few stations, mostly in north temperate latitudes: measurements from weather balloons, and readings from Dobson meters using the Umkehr technique, which gave a crude vertical distribution in roughly 5-km slices. Umkehr measurements depended on a calculated correction for stratospheric aerosols that was developed at one station and applied to the whole network, making it a persistent target of criticisms.[125] In the early to mid-1980s, the combined record of these sources suggested that ozone was increasing in the troposphere and decreasing in the stratosphere, but the two sources were not entirely consistent. Umkehr data showed significant ozone losses in the middle to upper stratosphere growing stronger in 1983 and 1984, in substantial agreement with 1-D models.[126] Both sources showed significant ozone increases in the troposphere, while balloon data—but not Umkehr—also showed significant decreases in the lower stratosphere.[127] This observed ozone decrease in the lower stratosphere, where models projected ozone should be increasing, was a serious discrepancy. Although these observations were ultimately shown to be correct, they were widely disregarded due to thin coverage and a general lack of confidence in these data sources.[128]

It was long hoped that satellite instruments, with their superior coverage and consistency, would make the problems of interpreting the Dobson network irrelevant. NASA sought to demonstrate the value of satellite instruments for the ozone debate after assuming leadership of U.S. stratospheric research in 1975,[129] but persistent operational problems, including instrument failures, gaps in coverage, and poor calibration with the ground network, limited their contribution until the mid-1980s. The first satellite ozone instrument, the backscatter ultraviolet (BUV), was launched in April 1970 but operated only intermittently after a June 1972 solar-panel failure, impairing its utility for global trends analysis.[130] Two more ozone instruments were launched in 1975, and another two in October 1978, the solar backscatter ultraviolet (SBUV) and total ozone mapping spectrometer (TOMS), which ultimately provided definitive clarification of global ozone trends. The SBUV instrument began registering ozone decreases soon after its launch, which were widely believed due to instrument degradation, but in 1981 NASA scientist Donald Heath reported a comparison by which he attempted to control for this degradation. He compared measurements of the vertical ozone profile from the first six months of BUV's operation (in 1970 and 1971) and the first six months of SBUV's operation (in 1978 and 1979), in effect assuming the two instruments were consistently calibrated soon after launch. From this comparison he identified significant depletion in the upper stratosphere, reaching about 0.5 percent loss per year around 40 km, despite a coincident stratospheric cooling trend that would tend to increase ozone.[131] Heath's claim attracted widespread popular attention and controversy in the United States and internationally, particularly after Mario Molina discussed it in Congressional testimony, but was not widely credited by scientists because the claimed corrections could not be verified.[132]

In total, several distinct claims had accumulated by 1985 that significant changes in global ozone were occurring, but each claim suffered from one or more identifiable weaknesses and they had not yet moved collective opinion away from the no-trend default. In 1985, however, the ozone trends debate experienced two shocks, which required reexamination of the entire debate and which were not resolved for three years: the report of the Antarctic ozone hole in June 1985, and a claim of global ozone loss in the satellite record that was circulated in late 1985.

The British Antarctic Survey (BAS) had conducted various geophysical observations, including total ozone, at its Halley Bay station in Antarctica since 1957, from early October until mid-March each year. The station was periodically threatened with budget cuts or closure, although its status become more secure with Britain's increased interest in a South Atlantic presence after the 1982 Falklands war.[133] Halley Bay was one of the many Dobson stations that did not send their data to the World Ozone Data Center, citing the poor quality of much of the data uncritically tabulated there.[134] The station's measurements of average ozone for the month of October began showing a weak downward trend in the late 1970s, which grew progressively stronger after 1980. The researchers initially suspected an instrument problem, but a newly calibrated instrument gave measurements consistent with the old one in 1982. When October ozone in 1983 and 1984 showed a continuing steep decline reaching 40 percent (from a normal October average of about 300 Dobson units to about 180), and large decreases were confirmed from a second British research station 1,000 miles to the northwest, the researchers decided to publish. In late 1984, they submitted their observations to *Nature*, where they appeared in May 1985, to the shock of ozone researchers worldwide.[135]

The observations were initially viewed with skepticism, because the extreme decline was so unexpected and because other data sources, including satellite instruments, appeared to show no such sharp decline. Within a few months, however, a review of archived data from NASA's two satellite instruments found that the extreme depletions were indeed present, but had not been reported. To simplify the processing of the enormous volume of data generated by the instruments, readings outside the range of 180 to 650 Dobson Units, which amply bounded any values ever previously observed, had been flagged and provisionally set aside as probable errors.[136] The reconstructed satellite data confirmed the BAS observations, revealing

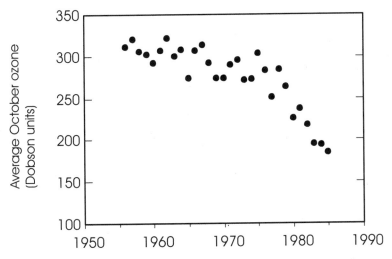

Figure 4.3 Discovery of the Antarctic Ozone Hole. Antarctic ozone in the month of October began declining precipitously in the late 1970s. Source: Data from J. D. Shanklin, British Antarctic Survey.

a huge region of extreme depletion that began appearing soon after the satellite's 1978 launch. October ozone losses over Antarctica averaged 40 percent in October 1983, increasing to 45 percent in 1984 and nearly 60 percent in 1985.[137] By the end of 1985, independent confirmation from satellites, Dobson meters, and balloons revealed that the seasonal reduction, which was soon called the "ozone hole," was real and covered essentially all of Antarctica.

In retrospect, a few other reports had prefigured these Antarctic losses, but had not given a complete enough picture to reveal the extremity of what was occurring. Satellite observations in 1980 found that the unique high-altitude clouds long observed over Antarctica covered a much larger region than previously known and persisted throughout the polar night, while another report noted extremely low levels of NO_2—both conditions that were later found to be associated with the extreme ozone losses.[138] In addition, Chubachi observed extremely low springtime ozone in 1982 and 1983 from the Japanese Antarctic research station Syowa, but lacked a long-term record to assess the observations' significance. Although he reported his observations at an ozone conference in 1984, he emphasized the subsequent recovery of ozone in November and December rather than the remarkably low initial values, and his observations attracted little attention.[139] A third Dobson instrument in Antarctica, at the U.S. South Pole station, also detected low October ozone levels after 1980, but these were less extreme than those at Halley Bay and were based on few observations due to later sunrise at the pole. These data had been reported to the world data center, but not yet been published.[140]

In addition to observations, the BAS paper included theoretical speculation linking the observed decline to CFCs and citing then-current hypotheses of nonlinear ozone depletion when chlorine exceeds nitrogen. But while their observations were of the greatest importance, it was quickly clear that their theoretical account could not explain the observed losses. Indeed, no available theory could do so. Several contending explanations were quickly advanced, of which only some linked the hole to CFC-induced chemical ozone depletion. A hastily assembled Antarctic observational expedition in September 1986 found suggestive support for a chemical explanation, but the cause of the hole remained intensely contested until resolved in late 1987, as discussed in chapter 6.

The second shock came later in 1985. NASA's Donald Heath had noticed downward trends in the SBUV data in the early 1980s, and began discussing them in late 1985 after verifying that the SBUV data confirmed the large Antarctic losses reported by the BAS.[141] He claimed that the data showed large losses between 1978 and 1984: about 1 percent annual decline in total ozone, with faster losses in the upper stratosphere (3 percent per year at 50 km), at high latitudes (1.5–3 percent per year), and in the Arctic springtime in a region centered over Spitzbergen.[142] The pattern of decline was consistent with model predictions in its variation by altitude, latitude, and season, but much larger than predicted. The SBUV data were known to be drifting downward relative to the Dobson network, at an estimated rate of 0.38 percent per year, but no one knew how much of this divergence represented problems with each data set and how much, if any, was a real trend.[143] Heath acknowledged that instrument degradation could account for some of the observed trend, as could declining solar activity and the anomalous events of 1982–1983, but argued that the trend was too large to be explained by any of these.[144]

This new claim of global ozone depletion, and the suggestion of possible Arctic losses similar to those in the Antarctic, attracted widespread public and political attention, particularly in Canada and Europe.[145] Scientists treated Heath's claim with extreme skepticism, however, as they had his earlier claim, principally out of concerns that he had not adequately corrected for instrument degradation and that the seven-year satellite record was too short to estimate a trend.[146] As in the earlier case, Heath's claim was later shown to be qualitatively correct in every respect, but substantially overstated in magnitude because his attempt to correct for instrument degradation was not adequate.

With diverse, ambiguous, and controversial claims proliferating about ozone trends, NASA launched a new assessment in the fall of 1986, the Ozone Trends Panel, to review all sources of data on ozone trends. The panel's goals were to determine what trends were clearly present, and to identify what portion of them could be attributed to natural causes, and to human activity. The 1986 NASA/WMO assessment, had addressed ozone trends, but this was one of the weakest parts of the assessment, and subsequent controversies and heightened policy attention called for immediate reexamination.[147] The panel worked through 1987 and reported its results in March 1988, as I discuss in chapter 6. During this time, the discrepancies between different ozone data sources, and the contending interpretations ranging from "marked downward trends, two to three times larger than theory" to "no obvious human-caused trends," remained unresolved.[148]

4.2 The Effects of Ozone Depletion

Although the anticipated effects of ozone depletion were the principal factor motivating public and political concern, little progress was made through this period in understanding or quantifying effects. This lack of progress reflected the intrinsic difficulty of the research problems, and also the consistently low priority and scant resources given to effects research. In both resources and progress, effects research contrasted starkly with atmospheric research.[149] In the early 1980s, when U.S. spending on atmospheric research exceeded $200 million per year and NASA's upper-atmospheric research program alone was about $25 million, *worldwide* spending on the effects of ozone depletion was less than $1 million.[150]

Most of the effects of ozone depletion arise from increased surface UV, which can cause skin cancers, eye damage, and suppression of immune response. It can also reduce agricultural yields, disrupt terrestrial and aquatic ecosystems, damage synthetic materials, and increase air pollution by increasing the photoreactivity of the troposphere. Potential benefits of increased UV have also been suggested, particularly nutritional benefits from increased vitamin D synthesis,[151] but have never received serious attention.

Since the beginning of the ozone debate, the most prominent effect has been cancer, particularly the two nonmelanoma skin cancers (NMSC), basal-cell and squamous-cell carcinoma, which together accounted for about 300,000 new cancer cases annually in the United States in the early 1980s.[152] McDonald galvanized the early SST-ozone debate with his 1971 estimates of NMSC increases, which he calculated from epidemiological data showing higher incidence at lower latitudes, among people spending more time in the sun, and on body parts more exposed to the sun. McDonald estimated an elasticity relationship between ozone depletion and NMSC of 6: each 1 percent ozone loss increases skin cancer incidence by 6 percent.[153] NMSC is so common and benign that tumors are often removed without pathological confirmation and reporting, but rates have risen for several decades despite this underreporting, by 15 to 20 percent in the United States since 1975.[154] Subsequent assessments retained the elasticity formulation, but varied the numerical value with increasing refinement of epidemiological analysis and study of direct mechanisms of cancer formation. Despite occasional estimates as high as 10, values have mostly ranged from 1 to 4, and have remained around 2 since the mid-1980s.[155]

Although NMSC is the best-known and best-quantified effect of ozone depletion, it may well be among the least important. The less common but more lethal cancer, melanoma, had about 32,000 new U.S. cases in 1991 and 6,500 deaths.[156] Like NMSC, melanoma shows rising incidence and declining mortality, but its association with UV is more complex and ambiguous. Risk is related to race, skin type, and latitude, but unlike NMSC, shows no correlation with occupational exposure or total cumulative exposure—in fact, most cases are in younger people—and some forms occur most often on body parts with little sun exposure. It has been hypothesized that the risk is associated with intermittent extreme sun exposure, or intense childhood exposure, but other factors are also likely associated.[157] Cataract formation, the best quantified effect other than NMSC, is also estimated to have a multiplier relationship with UV increase: 1 percent increase in UV-B is estimated to cause 0.6–0.8 percent increase in cataract incidence.[158] Suppression of immune response from UV radiation was first demonstrated in mice in 1976, in the skin and later systemically. UV has also been shown to impair contact allergy response in human skin. Systemic immunosuppression from UV in people was widely suspected but not demonstrated over this period.[159]

Initial studies on the effects of UV on productivity of agricultural crops showed a very small effect, while later studies showed higher sensitivity but great variability with species and conditions. For example, crops were less sensitive in the field than in growth chambers. By 1981, UV-chamber experiments on more than 100 plant species had found that about 20 percent were sensitive even to present UV levels,

20 percent were insensitive even to a fourfold increase, and the remaining 60 percent were of intermediate sensitivity—with effects in the field smaller than in growth chambers.[160]

Aquatic ecosystems were also thought likely to be sensitive to UV, but field studies were extremely difficult to conduct. By 1977, UV sensitivity had been identified in phytoplankton and in the juveniles of several ecologically and economically important fish and invertebrate species (e.g., anchovies, certain crabs and shrimp), but too little was known about these species' customary depths at specific times of day, and about how much they could adjust their depth in response to increased UV, to tell whether the effects would be significant in real ecosystems.[161] One experiment of enhanced UV in estuarine microecosystems found large losses in productivity, biomass, and diversity.[162]

Identifying UV effects outside experimental settings required detailed knowledge of patterns of surface UV radiation over space and time, but these data were extremely weak. Despite recommendations to establish a worldwide UV monitoring network as early as 1977, none was established over this period.[163] A network of UV monitors in the United States appeared to show decreases in UV intensity from 1974 to 1985, but the claim was widely disputed because the meters did not measure biologically active UV and were located in urban areas where increasing tropospheric ozone and other pollution were likely to offset UV increases from stratospheric depletion.[164]

The projected quantitative effects of UV enhancement also changed with model projections of depletion. When 2-D models began to suggest that the largest ozone losses would occur at high latitudes in winter, one consequence was that aggregate UV multipliers declined. On the other hand, UV penetrates the atmosphere most strongly under high-pressure, fair weather conditions, when total ozone is lowest and the ozone column is dominated by the middle- to upper-stratosphere—the region where, through the mid-1980s, depletion was expected to be greatest—suggesting that calculations of average depletion would underestimate UV effects.[165]

In addition to the effects of increased surface UV, ozone loss can have various climatic effects by altering the thermal structure of the atmosphere. Even when predicted changes in total ozone were small around 1980, large projected redistribution of ozone from the upper to the lower stratosphere would cause a major change in the atmosphere's heat balance. Many actors found general precautionary grounds to avoid making such a large change in the atmosphere,[166] and it was periodically suggested that too much attention was paid to global total ozone, and too little to possible redistributions and the resultant climatic effects.[167]

A direct climatic effect of CFCs was identified in 1975, in CFCs' infrared absorption and contribution to the greenhouse effect.[168] Ozone is also a greenhouse gas, and either reductions or redistributions could alter the atmospheric radiative balance, and hence the climate, although the effect was complex.[169] An ozone reduction would warm the surface through increased transmission of incoming visible and UV radiation, but would also cool it through reduced absorption of outgoing infrared radiation. Early calculations suggested that the total effect would be small, but also raised the possibility of larger climatic effects through changes in stratospheric circulation.[170] By 1985, the effects of ozone depletion and of anthropogenic greenhouse forcing were increasingly being considered together as "atmospheric change."[171]

Understanding of effects of ozone depletion in 1985 reflected little progress over ten years earlier. The list of potentially important effects had largely been identified by the mid-1970s, with only a few later additions. Few of the identified effects had even rough quantitative estimates, and these were likely not the most serious effects. Immune-system effects had the potential to be very serious, but had not been demonstrated to be so. Ecosystem effects were widely regarded as potentially serious, but were complex and little understood, and offered extremely limited opportunities for direct observation. Indeed, policy debate was pervaded by a general sense that effects did not matter: advocates of controls had reached an early consensus on precautionary action, rendering subsequent refinement of effects estimates irrelevant.[172]

4.3 Science for Policy: Scientific Assessments of Stratospheric Ozone, 1976–1985

The period from 1976 to 1985 was marked by a great volume of scientific effort and progress in understanding stratospheric ozone, and also by a large number of ambitious efforts to conduct official assessments of scientific knowledge—collective, deliberative, expert processes that sought to summarize, evaluate, synthesize, and interpret scientific knowledge to inform policy making and decision-making. The number, scale, and prominence of these assessment efforts reflected the influence of two views that were widely held by both policy actors and scientists throughout the period. On the one hand, policy decisions should be consistent with scientific understanding: what responses are appropriate depends in part on the likelihood, severity, timing, and consequences of ozone depletion, which can be addressed only by scientific inquiry. On the other hand, the content of scientific knowledge itself, as represented by published results and associated debates and arguments, cannot serve this need. In part, this follows from the general proposition that scientific knowledge alone cannot imply a unique optimal policy choice. It also reflects the practical constraint that most political actors are unable to process and evaluate primary scientific argument and make independent judgments of the present state of knowledge on questions of consequences and risks, or its implications for their preferred course of action. For these reasons, some form of intermediation between the domains of science and policy is essential in order to evaluate particular claims, summarize and synthesize present knowledge, and identify its implications of relevance to policy decisions. While policy actors may rely in part on trusted individual advisers to perform this mediation, expert assessment processes that are collective and deliberative are widely regarded as necessary to serve the broader functions of sharpening the terms of policy debates, clarifying and delimiting the basis of policy disagreements, and supporting the identification and adoption of preferred policy choices.

Despite widespread recognition of the importance of assessment to inform science-intensive policy debates such as ozone depletion, assessment processes remain acutely understudied. Consequently, the mechanisms by which assessment can contribute to policy making, and how these contributions are related to assessment organization, management, and methods, are inadequately understood.[173] This neglect has two serious consequences: theoretical understanding of policy outcomes on scientifically intensive issues has major gaps; and knowledge that could be available to guide practical decisions in the design and conduct of assessment is not being exploited.

To examine these mechanisms and relationships, this section reviews the most prominent assessment processes for stratospheric ozone conducted between 1975 and 1985: five assessments by the U.S. National Academy of Sciences (NAS), two by the U.K. Department of the Environment (DOE), annual assessment reports issued between 1977 and 1985 by UNEP's Coordinating Committee on the Ozone Layer (CCOL), and four by the U.S. National Aeronautics and Space Administration (NASA), which later gathered multiple sponsorships from both U.S. and international organizations.

The discussion examines variation in assessments' design, methods, and outputs. Significant characteristics include the scope of questions addressed, the number and types of participants and criteria for choosing them, the methods used to develop and state consensus and manage dissent, and the treatment of uncertainty and of new or unverified claims. In addition, assessments differ in whether they go beyond summarizing settled points, making judgments of the relative strength of competing claims or the likely resolution of outstanding disputes; and in whether and how they identify policy-relevant conclusions or make explicit policy recommendations. In a few cases, multiple assessments were conducted at roughly the same time, so their diverse choices, given similar underlying knowledge, can be observed. The discussion also examines the subsequent use of assessment outputs by policy actors, in an attempt to identify generalities about assessment effects, pathways of assessment influence, and conditions or choices that make assessments ineffective. Finally, the discussion attempts to distinguish between the policy influence of substantive scientific knowledge and controversy at a given time, and of the authoritative summary and synthesis of that knowledge in assessments.

The U.S. National Academy of Sciences

The U.S. National Academy of Sciences conducted its first assessment of ozone depletion from CFCs in 1975 and 1976, in the heat of the U.S. aerosol controversy. Soon after this assessment was completed, the 1977 Clean Air Act amendments required the EPA to commission additional ozone assessments from the NAS every two years. The Academy conducted four more assessments under this mandate, a small interim report in 1977 and larger assessments in 1979, 1982, and 1984. The 1976 and 1979 assessments and their role in U.S. policy were discussed in chapter 3. Following normal practice of the NAS, these assessments were all conducted by volunteer committees of about a dozen scientists, chosen for expertise relevant to the issue but with no strong prior stake in it. Most members were American, although committees often included one or two Canadian or European scientists, among them a Canadian chair of the 1977 and 1979 panels. All received professional support from NAS staff, with varying degrees of continuity in both members and staff between assessments. All took U.S. domestic policy makers as their primary audience, particularly the EPA and Congress.

The 1977 Clean Air Act amendments explicitly required these assessments to be comprehensive, covering the science of the stratosphere and any substances that might modify it, effects of stratospheric change, CFC uses and alternatives, and policy responses.[174] The NAS responded to this broad mandate by dividing the issue into broad areas addressed by separate committees with different expertise and responsibilities. Every NAS assessment had one body addressing stratospheric chemistry and physics and another addressing effects of UV radiation; some had superior bodies responsible for synthesizing implications and conclusions across all contributing bodies; and one assessment only, in 1979, had bodies addressing technological alternatives to CFCs and potential policy responses. Beyond these commonalities, there were substantial differences among the assessments, most conspicuously a major change in approach after 1979, between the first two full assessments and the last two.

The 1976 assessment included a panel on atmospheric chemistry and a committee to incorporate the panel's work, cover all other aspects of the issue, and synthesize results. In chapter 3, I discussed the unique conditions of this assessment, by which many actors committed in advance to base their decisions on its results and the IMOS Panel succeeded in establishing a presumption of regulation, leaving the committee the choice between accepting this presumption and explicitly rejecting it. The committee not only accepted it, but offered explicit policy recommendations, albeit in some cases worded very obscurely. They concluded that CFC restrictions would probably be required, after a delay for further research of no more than two years, and that "similar action by other countries should be encouraged by whatever appropriate means are likely to be effective."[175] They also offered expansive general guidelines for policy making under uncertainty, for example, that not all CFC uses should be eliminated, that those easiest to replace should be controlled first, and that delaying controls to learn more about what controls are needed can be the best course.

These broad didactic recommendations reflected much intelligence and common sense, but made the assessment highly vulnerable to attack. They were difficult to defend as grounded in the committee's expertise, with the important exception of the chair, the statistical decision theorist John Tukey. Moreover, they were based on a set of unexamined assumptions that were at best contestable, in some cases simply wrong: that the risk of ozone loss was not grave enough to consider eliminating all CFC uses (i.e., the committee assumed the results of a comprehensive benefit-cost assessment that they were unable to do); that delaying controls to learn more was acceptable if the resultant increase in depletion was small; and that the only feasible form of response was use-by-use restrictions. This last assumption, which they made after rejecting a CFC tax as ineffective unless impracticably large, easy to evade, and infeasible internationally, was particularly flawed. The committee provided no analysis and had no special expertise to support their claims about a tax, and it overlooked the possibility of comprehensive controls in some form other than a tax.

The conditions under which the committee worked may have forced them to draw broad policy conclusions—indeed, their tortuously worded recommendations may have represented their attempt to minimize their intrusion into government's domain—but this assessment can still be faulted for significant failings of both commission and omission. For example, saying specifically how long to wait before regulation (up to two years), without saying whether this delay was to apply to the commencement of rule-making or the implementation of controls, caused needless controversy and suggested the committee did not fully understand the implications of its recommendations.[176] Moreover, the committee missed an excellent opportunity to make their recommendations more useful by failing to state, even roughly, *what* new findings would strengthen or weaken the case for CFC controls.

After a hastily conducted interim assessment in 1977, which drew extensively on the recently completed first ozone assessment by NASA, NAS conducted its second full assessment in 1979. This effort, the first to be conducted under the broad mandate of the 1977 Clean Air Act amendments, sought to cover the mandate by establishing five separate bodies: an atmospheric panel; a committee to summarize the atmospheric report and integrate it with a discussion of health and ecological effects; a committee to assess CFC uses, alternatives, and policy responses; and two panels under it considering industrial technology, and socioeconomic impacts of regulation. The last two panels included academics and representatives from industry, labor, and environmental and consumer groups.

The atmospheric panel retained several members from 1976, although with slightly reduced non–U.S. participation, and took a very similar approach to its work.[177] They once again used calculations from one stratospheric model to produce a point estimate of steady-state ozone depletion (16.5 percent) and a range reflecting various sources of quantified uncertainty (5–28 percent). They also, very tentatively, presented depletion calculations under CFC reductions and growth. With a few qualifications, they expressed substantial confidence that large further changes in estimated depletion were unlikely, but as in 1976, this confidence was mistaken.[178] New results, including new values for reaction rates that the panel had not identified as important sources of uncertainty, caused large changes in estimated depletion over the next two years.

Reports of the committees on UV impacts and CFC alternatives appeared together, with a jointly prepared summary. The impacts committee, whose members were unchanged from 1976, summarized the atmospheric panel report with greater stress on the emission growth scenario, summarized the limited advances in knowledge of impacts, and drew policy conclusions. These conclusions were moderately strong, although substantially less expansive than those in the 1976 report. The alternatives committee drew on prior reports by EPA consultants, Rand, and DuPont, as well as consultations with industry and technical experts, to present a highly pessimistic assessment of prospects to reduce nonaerosol CFC uses that concluded substantial reductions would be impossible.[179] Compelled to rely on industry sources for much of their technical information about the availability, development status, performance, and costs of alternatives, this committee was both severely hampered in its ability to conduct an independent critical assessment, and misdirected in its basic framework, accepting presumptions that only fully developed and available alternatives should be considered, and that no significant cost or degradation of performance was acceptable in evaluating alternatives. The committee did note a surprising degree of disagreement among technical experts consulted, but did not note the extent to which industry's control over relevant technical information created a strong bias toward technical pessimism. Strikingly unsuccessful in their technological brief, the committee presented various broad policy conclusions, including a call for international aerosol controls and the assertion that U.S. leadership, together with measures to develop a stronger international scientific consensus, would likely induce other nations to follow the U.S. lead.[180] The resolution of this debate over technological control opportunities represented a decisive defeat for the advocates of further CFC controls.[181]

The third NAS assessment, which appeared in April 1982, took a sharp step back in both scope and ambition. Despite the legislative mandate for comprehensive assessments, this one excluded discussion of technological alternatives and policies. It was once again conducted by two committees, one on atmospheric science and one on biological effects of ozone depletion, whose reports were published together with a common summary. In contrast to the previous atmospheric panels, the newly re-formed atmospheric committee commissioned leading scientists to write six peer-reviewed background reports, which appeared as technical appendices. The committee did no new analysis, but simply summarized these reviews as its report. In many ways, this committee pulled back from the sweeping conclusions and policy recommendations of the prior NAS assessments. Where earlier committees had done new model calculations to project depletion, this one simply repeated estimates from a model comparison undertaken for a recent NASA assessment. The estimated range of steady-state depletion was 5 to 9 percent, reflecting new kinetics results that had decreased calculated HO and, consequently, ozone's sensitivity to chlorine. Where earlier committees had attempted to quantify and combine uncertainties from various sources in presenting ranges of projected depletion, this one made one limited attempt to do so, and even that attempt split the committee.[182] While the importance of depletion calculations with multiple perturbations was stressed, none was undertaken and there was no mention of either brominated chemicals or the potential for CFC emissions growth. Heath's recent claim to have observed upper-stratosphere depletion by comparing data from two satellites was mentioned, but not discussed or evaluated.

In contrast to the timidity of the atmospheric report, the report of the Biological Effects Committee was unusually forceful. The report reviewed developments since 1979 in molecular and cellular studies, ecosystem effects, and human health effects, based on papers presented to a July 1981 workshop. Although these fields remained full of gaps and unanswered questions—indeed, the report notes that many important questions identified by 1975 had received so little support that no progress had been made on them—this was where the significant contributions of this assessment lay: in clarifying previously known risks and identifying new ones not previously investigated. The committee presented several significant advances: newly identified interactions between biological effects at different wavelengths, including two cellular mechanisms that repair some of the worst UV damage;[183] and the first discussion of health effects other than skin cancer, including eye damage and immune system effects.[184] They proposed increasing the cancer-ozone multiplier to 2–5 for basal-cell carcinoma and to 4–10 for squamous-cell carcinoma, while they judged melanoma's causes too obscure to make a quantitative estimate.[185] In addition, the committee drew several forceful policy conclusions. They argued against the claim in the 1979 U.K. assessment that "other factors than UV cause skin cancer," noting that "it seems certain that more than 90 percent of skin cancer other than melanoma in the US" was associated with sunlight exposure. They argued for the first time that deaths from nonmelanoma skin cancers must be counted in ozone assessments, because while these cancers had low mortality (only about 1 percent), their incidence was so high that they caused nearly as many deaths as the more lethal melanoma. The committee also argued that UV effects on other animals and plants were likely to be as important as human health effects, but had been so little researched that they could not be quantified.

Despite the quality and forcefulness of this report, it gained little attention and had little policy influence. Press reaction to the assessment featured the reduced estimates of ozone loss from the atmospheric committee and ignored the increased estimates of cancer sensitivity—even though these were so large that they fully offset the reduced depletion and left projected cancer increases unchanged. Industry comment on the assessment ignored the impacts committee entirely, merely criticizing the atmospheric section for not being sufficiently complete.[186] Even the EPA's attitude to this assessment was ambivalent. In preparing their required report to Congress, the EPA first relied on NASA's 1981 assessment, which considered only atmospheric sciences, because the NAS assessment was not completed in time. Although the EPA acknowledged that their report was too narrow to meet their statutory mandate, they never submitted the promised fuller report, even after this NAS assessment provided them with the necessary information. Given the struggles in the agency at the time, EPA officials likely sought to avoid having to deal with the implications of these strong claims about health and biological effects.

The final NAS assessment of stratospheric ozone appeared in February 1984, again combining separate updates of atmospheric science and UV impacts by separate committees. As in 1982, the atmospheric section summarized major recent results in kinetics, modeling, and observations, with no independent analysis and no policy conclusions. New kinetic results had reduced steady-state depletion estimates to 2–4 percent, while multiple-perturbation scenarios with constant CFC emissions and modest growth of other pollutants gave nearly zero depletion.[187] Growth scenarios again received only brief mention, as did the recent suggestion of nonlinear depletion at high chlorine concentration. The report made no quantitative uncertainty estimates and criticized previous assessments for being too confident in doing so. The report on UV effects provided a modest update on the 1982 review. Prominent new results included another downward revision in the ozone-NMSC multiplier, back to 2; indications of increased UV effect in melanoma; and new evidence on immune-system effects suggesting that they were primarily suppressions, operated systemically, and occurred both in humans and in other animals.[188] Response to the report was muted and predictable. Industry and most press reports highlighted the steady decline of depletion projections since 1979, with one industry periodical calling the ozone issue nearly resolved.[189] Environmentalists and other advocates highlighted the large ozone loss projected from CFC growth, even with other pollutants increasing, a result that was present but downplayed in the report.[190]

The form, process, and participation of the four NAS assessments were largely set by standard NAS practice, and by the statutory mandate under which they were commissioned. The mandate hampered their effectiveness, particularly the requirement for comprehensive assessments covering every major aspect of the issue, without regard for the state of knowledge and the basis for consensus in different areas. While the mandate did not demand "integrated" assessment synthesizing knowledge from these disparate domains, it still sought to make the Academy resolve too many questions relative to the state of knowledge and the capabilities of an independent, elite scientific body like the Academy.

The Academy's four attempts to fulfill this broad mandate varied in approach, quality, and outcomes. Overall, the first two assessments were broad in scope and bold in identifying policy implications, while the latter two represented a major retreat in boldness and breadth. The one attempt to assess CFC reduction options was thwarted by structural problems associated with industry control over technological knowledge. The failure of the effort was no disgrace to the Academy: the same problems have consistently confounded attempts at independent technical assessments, and were not resolvable in the context of an Academy assessment. Attempts to assess UV impacts suffered from different systematic problems, associated with the weak state of knowledge in the field, the field's scant resources, and its limited progress. Moreover, even the strongest assessment in this field, that of 1981, attracted little public attention—less, indeed, than the less important and largely derivative atmospheric report of the same year. The reaction to these two assessments is one of the strongest instances suggesting some general handicap of effects claims in the general policy debate.

Not similarly handicapped by either structural limitations or weak underlying knowledge, the atmospheric bodies consistently produced high-quality reviews of the evolving state of knowledge, but even they never attained any significant influence on policy debates—with the anomalous exception, discussed in chapter 3, of the 1976 assessment.

The most notable feature of these atmospheric assessments was the vigorous attempt in the first two to conduct comprehensive, quantitative analysis of uncertainty in future depletion projections, followed by the equally vigorous retreat from such analysis in the last two. Such analysis has great value in principle, for supporting judgments of whether (and what) action is warranted and for directing research effort toward the most consequential uncertainties. But this exercise was widely attacked, by U.K. assessments, industry sources, and other scientists.[191] Indeed, such analysis is highly vulnerable to attack (whether partisan or sincere) for speculation and arbitrariness, because it necessarily lacks the foundation in agreed and published results from a broader research community that protects more conventional and passive scientific assessments. Such attacks may be unavoidable, and the appropriate response may be to manage the activity so attacks against it do not call other elements of the assessment into question, perhaps by separating them into different bodies. Alternatively, the analyses might have been ineffective in these assessments because of specific failings in implementing, explaining, and defending them. In each attempt, the panel highlighted the kinetic uncertainties for which quantification was easiest and most widely accepted. Other atmospheric uncertainties were treated, appropriately, by the panel's collective expert judgment, but were reported so tentatively that they were easily overlooked. Most seriously, by doing all uncertainty analysis on a range of steady-state depletion estimates, the panel consistently ignored uncertainty in the socioeconomic factors that determine future emissions, which were consistently larger than any geophysical uncertainties in their contribution to uncertainty over future ozone loss. Neglecting these severely impaired the panel's attempts to be policy-relevant.[192]

Other factors also limited the influence of these assessments. Because the assessment bodies were too small ever to include more than a few current leaders in the field, the quality of scientific review they achieved, while always very high, never attained definitive, reference-book status. Relying on a few writers, there was always room for minor idiosyncrasies in emphasis, interpretation, and particular points included or excluded that provided effective openings for subsequent attack. The reports also repeatedly expressed excessive confidence in the current state of understanding and made qualitative summaries of states of knowledge—for instance, judgments that theory-model agreement was "good enough within uncertainties of measurements"—that hindered comparison and provided no operational guidance for interpreting new results or observations. Other assessments consistently found the discrepancies more serious, but there was no basis for resolving such discrepant judgments. Finally, these assessments lacked a coherent vision of how and how far to proceed in identifying policy-relevant results, synthesizing implications, or drawing overt policy conclusions and recommendations. Rather, they first overreached with expansive policy advice, then overreacted by abandoning any attempt at synthesis, or even identification of the most important points. With the assessment bodies taking no coherent approach to scientific synthesis, reaction to them always stressed the new trends in steady-state depletion estimates—points that were easy to grasp, but arguably never the most important ones.

Policy actors used NAS assessment statements that supported their case, but did not use or trust the assessments more broadly. The only apparent instance of strong domestic influence was the 1976 assessment, but even this influence may be exaggerated, given the prior success of IMOS in shifting the presumption in the policy debate. Their use and influence internationally were even less, since they did not attain a sufficiently authoritative scientific status to avoid being identified with the U.S. government position.

The U.K. Department of the Environment

The U.K. Department of the Environment (UKDOE) sponsored assessments of CFCs and ozone depletion in 1976 and 1979. Both had a two-part structure. A short main report written by government officials summarized the issue and drew policy conclusions, while a longer appendix prepared by a committee of government, university, and industry scientists provided scientific discussion and background. The responsible officials and the committee membership were largely constant for these two assessments, and largely unchanged from the earlier assessment of stratospheric impacts of supersonic flight.[193]

The 1976 assessment represented a central component of the government's response to the independent Royal Commission on Environmental Pollution (RCEP), whose 1974 report had expressed strong concern over ozone depletion.[194] The scientific appendices provided moderately detailed tutorial discussions of the atmosphere, CFCs, ozone depletion, surface UV, and health and climatic effects, similar to those in the 1976 U.S. NAS assessment and drawing similar conclusions. Like the NAS assessment, this one used one model—in this case a British model—to obtain an estimate of 8 percent steady-state ozone loss.[195] The major differences from the NAS report are the prominence given to uncertainties, and the argument that 1-D models cannot be relied upon for policy analyses.

In addition to summarizing the scientific appendices, the short main report was an overtly polemical document that ranged far outside scientific discussion to make the case that no regulatory controls on CFCs were appropriate at the time. It cited new calculations showing the small Concorde fleet would have little effect on ozone, and implied—erroneously—that this refuted the original SST concerns raised by Johnston and Crutzen in 1971. It described the UV increase from projected depletion as similar to that experienced in moving from the north to the south of Britain, and also noted that UV radiation provided some benefits (principally a reduced incidence of rickets) and that skin cancer had many causes in addition to UV. It praised the utility and versatility of CFCs, and discussed potential alternatives in highly unfavorable terms. Finally, it argued that the global scope of the problem required global action. Although the IMOS report had made the same argument, the two drew opposite conclusions: where IMOS argued that the global problem called for U.S. leadership, the U.K. report stressed the futility of unilateral national action in advance of a global consensus. Its only action request was that British industry search for alternative aerosol propellants.

The second UKDOE assessment appeared in October 1979, with a similar two-part structure. This time, the first part was an official report of DOE's Central Directorate on environmental Pollution, presenting an interpretation of current understanding and explicit policy positions. The second part was a separate report by the Stratospheric Research Advisory Committee (STRAC), a 20-member body of government, university, and industry scientists.[196] As in 1976, the second part was a clear, didactic scientific summary that drew on essentially the same science as the concurrent NAS panel report and reached similar conclusions. Using slightly different kinetic data than the standardized set used in U.S. assessments, it reported steady-state depletion estimates from three U.K. models ranging from 11 to 16 percent, similar to that reported by the NAS panel the same year. It once again expressed strong skepticism about 1-D models, and also noted that since early 2-D model results showed the greatest depletion at high latitudes in winter, where surface UV is very low, the UV/ozone multiplier of 2, calculated from 1-D models, was likely to be too high: for example, one 2-D calculation with 12 percent global-average ozone loss gave only 13 percent global-average UV increase. Although stressing scientific uncertainties, the report argued that sources of uncertainty could not be quantified, and sharply criticized the 1976 NAS panel for trying to do so.[197]

While the tone of the STRAC report hinted at a competitive or adversarial stance toward U.S. science and its use in environmental policy, the first part of the assessment made this stance explicit. Written by government officials, as in 1976, this report provided a strongly partisan summary of scientific knowledge and drew strong, explicit policy conclusions. It criticized the United States (erroneously) for regulating aerosols against the advice of the NAS assessment,[198] and concluded that large remaining uncertainties continued to make any regulatory action on CFCs premature. Industry spokesmen in the United States made extensive use of this U.K. report to support their opposition to the EPA's 1980 proposal to control nonaerosol CFCs, but the passages they used were all drawn from the officially authored Part 1, not from the independent scientific report in Part 2. These were portrayed as showing serious enough scientific disagreement to reject regulation, even though the actual disagreements between the NAS assessment and the scientific part of the U.K. assessment were insignificant.

Soon after this report, STRAC was disbanded as part of a governmentwide drive to reduce outside advisory bodies. No similar U.K. ozone assessment body existed until a successor was established in late 1985. Despite STRAC's abolition, two of its members published an unofficial update two years later, which summarized new kinetics results reducing calculated depletion and provided an extensive discussion of uncertainty.[199] The update stressed the general need for better ways to analyze uncertainty, but was devoted principally to a detailed critique of the 1979 NAS panel's attempts to quantify uncertainty.

The U.K. assessments were curious hybrids of independent scientific reports and official government reports. The first part of each was entirely official, stating the government's policy position with a summary of supporting scientific and other argument. This purpose is clear in both reports, most strikingly in the forceful 1976 conclusion that the Concorde's effect on ozone was very small. Since this point was irrelevant to the report's charge to assess the CFC-ozone issue—and, moreover, the same kinetic changes that reduced ozone losses from SSTs increased losses from CFCs—it appears as a clear attempt to support the U.K. government in the campaign it was then waging to secure the Concorde's access to American airports.

These assessments were not exclusively advocacy for U.K. policy positions, however. Each report's second part was authored by an eminent group of British scientists, at least some of them independent of government and the CFC industries. Where these second parts diverged from the contemporary NAS assessments—as in their greater skepticism about using 1-D models for policy projections, and their refusal to attempt to quantify uncertainties—they reflected true differences in these expert groups' collective judgment about use of scientific knowledge in policy debate. Although these reports provided cogent reviews of scientific knowledge and recent results, the attempt to combine objective scientific review and partisan advocacy in one volume rendered their credibility suspect and their purpose obscure. These assessments provided comfort to opponents of controls, who used the partisan parts as if they were scientific reports to shore up their antiregulatory case, but only the scientifically ignorant were fooled. No opponent of the policy stance supported in these reports felt compelled to mount a serious response. The substantial scientific effort that went into these assessments was wasted as contribution to international policy debate, because the resultant report was tainted by its association with the U.K. government position.

The Coordinating Committee on the Ozone Layer (CCOL)

Attempts to conduct scientific assessment of the ozone issue at the international level began on an extremely modest scale. During the period considered here, the only international assessment body operating was the Coordinating Committee on the Ozone Layer (CCOL), established by UNEP after the 1977 Washington scientific meeting. Composed of representatives of ozone research programs, CCOL had a mandate that included coordinating research priorities, sharing results, and publishing periodic assessments. Its mandate covered the full scope of the ozone issue, including the ozone layer and its modification, effects of ozone depletion, and "socioeconomic aspects" (i.e., emission projections, alternatives to CFCs, and policy responses), but it did no research itself.

CCOL's founders sought to promote coordination of research and deliberation on its results and significance, in the hope that developing international consensus on scientific knowledge would in turn promote consensus on policy actions.[200] By involving participants with direct responsibility for research programs, rather than senior political decision-makers, CCOL sought to involve scientists who would command respect and who were motivated by technical and scientific considerations, not the interests of their institutions and political masters.

The CCOL's first meeting, in November 1977, established several basic aspects of its subsequent operations. Most important, it decided to let international organizations and nongovernmental organizations (NGOs) be members, including—over some objection—the industry-supported Fluorocarbon Panel. The reasoning for including these was that sharing and coordinating research most effectively required all research programs, not just national programs, to be members.[201] Having a research program was a condition of membership, so the Fluorocarbon Panel could join but environmental NGOs could not. National membership began with 14 countries, and gradually increased after 1980.[202] UNEP chaired the meetings, provided secretarial support, and published associated reports.

The committee met for a few days each year from 1977 to 1986, primarily, to compile and review reports from each participating research program. UNEP staff condensed these reports into a summary *Ozone Layer Bulletin* for wide circulation, and periodically prepared more extensive assessments that summarized research progress to inform national and international policy making. Drafts of these assessments were circulated at the annual CCOL meetings, then revised by UNEP on the basis of the committee's comments.

The first three years of CCOL's life saw a struggle over whether it would attempt to make strong, policy-relevant statements. In 1978, with the support of a group of national officials, UNEP's assessment tried to make such a statement, within evident diplomatic constraints, quoting a recent WMO statement that "the threat is real, and warrants continued action to control emissions," and urging policymakers to pay attention to the issue.[203] At the next meeting, U.S. and Canadian representatives argued that the 1979 NAS assessment should be taken as an international scientific statement but European officials rejected the proposal, calling the NAS report "U.S. science" and citing contrary statements from the 1979 U.K. assessment to rebut it.[204] The hard-fought language of this meeting's report stated that "considerations leading to the prediction of an ozone depletion due to anthropogenic emissions of halocarbons like [CFCs] are plausible and largely consistent," and endorsed efforts under way to limit CFC emissions, albeit in highly restrained terms.[205] These statements were the furthest advance in CCOL's ambition to draw policy-relevant conclusions. Subsequent meetings made no further attempts to address any statement to policymakers other than lists of research tasks, even after international ozone negotiations were established in 1982.[206] At the same time CCOL solidified a narrow view of its mandate, making no attempt to integrate its discussions of atmospheric issue and UV effects, and repeatedly declining negotiators' requests that it address "socioeconomic aspects" of the ozone problem, such as emissions projections, technologies, and response options.

Through the early 1980s, CCOL also took stances of increasing scientific conservatism on certain questions of high policy relevance. For example, on the question of what evidence would confirm ozone-depletion theory, CCOL reports beginning in 1980 stated with increasing force that confirmation required observing significant global ozone loss.[207] In 1983 they went even further, stating that replicating the present atmosphere was not sufficient to validate models, but that they must also correctly simulate an observed change in ozone and its time dependence.[208] Since such validation could be obtained only after a large ozone loss had already occurred, this was a stance even more extreme than industry representatives were advocating.

The CCOL faded away after 1984, its work overshadowed by the more ambitious and authoritative NASA/WMO assessment discussed below. CCOL did not meet in 1985, and met in 1986 only to review and endorse the report of the NASA/WMO assessment, after which two small ad hoc groups met to draft short summaries of other recent assessments on behalf of CCOL.[209] Presenting these two short documents to delegates as they convened in December 1986 to begin negotiations of the Montreal Protocol was the last act of the CCOL.

It was widely recognized that some body was needed to conduct assessments based on broad international scientific consensus, and UNEP staff and some national officials periodically tried to prod CCOL to do this job.[210] CCOL was fundamentally inadequate to achieve either useful synthesis or policy relevance, however, due to various limitations associated with its membership, leadership, procedures, and resources. CCOL was constituted of research programs, not individual scientists. Although it usually succeeded in attracting a few eminent scientists to participate on national delegations, most participants were program managers and officials who, like the programs they managed, were of uneven scientific stature. Although CCOL would not simply endorse an assessment by a national body (as the Canadians and Americans learned in 1979), it was also unable to produce international assessments that commanded any authority among either scientists or policy actors. However eminent a few of its individual participants, the group lacked the time, numbers, staff support, and collective stature necessary to clarify debates, resolve discrepancies or competing claims, or conduct synthesis. Its preparatory work was too limited, and its short annual meetings too dominated by the need to produce a report, to allow time for serious examination of substantive differences. Its reports were consequently little more than compilations of the re-search reports submitted by participating programs, even though these included much research of little relevance to the ozone layer and sometimes added highly tendentious interpretations and policy arguments. The CCOL reports drew no con-clusions or collective judgments, and presented conflicting views and interpretations without comment, sometimes using diplomatic language to obscure disputes or con-tradictions. They were, moreover, useless as scientific reviews, as they stated results with no criticism or evaluation, and included no references or attributions. More-over, even if CCOL's assessments had been strong and persuasive, it had no policy client. CCOL reports were submitted to national governments and UN agencies, none of which was obliged to receive, consider, and respond to them. Since all these limitations reduced the scientific interest of the proceedings and made them less attractive to top researchers, the CCOL's weaknesses of resources, participation, and authority were self-reinforcing.

Any hope that CCOL would produce useful policy-relevant syntheses was also hindered by the presence of the chemical industry's Fluorocabon Panel on the com-mittee. The panel was a curious combination of a partisan body and an independent scientific research program. Although the panel chose what areas to fund in part for the likelihood that they would vindicate CFCs, the research they supported in those areas was independent and high-quality, and the panel made no attempt to control the review of proposals, the conduct of research, or the dissemination of results.[211] But the panel itself issued occasional publications, reports, and policy briefs that, while supposedly based on the scientific research they supported, were highly selective and partisan.[212] As a member of CCOL, the panel naturally pressed for its interpretations of research to prevail. The extent to which CCOL reports in the 1980s mirrored some of the strongest components of industry's interpretation suggests that they often succeeded. Moreover, even when they exercised less sub-stantive influence, the panel's very presence on CCOL weakened any chance that it could be perceived as an independent scientific body.

Despite a few valiant efforts, the UNEP secretariat was also unable to provide the required authoritative synthesis. The periodic reports drafted by UNEP staff after CCOL meetings sought to provide the scientific synthesis needed to move the issue onto the international policy agenda. But these drafts were prepared without the necessary resources, technical knowledge, or scientific stature, and lacked the supporting scientific consensus they would need to be seen as anything other than bureaucratic documents. Although many activists had argued that the breadth and novelty of the ozone issue made it ideally suited for UNEP, it was not clear at this time that any issue was ideally suited for UNEP. It was a new and precarious organization, with no scientific status, about whose capacity many national officials had serious doubts. The result of all these handicaps was that despite great effort by some, the CCOL brought forward weak reports, achieved no advancement or integration of knowledge, and had no influence on policy. With the CCOL a failure, and with national assessment processes unable to gain international credibility, it appeared that no body was able to provide the authoritative international scientific consensus that all policy actors said was needed.

The U.S. National Aeronautics and Space Administration

In the United States, NASA conducted a series of ozone assessments in parallel with those of the NAS. While these began as purely domestic activities, NASA progressively broadened their international participation and institutional sponsorship, culminating with a massive 1985 assessment whose influence on the subsequent development of ozone policy was decisive.

NASA first promised the U.S. Congress to deliver an assessment of the effect of CFCs on stratospheric ozone by 1977, as a condition of the 1976 funding that established their upper atmospheric research program. The 1977 Clean Air Act amendments subsequently required NASA to deliver upper-atmospheric assessments biennially, a requirement parallel to that imposed on the EPA to commission assessments from the NAS. Unlike the EPA's requirement, however, NASA had complete freedom in how to do their assessments, operating under no specific instructions with no direct client—indeed, with little sign that anyone in Congress was actually interested in their product. NASA ran four ozone assessments under this mandate, in 1977, 1979, 1981, and 1985. With full control over the scope of each assessment, NASA organized them strictly as scientific reviews of the questions of upper-atmospheric physics and chemistry being pursued under their research program, making no attempt either to address the questions of UV effects and CFC uses and alternatives that EPA was instructed to consider, or to draw out explicit policy-relevant conclusions.

These four assessments all followed a similar model. Each was organized around a scientific workshop of a week or more, at which separate working groups of one or two dozen eminent scientists reviewed the state of knowledge in specific areas. Each working group received one or more background papers on selected important questions in its area, and sought to identify key problems, controversies, and research needs. A synthesis paper was then drafted, based on the position papers and the group's discussions. Each synthesis paper became one chapter of the assessment report, and was intended to be of scope and quality suitable as a reference for researchers in its area.

The first assessment workshop, held in January 1977, included 65 scientists organized in working groups on laboratory measurements, ozone measurements and trends, minor species and aerosols, one-dimensional modeling, and multidimensional modeling. Several industry scientists participated, including four atmospheric modelers from DuPont and four non-American scientists (three Canadians and Belgium's Guy Brasseur). The assessment took a narrow focus, considering only CFC effects on ozone rather than reviewing stratospheric processes and perturbations more broadly. While some of its outputs were of clear policy significance, the assessment made no attempt to highlight these, but only to attract scientific interest and credibility—at which it succeeded. The report was widely cited in research papers and other assessments, but received almost no mention in policy debates.

The most important contributions of this assessment were two services it provided to the research community, each an update of an activity previously conducted under CIAP: a tabulation of chemical kinetic data, and a model intercomparison. A standard set of kinetic data was essential to provide consistent inputs for model comparisons. The panel responsible for these data reviewed available measurements of all relevant reactions, discussed discrepancies and limitations, and provided recommended rate values. As this job required sustained effort, the panel began working before the workshop and continued afterward. They subsequently continued their work, providing periodic updates of recommended data for subsequent assessments and model comparisons.[213] Over time this group coordinated increasingly closely with an international body doing the same job, and beginning in 1981 the two bodies provided a single, joint report.[214] The model comparison exercise used this standard kinetic data to compare steady-state depletion estimates from nine 1-D stratospheric models. The assessment reported one depletion projection from each model, and an uncertainty range defined by the spread of estimates among models. Initially 5–9 percent, this range was revised to 10.8–16.5 percent when models were rerun using the faster rate reported for the reaction of NO with HO_2 in 1977.[215] In contrast to the NAS assessments, which analyzed uncertainty using variation in one model's results under different kinetic inputs, this assessment—like subsequent NASA assessments—stressed variation among models using fixed inputs.

The second NASA ozone assessment, conducted in 1979, followed a similar model with an expanded scale (90 scientists participating) and scope (seven topical working groups). Participants this time included four non-U.S. scientists, two each from Canada and Europe. Organizers' expansive description of this participation as "representing most of the institutions engaged in upper atmospheric research in the United States, Canada, and Europe" suggests that they were beginning to understand the importance of international participation for the credibility of the outputs. The assessment again incorporated a report of recommended kinetic data and an intercomparison of 10 substantially improved 1-D models, which now gave a range of 15–18 percent steady-state loss in total ozone and 38–55 percent loss in the upper stratosphere.[216] In addition to this model comparison, the discussion of uncertainty included one calculation of the effects of varying 65 uncertain inputs in one model, which gave a range of steady-state depletion from 7.5 to 29.8 percent.[217] The report stressed the importance of multiple-perturbation scenarios and reported early progress: preliminary calculations were available of effects of CO_2, tropospheric HO, and rocket exhausts separately, as well as one calculation that combined steady-state CFCs with doubled N_2O flux and doubled bromine concentration.[218] In contrast to the sharp criticism of 1-D models in the U.K. assessment the same year, this one criticized available 2-D models for their inadequate chemistry. Discussion at the workshop stimulated one important advance when it was noted that a lower level of calculated HO in the lower stratosphere would improve models' fit with NOx observations, and also would decrease calculated loss due to chlorine.[219] The fate of this assessment was very similar to that of 1977: widely treated as an authoritative scientific reference but unnoticed by policy actors.

The third NASA assessment, based on a May 1981 workshop, again followed a similar approach but with substantially expanded participation and the first co-sponsorship by other organizations, both U.S. and international. Non–U.S. participation increased to 25 of 109 participants, from 13 countries and 4 international organizations, with a non–U.S. scientist chairing a working group for the first time.[220] WMO and UNEP cosponsored the assessment, as did NOAA and FAA domestically. While NASA had limited success in its efforts to increase participation in 1979, international sponsorship and international participation reinforced each other this time, as WMO and UNEP involvement helped foreign scientists, particularly government employees, to justify their participation. On the other hand, NASA organizers initially feared that these organizations' sponsorship would harm the scientific credibility of the report, by fostering the perception that scientific uncertainty or dissent might be suppressed in a politically motivated drive for consensus. In a curious attempt to forestall that risk, the preface emphatically stated that the report was *not* a consensus document, but rather summarized the plausible scientific conclusions, even if they were multiple.

While the first three NASA assessments grew progressively in scale and ambition, the fourth one stands in a class of its own as the most important ozone assessment prior to the Montreal Protocol. Sponsored and led by NASA through 1984 and 1985 and widely known as the "Blue Books" for the color of its three-volume report, this assessment again followed a similar model but was unprecedented both in the number of participating scientists and sponsoring organizations, and in its aim to provide comprehensive, authoritative reviews of every atmospheric topic relevant to stratospheric ozone. More than 150 scientists participated, including many of the most respected people working in each area reviewed; they came from 11 countries—although the United States was still numerically predominant, and all participants were from OECD countries. Six other bodies cosponsored, including NOAA and FAA domestically, UNEP and WMO, the EU Commission, and the German Ministry of Research and Technology (BMFT). The assessment included reviews of chemistry, radiation, dynamics, and their interactions; couplings of the major chemical cycles, and perturbations of multiple pollutants; tropospheric as well as stratospheric processes; laboratory studies, field measurements, and models; long-term monitoring for ozone trend detection; and the role of multiple trace gases in climate change. The assessment's report exceeded 1,500 pages, the most voluminous contribution to the field since CIAP. The assessment was published as a WMO report as in 1981, but its international character extended much deeper: its major meetings were held in Europe, and 10 of 22 chapter chairs were non-U.S. scientists.

This assessment also made several important innovations in its process. Its extended work plan began with a weeklong conference in Germany in June 1984, at which 80 scientists heard a series of invited papers and reviewed the state of knowledge. This also gave organizers the chance to identify working-group chairs. Chapter groups met several times as they worked over the course of one year. Draft chapters were discussed at a second plenary meeting in June 1985, with a group of eminent outsiders attending as reviewers. In a major policy change, all chapters were explicitly authored by the working-group members, rather than produced by anonymous NASA staff based on working-group discussions at one meeting. The eminence and breadth of participation in each chapter was consequently advertised—and participants were credited for the heavy work demanded of them—on the title page. In addition, the report made more effort than prior assessments to summarize its main points in accessible form, increasing its value as a reference or tutorial: a Scientific Summary presented major conclusions of each chapter (many of them still research priorities rather than conclusions drawn from present knowledge), although there was no Executive Summary. Finally, the assessment was followed by a program of presentations to explain its main results to policy makers, particularly by two U.S. scientists who were later named scientific advisers to the U.S. delegation in the negotiations that began in late 1986.[221]

This assessment reported many significant new results, but no major surprises. The discussion of source gases noted what was by then widely accepted, that there were no significant tropospheric sinks for CFCs 11 and 12; that explaining observed trends in atmospheric concentrations required larger global emissions of CFC-12 than stated by the Fluorocarbon Panel; and that concentrations of halons were growing rapidly, and so could contribute significantly to ozone depletion.[222] The discussions of stratospheric chemistry summarized refinements of several reaction rates since 1981, but no major changes. The discussion of model results still relied extensively on 1-D models, but for the first time there were enough 2-D models with full chemistry to verify that their results were broadly consistent with those of 1-D models. The model comparison for the first time presented depletion projections for both steady-state and a range of CFC growth assumptions. Engaging the recent controversy over the feasibility of CFC emissions growth, the report stated that while Rand and the EPA had been mistaken in 1979 and 1980 in their predictions of significant short-term growth, recent experience had directly contradicted industry claims that significant CFC growth was impossible.[223] Projected steady-state depletion was 5–8 percent in 1-D models, and 9 percent in 2-D models, slightly increased over the prior two years by recent kinetics changes. In multiple-perturbation scenarios, continued growth of N_2O, CH_4 and CO_2 gave very small ozone loss if CFC emissions grew less than 1.5 percent per year, but large losses reaching 10 percent within 70 years if CFC growth was 3 percent.[224] Projected losses in 2-D models were two to four times larger at high latitudes than at the equator, with significant high-latitude depletion even under multiple perturbations with steady-state CFCs.

The report identified many important remaining uncertainties and discrepancies between theory and observations for nitrogen, halogens, and ozone itself. For nitrogen, despite good progress in understanding regional phenomena like the Noxon cliff, several observations remained inconsistent with theory and the required complete measurements of nitrogen species over the daily cycle were still not available. For halogens, stratospheric HCl appeared not to be increasing as expected, suggesting either that short-term variability was masking the increase or that there was some important error in theory. Most seriously, photochemical models continued to underpredict observed ozone in the upper stratosphere, where they should be most accurate, by 30 to 50 percent. Commenting on the seriousness of this discrepancy, the report noted that "this significant ozone imbalance in the photochemically controlled region of the middle atmosphere limits the confidence that can be attached to model predictions of future ozone changes in response to long-term increases in the atmospheric concentrations of source gases."[225]

Although no decisive advance had either strongly confirmed or refuted ozone-loss claims, the accumulation of many small advances fed a growing sense that the stratosphere was coming to be understood and that ozone-loss projections were increasingly well founded. This growing confidence was hard to account for, since the uncertainties and discrepancies identified in 1985 were about as numerous and serious as those in 1975. As so often in research, previously unrealized domains of ignorance were discovered as fast as previously realized ones were resolved.[226] Moreover, this confidence was undercut immediately on its release by the report of the Antarctic ozone hole and the claim of large global ozone loss in satellite data. As I discuss in chapter 6, resolving these controversies took two years of intense effort.

In contrast with the other assessments considered here, those by NASA began with no ambition to influence policy. Rather, they began as narrow-scope efforts to conduct authoritative scientific reviews of specific research areas and to provide researchers with certain public goods that required broad collaboration. The integration of these assessments with the largest upper-atmosphere research program in the world greatly helped NASA achieve these goals, by promising scientifically interesting events, motivating participation and effort from eminent researchers, and providing the resources to support specific new analyses, such as model intercomparisons. In turn, managers of the research program used the assessments to update their view of the state of the field, and to identify key problems and questions for further investigation.[227]

In this way, the NASA assessments built a reputation for scientific authoritativeness, first within the United States. Beginning in 1979, NASA sought to extend this reputation through increasing international participation and, from 1981, sponsorship. In large part, this international drive was motivated by the perception that differences of interpretation among the 1979 assessments had conveyed a spurious sense of wide scientific disagreements when they in fact represented interpretive and policy differences. By 1981, these assessments had attained a similarly authoritative status on questions of stratospheric science internationally, despite organizers' concerns that sponsorship by international policy organizations would taint the assessment's perceived objectivity. In the 1985 assessment, the authority of the assessment was so strong, its involvement of the top researchers in the field so nearly total, that it was able to reach modestly beyond purely scientific statements to make additional, common-sense statements, such as noting the potential for CFC growth. Consequently, it was able to draw synthetic conclusions of strong policy relevance without being attacked and without diminishing its authoritative standing. The most important such statement—that significant ozone loss was likely under CFC growth even if other emissions also grew—became the principal means by which the 1985 assessment subsequently exercised decisive policy influence.

4.4 The Policy Influence of Science and Assessment

After the bold stroke of the Molina-Rowland argument and its early qualitative verification, there were no subsequent developments of similar importance through this entire period. Rather, progress was achieved cumulatively through many incremental advances. Many of these came from progress in research instruments and methods, such as new kinetics techniques, field-measurement techniques and instruments that approached laboratory quality, new global satellite data sets, and advances in computing power and modeling algorithms.

On balance, scientific progress through the period tended to support those arguing that ozone depletion was serious and controls were warranted, although no clear confirmation of large losses was obtained. This support arose principally from the negative result that 10 years' pursuit of suggestions that would reject the claim or make it insignificant had not done so. Moreover, the large methodological advances that opponents had long argued were necessary to make ozone-loss claims credible were increasingly being achieved, and were providing no clear refutation of the claims. For example, the long-sought 2-D models with full chemistry gave global depletion projections similar to those of the 1-D models long denounced as not credible. Indeed, to the extent that results diverged, 2-D models projected somewhat larger losses.

Consequently, the atmospheric-science arguments on which to base a claim that ozone depletion was not a serious problem were largely reduced to two: increases in other pollutants would offset the effect of CFCs on total ozone globally, if other pollutants grew and CFCs did not grow much; and no global ozone loss had yet been detected. Even these arguments were being progressively weakened. The first begged the question of why other pollutants should grow but CFCs should not, and was undercut by evidence of resumed CFC growth after 1983. It was also threatened by 2-D model results that were beginning to show significant losses at high latitudes even under the multiple-perturbation scenarios that made global-average losses small. The second was being threatened by a mounting number of indications from diverse data sources suggesting that global-scale ozone loss was in fact occurring.

Two other types of arguments were available either to support or oppose CFC controls, which were based not on atmospheric science but on comparing the anticipated harm from ozone loss with the cost of measures to avoid it: that the impacts of ozone losses were sufficiently large (or small), and the cost and difficulty of reducing CFCs sufficiently small (or large), that such reductions were (were not) justified. Neither type of argument played any significant role in debates through this period, apparently for tactical reasons. Although it was a claim about effects of ozone loss on skin-cancer rates that began the policy debate in 1971, arguments about impacts played insignificant roles in policy and scientific debates thereafter. Activists did not need them, since widespread concern was established by the first suggestion of an increased cancer risk. Detailed arguments about quantitative effects risked complicating the debate and weakening their case. On the other hand, it was dangerous for opponents to argue that projected impacts did not matter. After early attempts to disparage ozone losses by comparing their impacts to moving 100 miles southward were forcefully rejected, opponents were unwilling to risk appearing reckless or eccentric by advancing similar arguments. Mainstream opponents did argue that vertical redistributions of ozone did not matter: they had to, since such redistributions were always projected even when projected total-column loss was small. Because activists never established vivid grounds for concern about vertical redistributions, this argument carried less risk for opponents, although it, too, was ultimately rejected on precautionary grounds. The disinterest of nearly all policy actors in precise arguments about effects likely contributed to the consistently inadequate funding that such research received.

Arguments about the cost and feasibility of technological options to reduce CFCs also played little role on policy debates, but for different reasons. The relevant knowledge and research capacity was strongly concentrated in private industry; no research capacity developed over this period that was independent of major firms in the sector, so industry groups had unchallenged authority to rule on the credibility of any such claims made in policy debates. Advocates of controls tried to claim that viable alternatives were available, but could not draw on technical information of sufficient detail and credibility to make the case, so soon withdrew. Opponents replied in general terms that alternatives were not viable, while revealing the minimum technical information necessary to win the argument. Revealing more than necessary could only empower their opponents and attract wider interest in questions that they wanted off the table quickly. The few attempts to conduct independent assessments of CFC reduction options, notably the 1979 Rand and NAS reports, were forced to rely on industry sources for most of their technical information. However pessimistic their assessment of reduction opportunities, they could not rebut industry charges that they were in fact too optimistic. After the fights over these two studies ceased, such arguments faded from policy debates until after the 1987 Montreal Protocol.

In the earliest ozone debates, both the SST and CFC issues reached policy agendas because individual scientific claims generated concern over a potential harm, which was communicated into policy circles by its original scientific authors or their associates. But in the period considered here, direct scientific claims had highly limited effects on policy debates and none on policy outcomes. The only use of direct scientific claims was the selective adoption by policy actors of results that favored their position. Since there was no decisive scientific advance in this period, most of the results so used were short-term noise in a gradually evolving debate, particularly the kinetic results that repeatedly increased and decreased ozone loss projections. It was predictable that advocates would use such results, as they would use any available arguments that favored their positions. But such selective use of scientific results had minimal influence on debates or outcomes: opponents may briefly have had to retreat, but could usually find (or wait for) another result that went the other way. Informed neutrals, recognizing the ephemeral character of such claims, were unlikely to be swayed by them.

With the direct influence of scientific results on policy debates so limited, many actors called for assessments to mediate this relationship and provide what benefits scientific knowledge could offer to clarify or advance policy debates. Many assessments were conducted, taking diverse approaches that reflected both strategic choices by assessment leaders and participants, and the resources and constraints associated with specific institutional contexts. Despite their diverse approaches, and however high the quality and objectivity of their scientific reviews, assessments over this period had essentially no influence on policy debates, with one striking exception. As with primary scientific results, partisans selectively cited assessment statements that supported their positions. But no assessment introduced new possibilities, formulations, or options to the policy debate; none moved any policy actor to change a position; none resolved any scientific questions that had become prominent in the policy debate; and none provided the focal point for convergence of any agreement—with the one exception.

Much of the assessment experience of this period illustrates potential pitfalls and conditions predictably associated with failure. Some failures simply reflected proponents' failure to recognize the need for assessment at all, or their underestimation of the challenge of the task. Just as the proponents of the 1970s ad hoc international negotiation process thought they could achieve policy agreement without assessment, so the proponents of the CCOL appear to have thought that a weak body coordinating national research programs and jointly reporting results could provide all the assessment that was needed. While the failure of the CCOL to influence policy was overdetermined, its experience illustrates a lower bound on the level of resources, participation, leadership, and collective authority needed to influence policy. Assessments that command little attention or respect by virtue of the collective stature of their participants; that draw no clear scientific judgments or conclusions about present knowledge except that more research is needed; that present no cogent new ways to understand the issue; and whose reports are both useless to scientists and inaccessible to lay persons, can expect to have no influence on policy, however high the quality of their work on other dimensions.

The early NAS and U.K. assessments illustrate two other clear pitfalls. First, when a scientific assessment is conjoined with strong policy conclusions that do not draw on either the assessment or the expertise of its participants, the assessment is readily dismissed as a partisan document. Second, for an assessment to gain credibility in international policy circles, it requires some minimum level of international participation and sponsorship. A related risk arises when multiple, parallel assessments are produced within a short time. The most striking instance of this was in 1979, when five ozone assessments were published by different institutions within one year. The result was widespread confusion, and exploitation of seeming differences between the assessments to claim that little was known on the topic. The sharpest argument concerned the NAS and UKDOE assessments, whose scientific statements were very similar but whose policy interpretations and conclusions were highly expansive, in opposite directions. These were selectively invoked by partisans, independent of nationality: U.S. industry relied extensively on the U.K. assessment to defend their policy positions and attack the policy conclusions of the NAS assessment, and even used the U.K. assessment's unofficial 1981 sequel, albeit more quietly, for the same purpose.[228] Even when assessments did not—as these did—go beyond their legitimate expertise in drawing policy conclusions, minor differences in inclusion, interpretation, or language could readily be exploited by partisans seeking to characterize the state of knowledge as more uncertain and less consensual than it actually was. In any of these cases, the ability of assessment to delimit policy dissent or its scientific bases was impaired, since partisan actors could shop for elements from multiple assessments, all with good credentials and seemingly authoritative, to support their preferred position or to characterize the state of knowledge as too weak to justify action.

The striking exception to this pattern was the NASA/WMO assessment of 1985, the "blue books." This assessment was unique not just in scale, ambition, and quality, but also in its subsequent influence, both on scientific research and other assessments, and on policy debates. NASA assessments, from the first one in 1977, had attained a status of authoritative scientific references that no other assessments did. In this capacity, the 1985 "blue books" assessment was widely cited by subsequent scientific publications, and frequently deferred to by other assessments.[229] Its stature and predominance were so great that it supplanted other assessments. Even other bodies obliged to produce their own assessments subordinated them to this one in various ways—by delaying to await its results, by presenting a report that met their legal requirement but was entirely derived from this one, or by simply announcing that this assessment took the place of theirs.[230] Since this assessment gained its definitive status in part by not being a product exclusively of NASA, it could no longer serve as NASA's required report to Congress, so even NASA had to produce a report separate from it.[231]

The same characteristics that secured this assessment's authoritative status with researchers and other assessments also contributed to its influence on policy. The number and stature of participants, the degree of effort they expended, and consequently the breadth and quality of consensus represented, all enabled it to make authoritative statements on the state of atmospheric-science knowledge at that moment that were beyond reproach in both scientific and policy circles. The scale of this effort had two practical implications for policy debates. First, it provided a strong signal of the scientific community's view of the seriousness of the issue. This could not have been contrived by the assessment's sponsors and organizers; despite its power as the major research funder, NASA could not have elicited such participation and effort if participants did not view the issue as important. Second, because the scale and authority of this effort supplanted all potential parallel assessments by making them scientifically superfluous, it eliminated the risk of other assessments advancing competing interpretations or generating confusion through minor differences of language. This made it impossible for any policy actor who might have wanted to advance a different view of scientific knowledge to mount a counterassessment that would command any credibility. By engaging a large fraction of the world's scientific resources in the field, this assessment achieved the status of an authoritative monopoly of atmospheric-science knowledge relevant to stratospheric ozone. Any policy actor seeking to support a position with scientific statements was compelled either to draw support from this assessment, or to rely on new results published later.

This authoritative status would have mattered little if the assessment had drawn no policy-relevant conclusions, but it did. In particular, the assessment's model comparison exercise this time included a full review of both 1-D and 2-D model results under various growth assumptions for both CFCs and other pollutants. This was the first time CFC growth was examined in the context of such a comprehensive review of multiple models and scenarios, and the first such attempt to escape attack for being speculative and unscientific. The report even cited recent growth to reject industry claims that little further CFC growth was possible. Highlighting the greater significance of uncertainty over future emissions growth put controversies over atmospheric-science uncertainties into perspective. This comparison was the central message of this assessment that influenced subsequent policy debates.

Two potential weaknesses of this assessment that might have impaired its influence appear not to have done so. First, the large continuing uncertainties and discrepancies that the assessment prominently identified, as well as the prominent new claims that appeared to undermine it, could readily have been used to argue that continuing scientific uncertainties made consideration of policy action still premature.[232] Instead, this assessment had nearly the opposite effect, strongly delimiting the policy debate by causing all actors to endorse a minimum policy position that CFC emissions should be held roughly constant. Second, as an independent initiative funded and led by NASA, this assessment had no official institutional connection to any policy-making body. Although the assessment met a clear need, no policy-making body was obliged to consider its results.[233] Although this might have allowed policy actors to use procedural objections to try to keep the assessment out of policy debates, its authoritative status and the strength of its central conclusions allowed it to condition the debate in a way that was beyond the reach of such objections. Chapter 5 will discuss the history of policy and negotiations through the 1980s that led to the signing of the 1987 Montreal Protocol and will lay out how this assessment exercised a decisive influence on policy.

5

Negotiations and Strategy, 1980–1987

By early 1980, the initial campaigns for international controls of ozone-depleting chemicals had failed. As these failures became clear, the officials who had sought international controls began pursuing the cause through other institutional vehicles, first with new attempts through the EC and the OECD. The traditional opponents of controls continued to resist these initiatives, and succeeded in blocking both. But negotiations that began under UNEP auspices in 1982 came to a different end, yielding the first international ozone treaties after several years work—the 1985 Vienna Convention and the 1987 Montreal Protocol. This chapter reviews the attempts to develop international cooperation to control ozone-depleting substances, from 1980 to the signing of the Montreal Protocol in 1987.

5.1 Early Maneuvers and the Establishment of Negotiations, 1980–1982

European Community, 1980–1982

In chapter 3, I discussed how Dutch, Danish, and German officials pursued ozone-protection measures in the EC but achieved only the weak council decision of March 1980. That decision required another review of the issue later in 1980, but the activists were unable to exploit this procedural requirement to force a stronger decision. To conduct the review, the commission engaged Belgian atmospheric modeler Guy Brasseur to review and compare the 1979 assessments of NAS and UKDOE. His review largely supported the conclusions of the U.K. report, highlighting uncertainty, criticizing 1-D models as inadequate for assessment, and suggesting that UV effects may be overstated. He concluded that further research was needed and that deferring action for a further five years' research would be acceptable. On considering the commission's summary of Brasseur's report, the Council of Environment Ministers decided against further action on June 30, 1980.[1]

Still the activists persisted. When the Dutch minister assumed the EC presidency for the first half of 1981, he sought to make further action on CFCs a priority of his term.[2] The commission sponsored a scientific workshop in January at which recent reductions in depletion projections were discussed, and submitted a report in June that again discouraged further action.[3] Although noting continued high aerosol use and rapid nonaerosol growth, as well as a recent CCOL statement that CFCs "may still pose a risk," the report recommended against changing the current "precautionary" EC policy. Rather, it proposed only quantifying the level of the production cap (which the 1980 decision had not done), developing voluntary emission-reduction measures, and supporting international replication of the EC capacity cap and 30 percent aerosol cut.[4]

In October 1981 the commission proposed to quantify the production-capacity cap at 480 Kte per year, nearly 60 percent higher than actual 1979 production.[5] The council failed to act on this proposal at its meetings in December 1981 and June 1982, while rejecting calls for further CFC cuts from Germany, Denmark, and the Netherlands.[6] This deadlock was not confined to CFCs: through this period, environment ministers' meetings were highly divisive and often failed to agree on anything. Ministers finally approved the proposal, and authorized the commission to collect CFC production data, in November 1982. In subsequent years, the commission continued to consult with industry on voluntary emission-reduction practices, and periodically reassessed the issue, each time advising that no further action was warranted. The voluntary practices, like the production-capacity cap that never came close to binding, were so gentle in implementation as to have essentially no effect.[7]

OECD, 1980–1982

The OECD had briefly considered the ozone issue in 1975, but soon ceded scientific assessment to UNEP and policy discussions to the ad hoc process. Neither of these bodies, however, gave any consideration to technologies and policies to reduce CFC emissions. In May 1980, after it was clear that the ad hoc process would produce no agreement, U.S. and Canadian officials proposed to the OECD Environment Committee that a comprehensive assessment of CFCs and ozone be done under the OECD. Prepared by an expert group drawn from the Secretariat and national governments, the assessment would include atmospheric science, UV impacts, present uses and policies, scenarios of future CFC growth, and potential controls, with one government responsible for each topic area. The OECD Environment Committee reviewed a draft of the report, including the CFC scenarios, in December 1980. Controversy arose, however, over the design and interpretation of scenarios. The initial proposal included just three scenarios, each corresponding to an assumed level of control policies by OECD nations: no additional controls, modest limits, and stringent controls. Through these scenarios, control measures were to be linked to resultant projections of emissions, and consequently to ozone depletion, impacts, and costs, in order to make a full integrated assessment. This attempt to couple emission scenarios specifically to both controls and impacts was abandoned, however, at a meeting in February 1981.[8] Instead, the 3 initial scenarios were replaced by 15, covering the entire range of CFC trends from 7 percent increase to 7 percent decrease annually. These scenarios were explicitly uncoupled from assumed controls; their only purpose was as inputs to atmospheric models.[9] Discussion of scenarios was separated still further from impacts and control options by removing scenarios to a separate, later report.[10]

When the expert group reconvened in September 1981, most members had been able to analyze only a few of the 15 revised scenarios, and none had examined the same set. Moreover, several modeling groups had not used previously agreed values for other model inputs. The discussion consequently lacked focus, and the comparability of model results—the principal purpose of the exercise—was limited.[11] Still, this was the only model comparison exercise for the next four years that considered CFC growth. OECD staff continued their efforts to produce an integrated assessment of CFC reductions for two years, but were frustrated by several factors: the prior decision to separate emissions from impacts and controls; industry and governments' refusal to divulge production, technology, and cost information; and weak knowledge of dose-response relationships for UV exposure. The reports of the exercise were of limited use and the, UNEP negotiations had superseded any possible OECD effort by the time they were finished.

UNEP, 1978–1981

UNEP's bid to be the forum for international discussion of CFC policies was rebuffed in 1977, leaving it responsible only for scientific coordination through the CCOL. The initial proponents of a broader UNEP role continued to pursue it, however, and resumed their efforts as the failure of the ad hoc process became clear. As early as UNEP's 1978 Governing Council meeting, the Norwegian delegation proposed that UNEP should also work to "harmonize actions . . . to protect the ozone layer . . . through the development of regulatory policies," but the effort gained little support.[12] A similar proposal in 1980 in the Governing Council gained a little more support, yielding an advisory decision that asked governments to find ways to reduce CFC use and emissions.[13] But when Sweden proposed a much stronger resolution in 1981, asking UNEP to convene negotiations for an international convention to protect the ozone layer, the decision passed despite widespread skepticism about UNEP's capacity to lead the activity, as did another authorizing UNEP to request national production and capacity data.[14] Delegations supported the measure for diverse reasons. Some saw it as the most effective route to international controls, or to preventing CFC production from shifting to countries without national controls; others saw it as an opportunity to delay calls for further action in other bodies.[15] The decision allowed UNEP to convene legal and technical experts to consider what principles and measures might be negotiated if future assessments showed the risk to be real, but not to negotiate for—or even formally to represent—governments.

Before this new ozone group first met, UNEP convened a group of senior legal officials for discussions of international environmental law. This meeting endorsed UNEP's ambition to establish a program to develop and review international environmental law, including facilitating international discussions about relevant legal guidelines, principles, and agreements. Ozone was identified as one of three priority issues for such discussions.[16] This meeting endorsed structuring international environmental negotiations and treaties using conventions and Protocols. For each issue, a convention or basic treaty would define the scope of the issue, structure the agreement, and state principles and objectives, while one or more Protocols—other treaties subsidiary to the convention, structured so as to be easier to modify—would contain specific provisions to manage the issue.[17] Sweden presented a draft ozone-layer convention including several specific proposals for CFC controls. They sought to have this negotiated and adopted rapidly, but other delegations were reluctant to move so fast.[18] The Canadian delegation reported that the Swedes did not understand recent scientific results that reduced estimated depletion, and argued that if forced to choose between a convention with tough CFC control measures and none, none was "more supportable at this time, scientifically and politically".[19] Only the Soviet Union openly opposed the proposal; the major EU nations and Japan were noncommittal, while the U.S. delegation expressed initial support but later retreated.[20] With the Swedish initiative rebuffed, ozone talks would begin without a specific control proposal on the table.

CFC Markets and Industry Strategy, Early 1980s

International ozone discussions resumed in the context of large shifts in international CFC markets from the late 1970s. Aerosol uses of CFCs 11 and 12 had contracted—precipitously in the United States, gradually in Europe—while markets for nonaerosol uses, other CFCs, and in other regions had continued to grow. U.S. use of CFCs in aerosols fell 95 percent from its 1974 peak to about 10,000 tonnes in 1980.[21] Other uses grew strongly and were vigorously promoted, particularly foams and auto air conditioning, but could only slightly offset the collapse of the largest market. Consequently, total U.S. production of CFCs 11 and 12 fell 45 percent from 1974 to 1980, declining from about 50 to 30 percent of world production. United States production capacity fell by only 18 percent, as the three largest producers each closed one plant and one small producer left the market. This modest consolidation of production left U.S. markets with substantial excess capacity, declining prices, and low margins that persisted through the early 1980s. DuPont had about half of this weak market, and was probably the low-cost producer.[22]

European markets showed similar but smaller trends, with aerosols starting as a larger share of total use and declining less. From a 1976 peak, European aerosol use declined 37 percent by 1982, partly offset by 42 percent growth of nonaerosol uses, so overall CFC consumption declined by 15 percent—with aerosols still making up more than half. With European exports holding steady, production declined by only 12 percent.[23] CFC production outside Europe and North America grew strongly through the late 1970s, nearly equaling the decline in Europe, so total non-U.S. production remained roughly constant. Aerosols were still the largest use worldwide, about 40 percent of world CFC 11 and 12 consumption in 1980.[24] Through the recession of the early 1980s, U.S. production of 11 and 12 remained flat while production elsewhere dropped about 5 percent.

The strongest market prospects were for CFC-113. Although only moderately successful when first marketed as a dry-cleaning fluid, CFC-113 was an effective solvent that was nevertheless gentle enough to clean plastics and other synthetics without damaging them. These characteristics made it ideal for cleaning grease, fluxes, and resins from electronic integrated-circuit chips and circuit boards, and its production grew strongly through the 1970s and early 1980s with growth of the electronics, computer, and aerospace industries. With its growth further spurred by substitution for other chlorinated solvents that were more toxic and smog-forming, and consequently coming under increasingly strict regulation, CFC-113's production tripled from the mid-1970s to the early 1980s, to become 20 percent of all CFCs by 1984.[25] Its production was more concentrated than that of CFCs 11 and 12. Like all two-carbon CFCs, it required a technically demanding two-step synthesis process. In the United States, only DuPont and Allied produced it; the smaller firms had tried, but never succeeded at making it cost effectively.[26]

5.2 International Negotiations to the Vienna Convention, 1982–1985

UNEP's ad hoc expert group on the ozone layer first convened at Stockholm in early 1982, with the major producer nations and a dozen developing countries attending.[27] Conditions were not promising for starting international negotiations. Extensions of the EC's 1980 measures had been repeatedly obstructed by deadlocks in council meetings. The U.S. delegation, particularly the EPA, was adjusting to Reagan administration appointees and their antiregulatory agenda. At industry in-itiative, Japan had recently announced an EC-style production-capacity freeze, but officials stated that this was only a response to other nations' measures, and that they regarded international CFC controls as premature and would not sign a con-vention containing them. Government submissions to the meeting showed that many had little understanding of the issue and its global character: several re-sponded that since they did not use CFCs, "their ozone layer" was not threatened.[28]

Since many delegates were unfamiliar with the CFC-ozone issue, much of the meeting consisted of tutorials. Recent scientific assessments by NASA, the CCOL, and the WMO were presented at the meeting, and experts gave presentations on ozone monitoring and trend analysis, atmospheric modeling, UV effects, and "al-ternative technologies and socioeconomic issues."[29]

The group was authorized as an expert working body, not a negotiating body, so delegations came without official instructions or the authority to commit their governments. While the supposedly informal and noncommittal character of the discussions had helped get them approved, delegates all understood that these ne-gotiations were serious, and that governments would be likely to endorse agree-ments reached there. With the group's mandate and authority both ambiguous, various participants struggled to pull discussions toward and away from negotiation of control measures. The Swedish chair exhorted delegates that their mandate was to consider how, not whether, to develop an ozone-layer convention, and that as independent experts they should be prepared to freely elaborate alternative possi-bilities—a fiction that none of the government representatives present believed.[30]

The Nordic states presented the draft convention they had been circulating in-formally for several months, which sketched several possible forms of controls: bans on nonessential uses in aerosols or other sectors; production capacity caps; and programs to reduce emissions, including specific technology requirements. Discus-sion of the proposal, unsurprisingly, revealed widely divergent views. The EC ar-gued that if any international measures were to be adopted, they should simply replicate the existing EC production-capacity cap.[31] Canada argued that controls were not presently warranted, but that a convention should be adopted to allow their timely enactment if and when they were needed.

The U.S. delegation did not even endorse international replication of their ex-isting domestic measures, and its members stated in informal conversations that if they had known in 1977 what they knew in 1982, they would not have banned aerosols.[32] In premeeting consultations, U.S. industry had argued that the convention should concern only scientific research, because even discussing potential regulatory programs, or "undue emphasis on data collection," would "result in regulatory momentum which is inappropriate at this time."[33] With British and French support, the U.S. delegation argued that discussing control measures or the contents of a convention was premature, and proposed cutting the meeting short.

Delegations also disputed how, if at all, to convene technical discussions of CFC alternatives and the cost of controls—questions that had never been treated adequately, and never examined at all internationally. Although these questions were part of its mandate, the CCOL had repeatedly refused requests to address them, arguing that they were fundamentally political, not scientific, so addressing them would politicize the committee and damage its scientific credibility.[34] The attempt to pursue these questions through the OECD in 1981 had deadlocked. The question of a body for technical assessment of responses was periodically revisited for several years, but no progress was made and no such body established until 1988.

At the end of the first session, UNEP was authorized to prepare a draft framework convention as a basis for discussion. This draft was circulated at the next session, in December 1982, as was a revised control proposal from the Nordic states, which combined non-essential aerosol bans with best-available technology requirements to limit emissions from other uses. UNEP deputy director Thacher argued that any such controls should as a technical matter be separated into annexes or Protocols to the convention, but negotiated at the same time.[35] The Nordic states, with Canada, Australia, New Zealand, and Switzerland, pressed for rapid adoption of a convention, while the Unites States, France, and the United Kingdom resisted. Some delegations and industry representatives objected even to nonoperational elements of the draft convention—such as its objective of protecting the ozone layer, its use of the language "likely adverse effects," and its "singling out" of CFCs when other compounds also affect the ozone layer—arguing that such language was prejudicial unless based on a comprehensive assessment of the issue.[36] UNEP officials proposed to do such an assessment of impacts and control options, but were rebuffed.

When negotiators next met, in April 1983, the Nordic nations again advanced the same control formula—a ban on nonessential aerosols and technology controls in other sectors—now proposed as an annex to the convention.[37] The British, French, and EC delegations continued to oppose any controls other than a production-capacity cap mirroring their own measures, while others supported the Nordic proposal in principle but with various specific differences.[38] The Canadians rejected immediate cuts as "scientifically indefensible," calling instead for measures to restrict future CFC growth and a framework for more stringent measures if and when needed. Despite continuing turmoil in the EPA, including the replacement of Administrator Anne Gorsuch by William Ruckelshaus in March and the forced resignation of all but one of the agency's senior staff, the American delegation arrived better prepared than previously, and was for the first time willing to discuss international aerosol controls. Although there was still no official U.S. position, other delegations interpreted this willingness as tantamount to official U.S. support for international aerosol controls, and some EPA officials informally stated that U.S. industry would likely support such an agreement.[39] This session also saw the first discussion of two-tiered obligations, with less stringent controls for developing countries. Despite these hints of progress, the continuation of negotiations remained precarious in terms of both authority and financing. Sweden, Switzerland, and the

Netherlands had sponsored the first three sessions, and with no other source of support evident, it was briefly proposed to hold the next session in Moscow in December, so it could be funded using some of UNEP's nonconvertible rubles.

Conflict over the ozone issue within the EPA peaked shortly after this meeting, leading to a sharp change in the direction of U.S. policy. Although never acted upon, the EPA's 1980 proposal to regulate all CFC uses posed a continuing uncertainty. Environmentalists, particularly the NRDC, periodically argued that the proposal comprised a "finding of harm" that obliged the EPA to proceed with regulations. For its part, the Alliance sought to forestall such action several times between 1981 and 1983 by having sympathetic members of Congress introduce bills reversing the strong precautionary mandate of the 1977 Clean Air Act amendments. These bills prohibited the EPA from further controlling ozone-depleting chemicals unless an international agreement was adopted or conclusive scientific evidence of ozone depletion was obtained.[40]

In this unstable environment, an official in the EPA's Toxics Office proposed in December 1982 that the 1980 regulatory notice should be officially withdrawn, citing reduced depletion estimates and the evident difficulty and ineffectiveness of further unilateral U.S. controls.[41] When the proposal was circulated for prepublication approval in June 1983, however, it attracted unexpected opposition from EPA's International, Policy, and Air offices. Officials in the Policy Office had become interested in CFCs while preparing a study on greenhouse warming, and concluded that the potential for resumed CFC growth made withdrawing the proposed regulation inadvisable. With the leadership of Toxics preoccupied in developing PCB rules and unwilling to defend the proposal, agreement was reached in August to move responsibility for stratospheric ozone from Toxics to the Air Division—an arrangement that was more consistent with statutory authority under the Clean Air Act, and that also gave the issue to officials more sympathetic to the need for further CFC controls. This shift of responsibility allowed the U.S. position in the international negotiations to change, since the Toxics Office had previously vetoed U.S. support for an international aerosol ban, and brought ozone depletion back as a priority issue in the EPA and the U.S. administration over the next two years. State Department officials opposed the new position, principally on the basis of general antiregulatory ideology, but were unable to block it.[42]

The new U.S. position supporting part of the Nordic proposal—an international ban on nonessential aerosols—although not its controls on other uses, was announced at the next meeting, in October 1983.[43] With this change, the United States returned to the position it had promoted under the Carter administration, an international aerosol ban replicating its domestic ban. In addition, the new U.S. position stated that all ozone-depleting substances should be included, that flexibility should be allowed to keep some aerosols if equivalent reductions were made in other uses, and that the controls should be placed in a Protocol that would be "integral to the Convention"—that is, states adopting the convention must also adopt the Protocol. The Nordic states quickly agreed to shift their proposed control measures from an annex to a Protocol, but they—like all other supporters of controls—wanted it to be optional.[44]

Like the two previous sessions, this one immediately followed a CCOL meeting, with substantial overlap in attendance. The CCOL provided a scientific update, and offered its services as a scientific assessment body to the convention. Until this session, a proposal to establish an assessment body with a broader mandate had been carried in the draft negotiating text. This "advisory body," composed of national representatives, would review scientific, economic, and technical information relevant to response measures—including technical assessment of potential CFC alternatives—and make recommendations to the parties.[45] The proposal was quietly dropped at this session, however, to avoid anticipated conflict over its membership and mandate.

Although the United States already had an aerosol ban in effect, U.S. industry organizations objected strenuously to the new position, arguing that supporting international replication of a ban implied that CFCs were still a problem, and so risked encouraging unwarranted controls on nonaerosol uses.[46] The air-conditioning industry even objected to the flexibility provision under which nonaerosol cuts could be traded off against aerosol cuts, arguing that if other nations used this provision to reduce air-conditioning use, that might create pressure for the United States to do the same—presumably by demonstrating that such reductions might be feasible or cost-effective. Industry groups were sufficiently alarmed that they tried to block U.S. participation in the negotiations through procedural objections, arguing that the change in negotiating position was made with insufficient public notice and without filing an environmental impact statement.[47]

At domestic briefings in early 1984, U.S. officials advanced several arguments to defend the new position. They asserted that the existing U.S. ban "logically required" support for an international ban; that U.S. experience showed the cost of an aerosol ban to be low; that an international aerosol ban would delay resumption of CFC growth, particularly in view of recent suggestions of nonlinear depletion at high chlorine concentration; that heightened concern over global climate change provided an additional reason to limit CFCs; and that the United States was not actually proposing an aerosol ban, but merely supporting a Nordic initiative.[48] The merit of these arguments was highly variable. Since only nonaerosol uses were growing strongly, a global aerosol ban—like the earlier national bans—would provide only brief respite from growth. Moreover, industry representatives argued with some justification that climate change should be addressed by examining greenhouse gases systematically.[49] Robert Watson, manager of NASA's upper atmosphere research program, supported industry calls for delay, arguing that three to five years' more research was needed to validate photochemical models for the upper stratosphere, while progress in understanding dynamic-chemical links in the lower stratosphere must wait longer, perhaps 15 years for the proposed upper atmospheric research satellite.[50]

Negotiations reconvened in January 1984, with proposed control measures now removed to a draft Protocol while the draft convention retained basic principles and framework for an agreement.[51] UNEP officials and activist delegates hoped that this session would conclude negotiations of the convention, but two unexpected conflicts arose here that delayed it another year: dispute settlement procedures, and the status of the EC Commission. For dispute settlement the United States sought compulsory and binding arbitration, while most others—most strongly the Soviet Union—wanted a weaker system. The delegation of the commission, which was seeking to expand its authority within the EC by obtaining increasingly independent standing in international bodies, proposed that it should be able to become a party on the same terms as nation-states—independently of whether any member-states joined, and with no requirement to submit a statement of its authority. National responses to this expansive bid varied widely, but all wanted at least some restriction on the commission's position: most delegations wanted to require that at least one EC member-state become a party for the commission to do so, and also sought a clear statement of the EC's authority and provisions to prohibit double voting.[52]

Little progress was made on the Protocol. The original supporters continued to press for controls, now including the Americans advocating a compulsory Protocol. The major European producers, now with the explicit support of the Soviet Union, Japan, and Chile, objected to even discussing controls. To support its new position, the U.S. delegation circulated an EPA background paper, which industry's Fluorocarbon Panel attacked on all fronts. The panel argued that vertical ozone redistributions were of no significance; that the apparent upper-stratosphere depletion in Umkehr observations was invalid because it relied on an unverified aerosol correction; that estimates of UV impacts were invalid because they were based on high exposures, when recent multiple-perturbation projections suggested UV increases would be very small; and that potential losses were distant enough to allow time for more research. As they had done since 1975, the panel suggested an additional three to five years of research were needed. As for CFC controls, the panel argued that the U.S. aerosol ban had been highly costly; that describing some aerosol uses as "nonessential" was subjective; that suggestions that CFC production would grow in the future were speculative; and finally, that CFCs' contributions to climate change should not be considered, since this issue was outside the working group's mandate.[53] The intensity of this response suggests that industry was primarily worried not about global aerosol controls, but about the risk of broader restrictions. The fact that the only U.S. concession to industry objections was to withdraw support for the "flexibility" clause, the most sensible control proposal yet advanced, also suggests that industry's main concern was avoiding attention to nonaerosol uses.

As the January session proceeded, opponents sought to weaken any agreement that might be adopted, and to delay decisions through procedural measures. Facing persistent opposition both domestically and internationally, the U.S. delegation dropped its insistence on a compulsory Protocol, but the activist delegations could not agree on how to weaken their proposal. Consequently the "second revised draft Protocol" that emerged from these discussions was heavily bracketed, essentially merging all proposals advanced, and too cumbersome to form the basis for subsequent negotiations. On the last day of the session, five European delegations and Chile proposed that UNEP's Governing Council be asked to suspend the group's mandate, call a diplomatic conference to finalize the convention immediately, and delay further discussion of Protocols for three to five years. The delay would be used to develop voluntary emission-reduction programs similar to those under development in the EC,[54] they argued, and for CCOL to conduct an assessment of all substances that might affect the ozone layer. The chair proposed a compromise— asking CCOL for an assessment but continuing the mandate of the working group—which prevailed. Still, advocates realized the United Kingdom and others would likely try again to decouple the convention and Protocol and revoke the working group's mandate in the Governing Council, where they would have to fight to continue control negotiations.

As expected, at the May Governing Council meeting the United Kingdom, France, and Germany proposed a decision calling for a diplomatic conference to finalize the convention after just one more working group session.[55] With negotiations in disarray, one session could clearly at best conclude the convention, so further discussion of Protocols would be postponed. With the United States, Canada, and the Nordic states pressing to continue the working group and delay the diplomatic conference, a compromise was adopted to hold the conference in the first quarter of 1985, after two more working group sessions. In an additional concession to the activists, the decision explicitly granted the working group a mandate to negotiate a draft Protocol on CFC controls, forestalling any further procedural attempts to block such discussion.[56]

At Governing Council, a group of activist delegations calling themselves Friends of the Protocol consulted informally on how to advance the negotiations. Recognizing that progress would require having clean Protocol negotiating text available before the next meeting, the group asked the Canadian and American delegations to take the lead in preparing it (the Nordics believed they had done so too often). The U.S. delegation reported that it could participate, but could not sponsor or lead this process because of domestic industry opposition, so the Canadian delegation offered to take the lead. Following informal consultations between Canadian and American officials, the group reconvened in Toronto in September to prepare draft negotiating text.[57] They resolved minor differences among themselves by presenting two alternative control measures in the text, of which parties could choose either one: eliminating "nonessential" aerosol use or cutting at least 80 percent of all aerosol use. The two alternative forms were designed to allow their U.S., Nordic, and Canadian sponsors to adopt one or the other without requiring any change in their existing domestic measures. A Canadian proposal to add a third option, cutting overall CFC use by 30–40 percent as an alternative to larger cuts in aerosols, was judged too complicated and likely to be unacceptable to other nations, so the draft included only aerosol controls.[58] Members of the group lobbied other delegations bilaterally through September and October to support the proposal, or at least to accept it as the basis for discussion. As arguments in support of controls, they used the as-yet unpublished "chlorine catastrophe" results of nonlinear depletion and the increasingly prominent linkages between CFCs and climate change.[59] The United States, Canada, and the three non-EC Nordic states sponsored the proposal under nominal Canadian leadership, while Australia offered support in principle.

When the UNEP working group reconvened in October,[60] the ozone issue had clearly risen onto industry's agenda. For the first time, three industry bodies attended as observers, but there still were no environmental groups.[61] An opening scientific briefing by NASA's Robert Watson reported on the recent scientific conference associated with the NASA/WMO assessment. The briefing stressed three major results: that model runs with multiple perturbations projected small ozone losses with CFC growth less than 3 percent per year; that nonlinear depletion was projected when stratospheric chlorine exceeded NOx, which could occur if CFCs grew at 3 percent for 75 years; and that statistically significant ozone decreases had been observed in the upper stratosphere.[62]

Despite EC procedural objections, the proposal from the Friends of the Protocol (which other delegates called the Toronto Group) was accepted as the basis for discussion. Although one U.K. official had previously provided input to the proposal,[63] the United Kingdom and other European nations attacked it harshly, counterproposing that a Protocol should follow their existing measures: a production capacity cap and a 30 percent aerosol cut. Other delegations noted that the EC's rigid position reflected deadlocked internal negotiations, both over controls and over the commission's authority, with the Netherlands and Denmark openly opposing the EC position and Germany wavering.[64] Seeking broader support, the

Toronto Group expanded its list of options to four, all intended to achieve similar levels of aggregate CFC reduction. The newly added options were a Swiss proposal to cut aggregate CFC use 20 percent (although this proposal, like the prior Canadian suggestion of comprehensive controls, was subsequently dropped) and a stronger version of the existing EC measures intended to attract the support of marginal European delegations—a production-capacity freeze with a 70 percent aerosol cut.[65]

A working group session in January 1985 provided one final chance to resolve differences between the Toronto Group and EC proposals, but the two sides instead continued to escalate their defenses of their proposals. Although the antagonists described the contending proposals as the "multi-option" (Toronto Group) approach versus the "single option" (EC) approach, their principal difference was their stringency: the three remaining Toronto Group options would reduce total CFC use by 35 to 50 percent; the EC proposal, by 20 percent at most. Each side also argued for the structural preferability of its approach to controls, with the Toronto Group focusing on consumption controls and the EC proposal on production controls.

Toronto Group delegations defended their proposals on the basis of possible nonlinearities in depletion and the prospect of continued growth. They defended their focus on aerosols because they were easy to reduce, and the multi-options approach because it "rewarded past actions." They attacked the EC proposal for its failure to achieve any near-term reductions and for locking in present production patterns, making it unlikely that developing nations would join.[66] For its part, the EC Commission argued that loopholes in the Toronto Group proposals would allow emissions to grow without limit. The options that controlled only aerosol use would not control rapidly growing nonaerosol uses; the option that controlled both aerosol use and production would allow unlimited imports for other uses; and the option limiting all use would allow unlimited production increases for export. In contrast, they argued, the EC proposal would limit total world production, and so would reliably avoid the prospect of nonlinear depletion. The European chemical association CEFIC, attending as an observer, circulated a paper that denounced the Toronto Group for hypocrisy and argued that CFC production could not grow due to firm limits on world supply of fluorspar ore and disposal capacity for by-product HCl.[67]

Each side's arguments were only partly sound. It was unreasonable for the Toronto Group to claim their proposal would prevent CFC growth, when nearly all growth was occurring in nonaerosol uses that their proposal did not control. But the EC proposal would also allow unlimited growth if nonparticipating countries expanded production for export to participating countries. Assuming the full global participation necessary to close this loophole in the EC proposal would also close the loopholes the EC attacked in the Toronto Group proposals.[68] Each side's criticism of the other's proposal as inequitable was also correct, for both were intended to be so. Despite pious calls for nations to move beyond their existing measures to protect the ozone layer, the Toronto Group's proposals were transparently designed so their proponents would not have to do so. They defended this aspect of their proposal as "rewarding past efforts."[69] The EC capacity cap was also intentionally biased, benefiting present producers over potential new entrants, and European over American producers.[70] The Toronto Group's proposals would provide larger immediate reductions, while the European ones would more effectively limit long-term growth. To the Toronto Group's credit, they portrayed their proposals as near-term measures, not a permanent solution. Controlling both production and consumption, an obvious solution in retrospect, was not proposed. Moreover, neither side considered the loopholes inherent in controlling only CFCs 11 and 12 when other CFCs that were just as ozone-depleting could replace them in some applications.

The session ended in stalemate, and both proposals were carried forward in bracketed Protocol text. Despite rumored threats that the activists would proceed without the EC, last-minute bilateral talks between U.S. and EC officials brought the two sides no closer together. Commission officials blamed their inability to move on the absence of a strong European environmental movement, although this can hardly explain their differences with the Toronto Group nations, where environmental groups also gave the issue virtually no attention.[71]

The diplomatic conference to conclude convention negotiations convened in Vienna in March 1985.[72] The same three industry groups attended as at the prior session, but broader public interest remained negligible: several environmental groups were specifically invited, but none came.[73] Tolba called last-minute informal consultations on the Protocol, but the two sides largely reprised the same arguments and no progress was achieved. European delegations added two new themes to their critique of the Toronto proposals: objecting to the claim that aerosols were frivolous uses when much U.S. consumption went to automobile air-conditioning and fast-food foam packaging, and invoking the new suggestion that depletion would increase nonlinearly at high chlorine concentrations to argue that staying below the nonlinearity threshold was the only reasonable policy goal, which the EC proposal would meet but the Toronto Group's would not.[74] Rather than let the conference deadlock, the Toronto Group finally agreed to drop their demand for an immediate Protocol in return for a seemingly insignificant procedural concession: a resolution authorizing UNEP to convene technical workshops on CFC growth and control options, and to reconvene Protocol negotiations, aiming for completion within two years.[75]

With the Protocol and its contentious control proposals off the table, delegates turned to the convention, which contained little of substance or controversy. Most points had been agreed upon in January 1984, and UNEP had been chosen as its secretariat at the May 1984 Governing Council meeting.[76] The two remaining points of conflict that had delayed agreement on the convention for a year, dispute resolution and EC participation, were resolved here. The United States, previously the most forceful advocate of binding and compulsory arbitration, had reversed its position after being sued in the World Court for mining Nicaragua's harbors and losing its procedural bid to avoid the court's jurisdiction.[77] With the most forceful former advocate of compulsory arbitration now rejecting it, the convention adopted a weaker, hybrid system: compulsory participation in a nonbinding process, with each party free to opt for binding third-party arbitration when it ratified. Sixteen nations that had supported the earlier U.S. position attached a declaration urging all parties to elect binding ratification and expressing regret at the weaker compromise adopted "at the request of one party."[78]

The EC Commission had continued to seek expansion of its authority in the environment and the right to become a party independent of its member-states.[79] Although the environment was not an area of EC competency prior to the 1987 Single European Act, the commission had successfully claimed the right to sign other major environmental treaties on the basis of their potential effects on European markets. Here, the commission again succeeded in expanding its participation, gaining the right to sign even if no member-states did. In return, it agreed to make a clear statement of its authority and accept rules to prevent double voting. European environmentalists and national delegations expressed frustration that EC negotiations were so concerned with these procedural matters that they gave inadequate attention to the substance.[80]

The Vienna Convention represented an exhausted compromise containing no substantive provisions beyond those agreed in the 1977 plan of action. The dispute-resolution procedure delayed agreement for about a year, but had been of no significance.[81] The Convention offered so little concrete benefit that the strongest argument the U.S. State Department could find to urge a favorable Senate vote was that it might encourage the Soviet Union finally to release their CFC production data. Weak as it was, however, the Convention still raised worries among some opponents of controls. Japan refused to sign it until 1988, when it ratified both the Convention and the Montreal Protocol. Even in the United States, an attempt by a senior State Department official to withhold U.S. signature was only narrowly deflected by last-minute representations to the secretary of state.[82] The only progress of the Convention, decisive in retrospect but not evident at the time, was the agreement to advance the subsequent agenda with workshops and resumed negotiations, and to empower UNEP to convene these. In informal consultations in late 1985, UNEP decided to hold two workshops, one on CFC growth projections and technical control options, and one on alternative approaches to control policies.

5.3 International Workshops and U.S. Domestic Initiatives, 1986

As preparation for the UNEP workshops and resumed negotiations was under way, a series of domestic events continued to strengthen the hand of a group of U.S. officials, principally within the EPA, who were strongly concerned about ozone. This group had conducted some research and analysis on the issue since it was moved from the Toxics Division to the Air Division in 1983. They gained significant early support from Lee Thomas, who succeeded William Ruckelshaus as EPA Administrator in January 1985, and were further strengthened by the filing and resolution of an NRDC lawsuit calling on the EPA to enact further CFC regulations.

The strong statutory obligation under the Clean Air Act provided the opening for such a legal challenge. The NRDC considered suing during Gorsuch's tenure but decided against it, fearing that she would react by withdrawing the ANPR. After Ruckelshaus replaced Gorsuch and CFCs moved to the Air Division, NRDC threatened to sue in 1983 but were persuaded to delay while early international negotiations held some promise of achieving control measures. When Protocol negotiations stalled, they filed suit on November 27, 1984. After the Alliance petitioned to intervene, the three parties worked through 1985 to negotiate a settlement, which was filed as a consent decree in October.[83]

The centerpiece of the settlement was the EPA's commitment to implement a "Stratospheric Ozone Protection Plan," a two-year program of research, policy analysis, and consultation to decide whether further regulations to protect the ozone layer were warranted. The plan included participating in the UNEP workshops, convening parallel domestic workshops prior to each, cosponsoring with UNEP an additional conference on health and environmental effects of ozone depletion and climate change, conducting a stratospheric ozone risk assessment for review by the EPA's Science Advisory Board, and announcing any regulatory decision with a draft Federal Register notice by May 1987 and a final notice by November.[84]

Different views of what this plan was, and should be, played out at the EPA's first domestic workshop, in March 1986, which examined CFC demand projections and technological alternatives. Lee Thomas carefully described the plan as a process to decide whether further regulations were warranted, not a commitment to regulate, but also stated forcefully that a precautionary stance was required and a decision to regulate did not require scientific certainty, or even confirmation that ozone depletion had occurred. Alliance chair Richard Barnett expressed concern that the plan was actually code for a decision already taken to regulate, and advanced a familiar set of arguments why CFC regulation was inappropriate: that it was unfair to "single out" CFCs when many substances affected ozone; that CFC production was likely to grow very slowly at most; that models continued to show no significant depletion for several decades, so time was available to improve scientific knowledge; and that while Antarctic depletion was clearly real, ozone levels returned to normal later each year and no mechanism could yet account for the observed depletion.[85]

Of many presentations by government, industry, and environmental groups at this workshop, one was particularly important. DuPont's Donald Strobach, speaking as science adviser to the Alliance, gave a summary of DuPont's alternatives research that was later credited as a major breakthrough. Highlighting several points that DuPont had announced in 1980 but which had been widely misunderstood, Strobach stated that DuPont had determined that several HCFC and HFC substitutes were technically feasible, but would cost two to five times more than CFCs; that roughly five years of process and applications development would be required to commercialize them; and that because the substitutes were not commercially viable under prevailing market and regulatory conditions, DuPont stopped its research program in 1981. This statement represented no change in either DuPont's or the Alliance's position—indeed, the only point that DuPont had not announced several years earlier was the termination of its research program. But because their earlier announcement had been so widely—and reasonably— misunderstood as stating that alternatives were technically infeasible, this presentation was widely perceived as a major revelation.[86]

Over the following months, other DuPont officials confirmed this account, and officials from the British firm ICI revealed that they had stopped their alternatives research program at the same time.[87] Both firms identified the same reasons—new kinetic results in 1980 that reduced calculated ozone depletion, and the recognition that the alternatives could not be marketed as cheaply as CFCs—but a calculation that the likelihood of CFC regulation had declined also surely figured in these decisions.[88] The reduced depletion estimates provided no reliable evidence for reduced risk; industry had stressed the unreliability of the high 1979 depletion estimates, and no important advances in understanding had made the lower 1980 estimates any more reliable. As for the cost of alternatives, while it would certainly be fortunate if ozone-safe alternatives could be found that matched CFCs' performance and price, to require this was to demand "no regrets" performance, assigning no value to avoiding ozone depletion. For producers, it was obvious that higher-cost alternatives could not be marketed without incentives to make them viable; but creating these incentives was the job of government, for which the higher cost of substitutes alone could not exclude regulations but must be weighed against the environmental harm the substitutes would avoid. As two DuPont officials wrote in 1986, "The rate of development of alternatives will correspond to the pressure applied; until now, there has been no pressure."[89]

UNEP's workshop on CFC growth projections and technical control options, which convened at Rome in May, achieved no progress. Officially designated as an informal meeting of national expert teams, the workshop included industry officials on several national delegations and the first representation of an environmental group.[90] The workshop illustrated how entrenched and rigid the standoff between the pro- and anticontrol factions had become: in separate sessions on CFC markets and projections, costs of present regulations, technical control options, and other ozone-depleting substances, no significant innovations were proposed and discussion deadlocked on all major topics. The most heated discussions concerned CFC growth projections. The EPA's John Hoffman summarized 11 growth studies, including several sponsored by his office and several independent ones, and argued that they suggested roughly 2.5 percent annual growth of CFCs 11 and 12 through 2050, with low- and high-growth scenarios of 1.2 and 3.8 percent, respectively. Other ozone-depleting chemicals, particularly CFC-113 and HCFC-22, were likely to grow faster. Since the recent NASA/WMO assessment had identified CFC growth as crucial in determining future depletion, growth projections had become important, and representatives from industry and European governments attacked these projections forcefully. They objected that the methods used were flawed and speculative, and that no growth rate should be specified for analysis, merely a range from 0 to 5 percent. In addition, DuPont's representative stated that U.S. industry was unaware of any plans to expand "CFC 11/12 capacity in the United States, Europe, or Japan"[91]—a statement that appears to have been a clear attempt to mislead, since it was true only because it was so carefully delimited; it was reported soon after the meeting that DuPont was planning to expand its CFC-113 capacity in Japan, and was negotiating to build CFC 11 and 12 capacity in China.[92]

The second UNEP workshop, held at Leesburg, Virginia, in September to discuss new approaches to control policy, was widely regarded as a major breakthrough.[93] Some aspects of the meeting were familiar. Some major European nations continued to leave the issue in the hands of industry (neither Italy nor Germany sent a senior government person), while several European presentations merely revisited the superiority of their capacity-cap approach to an international aerosol ban.[94] But many papers by U.S. and Canadian officials and NGOs suggested that they were no longer pursuing an aerosol ban.[95] Several presentations by U.S. officials and NGOs explored the consequences of alternative CFC emission paths using trends in stratospheric chlorine concentration as a policy indicator, and concluded that neither an aerosol ban nor an EC-style capacity cap would significantly reduce future chlorine growth.[96] The EPA's analysis posited the goal of holding stratospheric chlorine to its present level, and showed that because of long atmospheric lifetimes, even that modest goal would require an 85 percent cut in CFC-12 emissions. Although the same analysis had also been presented three months earlier at the Effects workshop, it attracted great interest here.[97]

The Leesburg workshop saw several other significant events. Soviet representatives suggested for the first time that they would be willing to release production data. In addition, sobering new results were presented from a dynamic 2-D atmospheric model that showed several percent ozone depletion at north temperate latitudes even in a multiple perturbation scenario with little CFC growth. This result called into question the contention that holding CFC emissions, at or near present levels would be adequate to avoid depletion.[98] The greatest interest was generated, however, by a Canadian proposal for a new comprehensive approach to global controls. The proposal would control aggregate CFC emissions, with each chemical weighted by its relative contribution to ozone depletion, rather than separately controlling individual chemicals or uses. Consequently, nations could choose how to distribute their control efforts among chemicals and among uses. Although similar approaches had been briefly proposed and discarded twice before, this proposal was perceived to be an important innovation. Within a specified global emissions limit, the proposal would distribute national quotas in proportion to some combination of GNP and population (e.g., 75 percent by GNP and 25 percent by population). National emissions would be approximated as the sum of production and net imports (imports less exports), but only exports to other parties would be included in this calculation, giving all nations an incentive to join the agreement. Members of the Toronto Group received the new proposal with great enthusiasm, and European officials described it as a marked improvement over earlier suggestions. Even Japanese officials suggested that such an approach might allow them to consider controls. The only unfavorable reaction to the proposal was from European CFC producers, who objected that it appeared complicated and would impose a paperwork burden.[99]

First Shift of Industry Position, 1986

The first major change of industry position was announced two weeks after the Leesburg workshop. By the early 1980s, U.S. industry had developed a sophisticated response to the ozone issue that combined the Fluorocarbon Panel's scientific research program—which tended to fund research judged likely to vindicate CFCs, but was conducted to high scientific standards—with continued calls to delay regulation pending better resolution of remaining scientific uncertainties. Since enough uncertainty always remained to justify delay, industry's opponents characterized this strategy as the "sliding three years." Less harshly, some observers suggest that senior industry scientists were confident that further research would discredit ozone-depletion claims, and they needed to delay regulation while awaiting the crucial finding that would vindicate CFCs.[100] Organizing and lobbying by the Alliance at the domestic level complemented this primarily science-based strategy at the international level.

Industry's substantive arguments against CFC controls changed little from 1980 to 1985.[101] In addition to the claim that scientific evidence remained too uncertain, their argument had two components. The first, that viable CFC alternatives did not exist for most uses, was seriously undermined by the details revealed about early alternatives research at the EPA's March workshop and later through 1986. The second, that CFC markets were likely to grow little if at all, was undercut when, after several years of stagnation, markets did resume growing in 1983, modestly in the United States and strongly elsewhere.[102] With 7–8 percent growth in world production of CFCs 11 and 12 in 1983 and 1984, and continued double-digit growth in CFC-113, world production surpassed its prior 1974 peak in 1984 and continued to grow sharply through 1985 and 1986. By 1985, there were also several significant investments in new and expanded CFC production capacity under way worldwide.[103]

This resumed growth put increasing pressure on industry, particularly in view of the conclusion of the 1985 NASA/WMO assessment that the difference between flat CFC production and even 3 percent growth was decisive for future ozone depletion. Pressure for a change in position was building in particular in the Alliance, where staff were growing increasingly embarrassed at issuing repeated corrections to their positions as CFC sales kept increasing. After attacking the EPA's Ozone Protection Plan in the March workshop speech as one of his first duties as Alliance Chairman, Richard Barnett reported increasing discomfort with the Alliance's strong stand, because he felt like "a shill for the chemical companies."[104] Finally, in late August the Alliance's Executive Committee made a comprehensive review of their position. Relying on the recent NASA/WMO assessment and on the evidence of resumed CFC market growth, they agreed after a day of intensive discussions that the Alliance would support internationally negotiated limits on future rates of CFC emissions growth.[105]

Announcing the Alliance's new position on September 16, Barnett stated that "on the basis of current information, we believe that large increases in fully halogenated CFCs . . . would be unacceptable to future generations," endorsed a "reasonable global limit on the future rate of growth" of emissions, and called for rapid development of substitutes.[106] In an accompanying interview, he stated that if further scientific findings supported a stronger international decision to limit capacity or production, the Alliance would also support that decision. Environmental groups praised the Alliance for the change of position but pushed further, stating that the new position indicated the need to consider still stronger controls on CFCs, perhaps even complete elimination.

Although DuPont officials had been centrally involved in the Alliance's change of position, DuPont also conducted a review of its own policy, and 10 days later announced a new position slightly stronger than that of the Alliance—supporting a global limit on CFC emissions, not just their rate of growth. More significantly, DuPont's announcement reaffirmed that CFC alternatives could be marketed in about five years if necessary incentives such as a global CFC production cap were provided. The statement continued to stress that CFCs posed no imminent hazard, describing the new position as a precautionary measure to respond to uncertainty about what CFC growth rate would be safe.[107]

In Europe, CFC industries had never faced a serious regulatory threat or formed a producer-user coalition similar to the Alliance. Industry's position was more strongly dominated by the chemical producers, who also enjoyed substantially greater influence over national and EC policies than their American counterparts.[108] When DuPont and the Alliance announced their support for limited international controls, the European producers made a rapid reassessment of their position and announced that they would support global production limits on CFCs 11 and 12 only. At the same time, ICI announced it would resume research into CFC alternatives, while expressing less optimism than DuPont about the prospects for success.[109]

These 1986 changes in U.S. industry's position have been widely mischaracterized. Many accounts have stated that DuPont endorsed international controls because it had achieved a breakthrough in developing CFC substitutes, and that this change in DuPont's position drove the U.S. delegation's subsequent pursuit of steep CFC cuts.[110] There is no evidence to support this claim, and substantial evidence to the contrary. The major CFC producers had all identified and partly developed the leading alternative candidates in the 1970s. The remaining barriers to marketing them were developing commercial synthesis processes, proving their suitability for specific applications, and testing their toxicity—all slow, incremental processes. Moreover, of these three projects, only synthesis development can be done in secret, and it is quite unlikely that DuPont achieved a decisive breakthrough here as early as 1986.[111] Developing and proving applications must be done in partnership with major customers, so cannot readily be concealed, while toxicity testing is a public good among firms, which producers undertook collaboratively. Despite rumors in 1986 and 1987 that one firm or another had the lead, the major producers all repeatedly stated the same estimated time to market substitutes (5 to 10 years), with toxicity testing accounting for most of the delay.[112] In fact, the first firm to market a substitute, albeit one of limited applicability, was Pennwalt.[113] Moreover, each CFC was used in so many diverse applications that depended on different properties that replacing each chemical with a single substitute for all its uses was highly unlikely, and in fact was not achieved. As late as 1992, DuPont's vice chairman complained that users' unrealistic hopes for perfect drop-in substitutes, which were unlikely for many applications, were delaying conversion away from CFCs.[114]

The claim of a secret DuPont breakthrough is also weakened by the fact that the Alliance changed position first. Although a large DuPont advantage in CFC alternatives could give them an interest in CFC controls, their customers and competitors would have opposing interests. Since the rest of the Alliance were either DuPont's customers or its competitors, attributing the Alliance position change to a DuPont breakthrough would require assuming that DuPont could trick them into supporting something against their interests. Finally, the claim that DuPont drove the aggressive U.S. negotiating position is refuted by officials' accounts of the development of the position, and by the rearguard action that U.S. industry, with DuPont participation, pursued through the spring of 1987 to pull back the aggressive U.S. position.

Rather, several other factors suffice to account for the change of U.S. industry position. The need to admit resumed CFC growth, together with the NASA/WMO conclusion that growth could bring large ozone loss, made the prior industry position untenable.[115] The DuPont statement in March that alternatives were technically feasible, while representing no change of position, sharply changed the terms of the debate by making clear that restricting CFCs would not require forgoing the services they provided, but just paying more for alternatives. Although the ozone hole was not yet attributed to CFCs, industry officials worried that the EPA would use it to promote stringent controls that were not warranted.[116] Endorsing international controls helped reduce the risk of more costly unilateral U.S. action. Finally, the position change in part reflected personal judgments of industry leaders that the risk was serious enough to require a response. Alliance chair Barnett was uncomfortable advocating an industry position that might be posing a serious environmental risk, and later reported suffering intense pressure from his family when a *New Yorker* article attacked him by name as a destroyer of the environment.[117]

Although U.S. industry made this move together, this was the first point at which the interests of producers and users, and of different producers, began to diverge. The Alliance had brought producers and users of CFCs together to resist further demand-side controls, which would harm them all. But who would be harmed by supply-side controls depended on the details. It is noteworthy that the three largest CFC producers—DuPont, Allied, and Pennwalt—endorsed the new Alliance position, but the two smallest producers—Kaiser and Racon—did not. This suggests a dawning realization by the larger producers that regulatory restrictions on CFCs, which were a weak business, might bring consolidation and more favorable market conditions—and by the small producers that such consolidation would come at their expense.[118]

Preparing for Negotiations, Fall 1986

Domestic and international maneuvering proceeded through the fall of 1986, preparing for the resumption of negotiations in December. Hoping to persuade European environmental groups to take the issue more seriously, a few U.S. environmentalists and officials conducted seminars and briefings in European capitals.[119] Germany established a new environment ministry in June 1986, whose position favoring 50 percent CFC cuts became the government's position "more or less by default," and when the EC Commission's November review of the issue recommended no policy change, Germany issued a formal protest.[120] The United Kingdom sponsored a new scientific assessment of ozone, and rumors circulated that resistance was softening even here, encouraged by a statement of the environment minister that decisions were needed that would make industry "uncomfortable."[121] France, Italy, Spain, and Greece remained firmly opposed to controls. EC environment ministers planned to set a negotiating position at their November 24 meeting, but could not because the Chernobyl accident and the Sandoz chemical spill dominated their agenda. Instead, they "agreed the current policy should be maintained until the outcome of the international negotiations is known."[122]

As the U.S. position was developed through the fall, the EPA highlighted worrisome new results, including model results with high temperate-latitude depletions and claims of global depletion in the satellite record, among them the "alleged" Arctic ozone hole.[123] But U.S. officials grew skeptical of the allocation scheme in the Canadian proposal as they examined its implications. Nations' CFC use varied widely relative to both GDP and population. Consequently, GDP-based allocations would give large surpluses to countries with low usage in relation to GDP, some of whom did not want them (e.g., the Nordic states) and some of whom the United States did not want to have them (e.g., the Soviet Union), while any population-based allocation would give huge surpluses to developing countries. The officials decided to adopt parts of the Canadian proposal that they favored—for example, the form of controls, treatment of trade, and joint regulation of all CFCs and all uses—into a new proposal with an allocation scheme more acceptable to the United States.[124]

Several U.S. environmental groups circulated control proposals ranging from 50 percent cuts to complete phaseouts. In contrast to the EPA's move toward comprehensive controls, all but one of these still prohibited specific uses. They also all included import bans on CFCs and broadly related products, intended to coerce other nations into enacting similar measures.[125]

Activity in the U.S. Congress to promote CFC controls also began in the summer of 1986. The Senate Subcommittee on Environmental Pollution held hearings on ozone and climate in June at which Lee Thomas announced that the EPA was considering seeking a CFC phaseout.[126] In an October letter to Secretary of State George Shultz, eight senators urged the United States to pursue complete phaseouts, and said that if the treaty fell short, they would introduce legislation to ban CFCs domestically and ban imports of CFCs and related goods.[127] In subsequent hearings, administration officials repeatedly urged against unilateral U.S. action, while legislators argued the need for a threat of unilateral action to advance international negotiations.[128]

The proposed U.S. negotiating position was first outlined in a State Department briefing paper on November 3. While industry and some officials had assumed the United States would propose to limit production at or near current levels, the draft position was much more aggressive. It called for a near-term freeze in production of all fully halogenated alkanes (CFCs 11, 12, and 113, and the two major halons were specifically identified), followed by a long-term phaseout of emissions subject to periodic review. The official position made this proposal specific. Circulated for interagency review in late November, it called for 95 percent CFC cuts in several steps.

This aggressive position shocked both the Europeans, who were still proposing only to cap CFC production capacity, and U.S. industry. Alliance chair Barnett denounced the suggestion of unilateral action, arguing that a unilateral approach had failed with the aerosol ban and would likely fail again. The analogy was weak, however, since the United States banned aerosols unconditionally and only later pressed others to follow, with no threat of sanctions if they did not. In an argument that anticipated the later domestic backlash against the U.S. position, Barnett argued against threatening unilateral action because the threat would be costly to carry out.[129]

5.4 Protocol Negotiations, 1986–1987

Protocol negotiations resumed in December 1986, with 25 governments, 3 industry groups, and 4 environmental organizations present.[130] There was little substantive negotiation in the session, as most of it was spent presenting and clarifying four proposals for CFC controls. The Canadian delegation pressed their proposal from Leesburg, without a specific numerical limit on world emissions but now revised to include five CFCs, two halons, methyl chloroform, and carbon tetrachloride, jointly controlled according to their ozone-depletion potential. The Canadians had not, however, addressed U.S. and other objections to their allocation mechanism, and their proposal gained little support.[131] The United States proposed cuts in CFC and halon consumption (calculated as in the Canadian proposal), to begin at present levels and drop in several steps to 95 percent reduction, contingent on periodic assessments. The choice of 95 percent reduction was described as eliminating all but essential uses with no substitutes.[132] The U.S. proposal weighted chemicals by their ozone depletion potential (ODP), as in the Canadian proposal; added trade restrictions against nonparties; and included CFC-114 to protect against its substitution for controlled CFCs. The United States signaled the strength of their position by a huge delegation—27 members, including several congressional staff advocating hard-line unilateral controls if international negotiations failed—and by midweek visits from the EPA administrator and deputy administrator. Other proposals were advanced by the Nordic states, who proposed to add immediate 25 percent cuts of CFCs 11 and 12 by industrialized countries, and by the Soviet Union, who presented a vague proposal for national limits on both production and consumption.[133]

Because the council of ministers had failed to consider the issue in November, the EC delegations arrived with the vaguest of mandates and no proposal. After two days of discussions, the three activist member-states led an EC caucus that produced a tentative outline of a proposal. The paper discussed limiting CFC production—not production capacity, a significant change—at 1986 levels or some other level, first for CFCs 11 and 12 and perhaps later for CFCs 113 and 114. It also endorsed some form of special provisions for developing countries, and stated that the European nations "viewed with great interest" proposals to weight chemicals by ODP, but they could not discuss any substantive provisions before the next council meeting on March 20. Other delegations were frustrated at the ambiguous status of these suggestions, and referred to the document—which the Europeans called a "provisional proposal"—as the EC's "nonproposal."[134]

In discussion of the proposals, it became clear that the advocates wanted controls to include both CFC-113 and halons, while Japan wanted to exclude 113 because they had just expanded production capacity to serve the electronics industry, and the Soviet Union wanted to exclude halons because they regarded them as essential for fire fighting in military installations and nuclear power plants—a grave concern so soon after the Chernobyl accident. In addition, Japan favored only a capacity limit, not a production limit. On the other hand, there was strong support for periodic scientific review and update, and substantial support for the U.S. proposal of trade restrictions against nonparties. Informally, U.S. industry representatives accepted the concept of freezing CFCs at current levels, while European industry argued that limits should be higher, to allow moderate further growth.[135]

The session closed with more procedural fights. The next session was planned for February, but the Europeans and Japanese tried to delay it until after the EC Council on March 20, while the Soviets proposed awaiting UNEP's Governing Council in June to clarify whether the group's mandate included halons. After blocking an attempt to table a consolidated draft text, the heads of the U.K. and EC delegations left the meeting early.[136] The scheduling conflict was referred to UNEP director Tolba, who held to the February schedule despite objections.

The February session opened with the first presentation by an environmental group in the negotiations. The European Environmental Bureau presented a statement, endorsed by 100 NGOs in 26 nations, calling for 85 percent reduction in chlorine and bromine emissions within five years and their eventual elimination.[137] This session also saw the first serious negotiation over alternative control proposals. While continuing to press for 95 percent cuts in four CFCs and two halons, the U.S. delegation also offered to explore options that traded off stringency of reductions against timing and scope. The Canadians and Nordics both abandoned their separate proposals and sought to coordinate with the United States, while still rejecting 95 percent reductions as scientifically unjustified. The Canadian delegation speculated that the United States was seeking such steep cuts to ensure the commercial viability of chemical substitutes, and suggested trading Canadian support of these for U.S. support of Canada's objectives on acid rain—the most contentious issue in the countries' bilateral relations at the time.[138]

After Belgium assumed the EC presidency from the United Kingdom in January, an unofficial meeting of the Environment Council had agreed to let delegates discuss a production freeze, on the understanding that they could take no formal position until the council met in March. This ambiguous instruction left European delegations at the February session uncertain what they could and could not discuss. A few delegates expressed tentative support for moving beyond a freeze to 20 percent cuts, but these suggestions—like those in the December EC paper—were informal, provisional, and surrounded by qualifications such as waiting a decade. These vague suggestions both frustrated the advocates of stronger controls, who did not know what they meant, and angered U.K. delegates, who had expected at least French and Italian support, and found themselves alone in their attempts to block such discussions.[139] Moreover, increasingly sharp conflicts among EC member-states were becoming evident. Denmark had recently moved ahead of EC policy by enacting an aerosol ban, while the Dutch had reduced aerosol use by 90 percent through a combination of labeling and voluntary measures.[140] Germany announced informally at the meeting that it planned to follow Denmark's lead and ban aerosols, and soon afterward threatened further unilateral action and denounced the EC proposals as inadequate.[141] Late in the meeting, Chairman Winfried Lang of Austria submitted "personal text" that attempted a compromise between the U.S. and EC positions, which included an initial freeze followed by cuts of 10 percent to 50 percent (20 percent being the most widely discussed figure, although with no date attached).[142] Other than this initiative, the only achievement of this meeting was to agree that the diplomatic conference to conclude a Protocol would be held in September.

In the closing press conference, after Chairman Lang had expressed optimism about continued progress, U.S. delegation head Richard Benedick vigorously criticized the results of the meeting and harshly attacked the integrity of some EC governments and the European chemical industries. Repeating criticisms he had made in the meeting, he denounced the EC's proposals as "ridiculous" and "totally unacceptable," and their proponents as "more concerned with industrial profits than with the responsibility for the environment and future generations."[143] Other U.S. delegates explicitly named the United Kingdom, Japan, and France as obstacles to progress, and blamed EC obstruction on their inability to surmount internal differences. Benedick also noted the risk that if negotiations failed to reach an acceptable agreement, the U.S. Congress would act unilaterally to restrict CFCs and impose trade sanctions on nations that did not do likewise.[144] His remarks antagonized many delegates and prompted a diplomatic protest from the United Kingdom.[145]

After the Vienna session, U.S. officials continued the intense bilateral diplomatic onslaught they had begun in January, with high-level scientific and diplomatic missions to major European capitals, the Soviet Union, China, Japan, and Korea, as well as discussion and question-and-answer sessions broadcast live worldwide by satellite television.

EC environment ministers reconsidered their position at a contentious council meeting on March 19–20. Member states were sharply divided over the stringency of controls and whether to broaden controls beyond CFCs 11 and 12. The three activists, Germany, Denmark, and the Netherlands, advocated a 50 percent cut, while the United Kingdom and France wanted at most a freeze. When France broke ranks to accept a 20 percent cut, the United Kingdom accepted an uneasy compromise: a production freeze of CFCs 11 and 12 within two years, followed by a 20 percent cut after four years. Europeans were also fighting the United States and others to apply controls to production of CFCs, not consumption. Anonymous officials had charged the United States with "pandering to an environmentalist gallery" and with hypocrisy, because U.S. firms were exporting CFCs to Europe, but the newly appointed director general for environment, Laurens Jan Brinkorst, tried to lower tensions with a conciliatory statement in late March.[146]

The February meeting had also seen continuing conflict over model projections of ozone loss. United States delegates had argued in a background paper that model projections showed reductions beyond a freeze were necessary to limit losses, while the U.K. delegation responded that large remaining disparities between models' projections meant that CFC cuts could not yet be scientifically justified.[147] UNEP and U.S. officials were convinced that the apparent disparities between model results were largely due to different input assumptions, and UNEP convened a small meeting of modelers to verify this by conducting comparative runs with common assumptions and emission scenarios. Hastily convened in a hotel room, the group included six scientists running five different 1-D and 2-D models.[148] UNEP's Peter

Usher brought sketches of a set of policy-relevant scenarios, which the modelers elaborated. Initial model comparisons using a few simple scenarios showed, as expected, that the five models' results were very similar. With this agreement established, one fast and simple model was used to generate depletion projections under a wide range of control scenarios—for instance, from a freeze to 50 percent cuts, from only CFCs 11 and 12 to all fully halogenated compounds, and with various assumptions about compliance levels and CFC growth in nonparticipating countries. These runs verified that alternative scenarios of future emissions contributed more uncertainty to future depletion than any model discrepancies or other known sources of scientific uncertainty. The principal result was that at least a "true global freeze" of fully halogenated compounds would be needed to hold global ozone loss under 2 percent by the mid-twenty-first century. Measures that were weaker on any dimension—for example, which excluded some chemicals, or had less than full global participation and compliance—would give larger depletion. Even 50 percent CFC cuts in industrialized countries would lead to 6–16 percent ozone loss by 2060, if other chemicals were excluded and compliance was not perfect.

The next negotiating session, at the end of April, saw increased participation, increased conflict over core control issues, and the first hints of possible agreement. Joining the negotiations for the first time, UNEP director Tolba set the tone with a forceful opening address. Referring to the results of the Wurzburg modelers' meeting, he stated that "no longer can those who oppose action . . . hide behind scientific dissent."[149] Indeed, since perfect participation and compliance could not reasonably be expected, stabilizing the ozone layer would require substantial reduction of emissions. A UNEP-sponsored scientific working group, chaired by NASA's Robert Watson, met with negotiators and provided support and elaboration for Tolba's statement. The group reviewed and endorsed the Wurzburg conclusions, provided recommended ozone-depletion values for various chemicals, and advised on the scope of chemicals to be included, concluding that if CFCs 114 and 115 were not controlled, they would likely replace 11 and 12 in many applications. The group also argued that because many uncertainties remained, sudden and unexplained changes in the atmosphere could occur, and the Protocol should provide for prompt response to new knowledge or observations. Despite activists' arguing for years that uncertainty implied true risks might be either higher or lower than present estimates, this represented the first official endorsement of this precautionary argument in the policy debate.[150]

This negotiation session was the most contentious yet. Ambassador Lang, recalled to Vienna on April 29 to deal with a domestic political crisis, was replaced by Ambassador Hawas of Egypt as chair, and effectively replaced by Tolba as the principal architect of a deal. Even before Lang's withdrawal, Tolba had organized small, closed meetings of key delegation heads, working in secrecy on an unofficial control-measures text. The group included the United States, Canada, Japan, New Zealand, Norway, the Soviet Union, and the EC Commission, plus the "troika" of current, prior, and incoming Council Presidents, Belgium, the United Kingdom, and Denmark.[151] In this session, negotiations were extremely blunt. The United States repeated its call for a near-phaseout and, although U.S. delegates consistently opposed congressional proposals for unilateral action when at home, raised this possibility to highlight the risks inherent in a weak Protocol.[152] The EC accepted in principle the possibility of moving to 20 percent cuts, but no further, while Japan remained at a freeze. German and Danish delegates, seeking deeper cuts, attacked their British colleagues, and Germany repeated its threats to break EC solidarity by moving unilaterally. Sharp differences over scope also persisted. The United States and Toronto Group members wanted broad coverage, while Europeans wanted only CFCs 11 and 12, and Japan and the Soviet Union sustained their adamant objection to including CFC-113 and halons, respectively. The Soviets tried to block discussion of halons on the procedural grounds that the group's mandate covered only CFCs.[153]

At the end of this session, Tolba circulated a "personal text"—a draft negotiating text for which only he took responsibility—that proposed a CFC freeze in 1990, a 20 percent cut in 1992, and "in-principle agreement" on a further 30–40 percent cut by 1994 to 1996. While the freeze and first cut were agreed to, the EC and Toronto Group still differed over the second cut: the EC wanted to require a majority vote to enact it, while the Toronto Group wanted it to be automatic in 1996 unless reversed by a two-thirds majority. Despite several delegations' reservations, the text showed the influence of the scientific working group in stating that controls should certainly include CFCs 11, 12, and 113; would include CFCs 114 and 115, "should scientific evidence confirm the need"; and would possibly include halons. Cuts would apply to both production and consumption, subject to an EC condition that they be treated as a single entity for trade purposes. Remaining points in the text were less contentious: restrictions on CFC trade with nonparties, special provisions for developing countries, and a requirement for periodic scientific assessments and review of control measures.

This draft text represented important progress, but despite enthusiastic reports by the press and some delegations that agreement had been reached,[154] the agreement was tentative, precarious, and incomplete.[155] Industry representatives expressed outrage at the stringency, scope, and speed of the proposed controls.[156] Several delegations initially expressed concern that opponents might attempt to revoke the negotiations' mandate at UNEP's Governing Council in June, although in fact both Japan and the EC moved beyond their earlier positions after this session. The Japanese Environment Minister announced his support for Tolba's text at Governing Council, while EC Environment Ministers met to consider whether to accept it on May 21.[157] Prior to this meeting, German officials had already announced that Tolba's text did not go far enough, and that Germany would begin talks with their chemical firms to seek near-total elimination of CFC production by 2000.[158] Concerned that the Toronto Group had gained advantages in the last negotiation session from knowing the outcome of council discussions, ministers kept their decision secret, although their press release hinted at the possibility of moving beyond their previous position in both stringency and scope.[159]

The U.S. Domestic Backlash, Spring 1987

While a group of U.S. officials in the EPA and the State Department were advancing an aggressive negotiating position through early 1987, a reaction was developing within the United States that attempted to restrain them. CFC producers and users began objecting as soon as the U.S. position for 95 percent cuts was announced in November. Although U.S. industry had dropped its long-standing opposition to any controls, they remained firm that scientific understanding—by which they meant the 1986 NASA/WMO assessment and continuing model results—supported only CFC growth limits or a freeze, not stringent reductions.[160] Many officials, in the United States and elsewhere, agreed.[161] Rather than attempting to intervene internationally, the opponents moved domestically to restrain the U.S. delegation, since it was clear that without vigorous U.S. activism, the international negotiations would quickly settle near a freeze. The opponents organized and gathered domestic political support through the winter, both in Congress and in executive agencies.

The first overt opposition to the U.S. position came in a series of letters from John Dingell, chair of the House Energy and Commerce Committee, to Lee Thomas and George Shultz, beginning on January 2, 1987. Acting on the basis of concerns expressed by the automakers over CFC alternatives in mobile air conditioning, Dingell inquired about the origin of the U.S. position, the extent of analysis supporting it, and the level of consultation undertaken, particularly with CFC user industries. He expressed particular concern about the November interagency approval of the position, asking what level of officials had approved it in such departments as commerce, the U.S. trade representative, and defense. He also sought complete access to the U.S. delegation for his committee staff.[162]

As European resistance to deep cuts began to soften after the February negotiating session, U.S. domestic opposition intensified. The Alliance launched a lobbying drive on February 20, asking its members to write to their legislators, opposing any treaty with a CFC phaseout. In March the Alliance wrote to several agency heads, requesting a comprehensive review of the U.S. position. But as opponents mobilized, so did congressional supporters of strong controls. As they had threatened in the fall, Senators Baucus and Chafee introduced bills on February 19 to cut CFCs and halons up to 85 percent and prohibit imports. Resolutions supporting the negotiations were introduced in both the House and the Senate, and similar bills were introduced in the House in early April.

Testimony at congressional hearings in March and May revealed strong disagreements over both the substantive and the tactical merits of U.S. demands for a near phaseout.[163] On substance, Rowland argued that evidence had been sufficient to justify steep CFC reductions since the 1970s, while the Fluorocarbon Panel's R. Orfeo argued that there was still no evidence of "imminent hazard" from current emission rates. Senior officials from State and the EPA uniformly opposed legislative threats of unilateral U.S. action, while several environmental groups argued that the combination of a forceful U.S. bargaining position and legislative threats as a backup was essential to move reluctant nations beyond a freeze.[164] Since industry representatives regarded nothing beyond a freeze as justified, they saw neither substantive nor tactical merit in the U.S. position. Since a Protocol with a production freeze would gain immediate support of all major producers, they argued that the United States should seek nothing more. Industry was especially alarmed at two aspects of the U.S. negotiating position: first, that uniform 50 percent cuts would greatly benefit Europeans, since they could achieve them easily by cutting aerosols; and second, that the proposal for weaker controls in developing countries would give them competitive advantages in key electronics sectors.[165]

At March hearings, Representative Dingell expressed concern that the U.S. delegation had negotiated "by the seat of its pants," lacked adequate technical and policy support, and gave away too much to domestic advocates of stringent reductions.[166] He pressed his objections in subsequent letters to Thomas, arguing that interagency authorization had been at too low a level, that the U.S. position lacked technical support, and that trade restrictions were added to the U.S. negotiating position in February without interagency authorization.[167] He also attacked the EPA's use of the consent decree from the NRDC lawsuit to retain their own discretion to act unilaterally if negotiations failed. The decree required EPA to issue a draft rule or announce that none was necessary by May 1, 1987. As international negotiations proceeded, EPA and NRDC agreed to delay this deadline to December 1, thereby holding over the negotiations a continuing executive-branch threat of unilateral U.S. action, in addition to the legislative threat. Dingell called on EPA to give up this threat, arguing that a unilateral rule was unlikely to be necessary, so EPA should make an announcement to that effect immediately. He noted that such an announcement would not preclude EPA from issuing a future rule, since they did not need a court-approved consent decree to authorize them to do so.[168]

The Office of Management and Budget (OMB) organized the requested review of the U.S. position in a series of bitterly contested interagency meetings over two weeks in April. Beginning with a series of presentations from scientists, industry, and environmentalists, the meetings reexamined every aspect of the position and its development. The OMB review finished with no resolution, just three days before the April negotiating session in Geneva. Still, NRDC and Senate advocates noted a weakening in U.S. delegation statements in Geneva, which appeared tentatively to accept the Tolba text with its maximum cuts of only 50 percent.[169] They charged that OMB had secretly demanded this weakened position, principally at the instigation of a group of midlevel political appointees in Interior and NOAA as well as OMB, whose hostility to environmental protection extended to the point of harassment and threats against U.S. delegation members.[170] These advocates argued that the United States must maintain both its strong negotiating stance and the threat of unilateral action to bargain effectively with the Europeans.

Because the OMB review had failed to resolve the conflict within the administration, the cabinet-level Domestic Policy Council (DPC) considered the issue at a meeting on May 20. Here, Interior Secretary Donald Hodel and Presidential Science Adviser William Graham criticized the draft treaty text, arguing that nothing beyond a freeze was justified even in an international agreement. When this meeting failed to reach agreement, a working group was established to review the policy once again, leaving the U.S. position and the delegation's authority ambiguous.

Based on a leak from the DPC meeting, the *Washington Post* on May 29 reported that in lieu of CFC controls, Secretary Hodel had advocated a "personal protection plan," which would encourage citizens to protect themselves from UV radiation with hats, sunglasses, and sunscreen.[171] This report, which may have been exaggerated or misattributed, nevertheless provoked a strong reaction against opponents of controls. A group of environmental organizations denounced Hodel at a May 29 press conference where NRDC's Doniger estimated that the proposed plan would cost ten times more than a CFC freeze.[172] This decisive tactical victory by advocates of controls rapidly drove supporters away from the backlash. Even Representative Dingell joined the criticism of the alleged Hodel proposal, introducing a resolution supporting U.S. pursuit of an international agreement that passed unanimously. A similar resolution in the Senate, which criticized the proposal and called for the administration to return to its near-phaseout position, passed by 80 votes to 2.[173] Key industry groups also quietly retreated from supporting the backlash, recognizing both that the extreme position now attributed to them was bound to lose, and that the alternative of unilateral U.S. action if negotiations collapsed or settled only on a freeze would be worse for them than a strong Protocol.[174]

Successive subcabinet level meetings under the DPC considered the issue in June, culminating in a decisive cabinet-level meeting on June 18. Calling in all sources of support, EPA and State officials contacted the foreign ministers of U.S. allies, requesting messages to the White House from ministers and heads of government supporting the strong U.S. position. The conflict was resolved shortly after this meeting, when President Reagan personally approved the strong position advocated by the EPA and State. The decision was kept quiet, reportedly to avoid embarrassing the losing side.[175]

International Consultations, June–July 1987

At the invitation of the EC, Tolba convened his key group of delegation heads for informal consultations in Brussels at the end of June. Informed just before the meeting of the presidential decision to maintain a strong U.S. position but ordered not to reveal any details, Benedick reports that this generated substantial confusion among other delegations over the solidity of the U.S. position.[176] Several remaining differences were narrowed or resolved here. It was agreed that controls would be measured from the baseline year 1986, and that developing countries would be granted a five- to ten-year "grace period" to meet control obligations. It was agreed that CFCs would be controlled jointly according to their ozone-depleting potential, a particularly crucial point for Japan because it would let them protect CFC-113. On the scope of controls, the EC agreed after brief resistance to add CFC-115, making five CFCs to be controlled jointly. Agreement was also reached that the two major halons would be included in the Protocol, but while the Toronto Group wanted to freeze them immediately at present levels, the EC wanted to defer this decision until after the first scientific review. Agreement was not reached, however, on the stringency of CFC controls: after the freeze and 20 percent cut in Tolba's text, the Toronto Group wanted the additional 30 percent cut to be automatic (or at least require a decision to stop it), while the EC wanted to require an affirmative decision to proceed and Japan and the Soviet Union did not accept it at all. The United States briefly tried to force the issue by demanding a third step to 80 percent reduction, but backed down.

The United States also advanced two new proposals here that were not resolved. The first—rumored to be a concession to the opponents who had lost the U.S. domestic conflict—required an extremely demanding level of participation for the Protocol to enter into force: ratification by countries representing at least 90 percent of world CFC consumption. The second U.S. proposal would grant CFC-producing nations additional allowance to produce for export to developing countries, helping to ensure developing countries had access to CFCs on competitive terms, and also benefiting U.S. producers by weakening EC domination of export markets.[177] At the end of this meeting, Tolba said he was 80 percent confident that a Protocol would be agreed upon at the September diplomatic conference; others said that he was being cautious, and that agreement was virtually assured.

One week later, a small group of lawyers met in The Hague to produce draft text for the diplomatic conference in September. Benedick reports that continuing uncertainty over the firmness of the U.S. position led European delegations to be surprisingly intransigent at both Brussels and The Hague, seeking to weaken several provisions of Tolba's text and gain further procedural concessions for the EC. As a consequence, the text forwarded to Montreal remained heavily bracketed.[178] United States officials tried to consult with the EC Commission later in the summer but were rebuffed; Commission officials did not want member-states to see them negotiating bilaterally with the United States. At a meeting on July 21, the Council of Environment Ministers heard charges that the United States had been probing member-states, seeking division, and decided not to adopt a final negotiating position until immediately before the September diplomatic conference.[179]

European environmental groups took their first major action on the issue in August, when Friends of the Earth UK launched a consumer boycott of aerosol spray cans. The boycott sought to end use of CFCs in personal care and household cleaning products by the end of 1988, and in all other nonessential uses by the end of 1989. Two weeks later, the Swiss Consumer Federation launched its own aerosol boycott.[180] The United Kingdom's new scientific assessment of the ozone issue was also released in August. Established in early 1986 to review and update the NASA/WMO assessment, this assessment's publication was delayed by nearly a year, making its scientific content slightly outdated. But its officially authored executive summary conveyed unreconstructed British skepticism about the seriousness of the ozone issue and the justification for any action, arguing that present CFC usage posed no immediate threat. Widely perceived as an attempt to justify a do-nothing stance, the report drew domestic and international denunciations, further weakening the remaining U.K. resistance to strong cuts.[181]

Montreal, September 8–16, 1987

Concluding negotiations of the Protocol took place in six days of officials' working groups, followed by final bargaining in the three-day diplomatic conference in September at Montreal.[182] The negotiations here settled details of a few items already tentatively agreed in Tolba's text, and resolved the few remaining areas of disagreement. Five CFCs and two halons would be controlled, in two separate ODP-weighted bundles. CFCs would be frozen at 1986 levels six months after the Protocol entered into force, followed by "automatic" and "semi-automatic" reductions of 20 and 30 percent at the end of 1993 and of 1998. The timing of the initial freeze was pegged to the date of entry into force, while the next two steps were specified by year.

Late negotiations in Montreal agreed on the "semi-automatic" status of the second reduction step, the control of both production and consumption, and freezing halons at present levels beginning in 1992. For developing countries, all control obligations would be delayed by 10 years, as long as their CFC consumption remained below 0.3 kg per capita. Production controls included two small elements of flexibility that consumption controls did not: first, the U.S. suggestion that producers be allowed to exceed their limits to export to developing countries for their "basic domestic needs" (by 10 percent until 1998 and 15 percent thereafter) was accepted; second, a Canadian suggestion that countries producing less than 25 kte be allowed to transfer their production quotas to others to maintain efficient scale of production, was also accepted. Finally, in order to allow the Soviet Union to finish building an already committed CFC plant, an exemption was crafted to apply to that plant alone. All parties were required to report annually to the secretariat their production, imports, and exports for each controlled substance, for a baseline year (1986 for the original signatories, otherwise the year a party joined the Protocol), and annually thereafter.

A series of trade restrictions on controlled substances and related products was included to give countries incentives to join the Protocol. Parties were forbidden to import controlled substances from nonparties after one year, and products containing controlled substances after about four years. Within five years, parties would determine the feasibility of banning imports of products *made using* controlled substances but not containing them. Based on consultations with trade lawyers, it was decided that nonparties who met all the same reduction and data-reporting requirements as parties would be treated as parties, in order to avoid an obvious GATT violation for trade discrimination.[183]

In contrast to these import controls, export controls were both weak and discriminatory. While industrialized-country parties were allowed to export controlled substances to nonparties, developing-country parties (henceforth known as "article 5 parties," because article 5 of the Protocol defined the special provisions granted for them) were not. Industrialized-country parties were under much weaker export controls. They were merely "discouraged" from exporting—and forbidden to subsidize exports of—technology that produced or used controlled substances to nonparties. The only measure to discourage exports of controlled substances was that the exports to nonparties would count toward both the exporter's production limit (like all exports) and their consumption limit—that is, exports to nonparties would be counted as part of domestic consumption. But this provision would come into effect only in 1993, so until that year industrialized-country parties could export controlled substances to nonparties with no penalty as long as they stayed within their production limit.

The two most contentious last-minute issues were the status of the EC (and, thereby, other hypothetical "regional economic integration organizations"—REIOs) and the ratification requirements for the Protocol to enter into force. In late negotiations, the EC proposed that it be permitted to meet all its obligations collectively, rather than each member-state meeting them separately. This demand provoked an intense debate that was resolved only in the final hours of the diplomatic conference. Other delegations perceived the EC as selectively seeking to be treated as a single sovereign state only in those respects that were advantageous: it was not, for example, offering to accept just one vote. There were valid substantive objections to the EC's attempt to pool all its obligations. Joint reporting, by pooling national data through the EC Commission, would make it impossible to tell whether any particular member-state was meeting its obligations. Defining the production obligation jointly would allow EC producers the benefits of full rationalization to maintain efficient production scale across all 12 member-states, a flexibility that was not granted to all producers. More fundamentally, meeting either production or consumption targets jointly would diffuse responsibility for compliance among multiple states and delegate it to the EC level, when other governments were not confident of either the EC's authority or its competence to deliver collective compliance.[184] For their part, European delegations suspected others of seeking to deny them their aspirations of economic and political union and the legitimate advantages that would follow from it. More pragmatically, Europeans declared that after implementing their single market, internal European trade would be as unmonitored as trade among U.S. states, so they would be unable to meet or document an obligation denominated in terms of intra-European imports and exports. In a last-minute compromise, it was agreed that once all member-states ratified, the EC could by unanimous declaration elect to meet their consumption obligations jointly, but not their production or reporting obligations.[185]

The American proposal to raise the threshold for entry into force to nations representing 90 percent of world CFC consumption also nearly scuttled the agreement at the last minute.[186] Delegations had worried about where to set this figure since February: putting it too low would risk leaving parties bound to treaty obligations without enough reciprocity, while setting it too high would give a few holdouts the power to obstruct the treaty. Tolba's draft text had set the figure at 60 percent of world consumption, which would require both U.S. and EC ratification, but not the Soviet Union or Japan. In contrast, the new U.S. proposal of 90 percent would give a veto to both the Soviets and the Japanese, neither of whom had been an enthusiastic participant in the negotiations. The United States held its position until the final hours of the negotiation against unanimous opposition, finally accepting a compromise that the Protocol would enter into force with 11 ratifications representing two-thirds of global CFC consumption. A related decision determined that a two-thirds majority of parties representing at least half of world consumption could modify the annexes setting the level and timing of controls on already-included substances, which were binding on all parties without ratification. With these last-minute resolutions the Protocol was completed, the culmination of 10 years' work and the first concrete accomplishment in international efforts to protect the ozone layer.

5.5 The Negotiation of the Montreal Protocol: Explanation and Significance

Early Failures in International Cooperation

The 1980–1982 initiative to conduct an integrated assessment of the ozone issue in the OECD and the early negotiations under UNEP shared a basic strategic approach. Both represented attempts by some activist national and international officials to overcome resistance to international negotiations by characterizing them as expert deliberations on nonpolicy matters. The attempt in the OECD failed. Opponents were able to ensure that the technical analyses lacked access to essential data, were under-resourced, broke no new ground, and achieved no relevance to policy. French officials were most forthright in their attempts to disable the exercise, but other national officials and some participating modelers also resisted connecting the work directly to policy choice. Indeed, doing what the activists sought—projecting future CFC growth, and associating alternative growth paths with consequences and with policy measures that might lead to them—would have been possible only as an exercise in integrated assessment of uncertainty. Such an exercise would have required knowledge and assessment tools that were not available, as well as a commitment to an exploratory assessment approach that may be infeasible in a political body like the OECD.

But the similar attempt in UNEP succeeded. The discussions—portrayed as expert discussions among legal and technical experts but transparently negotiations—were approved, the necessary parties came, and after briefly attempting to avoid talking, they talked. Several factors favored this attempt relative to that in OECD. UNEP was still widely perceived as incompetent, so opponents could hope to stall progress in this forum while using its existence to reject the establishment of others more likely to succeed. The supposedly "expert" questions of "relevant legal principles" that were posed to this group were more diffuse, posed no clearly impossible analytical challenges, did not depend on access to industry-held technical information, and yet could easily slide into policy discussions. Moreover, it is difficult to have a group's mandate revoked once it is established, although the Europeans tried, and it is too risky not to participate because of the risk that others will work around you to your disadvantage.

But while the UNEP initiative succeeded in establishing discussions that smoothly elided from discussing principles of a potential convention to drafting specific convention language, it did not succeed in achieving agreement on any concrete measures. The Vienna Convention was a vacuous agreement, containing only two advances of no concrete significance beyond what was already agreed in the Plan of Action eight years before: the dispute-resolution procedure and stronger participation for the EC. Much of the cause of this failure must be ascribed to substantively bad proposals. For three years, the two major factions each demanded that the other reciprocate their existing measures, so were stuck in endless arguments over which approach was preferable. As the initial proponents of international action, the nations of the Toronto Group hold primary responsibility for this waste of effort. Even as a matter of simple tactics, proposing an agreement that required action from their counterparts but none from themselves was unlikely to succeed. The corresponding EC position, taken in reaction to the Toronto Group position, may have been more transparently hypocritical, but the charges of hypocrisy leveled by both sides were correct.

Admittedly, the activists were in a difficult situation. They had already unilaterally taken the low-cost measures they were now seeking from others, and had failed to enact stronger domestic measures in part because domestic opponents pointed to the lack of international reciprocity for their earlier measures. They were stuck at the kink in their cost curves, and had no inducements to offer others in return for accepting cuts. Consequently, they were limited to just two kinds of argument: claims that the measures would be low-cost, which were ineffective since the firms that would bear any costs controlled the major European delegations; and moral exhortation, exposing themselves to charges of preaching and hypocrisy since they were offering nothing themselves.

A few attempts at compromise were made, such as the Toronto Group's adding an option similar to existing EC measures but stronger. Unable to avoid the basic asymmetry of the activists' proposals, however, these all failed. Moreover, on the few occasions that more rational approaches involving comprehensive CFC controls were proposed, they were quickly dismissed as posing insuperable negotiation complexities—an odd concern, in view of the deadlock that already prevailed. Industry representatives sought to reject these approaches precisely because they were more rational, and consequently posed a more serious threat, and in U.S. domestic negotiations were remarkably candid in stating so.

All participants in the debate used ostensibly science-based statements about environmental risk to support their preferred decisions. In negotiations before Vienna, however, while both the Toronto Group's aerosol ban proposal and the EC's production-capacity cap were consistent with a well-founded general sense of risk, there was no reasonable scientific basis to favor either proposal over the other. Rather, in their respective attempts to support positions that were shaped predominantly by their material interests, both sides made transparently selective, tendentious, and flawed arguments about the state of scientific knowledge and its implications for policy. While subtler and more sincere differences over the interpretation and use of policy-relevant uncertainty were also present, their effects on the debate at this stage were overwhelmed by straightforward attempts to use scientific argument to support positions determined on other grounds. The Toronto Group noted that the various aerosol-ban measures they proposed would achieve larger immediate reductions in CFC emissions, but they had no basis for arguing that large immediate reductions were warranted. Indeed, some of their supporting arguments were nonsensical, in particular citing the risk of future CFC growth as a reason to ban aerosols worldwide when nearly all growth was occurring in nonaerosol uses.

For their part, the EC defended their production-capacity cap as a "precautionary" measure that would prevent emissions from ever growing arbitrarily large, even though they applied the cap at a level so high as to make its effect distant and hypothetical. Moreover, they based their opposition to more concrete and immediate measures on a general stance of conservatism, arguing that unverified scientific speculation—their characterization of the state of knowledge—could not provide a legitimate basis for costly policy actions. But when a single, unverified scientific result was published (the "chlorine catastrophe" hypothesis) that precisely supported the EC's already-committed position, they leaped to state it as fact and claim that their position was vindicated. Certainly the relevant sets of U.K. and EC officials were heterogeneous in their views, and some no doubt sincerely subscribed to the official conservative stance. But these statements appear to have been opportunistic use of a single unverified scientific result, of precisely the same character as British scientists had spent a decade denouncing their American counterparts for indulging.

Rapid Movement to Controls in 1986 and 1987

From early 1986, the NASA/WMO "blue books" assessment changed the landscape of scientific statements available to support policy positions. This assessment's influence is evident in the extent to which two of its central statements—CFCs are growing again, and substantial growth in CFCs will bring large ozone depletion, regardless of the trends in other pollutants—were accepted by all parties in subsequent policy debate. It was these two pieces of information—the first increasingly available from multiple sources, the second provided exclusively by this assessment's model comparison exercise—that were responsible for U.S. industry's cautious acceptance of international CFC emission limits. Once U.S. industry had moved, the clear substantive basis for their change compelled their European counterparts to follow, however grudgingly.

This central message of the "blue books"—that present emission levels are likely acceptable, but growth would be bad—became the point around which nearly all positions converged in 1986. All the formerly recalcitrant opponents, including major industry actors, the United Kingdom, France, and the EC Commission, signaled their expectation that a freeze in production at or near present levels would be an appropriate agreement that responsibly reflected the available scientific knowledge. As the weakest level of control that could be justified on the basis of the evolving understanding, this became the minimum position necessary to maintain standing as a responsible participant in the policy debate. The force of the new consensus is shown by the number of former activists who also endorsed this level of controls, as did several scientists when they expressed policy opinions.[187] Not even the most long-standing activist delegations, the Nordic States and Canadians, thought the much stronger U.S. stance was justified.

Three scientific propositions circulating at this time supported a stance stronger than a freeze, but these were incomplete, provisional, and of weaker standing than the "blue books" consensus. Although evidence from the 1986 Antarctic expedition favored chlorine as the cause of the ozone hole, scientific spokesmen and institutions uniformly took the conservative stance that these observations were consistent with a chemical explanation but did not demonstrate it.[188] Several claims had been made that global ozone depletion had been detected, but each, for distinct reasons, was viewed skeptically. The newest results from one model showed significant depletion at high latitudes even with roughly constant CFC emissions, but these were not yet widely replicated.

Relative to this consensus, the U.S. negotiating position was extremely aggressive. The position was initially suggested by the simple analysis showing that an 85 percent emission cut was needed to stabilize atmospheric concentrations. The power of this "85 percent" result in framing subsequent discussions can be seen in all the places where the 85 percent figure appeared that were unrelated to the original calculation that produced it. The figure applied only to total emissions of CFC-12, but Farman, in denouncing U.K. obstruction, called for 85 percent cuts in all foam and aerosol uses, while U.S. NGOs demanded 85 percent cuts in all CFCs. United States pursuit of near-total CFC elimination also reflected Lee Thomas's view that, relative to quantitative limits, an outright ban offered simplicity in negotiations and policymaking, and maximal support for the investment needed to make alternatives viable.[189] United States officials did not use these pragmatic arguments to defend the position, however, but based it entirely on the 85 percent result and the need for a precautionary approach in view of the large outstanding uncertainties.

During the intensive negotiations of early 1987, no change in the scientific debate significantly shifted the support available for alternative policy positions. The only significant change through this period was a gradual increase in confidence attributed to the new 2-D model results showing significant high-latitude ozone losses even with small increases in CFC emissions.[190] The calculations done at the April Wurzburg meeting, though prominently featured in subsequent negotiations, represented no scientific advance. Rather, they simply limited delegates' confusion, by refuting spurious claims that model projections still disagreed, and by illustrating the obvious point that a "true global freeze" of emissions would require reductions, given realistic limits on participation and compliance.[191] Throughout this period, the scientific advice presented to senior U.S. decision-makers continued to state that either a freeze or "substantial" cuts were "consistent with the science." All records of the negotiations, and the scientific advice being offered to them, are consistent with the widely reported claim that negotiators did not consider the Antarctic ozone hole or the unverified claims of global depletion.[192] It would be far too strong to claim that these acute outstanding uncertainties had no influence on negotiations, of course. But while the stronger measures agreed to by splitting the difference between the activists and opponents were scientifically justifiable, so would substantially weaker measures have been. Given the strong preexisting consensus for a freeze, the influence of these uncertainties was most likely limited to shifting the balance between activists and opponents to reach somewhat deeper cuts than they would otherwise have agreed to. The rapidity with which negotiators pursued stricter controls after these uncertainties were resolved in early 1988 gives further support to this view.

Nor can any shift in long-standing arguments over attitudes to risk, or the standards of evidence needed to warrant policy action, account for the negotiated outcome. An early rhetorical shift in the U.S. domestic debate, rejecting industry's attempts to invoke a criminal-justice standard, supported the 1976 decision to cut aerosols but lacked the force to enact even the modestly stronger controls proposed in 1980. At the international level no such shift, rhetorical or real, had occurred by 1987. Activists repeatedly attacked the presumption that high confidence was needed to enact controls, but their opponents continued to invoke the need for "conservatism" and "caution"—that is, guarding against the risk of convicting innocent chemicals. Activists' declarations that the science was "solid enough" to take action were repeatedly met with simple denials. No progress was made in this debate through 1987—indeed, no progress was ever made at this level of abstraction; progress came only through eventual consensual resolution of specific evidentiary claims.

Accounting for this outcome consequently requires examining negotiating dynamics among major actors—their positions, supporting arguments, interests, and alternatives to a treaty—as they evolved through 1987. In this regard, the crucial factor was aggressive leadership by the group of U.S. officials who became the driving force for strong controls in 1986 and 1987 after many years of Nordic and Canadian leadership.[193] While there are some respects in which active U.S. backing, given its size and clout, is necessary to sustain any operational global agreement, the exercise of leadership by U.S. officials here went far beyond this necessary default role. NASA officials provided crucial scientific leadership by early recognizing the need for an authoritative international scientific assessment, and providing the resources and initiative to realize one. In addition, while U.S. negotiators developed few of the significant innovations in the Protocol, they were consistently the most energetic in assimilating good ideas advanced by others, modifying them as necessary to suit their purposes, and vigorously promoting the resultant proposal.[194]

The factor most immediately responsible for the agreed control level, however, was the promotion by an activist faction of U.S. officials of an extreme negotiating position and its maintenance through several months of increasingly intense domestic and international opposition. Markedly stronger than any previous proposal, this position took even the strongest supporters of controls by surprise. Whether it was the delegation's true goal or simply a bargaining gambit to capture attention and force the Europeans below a freeze, it embodied a crucial difference from pre-Vienna negotiations: the advocates of strongest action were now proposing to bear disproportionate costs by renouncing credit for their own prior actions. This change in orientation, which dated from the Canadian proposal at Leesburg, decisively signaled the activists' seriousness and provided iron-clad defense against the charge of hypocrisy.

The bargaining position gained still more force from the threat that if negotiations did not yield a strong agreement, the United States would control CFCs unilaterally, through either congressional or executive action, and impose trade sanctions on countries not doing the same. Although the U.S. delegation carefully followed administration policy in criticizing congressional proposals for unilateral action, this implied threat still helped them move other delegations by dramatizing the unattractive alternative that might prevail if negotiations failed. US industry groups opposed these threats of unilateral action, using the parallel of 1970s aerosol controls to argue that they would reduce, not enhance, other nations' incentives to negotiate, but the parallel did not hold.[195] The United States did not try to make others follow its aerosol ban, and likely could not have, since international trade in aerosol products was so small. This time, however, the serious threat to trade in CFC-related products clearly weakened the resistance of the Japanese and Europeans, and also helped weaken U.S. industry support for the domestic backlash after the "personal protection plan" debacle.[196]

The U.S. activists succeeded in maintaining uncertainty over how committed they were to their full-phaseout stance until late in the negotiations. Their retreat to 50 percent cuts may have been forced on them by the resolution of the domestic backlash (the details of which remain to be divulged), or may simply have been the anticipated concession necessary to move the Europeans below a freeze and reach agreement. The European opponents of strong controls were weakened in late bargaining by internal disagreement, with Germany threatening to break ranks, and by the Commission's predominant interest in using the negotiations to advance its own standing, rather than defending the interests of European CFC producers.[197]

In sum, the Protocol's 50 percent CFC cuts and halon freeze lay within the range of responses that could be reasonably justified by the consensual scientific knowledge of the time, but their particular value is best explained as a split-the-difference bargaining outcome between the near-phaseout advocated by the United States and the production freeze advocated by most other actors. Alternative resolutions of the outstanding scientific uncertainties would have favored one or the other of these positions—a freeze in the case that CFCs did not cause the ozone hole and global ozone was not really declining, and stringent reductions in the opposite case.

Evaluation of the Protocol

The Protocol was a remarkable achievement. Although neither perfect nor complete, it contained the first concrete and effective measures ever negotiated on the ozone issue, and embodied many important innovations. By abandoning sectoral use controls in favor of comprehensive controls of all chemicals and all uses, it accomplished two fundamental benefits. First, it targeted the control measures on the essential environmental problem being addressed, rather than on a set of activities partly related to the problem that happened to be available or vulnerable. Second, it accommodated national differences by allowing real flexibility in implementation, for example, allowing Japan not to adopt the hydrocarbon aerosol propellants they had consistently rejected out of fire-safety concerns. Extending the comprehensive approach to controls of multiple chemicals, rather than just multiple uses, provided still more flexibility, making it possible to adopt more stringent controls while allowing participants to protect ones they valued highly. Again the clearest example was Japan, which planned to protect solvent use of CFC-113 by its electronics industries. The scope of chemicals controlled appeared reasonable in view of available knowledge. All five of the major CFCs were included, avoiding the risk of demand shifting from controlled to uncontrolled ones, while halons were included because they were strong ozone depleters and growing fast. Other chemicals were briefly considered, but appeared to be secondary priorities.

Finally, the Protocol was visionary in its provisions for flexibility and adaptation, in two respects. First, the basic legal structure of the document was crafted to make it easy to adjust quantitative control obligations, by putting specific lists of chemicals and control levels in annexes subject to adjustment by simpler procedures than amending the treaty text. Second and more important, the treaty required that parties periodically support assessments of relevant developments in science, impacts, technology, and economics, then review the controls in force to consider whether these developments suggested changing them. This provision, although perhaps obvious in retrospect, was of fundamental importance to the subsequent success of the Protocol and reflected a substantial advance at the time. It reflected negotiators' response to the two forces squeezing them: the strong political need to achieve an agreement, and the great uncertainty over what stringency of controls was appropriate. Their esteem for the "blue books" assessment made it easy for them to decide how to respond to this uncertainty. The growth in sophistication of their understanding of the need for adaptive updating of the Protocol can be seen by contrasting this response to a proposal in the early negotiations that stated the convention must specify the nature of all future annexes.[198]

In contrast with the sensible structure of the Protocol's basic control obligations and the straightforward explanation of their specific level, many of the Protocol's details reflect an intense struggle between major CFC-producing nations to defend or expand the competitive positions of their national producers. On the stringency and scope of controls, CFC industries worldwide were willing to tolerate a freeze but wanted nothing stronger, and all lost. On the specifics of the form and implementation of controls, however, every detail would influence the competitive environment among CFC producers, particularly between those in Europe and North America, so maneuvering over these details took at least as much negotiating effort as the overall level of controls.[199] The sharpest negotiations concerned the issues of controlling production versus consumption, trade in controlled chemicals, and provisions for international rationalization of production capacity.

EC producers, who dominated export markets and had more effective excess capacity than North American producers because so much of their output was still going to aerosols, wanted the terms of the Protocol to help them maintain their export markets. North American producers wanted to weaken the EC's dominance of exports, or at least not have the Protocol strengthen their position.[200] This basic commercial conflict shaped many of the conflicts over design of controls. The Europeans wanted controls to be denominated in terms of production, which would let them reallocate production from domestic aerosols to exports. The North Americans wanted controls on consumption, which would allow all producers to reallocate production from declining domestic uses to compete for export markets. In the resultant conflict, each side argued that the other's proposed controls alone would leave large loopholes. Both were correct, and the agreed controls resolved the conflict through the obvious step of controlling both production and consumption.

North American officials were also more concerned about leakage of CFC production to nonparty developing countries, and sought two provisions to protect against this: to ban imports of both CFCs and products containing them from nonparties; and to define domestic consumption to include exports to nonparties, while exports to parties would be deducted from national production in calculating national consumption. The Europeans opposed this second measure, which would make their export trade depend upon their customers joining the Protocol. Finally, the Canadian delegation, at the request of DuPont's Canadian subsidiary, sought a provision allowing exchange of production limits between countries that included an arbitrary quantitative limit—only transfers from countries producing less than 25 Kte per year were allowed. The effect of this limit was to allow consolidation within North America but not within Europe. It was principally because of this provision that European industry officials scathingly called the 1987 agreement "the DuPont Protocol."[201]

The resolution of these issues gave some benefits to both sides, with the balance of advantage to the North Americans. Both production and consumption were controlled, giving some advantage to European producers, who could still shift production from domestic aerosols to exports. Counting exports to nonparties against consumption limits after six years limited this European advantage, however, as did the provision to allow production increases for exports to meet developing-country parties' "basic domestic needs." The base-year and industrial-rationalization provisions followed North American proposals, the latter conferring a significant advantage. More broadly, the controls substantially advantaged industrial-country producers by allowing them to export controlled chemicals but not allowing developing-country producers to do so. The most severe of these inequities were corrected in subsequent Protocol amendments.

The 1987 Protocol had three significant weaknesses. First, it gave little attention to the problem of reducing CFC use in developing countries. The only provision addressing the distinct needs of developing-country parties was the 10-year grace period in meeting their control obligations—a cheap, simple, and hastily conceived response, largely equivalent to delaying consideration of the question, that later experience showed to be misconceived. Second, the Protocol ignored CFC emissions, although it was emissions, rather than production or consumption per se, that did harm. Nonaerosol sectors presented many diverse ways to reduce emissions relative to consumption through recapture, recovery, and recycling measures. Controlling emissions would require detailed, use-specific national and subnational programs posing much more serious challenges of measurement and verification than aggregate controls on production and trade. But neglecting emissions in the Protocol implied a several-year delay in developing measures to reduce emissions from the large stock of chemicals already embedded in existing products. Third and most important, the Protocol was negotiated with little regard for the implementation of its obligations, how it would be verified, and how implementation problems would be addressed. Several decisions made in Montreal, either intentionally to resolve disputes or by neglect, subsequently posed serious problems for managing the Protocol and its implementation.

The Protocol was negotiated under intense uncertainty regarding both the risk of ozone depletion and the feasibility of the agreed measures, but much of the uncertainty was resolved rapidly. Major new research results and assessments within a year of the Protocol fundamentally changed the terms of subsequent debate. Both producer and user industry responded to the Protocol's stringent targets with an intense burst of research and technological innovation that within two years persuaded nearly all actors that not just 50 percent reductions, but total elimination of CFCs and halons, was feasible. As both atmospheric science and technology relevant to the Protocol advanced rapidly, the institutional structure adopted in Montreal to assess new knowledge and review and revise the Protocol was intensively used, and supported rapid maturation and strengthening of the regime.

6

Eliminating Chlorofluorocarbons

Science, Assessment, and Responses, 1986–1988

Two shocking observations of stratospheric ozone were reported in 1985: large springtime losses in Antarctica reported by the British Antarctic Survey (BAS), and a significant loss of ozone worldwide reported from a satellite instrument. Independent observations quickly verified that the extreme seasonal Antarctic losses were real but did not determine their cause, and a vigorous debate arose over whether it was a chemical phenomenon related to CFCs or a meteorological one. The claimed worldwide ozone loss could not be verified, and controversy arose over whether it was real, or simply an artifact of a deteriorating satellite instrument. These controversies remained unresolved through 1986 and 1987, and represented the most acute scientific uncertainties under which the Montreal Protocol was negotiated and signed. Both were substantially resolved over the next six months, however, adding sharply increased urgency to the pursuit of reductions in ozone-depleting substances that the Protocol began. Responding to a major new assessment of these results, DuPont announced in March 1988 that it would stop manufacturing CFCs and support strengthening the Protocol to phase them out worldwide. By late 1988, many other actors called for phaseouts, as a group of national and UNEP officials began planning to implement the Protocol's provisions for expert assessments of atmospheric science, effects of UV depletion, technology, and economics.

6.1 The Cause of the Ozone Hole

The report of the ozone hole was a shock to atmospheric scientists, not just because the losses were so extreme, but also because—in a strong setback to the growing confidence that stratospheric processes were coming to be well understood—it could not be explained. The observed losses were much larger than any existing theory could account for, larger than any existing model could generate even with aggressive manipulation to maximize losses. Moreover, the losses were occurring in the lower stratosphere, in spring, and only in Antarctica, while emerging 2-D model predictions said the largest depletions should occur in the upper stratosphere, in winter, and at high latitudes in both hemispheres. The problem attracted immediate attention from theorists, and by mid-1986 three distinct sets of explanations had been proposed: one based on natural nitrogen-catalyzed ozone destruction enhanced by the 11-year solar cycle; one based on changes in stratospheric circulation reducing the transport of ozone to Antarctica; and a few proposals for new chemical catalytic cycles based on chlorine or bromine. Only the third group of explanations implied that CFCs and related chemicals caused the hole.

The first explanation proposed that NOx created by solar radiation in the upper atmosphere accumulated in the lower stratosphere at the poles and accelerated ozone loss through the known NOx catalytic cycle. Solar maxima, such as occurred in 1980, should be associated with peak NOx production and hence, perhaps with a delay, with loss of Antarctic ozone. This explanation made several specific predictions. The hole should have been present shortly after previous solar-maximum years such as 1958 and 1969, and should have begun to reverse by the mid-1980s. Moreover, depletion should occur from the top of the stratosphere downward, and the region of depletion should contain elevated levels of NOx.[1]

The purely dynamic explanations for the depletion relied on the fact that ozone is not created in Antarctica, but transported there from lower latitudes. Consequently, a change in atmospheric circulation that reduced transport of ozone-rich air to the Antarctic stratosphere or increased its transport out could cause large reductions. Such a circulation change could be natural, or could be caused by greenhouse warming altering the atmosphere's radiation balance. Two variants of dy-

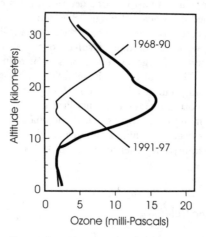

Figure 6.1 The Vertical Structure of the Ozone Hole. Most of the springtime ozone loss in the Antarctic occurs in the lower stratosphere, where ozone is naturally most abundant. Source: UNEP 1999a.

namic explanations stressed a weakening of general stratospheric circulation so that less ozone was brought to Antarctica; and a change in Antarctic circulation such that air was generally rising (rather than the normal sinking), so that low-ozone surface air was brought to the stratosphere.[2] Like the NOx-solar theory, these implied specific predictions. In addition to generally rising air under the depletion region, they predicted unusually cold temperatures, and a stronger and longer-lived polar vortex. Evidence on temperature was available, but mixed: October appeared to have grown colder, but not August or September.[3]

The chemical explanations proposed catalytic cycles different from those thought most important at midlatitudes, which depended on reaction with a free oxygen atom to complete the catalytic cycle. Since oxygen atoms are produced by solar radiation photolyzing molecular oxygen or ozone, they are most abundant where radiation is most intense—in the equatorial upper stratosphere. In the lower stratosphere in Antarctic winter, oxygen atoms are so scarce that the cycles depending on them could not account for the large observed losses.

Two catalytic processes were proposed in 1986. Both begin, like the midlatitude cycle, with a chlorine atom attacking ozone to form ClO, but they proposed different routes by which ClO regenerates Cl to close the cycle. One group proposed a cycle that depended on HO_2, in which ClO reacts with HO_2 to form an HOCl intermediary, which then photolyzes to release Cl.[4] Another group proposed a cycle based on atomic bromine being present as well as chlorine, by which ClO and BrO radicals react with each other to regenerate atomic Cl and Br, and molecular oxygen.[5] For either process to operate at the required rate, NOx concentrations had to be extremely low: otherwise, ClO would react rapidly with NO_2 to form chlorine nitrate, terminating the ozone-destruction path. To accomplish the required loss of NOx, both groups hypothesized heterogeneous reactions of chlorine nitrate, as suggested in 1984 by Rowland and colleagues, taking place on the surface of polar stratospheric clouds (PSCs) in the Antarctic night.[6] The reaction of chlorine with water would yield HOCl and HNO_3; since HOCl would then photolyze much faster than HNO_3, the resultant active chlorine would be unrestrained by NOx for a few weeks. Chlorine nitrate could also react with HCl to release active chlorine, or with HBr to release active bromine. These hypothesized reactions were first thought unlikely, because they required collisions between two molecules precisely at the surface of a cloud particle. But so little was known about the relevant surface processes that both the proposals themselves, and early skeptical reactions, remained highly tentative.[7]

The first attempt to resolve these competing explanations was a rapidly mounted expedition to the coastal Antarctic station McMurdo in August 1986, making various ground and balloon-based observations. The principal observations reported were that several NOx species and chlorine reservoirs were low in ozone-depleted air, while stable species such as N_2O and CH_4 were low and uncorrelated with ozone;[8] that ClO was present at elevated concentrations in daytime, as was OClO (which is formed from ClO);[9] that the region of greatest ozone depletion was in the lower stratosphere; and that stratospheric air appeared to be generally sinking, not rising.[10] Considered together, these observations essentially ruled out the NOx-solar cycle theory, because NO_2 was very low and ozone depletion was mostly in the lower stratosphere, when this theory predicted the opposite for each. Moreover, review of the historical ozone record showed no sign of similar large losses following previous solar-maximum years.

In discriminating between the purely dynamic and chemical hypotheses, the observations leaned toward chemical explanations but were not conclusive. The observations of generally sinking air and of low N_2O and CH_4 suggested that air came to the Antarctic stratosphere from the stratosphere at lower latitudes, not from the surface, while detection of ClO and OClO tended to suggest a chemical process. It also appeared that extreme meteorological conditions were somehow necessary to allow the large losses to proceed, however, and that rapid variations in ozone observed near the edge of the hole were predominantly dynamic in origin. More broadly, observations from one location in a huge affected region could not decisively rule out dynamic causes of the large-scale phenomenon. Despite the evidence favoring chemical explanations, the expedition leader's press conference statement that they "suspected a chemical process is fundamentally responsible for the formation of the hole" attracted wide criticism for drawing a stronger conclusion than the observations yet warranted, and other commentators were much more cautious in drawing the same conclusion.[11] As papers from the expedition and others elaborating dynamic explanations appeared through early 1987, the precise mechanism forming the hole and the relative contributions of chemistry and dynamics remained unresolved.[12]

In late 1986 and 1987, new theoretical and experimental work clarified the role of PSCs in the depletion chemistry, and proposed a third catalytic cycle that was subsequently found to be the most important. First, a mechanism was proposed to form PSCs containing both water and nitric acid, which would facilitate the hypothesized reactions occurring on cloud-particle surfaces.[13] Second, it was shown that the unlikely molecular collisions precisely at the ice surface were not necessary, because ice particles can scavenge HCl from the air and hold it at the surface, allowing it to react with a $ClONO_2$ molecule that collides with the ice later to produce Cl_2 and HNO_3. While the Cl_2 is released to the gas phase and photolyzes rapidly, the HNO_3 remains trapped in the ice.[14] This scavenging of HCl was unlikely if the clouds consisted mostly of water ice, but would proceed rapidly if the clouds were mostly nitric acid.[15] By tying up NOx, nitric-acid clouds would allow highly elevated gas-phase concentrations of HO and hence promote the rapid conversion of HCl into ClO. Consequently, nearly all atmospheric chlorine could be in active form, as opposed to about 1 percent under normal conditions. In addition, a third catalytic cycle was identified in which ClO reacts with itself to form a Cl_2O_2 intermediary, which then photolyzes to yield two atoms of chlorine and molecular oxygen.[16]

Despite this progress, more extensive observations were still needed to resolve the competing explanations. A second, larger expedition beginning in August 1987 combined ground observations from several sites with aircraft flights into the hole. Flying out of southern Chile, a high-altitude research aircraft went directly into the depletion zone in the lower stratosphere, while a converted airliner carried additional instruments through the upper troposphere below the depletion region. This expedition found the hole to be larger and deeper than ever previously observed, and provided several key observations that allowed a rapid resolution of its cause. Most decisively, an extremely strong negative correlation was observed between ClO and ozone, with ClO suddenly rising from a few ppt to 1 ppb and ozone simultaneously falling by half or more at the moment the aircraft crossed the boundary of the depletion region.[17] In addition, both total odd nitrogen and water were extremely low in the depletion region, and nitrate appeared to be incorporated into the PSCs as proposed.[18] In contrast to ClO, BrO was observed at only a few ppt, significantly lower than had been theorized for the Br/Cl depletion mechanism.

Finally, no evidence was found of sustained large-scale upwelling, the basic requirement of the exclusively dynamic explanations.

Delegates were concluding their negotiations in Montreal as this expedition was under way. Delegates had no access to news from the expedition, but its broad implications for the contending explanations of the hole were evident almost immediately and were reported on September 30, two weeks after the conclusion of the diplomatic conference. The main conclusion was that "The weight of observational evidence strongly suggests that both chemical and meteorological mechanisms perturbed the ozone. Additionally, it is clear that meteorology sets up the special conditions required for the perturbed chemistry."[19] The solar cycle theory was rejected again, even more decisively than the previous year. The balance between the meteorological and chemical explanations was more complex. It was clear that meteorology played a central role in the depletion, by creating the extremely cold conditions that allowed PSC formation and the associated removal of water and odd nitrogen from the air, and by isolating the air inside the polar vortex so chemical processes could proceed undisturbed by mixing from outside. Moreover, many details of the depletion, such as rapid short-term ozone changes over extended areas, appeared to be primarily meteorological in origin. But however tactful the

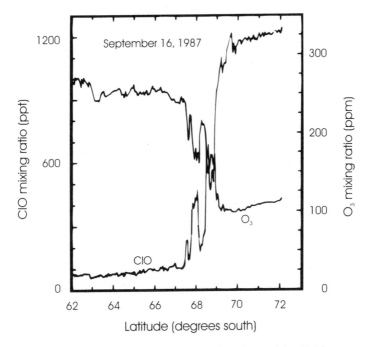

Figure 6.2 Chlorine and the Ozone Hole: The Decisive Evidence. The extreme negative correlation between chlorine monoxide and ozone observed as a 1987 research flight crossed the boundary of the ozone hole showed definitively that chlorine chemistry was responsible for the losses. Source: reprinted with permission from Anderson, Toohey, & Brune 1991.

language, the findings of this expedition dealt a decisive blow to the proposed exclusively dynamic explanations. There was no evidence of large-scale upwelling, as these explanations required. Most decisively, the extreme anti-correlation of ClO with ozone was widely described as the "smoking gun" that definitively showed chlorine to be responsible for the depletion. Additional chemical observations, including low HCl, extreme denitrification, and evidence of nitrate inside cloud particles, all supported the linked hypotheses that chlorine reservoir species reacted on cloud surfaces to release the active chlorine that then destroyed ozone.[20]

Many loose ends remained, of course. The relative contributions of the three proposed chemical processes could not be determined. The low observed levels of BrO argued against the Cl/Br cycle being quantitatively important, while the significance of the ClO dimer cycle depended on which of two possible ways the dimer photolyzed. Most important, the global implications of the Antarctic depletion, whether through hemispheric dilution of Antarctic ozone-poor air or through similar processes occurring elsewhere, were not clear.

The significance of these observations was discussed and clarified in a series of scientific meetings through the fall of 1987, particularly at a November meeting in Dahlem, Germany. Here, a working group on Antarctic ozone that included leading proponents of both the chemical and the dynamic theories, as well as participants from the recently completed expedition, produced a detailed consensus statement that summarized the contributions of chemical and dynamic processes to the depletion. It stated that dynamics provided the conditions necessary for anomalous chemistry to proceed, and explained specific features of the depletion, but that the fundamental loss process was catalytic destruction of ozone by chlorine species, which depended on heterogeneous processes taking place on cloud surfaces. The strong anti-correlation between ClO and ozone observed in the flights inside the hole was cited as the strongest evidence that chlorine played the major role in forming the hole.[21] This group also reported some progress in assessing the quantitative contributions of alternative proposed chemical mechanisms. The Br/Cl cycle could likely account for at most 10–20 percent of observed depletion, based on the low BrO concentrations observed. The ClO dimer process could explain observed depletion at 18.5 km within a factor of 2, subject to important remaining uncertainties about the rate of its formation and the products of its photolysis; however, the concentration of ClO dropped so rapidly below 18 km that it could not explain much of the depletion observed at lower altitudes. The contribution of the HO_2-dependent cycle could not be assessed because the abundance of HO_2 was still not known.

The group identified several important remaining uncertainties in addition to the photolysis products of the ClO dimer. The boundaries of the depletion region remained mysterious, as did the extent of global effect from Antarctic depletion. The rapid decline of ClO below 18 km made it difficult for any of the proposed chemical cycles to explain the depletion observed below about 16 km. And in general, quantitative models could not yet explain the observed pattern of depletion; even models with some heterogeneous processes could not replicate the observed pattern by increasing chlorine alone.[22] They also raised the speculative possibility of positive feedbacks between ozone dilution, temperature, and the size and stability of the polar vortex. With less ozone present, the polar stratosphere would absorb less solar radiation and so remain colder. Lower temperatures would increase the abundance and persistence of PSCs and create a more stable and long-lived polar vortex. These conditions, both seen in 1987, would enhance the processes destroying ozone.[23] Most seriously, the crucial role played by cloud surfaces suggested that similarly large losses might occur on surfaces of other stratospheric particles, such as the different types of PSCs that occur in the Arctic or the volcanic aerosols that are present throughout the lower stratosphere and increase sharply after major eruptions.[24] Consequently, the prospect of similarly extreme losses in other regions could not be ruled out.

The rapid progress of late 1987 in understanding the Antarctic losses clearly resolved the competing explanations in favor of chlorine chemistry, but still fell far short of a complete explanation. Important uncertainties remained, in particular the lack of a quantitative explanation of the observed pattern of depletion. In subsequent policy debates, however, it was never claimed that these remaining uncertainties weakened the status of the new evidence and understanding as a warrant for tightening CFC controls.

6.2 Declines in Global Ozone

The claim that global ozone loss had already occurred was the second prominent controversy that persisted through the Protocol negotiations. Although several such claims had been made from different data sources by 1985, the most contentious was the suggestion that the SBUV satellite instrument showed large ozone losses. Through 1986 and 1987 this claim attracted wide international public and political attention, but substantial skepticism from scientists, based on the short duration of the satellite record and the difficulty of correcting its data for known instrument degradation. The controversy extended to divisions among NASA scientists, including an occasion in March 1987 when two NASA scientists gave seemingly contradictory testimony before a congressional committee.[25]

While this controversy continued, one new analysis of Dobson data in 1986 presented a new approach to examining ozone trends that held the key to the resolution. Prior analyses of Dobson data suffered from two structural weaknesses: they combined records from stations of highly uneven quality; and by using deseasonalized data, they assumed that any trend would appear as a constant effect year-round. Harris and Rowland sought to avoid both these problems by examining data from selected stations with long, high-quality records, and by examining long-term trends separately for each month of the year. Starting with the station with the best long record—Arosa, Switzerland—they simply compared monthly-average data for the periods before and after 1970 (40 years and 15 years, respectively). They found significant decreases in winter, averaging 6 percent, and much smaller decreases, not statistically significant, in summer.[26] Applying the same approach to other stations, they found that *all but one* of the 22 stations north of 30 degrees showed significant midwinter losses over the last 20 years.[27] The winter trends had been obscured in earlier analyses, not just by being combined with summer data that showed no trend, but also by the larger interannual variability of ozone in winter than in summer. In estimating a year-round trend using data of such uneven variability, the least variable months are weighted the most and the most variable months the least. In this case, the pooling technique de-weighted data from the months with the strongest trends, although these monthly trends were highly significant when considered separately.

In late 1986, NASA established the Ozone Trends Panel (OTP) to assess the increasing number of claims that significant global-ozone declines were already under way, by comprehensively reviewing all sources of ozone data. Its purpose was to reconcile contradictions among data sources if possible, and to determine whether a judgment could be made about the existence of significant global ozone trends. The OTP was the second major international assessment conducted under NASA leadership. Initially, it was viewed as so closely connected to the prior 1986 assessment that it was referred to as "Volume 4," revisiting and extending one of the few areas whose treatment in the earlier assessment had been weak. The panel's 21 members were mostly U.S. university and government scientists, with one DuPont scientist and a few Europeans. At their first meeting, in December 1986, they identified working-group chairs for four priority areas—instrument calibration and performance, data-processing algorithms, comparisons among satellite instruments, and satellite-ground comparisons—that they planned to complete rapidly, reporting by mid-1987. In March 1987, however, the group greatly expanded its mandate, establishing several more working groups that brought this activity close to being a comprehensive update of the 1986 assessment. The added areas had seen significant progress since the 1986 assessment, so including them in this assessment would capitalize on the new opportunity to focus attention on the science of the issue. Unlike the 1986 assessment, however, the four groups actually working on ozone trends were conducting new analysis, not just reviewing and synthesizing existing knowledge. Their most important work was a laborious checking, rectification, and cross-correction of ozone data from multiple sources, which confirmed and greatly extended the identification of season-specific trends begun by Harris and Rowland.

Ozone data from ground stations and satellites both had serious problems, but of different kinds. Data from both instruments on the Nimbus-7 satellite (TOMS and SBUV) were slowly drifting relative to the ground network, at least partly due to degradation of a diffuser plate used by both instruments. Ground stations had highly variable quality in instrument operation, calibration, maintenance, and record-keeping, and consequently had random errors at each station, of unknown size and character but uncorrelated with each other. The panel exploited these known differences to check the two records against each other. They used short-term comparisons between ground and satellite measurements as the satellite passed over a station, and between nearby ground stations via the satellite passing over both, to diagnose problems in each ground station's data. They then retrospectively adjusted each station's data based on this diagnosis and the record of recalibrations in the station's logbook. This allowed them to construct a "provisionally revised" record of ozone data since the early 1960s from the 31 best stations, which they checked by verifying that the correction improved the station's agreement with nearby stations, with satellite overflights, and with stratospheric temperature records—which were known to be strongly correlated with total ozone. Finally, they performed a time-series analysis of this revised record including all known and suspected sources of variation in ozone, and tested for a linear trend since 1970.

Combined into broad latitude bands, the results showed significant ozone losses at northern midlatitudes (from 30 to 64 degrees) between 1969 and 1986: 1.7–3 percent decrease year-round and 2.3–6.2 percent decrease in winter, with the largest losses at higher latitudes. The largest monthly loss was 8.3 percent in January over the most northerly latitude band, 53 to 64 degrees. These observed losses were substantially larger than model predictions for these latitudes, which were only 0.5–1.0 percent in summer and 0.8–2.0 percent in winter. Some critics attacked the panel for revising the data, but these attacks were unfounded: the panel did the

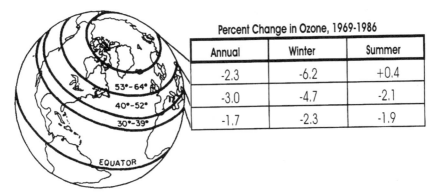

Percent Change in Ozone, 1969-1986		
Annual	Winter	Summer
-2.3	-6.2	+0.4
-3.0	-4.7	-2.1
-1.7	-2.3	-1.9

Figure 6.3 Northern Hemisphere Ozone Loss. The 1988 Ozone Trends Panel found statistically significant declines in winter ozone at north temperate latitudes, with the largest losses at higher latitudes. Source: UNEP 1991c.

analysis with both revised and unrevised data, and while the revisions improved agreement, the same pattern of significant winter losses, largest at high latitudes, was present in the unrevised data.[28]

The panel's analysis shed new light on two old controversies. It suggested that the atmospheric nuclear tests did contribute to the low ozone values of the early 1960s; and it provided an improved quantification of the effect of the 11-year solar cycle, suggesting that it would cause a 1 percent decline in global ozone from solar maximum to solar minimum. Examining data on the vertical profile of ozone, they concluded that the large upper-stratosphere losses in the SBUV record could not be confirmed, but that other data sources also showed substantial (though smaller) losses at that altitude—although these did not agree with each other.[29] Moreover, clear ozone increases in the troposphere suggested that losses in the stratosphere were actually larger than the observed losses in total ozone.[30]

The panel released their executive summary at a press conference with high public and political attention on March 15, 1988, one day after the U.S. Senate voted to approve the Protocol.[31] The main point stressed was that global depletion was real and worse than expected—roughly double what models predicted—and apparently was not due to natural causes (the summary text said the decline may be "wholly or in part" due to CFCs). Although Antarctic ozone was not their main focus, the panel also presented a slightly expanded treatment of the earlier Dahlem group's report, including many passages reproduced verbatim, prepared by a working group containing several members of the earlier group. This chapter restated the consensus that had solidified through the fall of 1987—that CFCs and other man-made chlorine compounds were the primary cause of the ozone hole—with only two additions: it further reduced the likely importance of the Cl/Br cycle, to 5–10 percent; and it reported the first evidence of larger-scale effects of the hole, 5 percent depletion year-round at all latitudes south of 60 South, likely caused at least in part by dilution from the hole.[32] Although the consensus on causes of the hole had been well known for several months, its restatement here drew great public and political attention. Speaking at the press conference, NASA's R. Watson called for stronger regulatory measures than those in the Protocol.[33]

6.3 Eliminate Chlorofluorocarbons?

Because the Protocol's central provision to cut CFCs by half was a bargained compromise between factions advocating a freeze and a total phaseout, debate resumed immediately over whether the measure was too strong or not strong enough. Calls for stronger measures gathered increasing support with scientific announcements through the fall, including the report of the Antarctic expedition and the Dahlem conference. In addition, an early analysis of the Protocol's control measures argued that rapid CFC growth in nonparties could quickly offset reductions made by parties.[34] Industry groups, which had supported only a freeze, countered that the Protocol's 50 percent cuts were a precautionary measure that protected the ozone layer with a generous safety margin. Opposed only by environmental groups and individual scientists, industry representatives made these arguments until early 1988.

The first more serious challenge to this industry stance came in February 1988, when three U.S. senators wrote to DuPont CEO Richard Heckert, arguing that the new evidence that CFCs caused the ozone hole made it time for DuPont to fulfill its long-standing pledge to stop making CFCs if "creditable scientific evidence" showed a threat.[35] Heckert's March 4 response, drafted by DuPont's Freon Products Division, said that no evidence yet suggested CFCs should be dramatically reduced and called the senators' request "unwarranted and counterproductive."[36] This reassertion of DuPont's opposition to further controls quickly became a serious embarrassment to the firm when the Ozone Trends Panel held its press conference on March 15. A DuPont scientist served on the panel, but like all members had been pledged to secrecy until the results were released. He immediately informed managers in the Freon Products Division of the report's significance, who quickly decided the situation needed a response from senior corporate management. On recommendation from line management, the Executive Committee decided on March 18 that DuPont would stop producing the CFCs and halons controlled by the Protocol within about 10 years. DuPont officials and other observers have noted that through three days of intense deliberations, discussion was concerned entirely with whether the new report represented the scientific evidence of harm that called for DuPont to fulfill its pledge, and not at all about the business implications of the decision.[37]

DuPont informed its major customers and competitors of its decision on March 23, and its employees and the press on Thursday, March 24.[38] In a second letter to the senators that reversed his stance of three weeks before, Heckert stated that the panel's conclusions—that CFCs may have contributed to an observed 2 percent global ozone loss since 1970, and were probably major contributors to Antarctic losses—had changed DuPont's view. Their goal was now an orderly transition to a phaseout of fully halogenated CFCs, and a shift toward alternative chemicals.[39] He noted that they had already made significant progress, citing a new HCFC-22-based blowing agent for food packaging foams, and stressed the importance of an orderly transition, since moving too fast could make the equipment that depends on CFCs (hundreds of billions of dollars) prematurely obsolete. DuPont's announcement did not include a precise timetable, but one official suggested that a 95 percent cut shortly after 2000 looked "reasonable," while a Senate proposal for 95 percent cuts in 1996 looked "difficult."[40]

DuPont's pledge to eliminate CFCs preceded such commitments by any other firm, government, or UNEP, and earned widespread praise. Criticisms of the move took two forms: that DuPont should have moved earlier; and that they moved because they had attained a breakthrough by which they expected to dominate alternative markets. DuPont officials argued that they moved when they did because decisive new information in the OTP report for the first time met their standard of demonstration that CFCs posed an environmental risk.[41] This claim has some merit, particularly for the global Dobson analysis that was the first authoritative verification of statistically significant total-ozone trends—trends that were moreover substantially larger than the best available models predicted. But much of the OTP report, including the verification that CFCs were the dominant cause of the ozone hole, had been known and thoroughly reviewed months earlier, particularly after the November Dahlem workshop. On Antarctic science, the OTP report contained little new substance, but merely repeated and modestly extended this already well-developed consensus in a more prominent setting. Similarly for the precise interactions of meteorology and chemistry in forming the hole, for the shares of depletion caused by each catalytic cycle, and for model depletion projections, DuPont's explanation for its change attributes substantial advances to the OTP report that were in fact available and verified over several prior months.[42] Even for global ozone trends, evidence of losses in both total ozone and the upper stratosphere had been accumulating for several years, although on this the panel did present the first authoritative confirmation. On the other hand, in at least one of its conclusions the panel went beyond what was well established, in attributing the reported *global* losses "wholly or in part" to CFCs.[43]

In sum, the substantive content of the OTP report was less novel and definitive than DuPont's rationalization of its policy change claimed. Rather, the OTP's status as a prominent, authoritative, international assessment was necessary for provoking the DuPont change in policy and for its other influence, in addition to the substantive advances it reported. In this respect the role of the OTP report paralleled that of the 1986 NASA/WMO assessment, whose influence depended upon its status and characteristics as an assessment rather than on any major substantive advance it reported. As a basis for criticizing DuPont's policy change, however, this argument gains only a little traction. Even granting that scientific evidence warranted such a change a few months earlier, large bureaucracies cannot make major policy changes instantly. Holding the earlier position for six months in the face of increasingly decisive evidence warranting a change does not appear to be unreasonably slow. Nor does it appear unreasonable to await a major assessment report and the explanation of their own scientist who worked on the report to react to the new results.

The other criticism advanced of DuPont's change of policy, that they expected to profit from a CFC phaseout due to some decisive breakthrough in alternatives, is as weak here as in explaining the earlier U.S. industry shift in 1986. All major producers were looking at the same small set of plausible alternatives, which they all knew how to make, although not yet cost-effectively at commercial scale. While a few diplomats and NGOs speculated that DuPont had a secret breakthrough, all major producers stated that such an advantage was impossible, since they were collaborating on the toxicity testing that was the slowest and riskiest step on the way to commercialization.[44] Capital markets appear not to have seen profitability in the decision, for DuPont's stock declined nearly 10 percent over the week of the announcement. Finally, one must credit the unanimous statements and recollections that business prospects did not figure explicitly in any of the crucial week's discussions. Denying these would require presuming an overt dishonesty totally at odds with DuPont's corporate identity and with the basic requirements of prudent management, in an uncommonly well managed firm.

But while such a direct claim of profit interest in the decision is not plausible, more diffuse claims of potential commercial opportunity are much more so. Whereas CFC and halon markets suffered from persistent soft prices and excess capacity, the proposed HCFC and HFC alternatives posed substantial technical challenges and offered the potential of some patent protection. Consequently, while alternatives markets posed many risks, it was also plausible that barriers to entry could make them more favorable than CFC markets for the largest and most technically sophisticated producers. This was surely understood by management in the Freon Products Division, even if not by corporate management. Indeed, by early 1988 the Protocol had already demonstrated the promise of such benefits for the strongest producers, helping to consolidate U.S. markets for CFCs 11 and 12 by driving out the two smallest producers.

Beyond any potential benefit in the new markets, DuPont also had broad corporate interests at stake. Heckert's aggressive response to the senators had put DuPont's reputation as a responsible and scientifically driven firm at risk. Defending a business line contributing only 1–2 percent to corporate revenues and income could not justify risking this corporate-wide asset. Indeed, this decision marked a turning point in DuPont strategy that elevated environmental responsibility to join worker safety as a first-rank corporate priority, a strategy change that Heckert's successor, Edgar Woolard, stated would be a hallmark of his administration when he assumed the chairmanship in April 1989.[45]

While the events of March 1988 pressed other major producers also to reconsider their CFC positions, none was under the same pressure as DuPont to announce a new stance immediately. Except for Pennwalt, which had committed to marketing substitutes in early 1987 and endorsed a global CFC phaseout the same day as DuPont,[46] other producers waited for two independent reanalyses of OTP data, which were released a few months later, in the summer of 1988. Using various statistical techniques and alternative representations of natural sources of ozone variation, these reanalyses all found significant winter losses between 1969 and 1986, ranging from 3.2 to 4 percent in the region 30–64 degrees North, confirming the main findings of the OTP. They also found the first suggestion of significant ozone losses in summer.[47] With CFC growth of more than 10 percent in 1986 and 1987 adding further urgency, a large number of government and industry endorsements of CFC phaseouts followed through late 1988 and 1989.[48]

Most major CFC industry organizations endorsed some form of phaseout over the following months, while continuing to stress that alternatives were not yet in hand and that the costs of an immediate phaseout could be devastating.[49] ICI, like DuPont, had been embarrassed by strong statements its officials made shortly before the OTP press conference, but resisted pressures to make an immediate change of policy. Testifying before the U.K. House of Commons Environment Committee on March 9, ICI officials argued that further cuts not only were unjustified, but could even obstruct substitute development, by forcing users to switch to immediately available but inferior non-CFC products before new alternatives were available. After the OTP Summary was released, a senior scientific advisory group denounced ICI's position before the same committee.[50] Although ICI resisted making an immediate change of policy, its senior management—like DuPont's—realized that actions taken by its Chemicals and Polymers Division to defend CFCs risked harming the reputation of the entire firm, although CFCs were a small share of corporate business. They changed policy in August, announcing they would also support an orderly CFC phaseout and rapid commercialization of alternatives, and also shifting control over environmental affairs to central management. Following this policy change, ICI mounted an aggressive public relations campaign in late 1988 to resist pressure to move further—seeking both to ensure enough time for an orderly movement from CFCs, and to defend alternatives from restrictions.[51] In the United States, the Alliance endorsed "additional control measures beyond 1998, with the ultimate objective of phasing out production of fully halogenated CFCs" in September 1988. By the spring of 1989, eight of the nine CFC producers in OECD nations—all except Atochem—had said they would cease CFC production when alternatives were ready to market.[52] Consistent with German public policy, the two German producers announced the most aggressive goal, to cease CFC production by 1995.[53]

Among governments, the greenest European states moved the fastest. As early as December 1987, the Netherlands, Denmark, and Germany had already pressed unsuccessfully in EC Council to adopt the "ultimate objective" of eliminating CFCs. The German government persisted, using the late days of its presidency to press the council to adopt stronger measures in June 1988. The council adopted two largely symbolic measures—calling for voluntary agreements with industry for stronger cuts, and announcing they would not attempt to claim an EC bubble—but rejected calls to endorse strengthening the Protocol, or require labeling of CFC-containing products.[54] The report of the German government's study commission on "the changing atmosphere," issued in September 1988, called for strengthening the Protocol to cut CFCs 85 percent by 1995 and to set firm dates, by 2000, to eliminate both CFCs and halons globally.[55] It also called for the EC to complete a phaseout by 1997, and for Germany to cut CFCs 95 percent by 1995. The Bundestag unanimously adopted the commission's recommendations in March 1989,[56] and the Environment Ministry began developing an ordinance to ban CFC uses and negotiating a voluntary production phaseout with industry.[57]

The EPA held to a qualified phaseout position through late September 1988, stating that a phaseout goal must be set, and that 95 percent cuts might be attainable in 10 to 15 years.[58] At the end of September, EPA administrator Lee Thomas sharpened this position, calling for total elimination of CFCs and halons, and a freeze on methyl chloroform.[59]

Considering its prior position, the movement of the U.K. government in 1988 was the most remarkable. At the December 1987 council meeting, the United Kingdom had led opposition to proposals for a CFC phaseout target in the EC, and argued for reinterpreting the Protocol to let the EC meet its production as well as its consumption limits jointly—the "EC bubble" that was a hard-fought point in Montreal, and that many regarded as seriously weakening the Protocol.[60] But this position changed sharply over the next six months, in parallel with the work of a new assessment by the United Kingdom's Stratospheric Ozone Review Group. In contrast to the same body's 1987 report, this one endorsed the OTP conclusions and went further, stating that it was "virtually certain" that CFCs caused the ozone hole and were "implicated" in midlatitude depletion; that 85 percent CFC cuts were needed to stabilize stratospheric chlorine; and that phaseouts of the major man-made carriers of chlorine and bromine were needed to prevent more severe ozone depletion.[61] Following release of this assessment's summary on June 14, 1988, the U.K. delegation joined the group seeking to strengthen the Protocol in the EC Council two days later.[62]

The United Kingdom's position consolidated through the fall of 1988, as part of its broader "greening," most prominently in a September speech to the Royal Society by Prime Minister Thatcher. The government adopted a goal of cutting CFCs 85 percent in October and sought to pursue the same reduction in the EC at the November council meeting,[63] two days after ICI had announced that it was leading the race to commercialize foam and refrigeration alternatives and endorsed strengthening the Protocol to phase out CFCs.[64] Despite the support of the commission and the original three green member-states, the council still rejected this proposal, with France and Spain in opposition.[65] The same week, the U.K. government announced it would sponsor a diplomatic conference on "Saving the Ozone Layer" in March 1989, as a forum to demonstrate that CFC alternatives were, or soon would be, available.[66] Although critics correctly noted that the new position served the United Kingdom's commercial interests, particularly protecting export markets for CFCs and alternatives, the change in policy was nevertheless of substantial importance in view of the United Kingdom's long-standing leadership of the opposition to CFC controls.[67]

6.4 Projecting Future Stratospheric Change: Halogen Loading versus Ozone Depletion

While the results of the 1987 Antarctic expedition and the Ozone Trends Panel settled two crucial and long-standing fights, they also cut the ground from under attempts to project future changes in stratospheric ozone. Existing models, whether using gas-phase chemistry or with the first attempts to integrate heterogeneous chemistry, could not replicate observed depletion either in the Antarctic or at mid-latitudes. Consequently, although concern about ozone loss was sharply elevated, no tools were available to help decide what to do about it through quantitative projections.[68]

A new approach proposed in the summer of 1988 promised to address this need. Instead of attempting to project the consequences of alternative future paths of emissions or controls for stratospheric ozone, the new approach examined their consequences for future stratospheric concentrations of chlorine and bromine. This was much simpler than projecting ozone loss because the processes by which active chlorine reached the stratosphere and eventually left it had long been known with much more detail and confidence than the processes by which they destroyed ozone while there, and the new discoveries had only increased this disparity. Stopping the analysis at the intermediate step of chlorine loading avoided all the complexity, uncertainty, and controversy associated with the actual ozone-loss processes.

Variants of this approach had been in use for several years. A simple variant lay behind the widely cited 1986 EPA argument that stabilizing atmospheric CFC-12 concentrations required an 85 percent emissions cut.[69] In their March 1988 announcement, DuPont also called for using a more basic indicator of atmospheric effects, such as chlorine loading, in view of models' inability to reproduce observed ozone losses. An EPA report of July 1988 first exploited the power of this new approach and first popularized it among policy actors. The report used chlorine-loading to compare multiple alternative assumptions about rates of CFC growth, levels of Protocol participation and compliance, and alternative breadth and stringency of controls. As the authors justified their approach, "chlorine and bromine abundances are currently thought to be the *primary determinants of the risk of ozone depletion*. Consequently, information about abundances of chlorine and bromine can be of use to the decision-making process without making final and certain conclusions about the quantitative relationship between their abundances and ozone depletion."[70]

The report presented many scenarios ranging from no controls (with average emissions growth of 2.8 percent through 2050), which gave stratospheric chlorine over 35 ppb by 2100; through several Protocol-like scenarios, yielding steady-state Cl of 8 ppb or more; through several increasingly stringent control scenarios showing that stabilizing chlorine at 1985 levels required eliminating all fully halogenated compounds, freezing methyl chloroform, and limiting growth of HCFCs. Finally, it presented a calculation of the share that each chemical contributed to excess chlorine, at present and under various future scenarios: these calculations strongly emphasized methyl chloroform and carbon tetrachloride, which contributed 22 and 26 percent of excess chlorine, respectively, but were not controlled in the Protocol.[71]

The new approach was extremely powerful. It met an urgent need to provide a comprehensible and defensible metric of stratospheric harm for policy makers. It allowed many alternative scenarios and choices to be compared, retaining the complexity of chemicals' differing atmospheric lifetimes while abstracting away from complexities of season and location of effect. Moreover, it was extremely evocative, conveying the strong sense that more stratospheric chlorine is worse. It was enthusiastically endorsed by policy makers and negotiators when they first saw it, and was later adopted by atmospheric scientists as a primary tool for communicating with policy makers. Its most important immediate consequence was to direct attention to methyl chloroform, which had been ignored because of its short atmospheric lifetime but which represented a large contribution to stratospheric chlorine and a significant opportunity to make near-term reductions.

6.5 Post-Protocol Planning, Establishment of Assessment Panels

In Montreal, Tolba had expressed willingness to accelerate the Protocol if scientific evidence merited it, and by the end of 1987 there were many calls to do so.[72] The increasing sense of urgency through late 1987 and early 1988 generated some differences over how best to proceed. Some advocated immediately reconvening negotiations to tighten the Protocol; others argued for first solidifying the Protocol's legal status by pursuing quick ratifications, so it would enter into force on schedule on January 1, 1989, before attempting any renegotiation of commitments, lest the controls already achieved be put at risk. After meeting with a dozen close advisers in Paris in January, Tolba adopted a compromise strategy: all efforts would initially be directed to gaining rapid ratifications, but the schedule after entry into force would also be advanced as much as possible, moving the first meeting of the parties from November 1989 to May and considering Protocol amendments in 1990 instead of 1991.

At the same meeting, Tolba and his advisers agreed to convene a series of early working group meetings to tie up loose ends in the Protocol. Although the Protocol embodied clear agreement on the core obligations to reduce ozone-depleting chemicals, it was incomplete in important respects. Even on the core obligations, important and potentially contentious points of defining terms and implementing commitments remained to be settled. On some other matters, the Protocol deferred fundamental decisions and provided only a sketch of the signatories' intentions.

The most important of these were the provisions requiring that parties periodically review the adequacy of the Protocol's control measures and consider amending them; and the associated provision establishing expert assessment panels whose advice the parties must consider in making these decisions.

The need for a vehicle for regular, official scientific advice had been recognized from the earliest attempts at ozone negotiations. Indeed, the CCOL had provided regular, although weak, scientific assessments directly to governments for several years even before the establishment of international negotiations. Nearly all proposed agreements advanced in early negotiations had included some body for regular scientific advice, either retaining the CCOL or establishing some new body consisting of official representatives of the parties.[73] By requiring that the parties periodically reassess the Protocol's control measures and consider changing them, however, and by linking this requirement to expert advice provided by assessment bodies, the Protocol conferred much greater influence on expert bodies than had previously been proposed. Both those who had sought stronger and those who sought weaker initial control measures had supported the requirements for assessment and periodic review, because the acute uncertainty under which they agreed on 50 percent cuts made it highly likely that further knowledge would reveal the appropriate control level to be substantially different from this, whether stronger or weaker. Although the Protocol language establishing the assessment panels lacked specifics, their establishment was inspired by the universally admired model of the 1986 WMO/NASA assessment, and it was widely expected that the new bodies would follow a similar model.[74] But the Protocol called for not just one assessment body but four, in widely divergent domains of knowledge: atmospheric science, effects of UV radiation, technology, and economics. The application of an assessment model developed for atmospheric science to these new domains of knowledge with inexperienced participants represented a substantial leap of faith, and required some adjustment.

A decision in Montreal had authorized a working group of the parties to meet in order to harmonize national production and consumption data. Following Tolba's January consultations, however, this group tacitly adopted a broader mandate when it met at Nairobi in March, and began working to resolve several matters left incomplete in the rush to complete the Protocol. This group also began discussing how to establish the assessment panels, and Tolba, in consultation with a few delegations, identified a tentative chair for each panel from among veterans of prior assessments and Protocol negotiations. After several months of informal consultations, the working group met a second time in October 1988 at The Hague. Here, in addition to continuing their elaboration of Protocol language, the group approved the initial design and mandate of the assessment panels and named tentative chairs.[75] Although many delegates had understood that there would be substantial political control over the panels,[76] Tolba proposed, in the interest of speed, that one "inter-governmental panel" would be the political supervisory body for four "reporting groups" composed of independent experts on each topic area. As the assessments proceeded, these four groups evolved into independent panels whose chairs coordinated among themselves to produce a synthesis report, and the political supervisory role of the intergovernmental panel was taken over—minimally—by the parties' working group.

In conjunction with this working group meeting, workshops were held to review the state of knowledge and plan the scope of the assessment reports for the science, effects, and technology panels.[77] The technology panel decided to proceed via separate expert committees for each major usage area. Although these would rely primarily on the expertise of members drawn from private industry, it was decided—with substantial controversy—that representatives of CFC producers would not be members, but would provide input to the assessment only by making presentations to the panel or serving as reviewers of the report. Principally Tolba's initiative, although supported by key national officials including Lee Thomas, this decision reflected the activists' mistrust of the CFC producers for their long history of obstruction, as well as their concern that the producers were too committed to their own chemical alternatives to participate with the desired objectivity.[78]

The drive for early ratification bore fruit late in 1988. The Vienna Convention entered into force in September 1988, after receiving its twelfth ratification. The Protocol needed eleven ratifications representing at least two-thirds of world CFC consumption to enter into force on schedule on January 1, 1989. Sixteen nations had ratified by the end of November, including the United States, Japan, and the Soviet Union but none from the EC, who were needed to meet the two-thirds consumption requirement.[79] Although an attempt by the Commission to have all member-states ratify together to demonstrate Community authority for environmental policy did not succeed, all did ratify by the end of December. The Protocol consequently entered into force on schedule, with 29 countries plus the EC Commission, representing 83 percent of world consumption.[80]

As the London conference approached, several further endorsements of phaseouts appeared. In February, Canada endorsed strengthening the Protocol to 85 percent cuts.[81] The EC reconsidered its position at a council meeting on March 2, 1989. Having unsuccessfully sought council commitment to stronger cuts at the November meeting, the United Kingdom was surprised to find itself upstaged here by others calling for even stronger cuts. The council agreed to cut the five Protocol CFCs by 85 percent as soon as possible, and eliminate them by the turn of the century.[82] The next day, in a speech to the National Academy of Sciences, President Bush officially changed the U.S. position, stating that 50 percent cuts might not be enough and calling for elimination of CFCs by 2000, provided safe substitutes were available.[83]

The London conference was held March 5–7. Although it had senior governmental participation, the conference had no authority to negotiate changes in the Protocol, but was intended to highlight the increasing availability of alternatives, the need for widespread participation by major developing countries, and the industrialized countries' willingness to do what was necessary to gain their participation. Coming in the wake of phaseout endorsements by most major industrialized nations, the conference saw essentially unanimous support for broad phaseout goals.[84] Indeed, calls for more rapid phaseouts continued to escalate here, with the EC environment commissioner advocating phaseouts by 1996–1997 and the German environment minister, by 1995–1996. A coalition of NGOs demanded still more forceful action, including immediate bans on CFCs, halons, methyl chloroform, and carbon tetrachloride, and limits to the period during which HCFC use would be allowed.[85]

The first formal meetings of the parties to both the Convention and the Protocol, held at Helsinki in late April, conducted several critical items of business to move the Protocol toward strengthening in 1990. Parties formally endorsed the mandate and organization of the assessment panels (although these had already been working for several months), and requested their reports by August.[86] They also established a subsequent negotiating body (the "Open-ended Working Group," or

OEWG) to review the panel reports and prepare draft proposals to amend the Protocol at the second meeting, at London in June 1990.[87] In addition, delegations here reaffirmed their commitments to phasing out CFCs by 2000, and for the first time engaged the details of what would be needed to persuade the major developing countries to join.

On the broad commitment to eliminate CFCs by 2000 there was extraordinary unanimity, although subsequent negotiations revealed significant remaining differences over the precise timing and extent of restrictions. All countries attending, both parties and nonparties (including the Soviet Union, which had demurred in London), signed the "Helsinki declaration." This declaration stated a political commitment to phasing out CFCs by 2000, with unspecified special provisions for developing countries; to phase out halons "as soon as feasible," although no specific date was stated; and to reduce or control "other ozone-depleting substances." Methyl chloroform and carbon tetrachloride were not mentioned explicitly, but were widely viewed as the prime candidates for these additional controls.[88] The U.S. delegation presented the analysis of future chlorine and bromine concentrations published by the EPA, which had already acted on its implications by notifying U.S. producers and users of methyl chloroform and carbon tetrachloride that controls on these chemicals were under consideration. The chlorine-loading analysis was still new to most delegations, and attracted great interest. Several delegations announced they would support including methyl chloroform and carbon tetrachloride in the Protocol on the basis of these results.[89] In addition, delegations asked the science panel, which had nearly completed its work, to take up the new approach in its report. The approach spread rapidly, and future concentration time-paths quickly became the primary means used by all advocates to support their preferred approaches.[90]

All parties present also recognized the importance of persuading major developing countries to join the Protocol. Only eight developing countries had signed in Montreal, and some major nonsignatories were planning rapid expansion of CFC production. China, for example, had announced plans to increase CFC production tenfold by 2000.[91] There was no agreement, however, over how much it would cost developing countries to eliminate CFCs, and how much funding should be provided to them through what arrangements—in particular, whether a new, dedicated fund should be established for the Protocol. The meeting deferred these questions, merely charging the subsequent negotiating body with exploring "modalities or mechanisms, including adequate financial funding mechanisms, which do not exclude the possibility of an international fund."[92] These discussions in Helsinki and London, although not conclusive, provided enough show of good faith to modestly advance the rate of Protocol ratification. Eighteen more ratifications arrived by the end of 1989, including 14 developing countries. But the largest developing countries held back, awaiting firm resolution of the crucial issues of finance and technology.

Parties also continued the work begun at the two previous working group meetings on several matters necessary to clarify and implement the Protocol. They agreed on a definition of "bulk substances" for purposes of implementing the Protocol's trade restrictions; they agreed that controlled substances used as chemical feedstocks, and recycled and recovered substances, would be excluded from the Protocol's production limits;[93] and they decided that the "basic domestic needs" of developing countries, for which the Protocol allowed interim increases in production and export from industrialized countries, did not include developing countries' export markets. The most contentious item concerned nations' reporting production and consumption data. Although the Protocol clearly appeared to require detailed reporting of disaggregated data, both the United States and the EC resisted this. Both sought to aggregate data across chemicals that were jointly controlled (e.g., reporting all CFCs together and all halons together), while the EC also sought to aggregate reports for all member-states. While many other parties thought disaggregated reporting was necessary to determine whether commitments were being met,[94] both the United States and EC claimed this would violate commercial confidentiality. In a compromise decision that prevailed over many objections and has since been widely criticized, it was agreed that nations would report separate data for each chemical to the secretariat, but that if the reporting party requested (as all have), the secretariat would aggregate chemicals before releasing the data. The disaggregated data would be held by the secretariat, not available for inspection by anybody except parties, and available even to parties only by "raising questions" about other parties' compliance.[95]

1989 Assessment Panels

Although their mandates were not formally approved until April 1989, the Assessment Panels began working after the Hague meetings in late 1988, continuing through August 1989. The four panels took diverse approaches to their respective charges—recent progress in atmospheric science, effects of increased surface UV radiation, technology for reducing CFC emissions, and economics. Two panels, atmospheric science and technology, exercised great influence over the international negotiations that followed; the other two exercised little.

Atmospheric Science Panel   The atmospheric science assessment concentrated on major findings since the major prior assessments, the NASA/WMO assessment of 1986 and the Ozone Trends Panel. It reported new results from further Antarctic observations, from Arctic observations seeking to identify whether ozone-loss processes similar to those in the Antarctic were occurring—or were likely to occur—there, and from further analyses of global ozone trends.

Further observations of the Antarctic depletion verified that the ClO dimer cycle was the most important contributor to the large lower-stratosphere depletion.[96] In addition, a previously proposed biennial cycle in the hole's severity was confirmed. The Antarctic stratosphere was a few degrees warmer in 1988 and the hole less severe, while the 1989 hole was stronger again.[97] Ozone loss also extended farther north in 1989 than had previously been observed, almost to Cape Horn, and this was likely due to in-situ chemistry, not transport of ozone-poor air.[98]

Looking for signs of similar ozone loss in the Arctic had been a priority from the moment the 1987 Antarctic expedition was completed. A small observational campaign was mounted in January 1988 with ground-based observations from Greenland and one stratospheric flight.[99] A more extensive campaign in the winter of 1988–1989 included coordinated ground observations from several sites, and multiple flights by two aircraft from Stavanger, Norway. Although extreme losses like those in the Antarctic were thought less likely in the slightly warmer Arctic stratosphere,[100] both years' observations, particularly those of early 1989, found highly perturbed stratospheric chemistry similar to that in the Antarctic. Temperatures were unusually cold, reactive chlorine and bromine were elevated, PSCs were present, and reactive NOx was very low. Large-scale ozone loss was not observed, although some scientists operating independently of the tightly organized NASA-sponsored expedition reported finding sharp narrow bands of depletion at specific altitudes.[101] The expedition reported no Arctic ozone hole, but described the Arctic stratosphere as "primed for ozone loss," likely to experience losses as great as 1 percent per day if these chemical conditions persisted long enough in cold sunlit air.[102]

In global ozone, all sources now showed consistent trends in the vertical distribution—significant decreases in both the upper stratosphere and, contrary to model predictions, in the lower stratosphere—but the panel treated these with great caution, based on short data series, sparse measurement networks, and continuing uncertainty over correcting for aerosols in Umkehr readings.[103]

Although existing gas-phase stratospheric models were known to be wrong in the Antarctic and increasingly believed to be wrong at midlatitudes, the assessment used them to project ozone loss and to calculate ozone-depletion potentials (ODPs) and global-warming potentials (GWPs) for various HCFCs and HFCs, generating some controversy over whether preliminary heterogeneous-chemistry models should be used instead. In addition, these analyses raised the new question of over what time horizon chemicals' environmental effects should be compared. All prior calculations had used "steady-state" ODPs, which compare contributions to depletion over the long term. Since HCFCs differ from CFCs principally in their short tropospheric lifetimes, however, their contribution to depletion is larger relative to that of CFCs if compared in the short term than in the long. This assessment again presented only steady-state ODPs, although some environmentalists argued that short-term ODPs should be used instead, a controversy that grew stronger over the next two years.

In its operations, the panel followed a strategy very similar to that of the 1986 assessment, with a few refinements. Once again, it presented a definitive review of atmospheric science, involving large numbers of the most eminent scientists in the relevant fields, many of whom had participated in the two prior assessments. Indeed, work on this assessment began so soon after the completion of the OTP report that it prompted concern that the atmospheric science community was being overtaxed doing assessments, diverting resources from needed research. Separate groups for each topic worked from January to June 1989, each producing a draft chapter for review at a weeklong plenary meeting in July.

Relative to the previous assessments that served as its models, the science panel made three significant changes in approach. First, it achieved reasonable progress in broadening international participation, especially from developing countries: scientists from 25 countries participated, including 8 developing countries, although OECD scientists still dominated numerically, particularly Americans. Since the Protocol required that this assessment be repeated periodically, its organizers also sought to consider succession in choosing participants, involving many relatively junior scientists whom they judged likely to benefit from the experience and take on increasing responsibilities in subsequent rounds. Second, the panel's official relationship to the Protocol required that it follow the Protocol's mandate in defining its scope. Consequently, the panel only reviewed topics of likely relevance to the required Protocol review—a mandate that excluded further consideration of trace-gas effects on climate—rather than conducting a comprehensive review of atmospheric chemistry and dynamics, as in 1986. Finally, in addition to scientific summaries, the assessment added an executive summary and a short section on policy implications. To ensure that these summaries represented legitimate consensus of participating scientists and did not become sources of controversy, they were drafted and reviewed word by word by all participants at the final review meeting, and not changed thereafter.

In response to parties' request at Helsinki, the panel presented projections of both chlorine loading and ozone depletion for several scenarios of halocarbon controls, which were chosen through informal consultations with officials and delegates, to be relevant to the range of options the parties were likely to consider.[104] Because the Protocol's targets now provided a logical baseline against which to compare future effects of alternative scenarios, the panel had a major advantage relative to all prior assessments, in that it did not need to state and defend an arbitrary baseline projection of future emissions in the absence of controls. The most permissive scenario considered froze CFCs and halons while allowing other ozone-depleting chemicals to continue growing, approximating the Protocol with substantial CFC growth in nonparticipating nations, and giving 9.2 ppb of chlorine by 2060. Intermediate scenarios cut CFCs and halons by 50 to 95 percent, progressively adding freezes of carbon tetrachloride, methyl chloroform, and HCFCs. The most stringent scenario, intended simply to study how fast stratospheric chlorine could be reduced, eliminated *all* halocarbon emissions, even HCFCs, by 2000. Even in this case, it was not until 2060 that stratospheric chlorine returned below 2 ppb, roughly the pre-hole level.[105] The scenario analysis stated a strong conclusion: "If substantial halocarbon emissions continue, the atmospheric abundance of Cl and Br will increase, and as a result, significant ozone decreases, even outside Antarctica, are highly likely."[106]

The report's summary highlighted four implications. First, the Protocol was based on projections of large ozone losses if CFCs continued to grow, but subsequent observations showed that significant depletion had already occurred. Second, chlorine-loading projections illustrated the inadequacy of the Protocol, which even with full implementation would let atmospheric chlorine grow from 3 to 9 ppb by 2100, reducing ozone by 0–4 percent in the tropics and 4–12 percent at high latitudes even without considering heterogeneous processes. Third, the risk that heterogeneous chemistry can occur on atmospheric aerosols worldwide, especially after large volcanic eruptions, suggests that global depletion may be larger than predicted. Finally, that the Antarctic hole appeared when Cl was about 1.5–2 ppb suggests that "To return the Antarctic ozone layer to levels approaching its natural state, . . . one of a limited number of approaches is a complete phase-out of all fully halogenated CFCs, halons, carbon tetrachloride, and methyl chloroform, as well as careful consideration of the HCFC substitutes. Otherwise, the Antarctic ozone hole is expected to remain, provided the present meteorological conditions continue."[107] This remarkable set of conclusions combined great caution in stating scientific conclusions on which significant doubts remained, with a set of conditional statements whose implications for policy action were extremely strong.

Effects Panel   The effects panel, a much smaller group, updated the EPA/UNEP 1986 effects study. Their major conclusion was that effects remained complex and underresearched due to inadequate support, and that more research was required, but that even present knowledge made it clear that some impacts posed significant threats. No increase in global surface UV-B was yet observed, although limitations of the network weakened confidence in this result: the network was sparse, used instruments unsuitable for detecting long-term trends, and was affected by cloudiness and pollution.[108]

The panel presented quantitative estimates of only two effects, both in the form of multiplier relationships: for nonmelanoma skin cancer (NMSC), the cancer-ozone multiplier was estimated as 3; for cataracts, the estimated multiplier was 0.6. Beyond these two, the panel identified many possible effects that provided grounds for concern, but were too complicated or not well enough understood to quantify. For example, they noted that UVB "can cause" immune-system suppression, which "might lead to" an increase in the occurrence or severity of infections or decreased effectiveness of vaccinations.[109]

UV effects had been shown on both terrestrial plants and aquatic organisms. About half of the plant species tested experimentally were found to be sensitive to enhanced UV: in some soybean varieties, the UV increase from a 25 percent ozone loss reduced yields by 25 percent. Harmful effects had been shown on many aquatic organisms, especially phytoplankton and zooplankton, and the juvenile forms of various crabs, shrimp, and fish, which may suggest broad negative effects on marine ecosystems. Other potential effects identified include increased chemical reactivity in the lower atmosphere, and hence increased smog, and damage to synthetic materials. Ecosystem threats from Antarctic ozone depletion constituted a large new uncertainty, though none had yet been observed.[110] As over the preceding 15 years, assessment of effects remained confined largely to listing possibilities, offering quantitative estimates for very few. Progress in research remained limited, due both to its intrinsic difficulty and to the limited resources directed to it.

Technology Panel   The technology panel had the most novel job, surveying current uses of ozone-depleting substances and available or soon-available alternatives in detail, to judge what reductions were technically feasible. Chaired by a Canadian delegate who had led pre-Montreal negotiations on technology and control options, the panel was organized in five "technical options committees" (TOCs), one for each major sector of chemical use. Each TOC included 15 to 50 technical experts, charged with assessing what level of reductions in their usage sector were technically feasible, by what time.[111] The definition of technical feasibility, approved by the parties, was "the possibility to provide substitutes or alternative processes without substantially affecting properties, performance or reliability of goods and services from a technical and environmental point of view"—notably, defined without explicit reference to the cost of alternatives.[112]

This panel's operations differed sharply from the others. Although participants included experts from universities and governments, most were technical experts from private industry, principally from user firms, engineering and consulting firms, and industry associations. Government officials provided most of the leadership and administration, however: three of the five TOCs were chaired by environmental officials, who with consultants took operational responsibility for producing the reports. Although participants were nominated by parties to the Protocol, chairs had substantial control over participation in practice, which they used to identify expert and energetic people committed to solving problems, with sufficiently wide representation from affected industries in each usage sector to produce high-quality and credible results.

The panel's conclusions were produced by aggregating sectoral and subsectoral conclusions from each committee, to project aggregate technical potential to reduce CFCs and halons. The conclusions were shocking, in that they stated that nearly all CFC and halon production could be eliminated by 2000. For CFCs, the schedule of technically feasible reductions identified was 50 percent reduction by 1993, and at least 95 percent by 2000. The remaining few percent, mostly required for servicing existing refrigeration equipment, was judged to be probably—but not cer-tainly—capable of elimination by 2000 through anticipated further technology de-velopment.[113]

The separate reports of each TOC provided backup, elaboration and qualifica-tions for this aggregate estimate of feasible reduction. For example, the foam TOC concluded that CFCs could be cut at least 95 percent by 1995, given the planned availability of HCFCs. CFC solvents could be eliminated by 2000, while aerosols could be eliminated *immediately* except for a few small medical uses. Refrigeration had the slowest feasible reductions and accounted for the 2–5 percent tail beyond 2000, because of existing long-lived equipment, particularly automotive air-conditioning, that needed continued servicing with CFCs.

The halon TOC was the only one that did not reach complete consensus on feasible reductions. The TOC concluded that while no perfect substitutes for halons yet existed, halons were used in many ways that were not necessary—for instance, vented for testing fire-protection systems or for training, or lost in leaks. They observed that conservation and improved management of the large "bank" of hal-ons stored in systems could allow a 60 percent cut in consumption within five years, even if alternative chemicals were not available. The TOC split, however, on the feasibility of reductions beyond this level. A majority judged a full phaseout feasible by 2005, while a minority wanted to make no such quantitative projection until experience was gained with alternatives. This difference produced what was later recognized as a rare instance of explicit majority and minority views in the TOC report.[114]

In this round, the technology panel and its TOCs established the pattern of multifunction work that they subsequently retained, to great effect. In addition to advising the parties on the specific questions of technological feasibility within their mandate, the TOCs served three other functions. First, they identified existing ca-pacity to reduce emissions of ozone-depleting substances that was not as widely known or exploited as it could be, and disseminated knowledge about it to the relevant sectors. Second, they brought together highly focused top technical experts to consider specific technical problems of reducing ozone-depleting chemicals, with the result that solutions to the problems were frequently identified. Third, their discussions also served to identify nontechnical barriers and blockages to accom-plishing the required reductions, which in some cases allowed government or in-dustry actors to organize to resolve the barriers. Because these last two functions had the effect of actually increasing capacity to reduce ozone-depleting substances, the panel was not merely advising the parties on a static set of feasible reduction options, but was actively expanding and diffusing the set of available options, so the extent of technically feasible reductions kept advancing.

Economics Panel   In contrast to the technology panel, the economics panel had a vaguely defined and ultimately impossible job. Unable to undertake any economic analysis of further reductions because it was working in parallel with the technology panel, it instead summarized EPA's earlier regulatory impact assessment and a con-sultant study commissioned by the government of Norway, to provide illustrative quantitative data on potential benefits of reducing ozone-depleting substances. In addition, it presented general guidelines for phasing down ozone-depleting sub-stances at minimum cost, similar to those provided by the U.S. National Academy of Sciences Panel on ozone in 1976.[115]

Synthesis Report   After the four panel reports were completed, the panel chairs met at Geneva in July 1989 to produce a "synthesis" report. In addition to summarizing the major conclusions of each panel, this report added newly available information, responses to new requests by parties, and integration and interpretation of the panels' results. In particular, it highlighted five alternative scenarios of further controls, each described by both its technical feasibility and its consequences for future stratospheric chlorine. From a baseline scenario of the Montreal controls only, successive scenarios added CFC and halon phaseouts; a freeze on methyl chloroform; phaseouts of methyl chloroform and carbon tetrachloride; and limiting HCFCs to 20 percent of the original CFC market. When analyzed in terms of resultant chlorine concentration, only the most stringent scenario brought an eventual future recovery.

6.6 Moving the Debate to Full Phaseouts: Explanation and Significance

In contrast to ozone assessments over the prior decade, both the Antarctic campaigns and the Trends Panel addressed highly specific, focused questions. The Trends Panel addressed the question framed more than a decade earlier by industry representatives opposing controls: Is there a statistically significant reduction in global ozone? Over the intervening decade, enough actors had accepted this observation as the ideal—or necessary—standard of justification for further controls on CFCs, that it became the most decisive piece of evidence.

Whether there is a significant trend in global ozone resembles an ordinary question in empirical research. Because of technical challenges in obtaining and analyzing the relevant data, only members of a small scientific community are in a position to make claims about ozone trends, and standard practice in addressing such questions—and among such communities—prescribes a strong presumption against novel claims. Claims of observed depletion were consistently held to the most rigorous scientific standards, even though they were generated from thin, diverse, ill-calibrated, inconsistent instrument systems that were extremely hard-pressed to meet the standard. Early reports of losses in Umkehr, balloon, and Dobson data, as well as Heath's two separate claims of trends in SBUV data, were all treated with extreme skepticism—to the point of seriously harming scientific careers—although all these claims turned out to be qualitatively correct, in some cases even quantitatively accurate.

But elevating this question to the status of the decisive evidence necessary to justify policy action, while retaining this standard of demonstration, represented the antithesis of a precautionary approach: do nothing until the observed harm is statistically distinguishable from the noisy global-ozone records. This approach to defining the observation necessary to warrant controls, and the evidentiary standard used to determine that the observation had been made, strongly biased the debate in favor of the opponents of controls. Once the specified standard was met with the analysis in the Trends Panel, however, most industry actors accepted both the result and its implications for action, when they might have tried to redefine the standard, for instance, by arguing that the observed trend was small, that the trends were largest where few people lived, or that knowledge of the effects of depletion was very weak.

Industry reactions were not just to the substantive resolution of the claim, however, but also to its prominent presentation in the assessment report of the Trends

Panel. The influence of the Trends Panel was far out of proportion to the novelty of its contents. On the Antarctic, all its contents but one relatively minor elaboration had been known, confirmed, and publicized for a few months. For global trends, several prior results had strongly prefigured its results, especially the Dobson analysis of Rowland and Harris, while the panel added the comprehensive extension, review and verification of this work. Even its results on this, its central analytic contribution, had been only imperfectly kept secret, in that the chair of the panel had informally announced its broad conclusions at the January 1988 Alternatives Conference.

But in addition to this substantive advance, the crucial contribution of the Trends Panel was to disseminate these results prominently, from an eminent, independent assessment body with wide official sponsorship. It was the announcement of the panel's results, given its stature and sponsorship, that seized attention and provided the crucial warrant for action, which was magnified by the drama of DuPont's ill-advised prior statement and subsequent rapid reversal. The response to the panel corroborates the observation from the 1986 NASA/WMO assessment that scientific results influence policy only through the intermediation of scientific assessments meeting certain conditions for effectiveness, which synthesize, validate, and publicize the underlying results and their policy-relevant implications. The experience of both these assessments also suggests the necessity of assessment processes and their leaders imposing some level of uniformity on the interpretation of results to extract strong, simple, bulletproof messages of clear policy importance. In some instances, this need to control interpretation of points likely to confuse or obscure the agreed message of the assessment represented a significant suppression of normally unruly scientific debate, which assessment leaders were able to accomplish as a consequence of the status of authoritative monopoly on policy-relevant scientific information that these assessments achieved.

Several important uncertainties and gaps in knowledge remained after the Antarctic expedition and Trends Panel. Important missing pieces included quantitative models of Antarctic depletion, good satellite records of global ozone, an explanation for the observed temperate-latitude depletion, and any basis for making quantitative projections of future depletion. Despite these enormous gaps in knowledge, nearly every significant policy actor reacted to the Trends Panel by calling for strengthening the Protocol, with most calling for complete CFC phaseouts.

Several explanations are possible for why no actors attempted to use these substantial remaining uncertainties as pretexts for delaying further controls while conducting more research. First, it may have been that after the Protocol was signed, no actor with sufficient influence over policy had an interest in arresting or reversing the movement away from CFCs, but only in limiting its speed. The small producers who could not compete in CFC alternatives were already being shaken out within months of the Protocol. Many individual users facing costly transitions would surely have wished to hold onto CFCs, but faced extreme collective-action problems and low likelihood of success in the face of a widespread presumption that further cuts were inevitable. The Alliance, the only organization representing all users, had shifted strategy under the influence of the producers and the largest users, seeking only to maintain an orderly retreat that would allow time to commercialize alternatives.

Second, it may have been that the remaining uncertainties, while of high scientific importance, had little policy significance. Observing that ozone loss had occurred, both in Antarctica and globally, and knowing that chlorine was causing the Antarctic (though not the worldwide) depletion, were sufficient to move policy. Beyond this threshold, the major remaining outstanding questions—such as whether the processes causing the Antarctic loss could operate worldwide, and how much ozone would be lost, by when, and where—served only to increase concern, not to provide a pretext to delay further action.

The widespread endorsement of tightening the Protocol to phase out CFCs did not resolve the operational questions needed for actual renegotiation of the Protocol. Moreover, the chorus of phaseout calls concealed substantial disagreement over questions of precisely what chemicals to phase out over what time period, which had to be resolved when negotiations resumed. It was here that the science and technology panels provided decisive assistance in framing the agenda for resumed negotiations and shaping the debates.

In atmospheric science, prior work, including prior assessments, had already decisively demonstrated the seriousness of the risk and the desirability of proceeding farther and faster. In this regard, the panel reviewed new progress to further develop the scientific basis for concern, but this job was mostly done and negotiators now had other needs. To help them resolve their specific questions and disagreements over what to do, negotiators needed some simple, understandable way to compare the atmospheric benefits of alternative choices. Time-paths of future ozone-depletion could have served this need, just as ODPs had done for some time, if scientific understanding had been adequate to produce them. Lacking the ability to produce these, however, the science panel adopted (after substantial refinement and improvement) the simple policy-relevant metric that had been advocated by the EPA and used it as their primary mode of communicating with policy makers. This work was presented to negotiators by a small subgroup after the panel's main work was completed. As with the simple model results developed at Wurzburg during Protocol negotiations, the full science group would have had little basis to object to these (but some did—and there were also some industry attempts to stir up objections to this new metric), but would likely have found them of no scientific interest. Crucially, the questions of intense scientific interest were those that were acutely unresolved—precisely the questions that this new heuristic sought to avoid.

The novelty and power of the technology panel's approach cannot be stressed enough. Despite several attempts, the problem of providing policy makers with credible technical information about ways to reduce CFC use had never been solved. This group accomplished four important results. They framed the questions in a sufficiently simple way that was both tractable to a technical group and relevant to the policy making audience. They structured participation and procedures so as to involve the required expertise, which resided principally in private industry, and structured the membership and procedures of subgroups to effectively control against both obstruction and insupportable technical claims. They exploited the policy and economic environment, and the institutional and professional ambitions of participants, to motivate an intense commitment to solving the problem of finding ways to reduce CFC use. Finally, they linked technical assessment to action, so that as the group resolved technical questions to advise the parties, it also achieved two other benefits: solving technical problems and facilitating innovations, thereby advancing the margin of what could be accomplished; and educating industry sectors about the necessity, feasibility, and benefits of emerging technical alternatives to reduce CFC use, thereby advancing the margin of what actually was accomplished. As I argue in subsequent chapters, this process of technological assessment was likely the most significant factor in accounting for the rapid reduction of ozone-depleting chemicals accomplished by the Montreal Protocol.

7

Industry Strategy and Technical Innovation, 1987–1992

7.1 Chlorofluorocarbon Markets after the Protocol

The 1987 Protocol exerted immediate pressure on industries producing and using CFCs worldwide, which grew more intense over the following two years as new scientific results and assessments led to a growing chorus of calls to eliminate CFCs. Firms responded with a flood of innovation to reduce use of CFCs and other ozone-depleting chemicals, which was substantially faster and more diverse than even the strongest advocates of controls had anticipated. At the time of the Protocol, CFC markets reflected trends that had persisted since the late 1970s. Driven by non-aerosol uses, world market growth of CFCs 11 and 12 continued after resuming in 1983, and passed 1 million tonnes in 1988.[1] Despite the resumption of growth, these remained low-margin commodity markets throughout the industrialized countries. The mix of uses differed substantially between the United States and the rest of the world. Aerosols were a smaller share of U.S. CFC use even before their drop in the 1970s, and by 1986 represented only a few percent of use in the United States and other countries that had banned aerosols, but 30 percent in the rest of the world and 50 percent in Europe. All other uses were proportionately larger in the United States, with an additional large increment for automobile air-conditioning.[2] Halon markets faced similar problems of soft demand and low margins in the 1980s, with prices falling 30 percent between 1980 and 1987. The only strong CFC businesses were CFC-113 and sales to developing countries, where markets grew rapidly through the 1970s and 1980s. While some CFC 11 and 12 were produced in developing countries, most of this growth was served by European producers, who dominated world exports.[3] Worldwide, about one-quarter of CFC use was in refrigeration (of which 40 percent was automotive air-conditioning), one-quarter in foams, one-quarter in aerosols, and one-quarter in solvents and miscellaneous uses. Contrary to popular opinion, home refrigerators were a tiny use, only about 1 percent.[4]

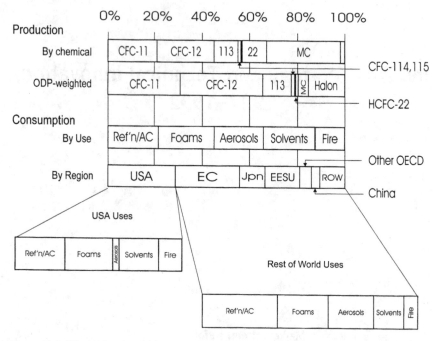

Figure 7.1 World Production and Consumption of Ozone-Depleting Chemicals. In 1986, CFCs 11 and 12 still represented more than two-thirds of world production of ozone-depleting chemicals (with chemicals weighted by ozone-depletion potential), although the contributions of CFC-113, methyl chloroform, and halons were growing rapidly. Aerosols were only a few percent of consumption in the United States and a few other countries, but more than 25 percent in the rest of the world. Source: author's calculations, based on data from AFEAS 1995; Midgley 1989; EPA 1988. UNEP 1999b; UNEP 2002.

The Protocol began to provoke large-scale reorganization of CFC industries and their political representation almost immediately. In the United States, the two smallest producers (Kaiser and Racon) announced their intention to leave the market.[5] At the same time, the Alliance began to assume a more prominent role as the voice of U.S. industry in international discussions, gradually eclipsing the CMA Panel, and the Alliance's executive director increasingly acted as spokesman for the organization, rather than the DuPont official who served as science adviser.

In Europe, the most immediate effect was a rapid drop in aerosol use, spurred by the consumer boycott initiated in August 1987, which reduced aerosols to 19 percent of world CFC use by 1988.[6] For nonaerosol sectors, however, there was no similar basis for confidence that CFC use could be easily and rapidly reduced. While the Protocol has been widely praised as precautionary because it was adopted under great uncertainty about the severity and immediacy of the ozone-depletion risk, it was even more striking for the extreme technological uncertainty under which it was adopted regarding the cost and feasibility of cutting nonaerosol CFC uses. In September 1987, no one knew what the technical challenges and costs would be of cutting CFCs by half in the nations that had already eliminated aerosols.[7]

The stakes were high, because many of the goods and services provided using CFCs, most notably refrigeration and electronics, were essential. Although CFCs were a small market, only about $2 billion worldwide, this quantity substantially understated their contribution to the world economy by an amount that was probably large, highly uncertain, and intensely contested. As industry argued with some justification, CFCs were intermediate goods that were incorporated into other products of substantially higher value that depended on them. Although the value share of CFCs in many final products was small, ranging from less than 0.1 percent in refrigeration and electronics to about 20 percent in insulating foams, the precise technical properties of CFCs were in many cases essential to the performance of the final products. Moreover, a great deal of capital equipment was optimized to, or dependent on, the precise properties of a particular chemical. In 1988, $28 billion of goods and services were produced with CFCs in the United States, using $135 billion worth of capital equipment ($385 billion worldwide).[8] CFCs were supplied competitively, with substantial overcapacity in the industry, so their producers could not capture more of the chemicals' contribution through raising prices. But potential limits in technical substitutability meant that the cost of cutting CFCs could substantially exceed their price, and in the extreme could approach the premature loss of all this capital. At this time, very little was known about the actual ease or difficulty of replacing CFCs. Some uses appeared to have very limited short-run substitution possibilities, but the experience of the 1970s aerosol cuts suggested that substitution difficulty might be overestimated in advance, while the extreme diversity of uses suggested that at least some should be easy to replace.

7.2 The Pursuit of Chemical Substitutes

The major CFC producers revived their programs to develop chemical alternatives to CFCs in 1986 and early 1987. Three promising candidates were already in commercial production (HCFC-22 at large scale, HCFC-142b and HFC-152a at small scale), although none of these was an ideal substitute for either CFC 11 or 12.[9] The substitutes whose thermodynamic properties made them the most promising— HCFCs 123 and 141b for CFC 11, HFC-134a for CFC 12—all required pilot-scale production for toxicity and process testing, and development of a commercial synthesis route, which were estimated to take about five years in total.[10] Early estimates suggested that the cost of HCFC-123 could be $1.50 to $2.00 per pound, and that of HFC-134a around $3.00, compared with about 65 cents per pound for CFC-12.[11] All proposed substitutes also needed varying degrees of development to make them work in existing applications, although an independent expert group convened by the EPA in late 1986 had concluded that this could be readily achieved under regulatory conditions that made the higher-priced substitutes viable.[12] The signature of the Protocol and the widely perceived risk of tighter restrictions to follow gave additional impetus to efforts to develop alternatives.

An important early step in developing alternatives was the formation of cooperative bodies to conduct toxicity and environmental testing of the most promising candidates. Since most proposed substitutes had not previously been marketed in commercial volumes, they required slow and costly testing for toxicity and environmental acceptability. Despite the intense competition under way to commercialize substitutes, it was easy to cooperate on these programs because the knowledge gained could not be withheld or used for commercial advantage. Thirteen CFC producers from the United States, Europe, and Japan announced the formation of a cooperative program for toxicity testing, the Program for Alternative Fluorocarbon Toxicity Testing (PAFT), in January 1988.[13] PAFT received authorization from antitrust officials in the United States, Europe, and Japan. In the United States, such protection was provided by the National Cooperative Research Act, which allowed competitors to collaborate on research serving the public interest. Testing under PAFT began in early 1988 with a two-year inhalation study of HCFC-123 and HFC-134a.[14] Two additional studies, each supported by a slightly different group of firms, tested HCFC-141b (a substitute for CFC-11 in foams) and HCFC-124 and HFC-125 (substitutes for CFCs 114 and 115 in refrigeration and sterilization).[15] In December 1988, 12 producers formed a similar cooperative group to test alternatives for environmental effects, the Alternative Fluorocarbon Environmental Acceptability Study (AFEAS). In contrast to toxicity testing, for which procedures were well established, environmental screening of new chemicals was (and is) a less well-specified task, so AFEAS had to develop criteria for environmental acceptability in a rapidly developing regulatory environment. After a small initial study, it was decided to assess each chemical by its contribution to ozone depletion and global warming, and the fate of its tropospheric breakdown products. AFEAS studied 12 chemicals—the same five as PAFT, plus three others that were already marketed and whose toxicology was known (HCFCs 22 and 142b, and HFC-152a) and four that were initially regarded as second-rank candidates (HCFCs 225ca and 225cb; HFCs 32 and 143a). AFEAS also took over the job of reporting CFC production when CMA's Fluorocarbon Panel ceased operation in 1990, and gradually expanded the scope of reporting to include HCFCs.[16]

While these new HCFCs and HFCs saw the most intensive development effort, there was initially also substantial interest in HCFC-22 as a CFC substitute. Its situation was unique, since it was a poor match for the thermodynamic properties of CFCs 11 and 12, but was already manufactured cheaply at large volume and its capacity could be expanded rapidly by converting existing plants. Although producers highlighted their efforts to develop new chemicals, several also initially undertook large expansions of their HCFC-22 capacity and promoted it as a near-term substitute, especially for aerosols and foam food packaging.[17]

The largest prize in this immediate competition appeared to be the European aerosol market. Through 1987, European producers rapidly expanded their HCFC-22 capacity, while DuPont conducted a major European sales drive for both HCFC-22 and nonfluorocarbon propellants.[18] In February 1988, however, ICI shocked its competitors by announcing it would not market HCFC-22 for toiletries, the bulk of aerosol markets, because of lingering concerns about its toxicity. Other producers were highly critical of ICI's decision, including DuPont—which had refrained from promoting 22 as a substitute for U.S. aerosols 10 years earlier out of similar concerns but had since decided it was safe enough. Still, all producers felt compelled to follow ICI's decision, eliminating European aerosols as a potential HCFC-22 market.[19]

This market became rapidly less attractive in any case, as Friends of the Earth had increasing success with its boycott of aerosol products. As had happened 10 years earlier in the United States, aerosol-product marketers could not hold a united front against this pressure in the face of their highly unequal dependence on CFC propellants. Two firms broke ranks in January 1988, and began labeling their non-CFC products "ozone friendly."[20] The next month, eight firms (representing 65 percent of the U.K. aerosols market) announced they would eliminate CFCs by the end of 1989. After an unsuccessful counterattack, the entire British aerosol industry followed in May 1988, agreeing to label non-CFC products immediately and cut CFC use 90 percent by the end of 1989. The aerosol industry in the rest of Europe followed one year later, in an April 1989 agreement with the EC Commission to cut CFC use 90 percent (a 40 percent reduction in total European CFC use) by the end of 1990. After 10 years of claims to the contrary, it was suddenly clear that CFC aerosol propellants were as easy to eliminate, and as politically vulnerable, in Europe as they had been in the United States.[21]

Efforts to promote HCFC-22 as a near-term substitute in foam food packaging had more success. In the United States, HCFC-22 received accelerated FDA approval in early 1988, first for fast foods and later for grocery display packaging, and was rapidly adopted in both markets.[22] Driven by this and other new markets, world production of HCFC-22 increased 50 percent between 1986 and 1992, then stabilized and began to decline. The decline reflected both specific limitations of 22 and a broader shift toward regarding all HCFCs as transitional substances, to be replaced in turn by nonozone-depleting options as they become available.

The more important substitutes were the small set of one- and two-carbon HCFCs and HFCs that had been identified in the 1970s. Although there are a few dozen such chemicals in total, a simple screening process based on each chemical's fractions of hydrogen, chlorine, and fluorine can identify those likely to meet required standards for nontoxicity, noninflammability, and short atmospheric life.[23] The few chemicals that pass this screen are roughly the same set that was deemed serious enough to be studied by AFEAS and PAFT.

In addition to being acceptable on environmental, health, and safety grounds, desirable substitutes must have thermodynamic properties that make them suitable to replace CFCs in specific applications, and an economical synthesis route. The major CFC producers raced to develop these few obvious candidates with crash programs that many participants described as the most intense competition of their careers.[24]

For CFC-11, different alternatives appeared most promising for its two major nonaerosol uses, blowing insulating foams and as the refrigerant in the large chillers of commercial air-conditioning systems. Despite its moderate inflammability, the most promising alternative for foams initially appeared to be HCFC-141b, and several firms quickly began to build or convert plants to produce it.[25] DuPont announced a patent in October 1988 for a new process to coproduce HCFCs 141b and 142b in a single step, and announced plans to convert a plant to the new process.[26] But long-term growth prospects for HCFC-141b were sharply curtailed in late 1988 when new model results suggested that its ODP was not 3 percent, as believed, but 10 to 12 percent, more than double that of other HCFCs.[27] As these reports were verified, flexible foam makers abandoned their plans to switch to HCFC-141b and producers began scaling back production plans.[28] Although the EPA had initially encouraged switching to HCFC-141b, it reversed its stance in 1993, banning its use as a solvent immediately and requiring that its production be phased out by 2003, nearly 30 years before most HCFCs.[29]

For refrigeration uses of CFC-11, the most promising alternative appeared to be HCFC-123. DuPont patented the first commercial synthesis route in mid-1988, and announced a large plant using the new process in June 1989, planned for completion in late 1990 in Canada. HCFC-123 production grew rapidly following its

Table 7.1 The Major Ozone-Depleting Chemicals, Their Uses, and Alternatives

Name and Formula	ODP (Steady-state)	World Production (Kte, unweighted)	Major Uses	Major Alternatives (partial list)
CFC-11 ($CFCl_3$)	1.0	407 (1986) 54 (1996)	Foams ~ 60% Aerosols ~ 30% (< 5% in US) Ref/AC ~ 8%	HCFCs 22, 123, and 141b; methylene chloride; hydrocarbons; HFCs 245fa and 365mfc; product reformulation
CFC-12 (CF_2Cl_2)	1.0	463 (1986) 119 (1996)	Auto AC ~ 20% (>40% in US) Other Ref/AC ~ 25% Aerosols ~ 35% (<5% in US) Foams ~ 13%	HFC-134a, HFC-152a, hydrocarbons, ammonia, CO_2, aerosol product reformulation
CFC-113 ($C_2F_3Cl_3$)	0.8	197 (1986) 6 (1996)	Solvents ~ 100%	Aqueous and semi-aqueous solvents, trichloroethylene, perchloroethylene, no-clean processes, HCFC-225
CFC-114 ($C_2F_4Cl_2$)	1.0	19 (1986) 0.7 (1996)	Foams ~ 77% Ref/AC ~ 23%	HCFCs 22, 123, 141b; hydrocarbons; HFC blends; ammonia; CO_2
CFC-115 (C_2F_5Cl)	0.6	12 (1986) 2 (1996)	Ref/AC ~ 100%	HCFC-22, HFC blends, ammonia, hydrocarbons, CO_2, water
Halon-1211 (CF_2BrCl)	3.0	15.5 (1986) 17.8 (1997)	Fire fighting ~ 100% (portable extinguishers)	Operational changes to reduce emissions, management of existing stock for critical uses, water, dry chemical, CO_2, inert gases

Compound	ODP	Production/Consumption (kt)	Application	Alternatives
Halon-1301 (CF$_3$Br)	10.0	11.2 (1986) 1.6 (1997)	Fire fighting ~ 100% (total-flooding systems)	
Methyl Chloroform (C$_2$H$_3$Cl$_3$)	0.1	707 (1989) 15 (1996)	Solvents ~ 100%	Aqueous and semi-aqueous solvents, trichloroethylene, perchloroethylene, dichloromethane, hydrocarbons, isopropyl alcohol
Carbon Tetrachloride (CCl$_4$)	1.1	~ 26 (1992) < ~ 5 (1996) (ex >95% feedstock use)	Solvents ~ 100%	
Methyl Bromide (CH$_3$Br)	0.6 (0.4 proposed)	46 (1986) 69 (1996) (ex <10% feedstock use)	Fumigation: soils ~ 75% Structures, durables ~ 16% Perishable goods ~ 9%	Heat, cold, solarization, chlorpicrin, 1,3-D, metam sodium, dazomet, phosphine, sulfuryl fluoride, integrated pest management
HCFC-22 (CHF$_2$Cl)	0.055	268 (1989) 470 (1996)	Ref/AC ~ 85%	HFC blends
HCFC-141b (CH$_3$CFCl$_2$)	0.11	0 (1986) 121 (1996)	Foams ~ 88% Solvents ~ 9%	HFC-152a, HFC-134a, CO$_2$, hydrocarbons, HFCs 245fa and 365mfc
HCFC-142b (CH$_3$CF$_2$Cl)-	0.065	7 (1986) 38 (1996)	Foams ~ 92%	HFC-152a, HFC-134a, CO$_2$, hydrocarbons, HFCs 245fa and 365mfc

introduction in January 1991, despite some continuing toxicity concerns and a 1989 suggestion that its atmospheric breakdown products could be environmentally harmful.[30]

To replace CFC-12 the clear leading contender was HFC-134a, a zero-ODP chemical that matched CFC-12's performance characteristics closely.[31] HFC-134a was the largest prize in the race for new chemicals because it was expected to serve the large automobile air-conditioning market, and several firms pursued it via multiple synthesis routes. ICI made an aggressive commitment to HFC-134a in early 1988 and opened their first commercial plant in October 1990, two months before DuPont.[32] Through aggressive early development and marketing, as well as engineering work to help automakers solve problems of incompatibility with seals and lubricants, ICI claimed a substantial fraction of this market in the United States as well as in Europe.[33] By 1991, DuPont and ICI each planned to have 134a plants operating in Europe, North America, and Asia by 1995. Allied, Atochem, Hoechst, Montefluous, Showa Denko, and Daikin all had plants under construction, while several other firms had announced building plans.[34] Both DuPont and ICI announced important catalyst breakthroughs in 1992, which roughly doubled their capacity.[35] Although rapid demand growth was projected, reaching 100 to 200 million pounds per year by the late 1990s, this building boom created an initial surplus in the face of softer than expected demand for the first few years.[36]

Despite intense development efforts, substitute chemicals faced several delays in coming to market at large scale, and often were not available in large enough quantities to support the transition of major uses until long after their availability was announced. As late as mid-1991, the only alternatives commercially available were the three that were marketed before the Protocol, HCFCs 22 and 142b, and HFC-152a. Several others were very close to commercial availability (e.g., HFC-134a) or were awaiting results of toxicity and environmental testing (e.g., HFCs 32 and 125, HCFCs 123 and 141b).[37] By late 1991, it was also becoming widely accepted that HCFCs would be used only as transitional substances. Attention increasingly shifted to HFCs, particularly to chemical blends that could match CFCs' properties better than any single alternative. HFC-32 emerged as an especially important component in refrigerant blends, some approaching drop-in performance, although its inflammability and high compressor discharge temperature made it problematic to use alone. Both DuPont and ICI opened HFC-32 plants in the summer of 1992,[38] and by 1993, DuPont, Allied, ICI, and Atochem were all marketing various patented refrigerant blends.[39] After a few years of offering distinct, patented, but similarly performing blends that competed for the same applications, producers began cross-licensing each other's blends that had come to dominate particular applications by the mid-1990s. As production of chemical substitutes increased, manufacturers began consolidating and closing their CFC capacity in 1990.[40]

The race to develop and market these substitute chemicals posed large technical and commercial risks. Nearly all producers committed to ceasing production of CFCs before they had alternatives in hand, and committed to commercial-scale production of promising alternatives before receiving final toxicity results or completing all process and application research.[41] Although the industry cooperatives for testing toxicity and environmental impacts pooled some of these risks, the uncertainties associated with developing cost-effective, large-scale synthesis processes for new chemicals, and demonstrating their suitability in specific applications, represented large remaining risks. The producers that lacked the resources to pursue all promising chemicals simultaneously had to make large gambles on the success of particular ones. This competition favored the largest producers and those with the strongest research capability, as the pattern of smaller producers being sold or closed within a few years demonstrated. The three smallest U.S. producers—including Pennwalt, which tried to compete in new refrigerant and foam markets—were all sold by 1989.[42] Atochem's acquisition of Pennwalt and Racon, as well as ICI's construction of new U.S. plants, brought European producers a substantial fraction of U.S. production capacity for the first time, reaching about 25 percent by 1993.[43]

But even the largest producers faced serious risks in this transition. These were not limited to the technical risks inherent in learning how to produce the new chemicals profitably and make them work in existing applications, but included diverse market, regulatory, and political risks. For most applications a few different alternative chemicals were proposed, which competed on performance, ease of substitution, cost, and health and environmental impacts. The eventual cross-licensing of refrigerant blends to eliminate proliferation of similar products reduced one such competitive risk, but others were more subtle and systematic. For example, in some applications near-term HCFC substitutes (which were relatively cheap and easy to substitute in existing equipment) competed with HFCs or other longer-term substitutes. Where expensive capital equipment such as refrigeration systems had to be designed around a chemical's properties, this competition among chemicals was mirrored in competition among equipment designs.[44] Producers also faced the risk that chemicals they expended large efforts to develop would later be judged environmentally unacceptable or too toxic for the proposed application, as happened to HCFCs 22 and 141b.

The most acute risks were regulatory. Just as creating viable markets for higher-cost alternatives depended on the incentives created by restriction of CFCs, so details of the regulatory treatment of CFCs and substitutes would greatly affect the profitability of all substitute markets, and could confer large advantages on one proposed class of substitutes or another. Weak or inadequately enforced CFC controls would gravely harm the viability of all substitutes, while too rapid a clampdown on CFCs could prevent an orderly transition, drive users into inferior alternatives that were immediately available, or impose large costs of premature capital retirement.[45] Regulatory treatment of the new alternatives also posed serious risks, particularly in the United States, where the Significant New Alternatives Program (SNAP) under the 1990 Clean Air Act amendments gave the EPA authority to rule proposed CFC alternatives acceptable or unacceptable for specific applications, based on a broad set of environmental and safety criteria. Industry was harshly critical of this provision, especially EPA's refusal to promise that an alternative, once judged acceptable, would not subsequently be ruled unacceptable, arguing that the resultant uncertainty deterred the investments that were needed to eliminate CFCs.[46]

As a class, HCFCs suffered from acute regulatory uncertainty for several years, in the United States and internationally. Although HCFCs were the first alternatives available and destroyed much less ozone than CFCs, they still destroyed a little—about 2 to 10 percent as much as CFCs. Their intermediate level of harm made the right treatment of them unclear: Should uses be switched to HCFCs as fast as possible, or wait for entirely nonozone-depleting options? Both producers and users worried that regulators would push them rapidly into HCFCs, then cut HCFCs as soon as CFCs were gone, not allowing long enough product lives for HCFCs or associated equipment to provide an adequate return on investment.[47] And in fact, calls to restrict HCFCs began to circulate in both the U.S. Congress and Protocol negotiations in early 1990. These proposals sought to restrict either the quantity or the duration of global HCFC use, some calling for product lifetimes as short as 10 years. By 1993 the regulatory treatment of HCFCs had been clarified as "transitional" chemicals, to be allowed product lifetimes sufficient to support both producer and user investments, but eliminated thereafter.

Despite all these risks, the new markets promised large rewards to the producers who picked the right chemicals, developed them successfully, built capacity in time to capture market share, and avoided early regulatory restrictions. These producers would enjoy strong positions in new markets with stronger barriers to entry and higher prices and profitability than for CFCs. Although the physical volume of the new markets would be smaller, prices of HCFCs were projected to be two to three times higher than pre-Protocol CFCs, and patented HFC blends as much as 10 times higher.

Managing these risks, in particular the interactions between the expected success of new chemicals, their likely environmental impacts, and regulations, became essential elements of corporate strategy. Producers naturally sought to develop chemicals that were likely to escape strict regulation, lobbied to secure favorable regulation of those they decided to pursue, and attempted to exploit regulatory loopholes.[48] All producers, for example, resisted 1990 calls for early HCFC phaseouts, but DuPont—which had the strongest commitment to the chemicals—took more forceful measures. DuPont suspended construction of four HCFC plants worldwide in June 1990, and only resumed work two months later, when satisfied that Protocol negotiators and U.S. regulators would not cut HCFC lifetimes too short.[49] After spending three years defending long HCFC product lifetimes, ICI reversed course in 1991, citing regulatory uncertainty in the EC, and announced— as did Hoechst the same year—that they would concentrate on HFCs and not attempt to commercialize any new HCFCs.[50]

Producers also sought to influence evolving regulations to cultivate their environmental images and to shape markets to their advantage. For example, once ICI had committed to a predominantly HFC strategy, it joined environmentalists and European regulators in advocating strict controls on HCFCs. Equipment manufacturers with little or no exposure to HCFCs also joined in advocating their early phaseout.[51] Similarly, as DuPont reduced its CFC production, it twice advanced its CFC phaseout date—to year-end 1996 in an October 1991 announcement, then to 1994 in a 1993 announcement—thereby keeping its own schedule ahead of that enacted in successive Protocol amendments.[52] These repeated advances angered equipment manufacturers and other users, who claimed that DuPont was polishing its environmental image and yielding to unwarranted pressure from environmentalists, while they were bearing the costs in premature equipment conversion.[53]

In fact, the costs and risks borne by many users in the transition to new chemicals were significantly greater than those borne by CFC producers. For producers, the anticipated higher prices and greater concentration of alternative markets represented a benefit that might offset the risks of the transition, but for users these factors only made matters worse. Moreover, no viable alternatives were apparent for some sectors: the only plausible fluorocarbon alternatives for solvent uses of CFC-113 all had serious toxicity problems or were judged too ozone-depleting.[54] Users also bore much of the risk of market disruptions, and of uncertain availability, cost, and performance of alternatives. These risks were dramatized in mid-1992, when producers' announcements of rapid CFC production cuts before most alternatives were in commercial production provoked widespread hoarding, price spikes, shortages, and plant shutdowns.[55] Although shortages eased later in 1992, users still faced the unattractive choice between remaining dependent on high-cost CFCs of uncertain availability, or moving rapidly to immature alternatives of uncertain performance, availability, cost, and regulatory lifetime. The next section discusses users' responses to these risks, which came to represent the largest risk to producers as well: the large-scale loss of CFC markets to nonfluorocarbon alternatives.[56]

7.3 Mobilization of Chlorofluorocarbon User Industries: The Engine of Innovation

Prior to the 1987 signing of the Protocol, industries using CFCs had played little role in national or international policy debates. Although the Alliance represented U.S. producers and users and gained political influence from this breadth, its agenda was primarily shaped by a few large firms, the CFC producers and the manufacturers of refrigeration equipment. Between 1980 and 1987, discussion of CFC alternatives concentrated almost exclusively on chemical alternatives, with the strong presumption that desirable alternatives should resemble the CFCs as closely as possible in all relevant respects, and would come from current suppliers. The extremely low price elasticities estimated for CFCs as late as 1986 reflected the widespread assumption that users had few options to reduce their consumption.[57] Although regulators began suggesting in 1986 that these assumptions and the weak participation of users in policy debates posed the risk of neglecting other alternatives, the only user group to participate in pre-Protocol debates was the American Electronics Association. This group opposed the Protocol, arguing that there was no alternative for the CFC-113 solvents on which its members depended.[58]

In late 1986, however, a few major users began seeking ways to reduce their consumption and emissions of CFCs through in-house research programs and task forces.[59] The Alliance provided early support to efforts, beginning with a November 1986 conference at which 500 producers and users began to investigate in detail where CFCs were used, where they were emitted, and how emissions could be reduced.[60] Although the focus was on reducing emissions, these inquiries naturally revealed many ways to reduce use as well.

Within a year, as the desperate situation of many CFC users became clear, these preliminary investigations grew into a flood of development work by both current users and third parties seeking to market new solutions. The restrictions already agreed upon in the Protocol and the risk of further restrictions on CFCs and alternative chemicals gave users an interest in reducing their dependence on all these chemicals as quickly as possible.[61] Further motivation to avoid fluorocarbons arose from the uncertainty and cost associated with the alternatives being developed by the CFC producers, and the perception that the CFC suppliers, in endorsing the movement to new chemicals that would harm users but might benefit them, had betrayed their customers. While some sectors were already reducing CFC use for unrelated reasons, these factors combined to spur an intense search for CFC alternatives, particularly nonfluorocarbon alternatives.[62] By early 1989 most users were not just trying to meet the Protocol's 50 percent target, but to reduce their dependence on CFCs as rapidly as possible, while many were also trying to eliminate HCFCs.[63] An early landmark in the rush to substitutes was the first CFC and Halons Alternatives Conference, jointly sponsored by the EPA, Environment Canada, and the Conservation Foundation in January 1988, at which several critical break-throughs were announced. Repeated annually with UNEP and the Alliance joining as cosponsors, the Alternatives Conference became the focal point for exchange of technical information about both fluorocarbon and nonfluorocarbon alternatives for each specific use.

These efforts made CFC elimination much faster and easier than projected, and made markets for HCFCs and HFCs substantially smaller than projected, although the pace and character of developments differed substantially among uses. Many important alternatives had been known or used at small scale for decades, while others required significant innovations. In some uses attractive non-CFC alternatives appeared so fast that CFCs disappeared in a few years, while eliminating CFCs in others proved much more difficult than initially expected.

Aerosols, the largest remaining global CFC use in 1987, were eliminated the most easily and predictably. Despite a decade of claims to the contrary, the hydrocarbon propellants and alternative packaging formats that a few countries had adopted in the 1970s were immediately available in the rest of the world. Where hydrocarbon propellants were restricted because of inflammability or smog formation, dimethyl ether and HFC-152a were also immediately available, although at substantially higher cost.[64] Metered-dose medical inhalants, a specialized class of aerosols, appeared much harder to substitute, but represented only a few percent of total aerosol use. The national aerosol bans of the 1970s had exempted these as essential uses, a precedent that the Protocol followed until substitutes were developed and tested in the late 1990s. As had been widely predicted, eliminating CFCs in aerosols—except for the last few percent—was easy, fast, and cheap.

Plastic foams used 267 Kte of CFCs in 1986, about one-quarter of world use.[65] Foams were a diverse group of uses, comprising three large types (flexible foams, rigid polystyrene, and rigid polyurethane) and several smaller types, which used different blowing agents and differed markedly in their ease of reducing CFCs. Some eliminated all fluorocarbons within three years, while others experienced persistent difficulties for more than 10 years. Flexible foams, used for cushioning in automobiles, mattresses, and furniture, were blown principally with carbon dioxide but used either CFC-11 or methylene chloride as a secondary blowing agent to make the foam soft. For these foams, two approaches to reducing CFC use were evident by 1989: process changes to replace CFC-11 with methylene chloride, and eliminating the secondary blowing agent entirely and accepting a slightly firmer foam. Union Carbide announced a new non-CFC process in September 1988, while British foam manufacturers agreed in May 1989 to switch entirely to methylene chloride by 1993. European manufacturers reached a similar agreement with the EU Commission in August 1990. Although flexible foams were initially expected to be a large market for HCFCs 123 and 141b, the U.S. and European markets both essentially disappeared by 1994 as all users switched to nonfluorocarbon blowing agents.[66]

Rigid polystyrene, used in several forms to package food and other products, was mainly blown with CFC-12, but some products also used hydrocarbons. Several forms of polystyrene switched quickly to HCFC blowing agents.[67] One form, rigid polystyrene sheet, initially faced a more uncertain conversion because it needed regulatory approval for use in fast-food packaging and food-store display trays. Although it was only 3 percent of CFC use, this product's high visibility made it a target for consumer pressure as early as 1987.[68] After McDonald's, the largest user, responded to early pressure by announcing in August 1987 that it would require all its packaging material to be CFC-free within 18 months, several packaging manufacturers worked rapidly to change their products. One firm had already been testing new blowing agents, and within two weeks announced a new process that could be implemented rapidly with little change of equipment, using a blend of HCFC-22 and pentane. At the urging of EPA officials, the FDA gave expedited review to the new process and approved it for fast-food packaging—although not for food-store packaging—in October 1987. In January 1988, DuPont began marketing a highly purified HCFC-22 for food packaging, and the largest U.S. maker of disposable food packaging announced it would switch all its operations to the new process within 60 days.[69] EPA officials then convened a negotiation among packaging manufacturers and environmental groups in February, to promote rapid industrywide adoption of the new processes.

After the FDA's March approval for food-store packaging removed the largest remaining obstacle, the food-service packaging industry (FSPI) announced an industrywide voluntary commitment to end CFC use by the end of 1988. Because the new processes raised costs significantly, no firm could make the move alone but the whole industry could readily do it together. Three major environmental groups endorsed the plan, on the condition that the industry would switch to fully nonozone-depleting blowing agents when available.[70] The new technologies were shared domestically through the FSPI, and internationally through the Foams Technical Options Committee of the Protocol's Technology Assessment Panel, although their approval in food packaging was delayed in several countries due to lingering concerns over potential carcinogenicity. The promised second switch was completed within a few years, when manufacturers converted to hydrocarbons, HFC-152a, and carbon dioxide.[71] The polystyrene foam sector was anomalous in that it easily changed technology twice within a few years, since the cost of each change was so low.

The largest foam sector, and the largest single use of CFC-11 in countries that had banned aerosols, was rigid polyurethane foam, used as insulation in buildings, appliances, and refrigerated transport. Eliminating CFCs here was most difficult, because CFCs represented a large share of total product cost and because alternative blowing agents were substantially less effective insulators. Consequently, switching to non-CFC blowing agents would either raise the foam's cost or degrade its insulating value so much that the industry risked losing market to other insulating materials. Through 1988 there were serious concerns that this industry might not survive the wait for HCFCs to be commercialized; their difficulties were so acute that they were the only usage sector granted a partial exemption from the tax imposed on CFCs in the United States in 1990. By late 1989, however, it began to appear feasible to reduce CFC use in two stages: first by modifying blowing processes, then by switching to HCFCs, principally HCFC-141b, as they became available in the mid-1990s. When HCFC-141b became available faster than projected, it largely replaced CFCs by 1993, although the industry struggled for several years to maintain insulation effectiveness by modifying blowing technology to obtain smaller cells. The challenges of this sector persisted, as it later also faced the greatest difficulties in moving away from HCFCs.[72]

The refrigeration and air-conditioning sector, like the foam sector, comprised several distinct uses with diverse needs. This sector faced several unique difficulties in reducing CFCs: the amount and variety of long-lived, CFC-dependent capital equipment in service; the need for refrigerants to be compatible with machinery, lubricants, seals, hoses, and gaskets; and the need to meet regulated energy-efficiency standards. Long equipment lifetimes posed particularly hard choices, between making an immediate long-term commitment to HCFCs that might not be available for servicing through the equipment's life, and staying with CFCs—which were highly likely to become unavailable—while awaiting nonozone depleting solutions.[73]

The main categories of refrigeration uses were automotive air-conditioning, household refrigeration, commercial refrigeration, and the chillers used to air-condition large buildings. The clear early leader to replace CFC-12 in automobile air-conditioning was HFC-134a, although the need for equipment redesign and new lubricants and seals made it uncertain how soon a large-scale switch could be accomplished.[74] The major automakers regarded the conversion problems as serious enough that they pursued four other backup options before committing to HFC-134a in 1991.[75] They produced the first cars with HFC-134a systems in 1992 and converted all models by 1994. In contrast to the early view that HFC-134a systems would carry a 5–15 percent energy penalty, optimization of the new systems that took advantage for the first time of the thermodynamic properties of the lubricants gave efficiencies better than those of conventional systems by 1992.[76] A diverse set of nonfluorocarbon systems was proposed that used hydrocarbons, helium, carbon dioxide, metal hydrides, and other refrigerants, but none of these received serious consideration. The reasons advanced for dismissing these alternatives were all legitimate concerns—for instance, inflammability, energy efficiency, retooling needs, and high electrical loads—but provided no clear basis for rejecting them out of hand in comparison with HFCs.[77]

The problem of eliminating CFCs in auto air-conditioning was not limited to new cars, but included servicing the stock of 140 million vehicles on the road with CFC-12 systems. Most automotive refrigerant was used not to charge new systems but to refill existing systems after leaks, service, or accidents, and it appeared that the supply of recycled CFC-12 could meet less than half this need. Since HFC-134a could not be used in older systems without hundreds of dollars of system refitting, many "drop-in" replacements were quickly marketed to fill this gap.[78] These included a few HCFC and HFC blends that matched the performance of CFC-12 very closely, and many products of high inflammability or suspect performance.[79] Although the EPA eventually named only two acceptable alternatives for automobile air-conditioning—HFC-134a and one blend—delay in issuing the rule allowed many potentially dangerous and damaging substitutes to be marketed, provoking enough concern that the Society of Automotive Engineers petitioned the EPA to restrict the sale of alternatives. A few years of technical refinements eased the problem of converting existing systems to HFC-134a, with retrofit costs falling from about $1,000 to $100. Trends in coolant prices also helped ease the shift, as CFC-12 rose from $10 to $50 per kilogram between 1992 and 1996 while HFC-134a dropped from $25 to less than $10.[80] Still, concern about shortage of CFC-12 to service the existing automobile fleet was strong enough that in December 1993 the EPA asked DuPont to continue producing CFC-12 through the end of 1995, although the company had planned to end production one year earlier.[81]

Household refrigerators were a small CFC use but, like foam food packaging, a highly visible one. Conventional home refrigerators used CFC-12 as a refrigerant and were insulated with polyurethane foam panels blown with CFC-11. The major U.S. manufacturers introduced new HFC-134a models in 1993 and 1994.[82] Although the change of refrigerant and the loss of CFC-blown insulation initially posed large challenges to meeting energy efficiency targets required in 1993, design changes allowed the targets to be met and exceeded.[83] In fact, a competition sponsored by the EPA and 24 electric utilities stimulated development of competitive new CFC-free models using 25–50 percent less energy than even the new standards required. The competition's $30 million prize went to Whirlpool for a model using HFC-134a refrigerant and HCFC-141b in the insulating foam.[84] European household refrigeration went a completely different route, led by the efforts of the environmental group Greenpeace. It was widely known that light hydrocarbons such as propane and butane were feasible alternatives to CFC refrigerants, but had two disadvantages: pure hydrocarbons would require equipment redesign in most applications, and they were inflammable. Hydrocarbons were not seriously considered in the United States mainly due to concern about fire risk, although this may have been exaggerated in view of the small quantities used. But a Greenpeace campaigner in Germany learned of an independent inventor's hydrocarbon design, and in 1992 funded a failing East German appliance manufacturer, DKK-Scharfenstein, to produce a household refrigerator based on the design. The refrigerator used a mixture of propane and isobutane as a refrigerant, and foam insulation blown with cyclopentane (in thicker-than-normal walls, to compensate for the lower thermal insulation value of this foam). Greenpeace and Scharfenstein introduced the refrigerator, which they called the Greenfreeze, at a July 1992 press conference and solicited direct orders from consumers. Although the initial model was a small refrigerator with no freezer, its promoters received 65,000 orders within four weeks. This immediate success averted the planned closure of the firm and led to its purchase by an international group of investors in November 1992.[85] In March 1993, Greenfreeze won the German Blue Angel award for environmentally superior products.[86]

Although the major appliance manufacturers had declined Greenpeace's initial proposal and subsequently attacked the Greenfreeze as unsafe, consumer and public pressure soon forced them all to introduce their own hydrocarbon models.[87] Bosch-Siemens and Liebherr introduced hydrocarbon models in 1993 to compete with Greenfreeze. By the end of 1993, several firms had made design improvements that allowed larger hydrocarbon refrigerators with freezers, which were more energy-efficient than CFC-based models. After Greenpeace attacked them for releasing only a new HFC-134a model in February 1993, Electrolux introduced a hydrocarbon model, and announced in August 1994 that it would convert its entire European line to hydrocarbons by the end of 1995.[88] The other major manufacturers on the European continent, although not in the United Kingdom, followed with similar targets to switch most or all of their lines.[89] Rapid penetration of household refrigerator markets by hydrocarbons followed in much of the industrialized world, and in major developing-country markets.[90] By the late 1990s, more than 12 million hydrocarbon refrigerators had been sold worldwide, and more than 100 hydrocarbon models were offered. The only market not using them was North America, where larger refrigerators posed harder technical problems and U.S. litigiousness made manufacturers more cautious about even minimal fire risk.

Commercial refrigeration systems in food stores and warehouses used three refrigerants, CFC-12, HCFC-22, and R-502 (a blend of CFC-114 and HCFC-22), depending on the required operating temperature. The sector first expanded use of HCFC-22, but a replacement for R-502 in the coldest applications remained unavailable until 1993—a situation that led the Bush administration in its last days to reconsider its earlier request for rapid voluntary CFC production cuts.[91] As in many other sectors, however, substitution eventually went faster than expected, following the introduction of several near drop-in blends in 1993. At first there were only HCFC blends for many applications, but HFC blends were commercialized for most applications by 1995 and HCFC use in the sector began to decline thereafter.[92]

Nonfluorocarbon refrigerants, such as hydrocarbons and ammonia, are also feasible in commercial systems, although the large charges of these systems and the need for charging in the field make fire and toxicity more serious concerns than for small household systems. Environmental groups promoted these options vigorously, but industry interest only began to grow only in the mid-1990s with the introduction of systems using a tightly confined ammonia loop and a secondary loop of brine to carry cooling long distances.[93] In 1994 and 1995, prompted in part by Greenpeace attacks for using HCFCs and HFCs, two U.K. supermarket chains announced plans in 1995 for stores using ammonia or hydrocarbon systems for all refrigeration.[94] Secondary-loop systems using ammonia also saw revived interest in shipping by 1994.[95]

The final major refrigeration use was CFC-11 in chillers, which chill water to provide cooling in commercial buildings (home air-conditioners run on HCFC-22). After initially investigating multiple alternatives, some chiller manufacturers switched their lines to HCFC-123, others to HFC-134a. Both choices posed moderate technical problems—HCFC-123 was not compatible with standard seals and gaskets, while HFC-134a would not mix with standard mineral oils—but chiller manufacturers solved these, and by 1992 were offering designs that could operate with either conventional or new refrigerants.[96] Some of these new designs represented major advances, offering improved performance, increased energy efficiency, and unprecedented low leakage levels.[97] However, broader building design changes that would have integrated lighting improvements to reduce cooling loads, or more ambitious schemes to integrate heating, cooling and cogeneration applications with absorption-cycle cooling cycles, did not make significant early inroads.[98]

In sum, nearly all refrigeration and air-conditioning applications went to HCFCs and HFCs, including several blends. After an early proliferation of untested options caused confusion and delays in some sectors, the industry association issued codes, a numbering system, and standards for new refrigerants in mid-1992.[99] Mobile air-conditioning and home refrigerators in the U.S. switched to HFC-134a; commercial refrigeration went to HCFC-22 and several HCFC and HFC blends; and chillers went to HCFCs 22 and 123, and HFC-134a, depending on their operating pressure. Nonfluorocarbon alternatives did not receive serious consideration—except in household refrigerators, where Greenpeace's aggressive promotion of a hydrocarbon design eventually transformed the market everywhere except in North America. Nonfluorocarbon options began to gain renewed attention in the mid-1990s, however, as potential alternatives for some uses that appeared to be the most dependent on HCFCs.

The solvent sector provides the most striking instances of users innovating around fluorocarbons, turning what was initially viewed as the hardest sector from which to eliminate CFCs into the easiest. CFC-113 is an effective solvent, but is mild enough to use on plastic and rubber parts or for cleaning fine leather and suede, and sufficiently nontoxic to use for hand cleaning of precision metal parts. Methyl chloroform was a good substitute for many applications, and was widely used for metal degreasing, but also contributed to ozone depletion (although only about 10 percent as much as CFCs). Replacement of other solvents such as perchloroethylene and trichloroethylene, which the EPA listed as hazardous air pollutants in 1986, had further increased growth of both CFC-113 and methyl chloroform.[100]

Early attempts to find nonozone-depleting alternatives to fluorocarbon solvents had limited success. HCFC-132b initially appeared promising, but was too toxic. DuPont announced a few new solvents in 1988 that reduced CFC-113 use by blending it with other chemicals, but the reductions were small (ranging from a few percent to 37 percent for various applications).[101] In March 1989 both DuPont and Allied introduced several new blends based on HCFCs 141b and 123, but none of these adequately matched the performance of CFC-113 for all applications. Most were stronger solvents, and so could not replace CFC-113 in important electronics applications.[102] Through 1987 and 1988, electronics and chemical firms continued to state that the prospects for replacing CFC-113 solvents were poor, and that the EPA had overestimated them in enacting regulations.[103]

The first major change in this picture was announced at the January 1988 Alternatives Conference. AT&T had been working to reduce use of all chlorinated solvents since the late 1970s, for reasons of environment, cost, and workplace health, and had replaced them with aqueous cleaning of chips and boards at several plants by 1984. Aqueous cleaning was known to be suitable for nearly all nonsurface-mount applications, but was not suitable for the more demanding surface-mount applications. At the 1988 Conference, AT&T and a small firm, Petroferm, jointly announced a new, environmentally benign solvent, based on terpenes, that cleaned surface-mount applications better than CFC-113 for about the same price.[104] The two firms had jointly developed processes for large-scale electronics cleaning after an AT&T engineer saw the solvent advertised in a petrochemical-industry supply magazine.[105] AT&T announced that the new solvent would let them cut their use of CFC-113 by one-third, or about 450 tonnes per year.[106] Petroferm also developed terpene solvents for a wide range of other plastic- and metal-cleaning applications.[107]

The Solvents Technical Options Committee (TOC) under the Protocol was a major engine of innovation in reducing CFC solvent uses. Stephen Andersen, who was responsible for industry cooperative programs in the EPA's Stratospheric Protection Branch and was later named chair of the Solvents TOC, began approaching major solvent-using firms at the January 1988 Alternatives Conference. He contacted both AT&T and Nortel, competing manufacturers of telecommunications equipment, to request their participation in the Solvents TOC and challenge them to reduce CFCs in their own operations. Both firms were already concerned about tighter CFC restrictions and were trying to reduce their use rapidly. Nortel committed publicly in July 1988 to cutting their worldwide CFC use in half by year-end 1991, and later revealed that their actual goal was to eliminate them completely by that time. AT&T announced in 1989 that it would eliminate CFC emissions by the end of 1994.[108]

The Solvents TOC brought together major solvent-using firms for its work during the first half of 1989. With top technical experts on specific solvent uses collaborating, the TOC found that in addition to identifying existing reduction opportunities, they were able to solve technical problems that had obstructed implementation of other reduction opportunities and even identify new opportunities. In one striking example, a group of TOC experts on a March 1989 site visit observed an innovative German machine that soldered inside a controlled-atmosphere chamber. Although the machine's operation was problem-ridden, the group recognized that the concept of soldering in an inert atmosphere could control oxidation and reduce the need for fluxes, thereby making it unnecessary to clean components after soldering and allowing huge reductions in solvent use. Participating experts from AT&T, Nortel, and Ford immediately arranged to have their firms purchase the machines, and began collaborating to perfect the concept.[109]

From their experience on the Solvents TOC, officials from AT&T and Nortel became convinced that pooling effort and sharing technical information could greatly advance efforts to reduce CFC use. Working with the EPA's Andersen, they identified counterparts from other firms to invite to an organizational meeting in October 1989, immediately before the second Alternatives Conference, where they discussed forming an industry cooperative to continue such cooperation. Fourteen firms attended and nine agreed—despite suspicion about the EPA's intentions—to establish the cooperative, named the Industry Cooperative for Ozone Layer Protection (ICOLP).[110] The EPA had already supported the formation of similar research consortia for halon reduction, for lubricant and equipment compatibility of automotive air-conditioning alternatives, and for toxicity and environmental testing of new fluorocarbons, and so was able to guide the formation of ICOLP under the National Cooperative Research Act.[111]

By 1992, ICOLP had expanded to 15 corporate members and had developed affiliate memberships for government agencies and associations, including the U.S. Air Force—whose support was crucial when the group realized that more than half of world CFC solvents were used because U.S. military specifications required them.[112] ICOLP's approach was based on R&D collaboration and free sharing of knowledge and technology with both members and outsiders. In addition, member firms produced technical manuals and an on-line technology database to publicize their experience, and ran technology-transfer workshops and provided consulting services to help eliminate ozone-depleting solvents in developing countries. Members reported gaining several benefits from participating: consultation with top experts from multiple firms to help solve their hardest reduction problems; access to specific technologies, including some from their strongest competitors; pooling the risk of being unable to eliminate certain uses; and favorable publicity and access to government policy makers.[113]

Viable solvent options proliferated rapidly, spurred by the research initiatives of large, sophisticated firms like Nortel and AT&T, the review of promising technologies conducted by the Solvents TOC, and the collaborative research and free exchange of results promoted by ICOLP. By 1990, many diverse options (including aqueous, semi-aqueous, hydrocarbon, and no-clean) were available to replace virtually all CFC-113 use. Although methyl chloroform was initially considered the best alternative to CFC-113, many of the new nonfluorocarbon alternatives were also effective replacements for it.[114] The corporate leaders who had pledged to eliminate CFCs from their operations all met their goals between 1991 and 1993, spurred by a May 1993 EPA deadline for warning labels on products made with ozone-depleting substances. By late 1993, more than 60 major firms based in five countries had ended CFC use in their worldwide operations.[115]

This rapid innovation by solvent users took CFC producers by surprise. The one fluorocarbon that appeared promising as a CFC-113 substitute, HCFC-225, faced a long development time during which user innovation rapidly eroded its market.[116] ICI, one of only two firms that marketed both CFC solvents and solvent cleaning equipment, recognized the inevitable in 1990 and began undercutting its own fluorocarbon business by selling terpene cleaning equipment.[117] The solvents sector, which in 1987 was thought the hardest place to eliminate CFCs, turned out to be among the easiest, and the fastest to move away from fluorocarbons altogether.[118]

The final usage sector restricted by the Protocol was the halons, which the 1987 Protocol froze at present levels beginning in 1992. Halons were highly effective fire extinguishers, but were substantially more ozone-depleting than the CFCs.[119] Although ozone depletion from bromine-containing chemicals had been periodically discussed since 1975, halons were added to the Protocol's agenda only in the final months. Learning of their involvement in ozone depletion very late, the firms that made halons but not CFCs first tried, without success, to raise the same scientific questions about bromine—such as possible tropospheric sinks, uncertain stratospheric chemistry, and the need for years of measurements to verify atmospheric lifetimes—as had been discussed for chlorine for 15 years.[120]

Halon users took a different approach. The fire-protection industry association (the National Fire Protection Association, or NFPA) learned that halons were ozone depleters only in 1986, when the EPA contacted them as they were about to revise the fire code to require full-discharge testing of all halon systems. Gary Taylor, chair of the NFPA Standards Committee, pressed successfully to stop the proposed change, and the NFPA joined the EPA, the U.S. Air Force, the fire-prevention equipment manufacturers, and three trade associations in establishing a group to identify and reduce unnecessary halon emissions. The EPA and the U.S. Air Force established a parallel government interagency group. These new networks, both established before the Protocol, provided technical support for the effort to include halons in the Protocol.[121]

These groups initially sought to limit halon emissions as much as possible to fighting fires. They quickly learned that although only 20 percent of annual halon production was released (as opposed to 85 percent of annual CFC production), less than 5 percent of releases went to fight fires. The great majority of emissions came from discharge-testing of building fire systems. Even where fire codes did not require discharge testing, insurance companies often did, both to test system performance and to ensure that rooms were sealed tightly enough to hold the required halon concentrations. By early 1988 the two groups began developing and promoting alternative testing methods that did not require releasing halons, such as discharge-testing with alternative chemicals or using blower-doors to create a pressure differential.

The U.S. military was both a large halon user, consuming 35 percent of the U.S. market, and a strong leader in reducing it.[122] In early 1989, the U.S. Air Force issued a new policy that limited halons to critical uses (i.e., in combat or other situations where evacuation was impossible) where no alternative provided adequate fire protection. The U.S. Marines, Navy, and Air Force then collaborated to develop the first halon recycling equipment, which was subsequently commercialized by private firms. The U.S. Army, which needed halons to protect armored-vehicle crews in combat fires, adopted a strategy that combined replacing halons in noncombat uses and halon recycling, for which they led a government-industry effort to develop a purity standard. The U.S. military also led international reductions, sponsoring two NATO conferences to share knowledge with other armed forces and to seek their support for accelerated phaseouts.[123]

These efforts rapidly identified several ways to make large reductions in halon emissions. First, large reductions were available simply by avoiding unnecessary releases, through such measures as alternative system testing methods. Second, sharing information among users revealed that a great deal of halon was used unnecessarily due to mistaken beliefs that other extinguishing agents would damage valuable equipment. For example, a December 1989 workshop revealed that both IBM and NASA used water and CO_2 systems to protect their computers without damage, while many other users believed that computers could be protected only with halons.[124] Third, it was recognized that the large halon "bank," 200–350 Kte sitting in tanks in existing systems, could be jointly managed to supply critical systems for many years without producing any more.[125] Finally, several chemical halon substitutes were proposed beginning in 1990, although none fully matched halons' performance. Great Lakes Chemical announced a substitute in May 1990 that could serve most applications, but was too toxic for total flooding of occupied rooms and had an ODP of 0.5.[126] In June 1990, DuPont proposed HCFC-123 and HFC-125 as somewhat costlier and less effective replacements for halons 1211 and 1301, respectively, pending toxicity testing and EPA approval under the SNAP program.[127]

It was because halons were believed extremely difficult to reduce that they were only frozen, not reduced, in the 1987 Protocol. These innovations in reducing unnecessary discharge and better managing of existing stocks reduced the need for halons so rapidly, however, that they were the first group of ozone-depleting chemicals to be eliminated when production ceased at the end of 1993. Halons' unique usage patterns, the large existing stock, and the small fraction of consumption actually used where it mattered, allowed production to be eliminated long before chemical alternatives were fully commercialized. In that sense, the phaseout of halons was an even more extreme success story than that of solvents.

Although this user-driven innovation occurred worldwide, it was most vigorous in North America—as was the movement away from CFCs in general. In Japan, the major initiatives were developed by industry associations with government guidance. In coordinated 1989 announcements, the automakers stated that they would switch to HFC-134a, the cosmetics firms that they would switch to hydrocarbon propellants, and the electronics firms that they would phase out CFC-113, although the means to do so were not yet clear.[128] European nations remained well ahead of their Protocol deadlines, principally due to rapid aerosol cuts, but European firms remained far behind their North American counterparts in the more challenging sectors, with Germany a partial exception.[129] While the major U.S. electronics firms had aggressive CFC-elimination programs in place by 1989, surveys of U.K. electronics firms in 1990 and 1991 found that few had near-term reduction goals and 20 percent planned to keep using CFCs as long as possible.[130] Speaking at the 1992 Alternatives Conference, the German Environment Minister criticized German appliance firms for resisting his calls for a 1993 phaseout and announced a series of government-industry consultations to determine how fast a phaseout was technically feasible.[131] In December 1992, the U.K. Department of Trade and Industry was so concerned that firms were not adequately preparing for the impending EC cuts—an 85 percent CFC cut in January 1994 and a phaseout one year later—that it launched an industry awareness program to alert them.[132] European industry lagged so far behind in 1993 and 1994 that an ICI spokesman criticized European refrigeration manufacturers for their lack of progress.[133] The U.K. dry-cleaning sector made so little advance preparation to abandon CFC-113 that firms found themselves in a last-minute panic on the eve of the phaseout in 1994.[134]

7.4 Regime Formation and Industry Strategy

In addition to being a time of rapid regime formation, the period from late 1986 to early 1988 marked a sharp divide in industry's response to the ozone issue. After 10 years of unchanged policy positions and no significant progress in identifying ways to reduce use of ozone-depleting chemicals—despite sustained environmental concern—this period began a burst of development and innovation from producers and users of ozone-depleting chemicals that made their early elimination appear plausible by 1989, and confidently foreseen by 1991. This shift to rapid progress was not spurred by any particular technological breakthrough. There was no technical reason that the burst of innovation that began in 1986–1988 could not have happened several years earlier. Indeed, many of the alternatives that allowed rapid reductions were partly known 10 or more years earlier, but needed sustained attention to refine them and solve associated problems before they could be viable. That this did not happen earlier represented a serious lost opportunity. That it happened at this time is an indication of the power of regulation and institutions to influence the rate and character of technological change.

Before negotiation of the Protocol, there was no market or regulatory pressure to promote the search for ways to reduce CFC use. The major CFC producers had investigated chemical alternatives in the 1970s as a defense against the risk of regulation, but abandoned the efforts when the immediate risk declined and they concluded that the new chemicals could not compete with CFCs in price. Their strategy was defensive, seeking to learn enough to identify alternatives in case of regulation, but to avoid revealing—or even learning—so much about alternatives as to make regulation more likely. Users of CFCs, regarding stringent restrictions as unlikely and trusting the producers to develop alternatives if cuts were imposed, had no interest in pursuing CFC reductions unless the same measures served more immediate competitive purposes, such as improving product quality or reducing costs. In aggregate, regulators never seriously considered controls, in part because they believed they would be too costly, while the firms best able to develop alternatives made no effort to do so because they believed regulation was unlikely. Their lack of effort in turn ensured that controls continued to appear too costly—a perception to which they were happy to contribute.

The linked conditions sustaining this no-progress equilibrium were shattered by the Protocol's control measures—indeed, by the prior widespread shift in late 1986 to the belief that controls were likely. The first shift in industry's policy position in the fall of 1986 was pushed by rhetorical factors. The justifications on which industry had founded its opposition to controls were undermined by revelations about the 1970s alternatives research programs, resumed CFC market growth, and the 1986 assessment's conclusion that continued growth would likely bring large ozone loss. These factors compelled a change in industry groups' positions if they were to maintain their standing as responsible participants in the policy debate. But once this change was made, it contributed—along with other actors' responses to the same information—to a transformation of the strategic landscape for CFC producers and users. For the first time, it appeared likely that serious restrictions on CFCs might be enacted.

This shift fundamentally reshaped the interests of firms dependent on CFCs. The producers immediately revived their alternatives research programs, to defend against the risk of CFC restrictions and to pursue opportunities in alternatives markets once CFCs were restricted. The shift in users' interests was even stronger, although differing among sectors. Users faced acute risks from the CFC cuts already agreed upon, and from the prospect of more stringent and broader controls. All users faced threats of disruption to their operations, degradation of product quality, and increased costs: for some, the survival of their business was threatened. With serious disruption unavoidable, users had an interest in reducing their dependence on CFCs as rapidly as possible, and the opportunity and incentive to reexamine their products and processes comprehensively to achieve this reduction. It was by no means clear that the best way to achieve this reduction was to adopt the chemical alternatives being promoted by the CFC producers. The unavoidable disruption of normal operations made it feasible to search more broadly for alternatives, and users had little reason for continued confidence that the CFC producers would protect their interests. The chemical alternatives were clearly going to cost more than CFCs, their performance and availability were uncertain, and they also faced risks of future restrictions. Consequently, many users tried to reduce their dependence not just on CFCs, but on all ozone-depleting chemicals. The interests of producers and users, which had been aligned as long as the status quo could be reliably defended, now diverged—because the risks to users associated with alternative chemicals compelled them to look out for themselves, and because the prospect of consolidation in alternatives markets introduced directly opposed interests between producers and users on the issue of price.

The rapidity of innovations to reduce CFCs, the central role of user-driven innovation, and the widespread development of nonfluorocarbon alternatives were all facilitated by domestic and international policy choices. The organization of the Protocol's Technology Panel encouraged innovation in all usage sectors, and its exclusion of CFC manufacturers differentially directed this advantage toward nonfluorocarbon alternatives. United States regulators were instrumental in alerting user firms to the coming restrictions, challenging them to take leading roles in reducing their usage, and facilitating the formation of collaborative organizations for firms to work together on shared technical problems. These bodies allowed joint technical work among competitors without the risk of antitrust charges, and also allowed groups of competitors to take decisions in parallel that carried cost penalties, not imposing a competitive disadvantage on any one for moving first. Crucially, officials also worked with industry to identify and remove barriers to the development of alternatives that arose from government regulations and practices.

The results of these efforts differed strongly among usage sectors, in the ease and rapidity of the initial movement from CFCs, the share moving to other halocarbons—and later, in the ease of those that initially went to halocarbons in moving beyond them. In aggregate, reducing CFC use was much faster and easier than had been projected. By late 1990, it was evident that nearly every sector had numerous technical opportunities available to reduce use, and that new options were appearing monthly. By 1992, it was clear that CFC use could be reduced virtually to zero in all sectors without serious disruption.[135] The solutions that enabled this were diverse, including conservation measures, new fluorocarbons, nonfluorocarbon chemicals, and process changes that made the services formerly performed by CFCs unnecessary. Alternatives of all these kinds were available earlier and performed better than projected, but the sectors that appeared most difficult shifted over time: first solvents and halons, then insulating foams, and finally certain refrigeration equipment.

The degree of cooperation among firms also differed strongly among usage sectors. Solvents and halons, the sectors that first appeared most difficult, had two important advantages: firms had substantial capability to solve technical reduction problems if they cooperated; and because these are standards-dominated industries, firms were willing to cooperate. Because firms needed to meet well-defined performance standards but did not compete to outperform the standard, it was relatively unlikely that a firm could gain an important competitive advantage by finding an alternative to CFCs and not sharing it. In renouncing CFC alternatives as a dimension of competition, firms gave up only modest commercial opportunities tangential to their main lines of business. Similar circumstances facilitated cooperation in food-packaging foams. The industry was standards-dominated because of the need for FDA approval, and feasible alternatives were quickly and obviously available, at a cost penalty that was not so large as to make the entire industry uncompetitive with other products. Consequently, firms could readily agree to make the move together, with minimal competitive effects. In contrast, there was little cooperation among firms in the refrigeration and insulating foams sectors. In both these sectors, alternative chemicals posed large, important uncertainties about product performance, which could confer decisive competitive advantage on a firm that made a significant breakthrough. Moreover, for insulating foams, not even the whole industry could move together to adopt an alternative with a cost or performance penalty, because the penalties appeared large enough to risk loss of the entire market to other insulating materials.

The intensity of users' rush from CFCs and the flood of nonfluorocarbon alternatives developed by users and third parties greatly exceeded producers' expectations. The largest chemical producers expected to profit from the switch to chemical alternatives that they would control. In DuPont's successive estimates of the shares

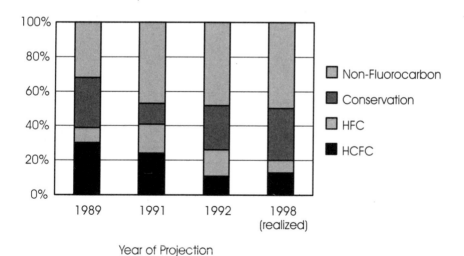

Year of Projection

Figure 7.2 Projected and Realized Replacements for CFC Uses. It was initially projected that about 40 percent of the CFCs used in industrialized countries would be replaced by HCFCs or HFCs. As the phaseout was implemented, however, attractive new nonfluorocarbon alternatives to CFCs were repeatedly identified. Ultimately, HCFCs and HFCs replaced only about 20 percent of former CFC use. Source: UNEP 1999b.

of former CFC markets going to various alternatives, the share predicted to go to other fluorocarbons stood at 39 percent in mid-1989, rose slightly to 41 percent in 1991, then plummeted to 26 percent in 1993. In the actual replacement of CFC uses realized by 1998, only 20 percent went to HCFCs and HFCs.[136]

Producers responded to these changes in various ways. While some exited the business, others shifted to smaller volumes of higher-margin chemicals or broadened their business to sell related products and services in addition to chemicals, such as equipment, technical advice, and recovery and recycling facilities.[137] Those that remained in the fluorocarbons business needed several elements in their post-Protocol strategy. They of course tried to pick the most profitable alternatives to develop, to produce them cheaply, to market them effectively, and to defend a reputation for good corporate citizenship. But in addition, once they had invested in producing particular chemicals, they had to defend these as long as possible against both regulatory restrictions and competition from other alternatives. In part this required attempting to elicit regulatory commitments before making investments. Since large-scale technology choices were being made both in markets and in regulatory arenas, it also sometimes required attacking the safety or environmental acceptability of other proposed alternatives, as well as their performance—especially those that appeared the most threatening.[138] Producers who chose not to invest in particular alternatives had clear interests in joining environmentalists or governments in highlighting their environmental harms and calling for their stringent regulation, particularly when marketing competing products.

The political strategy of industry associations such as the Alliance also had to change. The Alliance's strategy of uniting producers and users around shared interests in the original chemicals became infeasible as soon as the chemicals came under serious threat. This approach remained valid only for a narrow set of issues on which the interests of producers and users remained sufficiently aligned, such as defending adequate product lifetimes for fluorocarbon alternatives and, more delicately, defending continued availability of CFCs for long enough to let all usage sectors make an orderly transition to new technologies. Both these stances implied resisting the most aggressive environmentalists and regulators, whose interest in reducing ozone depletion as rapidly as possible made them willing to risk premature abandonment of early investments in interim technologies as less ozone-depleting ones become available. But having retreated from their long-standing opposition to any restrictions, the Alliance and other industry bodies faced the subtler challenge of continuing to defend their members' interests while maintaining their standing as responsible participants in continuing regulatory debate.

8

Building an Adaptive Regime

The Protocol Evolving, 1989–1999

The most important innovation of the 1987 Protocol was the set of linked measures and institutions for the repeated reassessment and revision of controls on ozone-depleting chemicals. Going far beyond conventional provisions for treaty amendment, the Protocol established a complex set of requirements, deadlines, and supporting procedures and institutions to review control measures periodically and consider modifications in view of scientific, technological, and economic developments. Established as responses to the acute uncertainties, both scientific and technological, under which the 1987 Protocol was negotiated, these provisions served the interests of both the advocates and the opponents of strict controls, by making it possible—at least in principle—to respond to new information by either tightening or loosening control measures.

Since 1987, four complete cycles of assessment have been conducted and the Protocol has been revised five times. Each revision has tightened the control measures, through some combination of larger reductions, earlier phaseouts, and controlling additional chemicals. In addition, the Protocol's institutions and the process of assessment and revision itself have been tuned and elaborated over this time, with new bodies established and the jobs of existing bodies modified to meet evolving needs.

8.1 Negotiating the First Protocol Revision, 1989–1990

After more than a year of informal discussions, ozone negotiations resumed in August 1989 in the Open-ended Working Group (OEWG), aiming to finalize Protocol modifications at the parties' second meeting, planned for London in June, 1990.[1] The working group was "open-ended" in the sense that nonparties—which still included such important nations as China, India, and Brazil—participated on essentially the same terms as parties. In contrast to pre-1987 negotiations, there was also consistently strong participation from both industry and environmental groups. Two major areas of negotiation were pursued in parallel working groups: modifying the Protocol's control measures, and establishing mechanisms for financial and technical aid for developing countries. These negotiations were more complex and, particularly on finance and technology, more contentious than those of 1987.

Moving to Full Phaseouts

The assessment panel reports strongly set the agenda of the resumed negotiations. The science panel had reported that every important result since 1987 had increased concern that CFCs, halons, and other halogenated chemicals were harming the ozone layer,[2] while the technology panel had concluded that use of these chemicals could be reduced at least 95 percent by 2000. Even before negotiations resumed, these results had prompted many governments and industry actors to endorse phasing out the original five CFCs by 2000, while several governments sought even earlier phaseouts. Strengthening controls on other dimensions, such as extending the phaseout to the 10 other fully halogenated CFCs not initially controlled, also had unanimous support. Since these were not commercially produced, this simply forestalled switching to them from the controlled CFCs. There were significant disagreements, however, on other aspects of controls. Even the broad agreement on a CFC phaseout concealed important differences over the phaseout date, interim reductions, and keeping a small supply available to service installed equipment or meet essential uses. Stronger differences were present over the treatment of halons and of other ozone-depleting chemicals not initially controlled: methyl chloroform (MC), carbon tetrachloride (CT), and HCFCs.[3]

The August meeting heard reports from each panel and reviewed a draft synthesis report of the four panels' findings.[4] The science panel cochairs summarized recent findings that available theory and models could not explain observed ozone losses, either at the poles or in midlatitudes. While the 1987 Antarctic expedition had persuasively demonstrated the central role of chlorine in causing the ozone hole, its role in midlatitude depletion remained uncertain. Of course, no known natural processes could account for observed midlatitude losses either, so anthropogenic chlorine was widely assumed to be responsible, even if this had not been demonstrated. But since models could not replicate the observed depletion, neither could they project the effects of alternative control measures on future ozone with any credibility.

In view of these clear weaknesses of models, the science panel cochairs also presented new calculations using chlorine-loading projections, as parties had requested in 1989 when they saw the EPA's preliminary analyses using the approach. The new analysis refined and improved the earlier EPA work in several ways, but reaffirmed its main claim that stratospheric loading of chlorine and bromine was "the key ozone-loss indicator."[5] Several control scenarios were examined. The weakest scenario, the controls in the original Protocol, gave 8 ppb chlorine by 2050 and 10 ppb by 2100. Others phased out CFCs, then added freezes and phaseouts of MC and CT, and finally a limit on HCFCs. Only the final scenario reversed the growth of stratospheric chlorine, returning it to 2 ppb around 2075.[6]

The most obvious conclusion of the new analysis was that only a stringent policy phasing out all major ozone-depleting chemicals and limiting HCFCs would return stratospheric chlorine to its pre-ozone-hole level—a result that strongly steered the negotiating agenda toward such stringent controls.[7] The analysis also vividly illustrated the effect of CFCs' long atmospheric lifetimes, in that small delays in controlling them caused substantially longer delays in atmospheric recovery. Finally, the scenarios showed how control decisions could vary two distinct dimensions of policy choice, through the stringency of controls on short- and long-lived gases. Even with the strictest controls, stratospheric chlorine was bound to increase for another decade or two before peaking and then declining. Since emissions of any gas are nearly gone from the atmosphere after about three atmospheric lifetimes, long-lived gases like CFCs and halons would determine the time of atmospheric recovery. Stopping these emissions earlier was consequently the most effective way to advance the return of stratospheric chlorine to low, pre-hole levels. In contrast, the most effective way to reduce the near-term peak of chlorine concentration would be stricter controls on short-lived gases, which would have the fastest atmospheric effect. After presenting these analyses in August and revising them in response to parties' suggestions, the authors returned to the November OEWG meeting, where they analyzed more proposed control scenarios in real time.[8] Although these analyses were done by a few scientists independently of the science panel's work, the report was included with the parties' permission as an annex to the synthesis report. Chlorine-loading analyses had been circulating for a year, but these presentations solidified the approach as the principal means of comparing policies, and had substantial influence on subsequent control negotiations.[9]

After the August session, negotiations continued through six more OEWG meetings and a dozen-odd smaller and informal meetings through June 1990. Throughout the negotiations, parties relied extensively on UNEP Director Tolba to probe informally for promising proposals and bring suggested deals forward as his own personal suggestions. Tolba's first such package, presented to the August meeting, proposed to eliminate CFCs by 2000 with a 5 percent consumption "tail" through 2005 to service installed equipment; cut halons 50 percent by 1995, with an unspecified phaseout date; control MC and CT with an unspecified schedule; control any HCFCs with an ozone-depleting potential (ODP) greater than 10 percent; require data reporting for all HCFCs and HFCs; and extend trade restrictions to ban exports of controlled substances to nonparties and imports of products containing them from nonparties.[10] After consultations with Tolba, the Bureau of the Parties in September brought forward a slightly different proposal, which included interim targets for CFCs, CT, and MC but no halon phaseout.[11]

All major producers except Japan and the Soviet Union supported eliminating CFCs by 2000. The Soviets argued that long-range production plans already committed to might preclude a 2000 phaseout, while Japan sought to let consumption (not production) continue until 2005, a position consistent with the technology panel's conclusion that at least 95 percent—but not necessarily 100 percent—of CFC use could be eliminated by 2000.[12] Both dropped their objections by November, yielding broad agreement to end CFC production and consumption by 2000, with some delegations still seeking earlier phaseouts. As in 1987 there were proposals in the U.S. Congress to move faster, but lacking the threat of unilateral trade sanctions to back them up and the will of the U.S. delegation to exploit them for diplomatic leverage, these did not force international negotiations ahead as they had in 1987.[13] Interim cuts were a persistent conflict. Europeans, who could cut CFC use in half immediately by restricting aerosols, advocated sharp early interim cuts, as did the Nordic states and other activist nations (some of whom lacked this easy way to achieve rapid cuts); the United States, which had few such opportunities for rapid, easy reductions, as well as the centrally planned economies, wanted only a phaseout date with no interim targets. This disagreement persisted until resolved in the final compromise in June.[14]

Positions on halons moved rapidly soon after negotiations began. Several reports in early 1989 had increased concern about halons, suggesting that more production was emitted than had been thought, that atmospheric concentrations were increasing sharply, and that bromine might contribute a significant share of present ozone depletion.[15] Although both Tolba's and the Bureau's initial proposals were weak on halons, by November most delegations accepted a phaseout between 1995 and 2005—with the conspicuous exception of the Soviets, who supported only 10 to 50 percent cuts by 2000.[16] Between January and March 1990, delegations increasingly accepted the technology panel's argument that the existing bank of halons could meet priority needs for many years without new production, and several nations advanced their proposed phaseout dates—the EU from 2005 to 1999, Canada and the Nordics to 1995.[17] By March, there was near-consensus for a 50 percent cut by 1995 and a phaseout by 2000, with exemptions for essential uses. In May, the Soviets accepted that the essential-use exemption provided adequate insurance and agreed to this formulation, even while technology was advancing so fast that many expected there would in fact be no essential uses. In contrast to CFCs, parties did not agree to ban the 46 other halons that were not commercially produced.

Carbon tetrachloride (CT) and methyl chloroform (MC), two ozone-depleting chemicals not controlled in 1987, were central in these negotiations. Once widely used for dry cleaning, grain fumigation, and other solvent applications, CT had been banned for its toxicity in most industrialized countries. Although known to be as strong an ozone depleter as the CFCs, it was not considered for control in 1987 because delegates presumed—without global data—that its only use was as a feedstock to make CFCs and other chemicals, so its emissions would be minuscule. Although correct for the OECD nations, this assumption was not correct for the rest of the world. New atmospheric measurements in 1989 revealed that CT's concentration was substantially higher than expected and increasing about 1 percent per year, indicating continued large-scale use as a solvent in Eastern Europe, the Soviet Union, and developing countries. On learning of these new observations, most delegations pressed for a rapid phaseout, with the most aggressive seeking elimination by 1995.[18] Japan and the Soviet Union initially resisted, but only briefly. The bureau's September proposal to phase out CT in 2000 was accepted, although interim cuts remained controversial through May.

The situation of methyl chloroform (MC) was unlike that of any chemical previously considered in the negotiations. Because of its short atmospheric life, MC was only a moderate ozone depleter, but it was used in large volume as a solvent, particularly for metal degreasing and dry cleaning.[19] Although identified as a ozone depleter in the 1970s, MC was long given lower priority than CFCs or halons because of its low ODP.[20] It was even considered as a less ozone-depleting substitute for CFC solvents. Very limited data on its production and use was available until an industry survey published in late 1989 showed that world production had quadrupled from 1970 to 1988, with rapid growth in the late 1970s and slightly slower growth in the 1980s. Chlorine-loading projections showed two opposing implications of alternative MC controls, both due to its short atmospheric lifetime. On the one hand, continued near-term use would not delay atmospheric recovery to 2 ppb chlorine, because current emissions would be gone from the atmosphere within a few decades. On the other hand, as a short-lived gas produced in such large volumes, MC offered the largest opportunity to reduce near-term peak chlorine concentration—by as much as 0.4–0.5 ppb with a rapid phaseout.

These implications of cutting MC, as well as its continued high growth rate, made it a point of sharp controversy. Predictably, environmental groups and some activist officials highlighted the ability of sharp cuts to reduce near-term atmospheric chlorine, while MC producers stressed the irrelevance of such cuts for eventual atmospheric recovery. Environmentalists also highlighted the risk of large-scale substitution of MC for CFC-113 driving another growth surge. Producers countered that past MC growth represented one-time substitution for more polluting chlorinated solvents and production would soon level off; that even large-scale substitution of MC for CFC-113 would be environmentally beneficial, since a solvent with ODP 0.8 would be replaced by one with ODP of 0.1 to 0.15; and that solvent uses offered so many ways to cut emissions through enclosure, vapor capture, and recycling that production and consumption controls were unnecessary.[21]

The technology panel and its solvents committee had given only limited consideration to MC and CT in their initial work, but more extensive assessments of both were added to their mandates late in the process. This belated addition posed serious challenges, and the panel's work on these chemicals was less detailed and well supported than their work on CFCs and halons. For CT, the solvents committee simply noted that it had been successfully eliminated in OECD countries, and on that basis concluded, without a detailed examination of usage elsewhere, that it could be eliminated in all industrialized countries by 2000. For MC, the new industry survey provided data on aggregate use and trends, but little information was available on patterns of use in specific applications. Consequently, the panel drew on the recent experience of rapid development of alternatives for CFC-113 solvents, and on a review of the thin published literature, to conclude—largely by analogy—that substitutes were probably available for 90 to 95 percent of MC use. Industry criticized this conclusion harshly, arguing that usage patterns were in fact quite different for MC and CFC-113, and that many firms were counting on using MC to reduce CFC-113. Nevertheless, while noting that an orderly transition away from MC would require some time, the panel concluded that MC could be "frozen or substantially reduced" without obstructing a timely phaseout of CFC-113.[22]

Controversy over the technical feasibility and economic implications of controlling MC persisted through 1989. The producers—a different group than the CFC producers, not previously involved in ozone negotiations, dominated by Dow Chemical with 40 percent of world production—organized a producer-user group similar to the Alliance in late 1988, after EPA officials began calling for MC and CT controls. In 1989, this group tried to repeat the Alliance's 1980 success by mobilizing their many small users—about 73,000 U.S. firms—for an antiregulatory lobbying campaign directed at the U.S. Congress.[23] Presenting chlorine-loading analyses in congressional testimony in May 1989, science panel chair Robert Watson faced hostile questioning from Congressman John Dingell, who argued that substituting MC for CFC-113 would substantially reduce ozone depletion and that alternatives to MC had not been identified.[24] In response to their persistent criticism of the technology panel's treatment of MC, the industry association was granted the unusual opportunity to make a presentation to the November OEWG meeting, where it argued that the maximum feasible MC controls by 2000 would be a 20 to 25 percent cut in *emissions*.[25] Addressing the same meeting, a representative of the technology panel responded by strengthening their original conclusion, stating that a total MC phaseout was feasible by 2000.[26] Detailed analysis of MC uses and alternatives to support either conclusion remained unavailable, however. The MC industry tried to support its claim that growth was slowing with a survey of production, but results showed substantial continuing growth, with a 5.8 percent increase in 1989.[27]

Understandably in view of this sustained controversy, delegations' proposals on MC were far apart. In contrast to its positions on other chemicals, the Soviet Union proposed the strictest MC controls, a phaseout by 1992.[28] The most activist OECD delegations sought a 2000 phaseout with a 50 percent cut by 1994, close to the bureau's September proposal. Japan, which had planned to use MC as a major substitute for CFC-113 in electronics, proposed that MC controls only be studied. Although the EPA followed the implications of their chlorine-loading analyses and sought a rapid phaseout, the U.S. position was much weaker, reflecting interagency disagreement: a freeze, followed by cuts of 25 to 100 percent.[29] The EC endorsed a near-term freeze, but proposed waiting until 1994 to identify a reduction schedule.[30] Positions converged somewhat by November, when the United States, EC, and Japan all agreed on a near-term freeze; but while Japan and the EC (particularly France and the United Kingdom) wanted to go no further than a freeze, the Nordics and the EFTA still sought a 2000 phaseout and the United States still sought cuts of 25 to 100 percent, based on future assessments. Despite some further softening of positions through the spring, substantial differences remained through May.

The intermediate status of HCFCs, whose ODPs ranged from 2 to 15 percent, made their treatment also controversial. Environmentalists and some activist delegations advocated moving to zero-ODP alternatives as soon as technically feasible, while some went still further to argue that even HFCs, which have zero ODP, should also be treated as transitional substances because of their contribution to greenhouse warming. In 1989 and 1990, however, HCFCs were the most readily available substitute for many CFC applications and the only substitute for some. Consequently, rapid CFC cuts required switching some applications to HCFCs. But since large investments were needed to produce HCFCs and redesign equipment to use them, industry argued that they could justify these investments only if the chemicals were assured long enough lifetimes—DuPont argued for 30 to 40 years—to give adequate return on investment.[31] The most aggressive HCFC controls being discussed would very likely require premature abandonment of the investments made to produce and use them. DuPont representatives reinforced this point by noting that even in late 1989, less than 5 percent of the capacity needed to complete the switch from CFCs was committed worldwide, and continued uncertainty about the viability of these investments would only prolong the use of CFCs.[32] The Alliance argued that HCFC controls should not be considered until the 1992 or 1994 Protocol revision, supporting the position with their own chlorine-loading projections, which showed that if HCFC restrictions delayed the switch out of CFCs by even a few years, the net effect would be to delay atmospheric recovery.

National positions on HCFCs also diverged. At one extreme, European industry, most EC delegations, and Japan opposed including HCFCs in the Protocol, even for data reporting. At the other extreme, the Nordics, Australia, and New Zealand proposed phaseouts by 2010 to 2020. The Nordics also proposed restricting specific chemicals and uses, limiting the most damaging HCFCs—those with ODPs above 1 to 2 percent, atmospheric lifetimes over 10 years, or significant global warming potential—to specific "critical uses" to be identified.[33] Germany also proposed treating the most ozone-depleting HCFCs more strictly, in particular HCFC-22, by restricting them to essential uses and banning production by 2000.[34] The chemical firms forcefully opposed these proposals, and both ICI and DuPont threatened to abandon their HCFC investments if tight controls were enacted.

In November, the United States proposed a two-stage phaseout schedule for HCFCs that was fairly close to the position of U.S. industry, eliminating them in 2020–2040 for new equipment, and in 2035–2060 for servicing existing equipment.[35] By this time DuPont had accepted the eventual phaseout of HCFCs, but argued that no interim controls should be imposed and phasedown should not begin before 2030. United States industry also opposed the unilateral U.S. cuts in HCFCs in proposed Clean Air Act amendments, arguing that these would limit the delegation's flexibility. As the deadlock persisted through the spring and advocates saw little hope for binding HCFC controls, they shifted their efforts to developing a nonbinding resolution for an HCFC phaseout.

A Mechanism for Financial Assistance

The primary interest of the developing countries in these negotiations was the establishment of a financial mechanism. Although delegations in 1989 had supported the principle of financial assistance to offset developing-country costs of eliminating ozone-depleting chemicals, this broad agreement concealed sharp differences over the size and manner of the payments, and how they would be controlled. In contrast to negotiations over chemical controls, these were basically bilateral negotiations between the industrialized countries that would make the payments and the developing countries (called "Article 5 Parties" for the section of the Protocol that specified their special treatment) that would receive them. Occasional aggressive posturing by both sides in these negotiations concealed substantial uncertainty over the strength of their relative bargaining positions. The greatest uncertainty concerned what would happen if terms could not be agreed upon to bring the developing countries into the Protocol: Could they establish a viable regime outside the Protocol, producing and using banned chemicals and related products, and exchanging them among themselves; or would they soon be forced by market forces or the Protocol's trade restrictions to make the same reductions without compensation? Although negotiations were constrained by prior agreement on the principle of compensation, these starker possibilities underlay each side's calculations. A global phaseout required participation of developing countries, some of which were planning large-scale expansion. But staying outside a regime that was driving a transition to CFC alternatives carried clear risks of sanctions and of losing trading opportunities and spillover benefits from this technological shift.

In informal discussions called by Tolba before the first working group meeting, two issues were identified as crucial for developing countries: assessing their costs of joining the Protocol and designing the institutional arrangements for a financial mechanism. Early cost estimates were very weak, and obtaining reliable estimates was complicated by the great variety of situations of developing countries: some produced CFCs, some produced capital equipment or products dependent on them, and some only imported. In early discussions the Indian Environment Ministry said it would require compensation of $2 billion to eliminate CFCs, although this number was a clear overestimate.[36] The only published cost estimate, commissioned from a consultant by the Dutch government, was $400 million per year for 10 years for all developing countries, but this was also an ad hoc estimate based on little analysis.[37] After developing-country delegates attacked this estimate at the first meeting as much too low, it was proposed that industrialized countries sponsor case studies of CFC use patterns and phaseout costs in individual developing countries. The Netherlands, the United Kingdom, Canada, and the United States all sponsored such studies, which became available through the first half of 1990. Based on studies under way, the EPA in February estimated the cost as $100 million per year for three years with present parties, plus a further $100–$200 million if China, India, and other major nonsignatories were to join. By June, further studies suggested total costs of $220–$260 million over the first three years.[38] As costs became better understood and estimates declined, it became increasingly clear that industrialized-country parties could pay them without difficulty.

Discussions of how the financial mechanism would be organized and controlled were even more contentious than those on how much funding was needed. At the first meeting, a group of developing countries stated four basic requirements: funding would come via a separate trust fund under UNEP, not through existing, donor-controlled institutions like the World Bank; providing funding would be a legally enforceable obligation of the Protocol; funds would be new and additional relative to existing aid flows; and "free access and non-profit transfer" of ozone-friendly technologies would be guaranteed. In contrast, the industrialized countries generally favored a funding vehicle associated with the World Bank. Over several arduous sessions, an approximate consensus emerged that the fund would be governed by a small body accountable to the parties, with equal representation from industrialized and developing countries. This body would draw on the expertise of existing institutions including the World Bank, the United Nations Development Programme (UNDP), and UNEP for specific implementation jobs. Three issues were the most persistently difficult: whether contributions would be obligatory; whether developing countries' obligation to comply with control measures would be formally conditional on receiving adequate assistance; and the terms used to describe access to alternatives technologies, all of which would likely be privately owned in industrial countries.[39]

A more serious last-minute obstacle arose over American aversion to the basic principle of committing new funds through a new institution. Although all delegations had formally approved the "principle of additionality" in February, the U.S. administration was divided and the delegation had in fact maintained a reservation on the meaning of the principle—in effect reserving judgment on whether the United States actually supported additionality. In early May, after the White House Conference on Climate Change raised senior officials' concerns that an ozone fund might be a precedent for a much more costly global climate fund, the U.S. delegation "clarified" its position by stating that any funding for the Protocol should come through the World Bank from existing resources. The clarification drew uniformly harsh reactions, domestically and internationally. In addition to denunciations from many other delegations, a World Bank official said the Bank would participate in the fund only if the resources were additional. Domestic reaction included a call to the President from the chairman of DuPont, stating the firm's support for the fund. Prime Minister Thatcher reportedly implied to President Bush that if the new U.S. position scuttled the upcoming London meeting, she might reciprocate at the G-7 summit to be held in Houston one month later.[40]

Final Negotiations for the 1990 Protocol Revisions, London

One month before the London meeting, Tolba circulated a revised set of "personal" proposals that many described as unusually timid for him. CFCs, CT, and halons would be phased out by 2000, with interim cuts and halon essential-use exemptions to be decided in 1992. MC would be cut 50 percent by 2000, with a nonbinding resolution to phase it out by 2010. HCFCs would be added to the Protocol for reporting and would be phased out by 2040, with an additional nonbinding resolution to eliminate them "if possible" by 2020. Finally, for halons other than the three initially controlled, most of whose properties were unknown, a nonbinding resolution would call on nations to report data and use them only as transitional substances.

By June, eight producing nations had stated their support for CFC phaseouts by 2000. On June 6, the EC Council endorsed Tolba's proposal as a minimum position, but their statement was ambiguous on a faster phaseout, reflecting internal disagreement.[41] In the United States, the House and Senate had both passed versions of Clean Air Act amendments that phased out CFCs, halons, and CT (by 2000), MC (by 2000 or 2005), and HCFCs (by 2015 to 2030).[42] Among both government and industry positions, the Germans were the strongest: a May 30 Cabinet ordinance, based on negotiated agreements with chemical producers, eliminated CFCs by 1995, halons by 1996, and MC and CT by 1992, with additional restrictions on specific uses and on HCFC-22.[43] The major CFC producers also supported CFC phaseouts by 2000, and both DuPont and ICI announced plans for large new HFC plants on the eve of the London meeting.[44] The Alliance took a more cautious position, seeking to protect CFC users against the risk of delayed alternative availability for some uses by keeping a small "tail" of 5 to 10 percent of CFC production and use beyond the proposed 2000 phaseout date.

Environmental groups criticized Tolba's proposal as too timid, arguing that if Germany could eliminate CFCs faster than he proposed, all industrialized countries could do the same. They especially objected that he proposed only a 50 percent MC cut, citing the technology panel's conclusion that cuts of at least 90 percent were feasible. In a series of briefing papers using chlorine-loading calculations, they compared the consequences of Tolba's proposal with the strongest position of any national delegation on each chemical—eliminating HCFCs by 2010, MC by 2000, and CT by 1995—and also with a still stricter schedule that they advocated, which advanced CFC and halon phaseouts to 1992, and HCFCs to 2010.[45]

A crucial U.S. statement on June 15 broke the deadlock on finance. Released by White House chief of staff John Sununu, the statement "clarified"—or, rather, reversed—the May statement that funding must be through the World Bank with existing resources. In doing so, it stressed the limited and unique nature of the ozone negotiations, the strong scientific basis of the issue, the predictability of the required funding amounts, and—in particular—the nonprecedential character of this financing arrangement for any other global institution.[46] United States insistence on preambular language stressing the non-precedential character of this fund was later used to force U.S. support for an MC phaseout.

A final negotiation session of the working group was held the week before the diplomatic conference. Here, agreement was reached that CFCs would be phased out by 2000, although some delegations still sought earlier elimination and differences remained over interim reductions; that halons would be phased out by 2000, with an essential-use exemption; and that CT would be eliminated by 2000, with an 85 percent cut by 1995. On MC, agreement was reached only to freeze in 1993 and reduce 30 percent in 1995; beyond that, proposals ranged from a 40 percent cut by 2000 (from the EC) to an 85 percent cut by 2000 (from the Nordics and Australia, who had retreated from their earlier proposal for a 2000 phaseout). On HCFCs, several delegations still sought phaseout commitments by 2010 to 2040, but the EC persisted in its opposition so it was agreed to add them to the Protocol only for data reporting, and to treat their reduction in a nonbinding resolution this time.

The diplomatic conference was thus left to resolve three groups of issues: treatment of CFCs prior to 2000; treatment of MC; and the contentious issues of finance and technology for developing-country parties. Hard negotiations, mostly conducted in closed contact-group meetings, lasted to the final moments of the conference, and were described by Tolba as the most difficult and complicated of his career.[47] On CFCs, the remaining differences between the EC and United States were resolved with a compromise that kept the phaseout at 2000 and added interim cuts of 50 percent in 1995 and 85 percent in 1997.[48] Thirteen delegations also signed a nonbinding resolution committing to advance their national CFC phaseouts to 1997. As agreed in the final working-group session, all delegations endorsed a nonbinding resolution to end HCFC use by 2040, and by 2020 "if possible."[49] DuPont announced at the meeting that a 2040 phaseout might suffice for them to resume construction of their HCFC plants, which they did three months later.[50]

A decision to eliminate MC by 2005 was the surprise of the London meeting. Coming into the session, the strongest proposal on the table was the Nordics' 85 percent reduction. The EC and Japan supported only Tolba's proposed 50 percent cut, as did the U.S. delegation despite the stronger measures coming forward in Clean Air Act amendments. Benedick and other participants have described the sequence of cascading bargains by which delegates agreed on a 2005 MC phaseout. It was crucial for this outcome that the MC producers, as newcomers to the issue, lacked both the experience and the growing reputation for responsible participation that CFC industries were beginning to enjoy. Senior management of Dow, the largest MC producer, reportedly refused a last-minute request to identify a control level they could accept, so the U.S. delegation was not particularly concerned with protecting the company. Late in the meeting, the Norwegian delegation pressed the United States to support a phaseout by threatening to withhold approval of the non-precedential preambular language on which U.S. acceptance of the fund depended. When the U.S. delegation telephoned the White House for instructions, Chief of Staff John Sununu agreed to the trade. Having joined those supporting the phaseout, the U.S. delegation then brought the EC on board, with the support of Germany, by making U.S. support for the "industrial rationalization" and joint reporting provisions that were the top European priorities conditional on European support for the MC phaseout. Since the Soviets did not use MC and had supported strict controls from the start, this left Japan as the only major producer resisting a phaseout, and its delegation would not stand in isolation. The meeting agreed to eliminate MC production and use by 2005, with interim cuts of 30 percent in 1995 and 70 percent in 2000. The baseline year for newly controlled chemicals was set at 1989.[51]

The U.S. retreat had allowed negotiations on finance to proceed. In addition, preliminary results from reanalysis of satellite data, discussed in several presentations at London, appeared to raise the stakes for developing countries by showing ozone losses over virtually the entire globe, including a 3 percent decline over 10 years at equatorial latitudes.[52] Still, several difficult last-minute issues remained. At the diplomatic conference, delegates agreed to establish an interim fund, which would finance specifically identified categories of developing countries' "agreed incremental costs" of meeting their targets. As the United States insisted, the decision stated that the mechanisms adopted here were "without prejudice" (i.e., without precedent) for other issues such as climate change. The fund would be governed by a 14-member Executive Committee of the parties, seven each from industrial and developing countries, with informal regional rotation of representation within each group, and would use a voting scheme (when consensus failed) that accorded strictly equal power to each group.[53] The committee would be supported by a secretariat based in Montreal, and would work through existing "implementing agencies"—initially the World Bank, UNEP, and UNDP, each with assigned roles—to do its actual administrative work. Reflecting the relatively low phaseout costs identified by national case studies, delegates set interim funding at $160 million over the first 3 years, to be increased by $40 million each if India and China became parties. Contributions by industrialized countries would be set according to the UN scale of assessment,[54] and were to be additional to other funding—a slogan, not an measurable objective—but parties were permitted to meet up to 20 percent of their obligation through bilateral projects that advanced the Protocol's objectives.

The most contentious matters, which concerned the precise language used to describe several central provisions of the agreement, were all resolved through artful ambiguity of drafting. Conflict over whether industrialized countries' payments to the fund were voluntary or obligatory was resolved by using the word "assessed" in one place and "voluntary" (which they in fact were) in another. Conflict over how to describe the terms on which CFC alternative technology would be provided to developing countries—with the donors favoring "commercial" or no description, and the recipients, "preferential and non-commercial"—was resolved through use of the largely meaningless phrase "fair and most favorable." Although delicate language partly obscured the meaning in each case, the substantive resolution of both issues met donor-country demands, as was necessary to gain agreement on a fund at all. The final conflict concerned the relationship between financial and technical aid and the developing countries' obligations to comply with the control measures. Since the controls were binding treaty obligations but the financial contributions were not, developing countries sought to make the existence of their control obligation depend on their receiving adequate assistance. Industrialized-country delegations rejected this contingency, believing it risked letting countries declare unilaterally that they had not received enough aid and so were not bound by their control obligations. The issue was resolved in last-minute bargaining, with Indian environment minister Manika Gandhi overreaching with aggressive press statements and losing the support of China and other major developing countries. As concluded, the agreement made a statement of fact, that parties' "capacity to fulfill" their commitments—rather than the obligation to fulfill them—would depend on receiving assistance. In practical terms, parties also agreed that any developing country concerned it might be unable to comply would notify the secretariat, which would convene parties to consider the situation and discuss responses.[55]

In addition, a few points of implementation were refined. Two measures were changed at European insistence: exchange of production quotas for industrial rationalization was allowed among any parties, not just those making less than 25 Kte; and joint reporting of EC consumption (but not production) was allowed, to reflect the incompatibility between the single European market and the reporting of national-level trade necessary to calculate national consumption under the Protocol. It was clarified that parties not accepting the new amendment would be treated as nonparties for trade restrictions in the newly covered chemicals. Discrimination between industrialized and developing countries was eliminated in the rules for voting adjustments to the Protocol's controls (voting weighted by CFC consumption was replaced by the system developed for the financial mechanism, a two-thirds majority overall with simple majorities of both industrialized and developing parties) and in trade measures (exports of controlled chemicals to nonparties were banned from all parties, not just developing countries). Changes in the extra 10 percent of production allowed for export to meet developing countries' "basic domestic needs" were discussed, but the developing-country parties were divided—those producing CFCs wanted the allowance reduced, while importers wanted it increased—so it was left unchanged. With agreement on the financial mechanism, these changes allowed the major developing countries to endorse the Protocol. At the end of the meeting, the delegations of China and India announced that they would advise their governments to ratify. Many other developing countries also ratified over the following two years.

These Protocol revisions made large changes in projected atmospheric concentrations of chlorine and bromine, both through the increased stringency of controls in industrialized countries and through bringing the major developing countries into the control regime, albeit with a 10-year delay. Whereas the original Protocol would have let stratospheric chlorine rise to steady state at about 11 ppb after a century, under these revisions chlorine was projected to peak around 4.1 ppb soon after 2000, then decline to below 2.0 ppb by 2060 to 2075. But the legal mechanism by which the London amendment achieved these advances raised questions of subsequent complexity and difficulty, which were anticipated in the report of a working group on implementation.[56] The stringency or timing of controls on already controlled chemicals could be changed by "adjusting" the Protocol, so the revised controls would bind all parties without ratification. But adding new controlled chemicals required amending, rather than adjusting, the Protocol, so parties would become bound by the new measures only when they ratified the amendment. Adding new chemicals by amending the Protocol raised the possibility of a nested series of instruments with different parties to each holding different obligations—posing difficult problems regarding obligations to report data, and the status of different states as parties or nonparties, complying or not, for purposes of the trade provisions. The issue of nested membership became increasingly pressing after 1990, as controls were tightened and new chemicals were added without increasing the size of the multilateral fund, in further amendments that many developing countries refused to sign.

8.2 Second Round of Protocol Revision, 1991–1992

Calls for further tightening of controls resumed within months of the London meeting. At the national level, several governments that had advocated faster phaseouts than were agreed to in London quickly moved to enact stronger measures domestically. The EC Council discussed faster reductions in August 1990, then enacted a regulation in January 1991 advancing its CFC and CT phaseouts to mid-1997.[57] The German ordinance advancing the national phaseouts to 1993–1995 was enacted in May 1991, after several months of delay for review by the commission.[58]

At the international level, the London meeting set in motion a second round of assessment and review of the Protocol's control measures on an even faster schedule. Informal consultations with delegates to consider instructions to the assessment panels, which in the first round took place more than a year after the Montreal meeting, began at the London meeting. The panels reported in mid-1991, after which negotiations resumed, aiming to complete further Protocol revisions at the fourth meeting of the parties, in the fall of 1992.

The mandates and instructions of three panels, all except the effects panel, were revised for this round. The science panel retained its broad mandate to review progress in atmospheric science, and was explicitly asked to conduct further chlorine-loading analyses of control scenarios and to estimate the ODPs and global-warming potentials (GWPs) of CFC substitutes. In addition, parties accepted the panel's request that they be asked to examine stratospheric effects of high-flying aircraft, rockets, and the space shuttle, which had not been assessed since the 1970s. The technology panel, in addition to its continuing mandate to assess what further reductions would be feasible, was asked to report on progress toward meeting the Protocol's targets and to address a few specific questions that parties expected to be important in the next round of negotiations: What is the earliest possible date to phase out MC; how much of controlled substances are developing-country parties likely to need, and are they likely to be available; and what quantities of HCFCs will be needed in specific applications, for how long? In response to the difficulties faced by the Economics Panel in 1989, it was merged with the technology panel to form the new Technology and Economics Assessment Panel (TEAP).

Atmospheric Science: New Results and a New Assessment Report

As had occurred after the 1987 Protocol, new results after the 1990 London meeting continued to indicate a more serious threat to ozone worldwide. The most important advance in ozone measurement was a breakthrough in analyzing data from the Nimbus-7 satellite instruments, which finally allowed a reliable correction for instrument degradation. The correction exploited differences in the rate of instrument drift at different wavelengths to construct a correction that did not depend on using the ground-based data.[59] With the correction, the satellite and Dobson records for the first time provided independent sources of ozone data, which both confirmed earlier reports of midlatitude winter losses accelerating in the 1980s. The corrected satellite data also showed significant midlatitude losses in spring and summer as well as winter.

Another refinement of the satellite data released in April 1991 increased the estimated rate of midlatitude ozone loss to 4–5 percent per decade year-round, 2 percent in summer and 8 percent in winter, roughly double the rate estimated by the trends panel in 1988.[60] Calling the new report "stunning and disturbing," EPA administrator Reilly said the new observations meant that steps taken so far "may well turn out to be inadequate," as the EPA doubled its estimate of extra U.S. skin cancer deaths over to 200,000 over 50 years. On release of these new data, more than 30 U.S. senators wrote to President Bush, asking that the United States pursue an accelerated phaseout, but the administration declined to change the U.S. position.[61]

The new data held another shock in the vertical distribution of ozone. While theory still predicted that midlatitude losses would be largest in the upper stratosphere, the new observations clearly showed that the largest losses were in the lower stratosphere, around 12–25 km, with a smaller decline around 40km and a region of no loss in between, suggesting the possibility of two separate loss mechanisms.[62] Although several indications of lower-stratosphere losses had been accumulating for a few years, these had been treated very tentatively because of weaknesses in each source and differences in the magnitude of apparent losses between sources.[63] Stratospheric models, still unable to account for the observed decline in total mid-latitude ozone, were also unable to explain this preponderance of lower-stratosphere losses. A few results suggested the cause might be heterogeneous chlorine-ozone chemistry acting on the surfaces of naturally occurring sulfate aerosols—the aerosols were most abundant at the same altitudes as the observed losses, and laboratory experiments showed that such aerosols increased the rates of the relevant reactions[64]—but these suggestive results were far from a confirmation. In mid-1991, the first model calculations that included estimates of these heterogeneous processes suggested they could account for about half the observed losses, while hemispheric dilution of larger losses in polar regions might account for the other half.[65]

This preponderance of lower-stratosphere losses also brought a crucial change in understanding of linkages between ozone depletion and climate change. Because ozone is a greenhouse gas, ozone loss in the lower stratosphere would cool the surface and troposphere by nearly as much as the direct greenhouse effect of CFCs would warm them. This result sharply reduced CFCs' total contribution to warming: instead of the 10–15 percent of anthropogenic greenhouse warming estimated in 1990, their true effect could be near zero, or even negative.

Increasingly serious ozone losses also continued in polar regions. The Antarctic ozone hole continued to appear each year with increased intensity and extent, breaking a former pattern of alternating years of more and less severe losses as increasingly large, deep, and long-lived holes appeared in 1989, 1990, and 1991.[66] The dynamic processes causing the earlier oscillation appeared to be overwhelmed by the intensity of chemical depletion processes operating. In the Arctic, analyses of ozone losses in the winter of 1989–90 released in September 1990 suggested losses inside the vortex averaging 25 percent in the lower stratosphere and 6 percent in the total column, substantially larger than initially reported.[67] The next winter, Scandinavian scientists reported localized losses in excess of 40 percent, which they called "mini-holes."[68]

Against this backdrop, the science panel followed essentially the same process as for the 1986 and 1989 assessments. Participation was slightly expanded, now with 25 countries represented, including nine developing countries.[69] Groups for each topic worked from April through October 1991, and the assessment's scientific and executive summaries were released in late October. The panel reviewed recent advances in polar observations, global ozone trends, and model projections. It described recent Antarctic work as beginning to yield a "semi-quantitative understanding" of the loss process, but recent Arctic observations as yielding only slight changes in understanding.[70] Although large chemical perturbations and large localized ozone losses had been observed, and it was widely believed that a long, cold winter could bring large Arctic-wide losses, the panel reemphasized that observed Arctic losses were not comparable to those of the Antarctic in depth or extent. Regarding mid-latitude losses, the panel endorsed the hypothesis that heterogeneous processes on the surfaces of natural sulfate aerosols were responsible, stating that "although other possible mechanisms cannot be ruled out, those involving chlorine and bromine appear to be largely responsible for the ozone loss and are the only ones for which direct evidence exists."[71] The June 1991 eruption of Mount Pinatubo in the Philippines offered both the opportunity to test this hypothesis and a substantial near-term risk, because if aerosol-surface chemistry was causing the observed losses, the eruption's large injection of aerosols would sharply increase losses.[72]

As in prior assessments, a central piece of the panel's work was a quantitative projection of the atmospheric consequences of various scenarios of future emissions and controls. Although they presented ozone-loss projections, they emphasized the simpler and more widely trusted chlorine-loading projections. In a further extension of the chlorine-loading approach, the panel defined four target measures derived from chlorine time paths on which they scored emission scenarios relative to a baseline scenario defined by the London agreements: the peak concentration of stratospheric chlorine (which would reach 4.1 ppb under the London baseline); the date of decline to 3 ppb (2027); the date of recovery to 2 ppb, likely to restore the Antarctic hole (2060); and the time integral of chlorine concentrations above 3 ppb, a combined measure of the peak and the rate of recovery. In terms of these target measures, the scenario analyses showed that further advancing phaseout dates would reduce both peak concentrations and time to recovery. These analyses also clarified important differences among HCFCs, depending on their atmospheric lifetimes: tight controls on short-lived HCFCs would decrease peak chlorine slightly (by .01–.02 ppb) but would not speed recovery to 2 ppb, while cutting longer-lived HCFCs would reduce both the peak and the recovery time.

Finally, the scenarios brought to prominence a chemical not previously considered seriously for controls, methyl bromide (MeBr). Methyl bromide is a biocide, used for extermination, sterilizing goods for shipping, and sterilizing soils in preparation for planting specialty crops. Soil sterilization represented about 75 percent of total use, with nearly half of this used on just two crops, strawberries and tomatoes. The largest user of MeBr was the United States, where usage had increased sharply after the EPA banned the similar pesticide ethylene dibromide for toxicity in 1984. Although occasionally identified as a possible contributor to ozone loss as early as 1975, MeBr had never been proposed for control, principally because of acute uncertainty over the size of anthropogenic releases relative to natural sources and sinks. The 1986 NASA/WMO assessment had reviewed MeBr and suggested that about half of emissions might be anthropogenic, but concluded that data were not sufficient to establish an atmospheric concentration trend and speculated—erroneously, as it turned out—that significant growth in its use would be precluded by its high toxicity.[73]

Methyl bromide appeared to be more important in part because of a new method the Panel developed to estimate ODPs that used observed correlations of source gases with ozone loss to constrain model results, since models could not account for observed losses. The new method substantially increased the ODPs of brominated chemicals, since heterogeneous chemical processes increased bromine's relative contribution to depletion. The new results suggested that controlling MeBr was a high-payoff opportunity to reduce ozone loss: under certain conditions, each 10 percent reduction in MeBr emissions would achieve as much as a three-year advance in the CFC phaseout. As with MC in 1989, the panel's analysis forced MeBr onto the control agenda with substantial accompanying controversy. The crucial statement of the high payoff from MeBr controls appeared in a summary section on "implications for policy formulation," while only weaker and more general statements appeared in the body of the report—suggesting that the conclusion was formulated after the bulk of the panel's work was completed.[74]

UV Effects

As in previous assessments, knowledge of the effects of increased surface UV remained ambiguous, scattered, and weak, and the programs pursuing it inadequately funded. The effects panel reviewed recent work, highlighting the few significant new results by juxtaposing the summaries of each chapter from their 1989 and 1991 reports. Although observations at individual sites both in the Antarctic and in temperate latitudes showed the expected association between low ozone and high UV in short-term fluctuations, there was still no large-scale trend in surface UV evident. Known inadequacies of the standard UV meter and inappropriate siting of stations near polluted metropolitan areas made the UV measurements conducted in the United States between 1974 and 1985 unsuitable for detecting a long-term trend, however; improved instruments were installed at a few stations worldwide after 1988, but had too short a record to establish long-term trends reliably. Quantitative estimates were still available for only two effects, nonmelanoma skin cancer and cataracts. New epidemiological work suggested a reduction in the ozone-cancer multiplier to 2.6, so a 10 percent ozone depletion would generate a 26 percent increase in cancer incidence, or 300,000 additional cases per year worldwide. The same loss was projected to cause a 6 to 8 percent increase in incidence of cataracts, or 1.6–1.75 million additional cases worldwide per year. Additional ocular and immune effects had been demonstrated. In the first observation of a real ecological effect of ozone loss, the biomass and productivity of phytoplankton in Antarctic waters were found to decrease under the ozone hole, where UV radiation was also found to penetrate deeper than expected in the clear waters, up to 65 meters.[75] Prospective studies of aquatic UV-B sensitivity were still hindered, however, by interspecies differences in sensitivity, adaptive strategies, and UV-damage repair mechanisms, some of which appeared to have sharp thresholds.[76] The panel still judged ozone depletion likely to have substantial effects on tropospheric pollution and damage to materials, but neither effect was better understood since the last assessment and neither could be quantified. The panel's report attracted little attention or response when it was published in November 1991.

Technology and Economics

Although the technology and economics panels were now combined, the combined panel closely followed the successful strategy of the 1989 technology panel. Now jointly chaired by environmental officials from the United States and the United Kingdom, the panel summarized progress in phasing out ozone-depleting chemicals and assessed the potential for further reductions. Technical options committees for each major usage sector were once again staffed principally by industry experts, now including experts from the CFC manufacturers and somewhat broader participation of developing-country experts.[77] All were all jointly chaired, in most cases by the 1989 chair with a developing-country technical expert added as cochair.

As in 1989, the panel reached strong conclusions about the feasibility of further reductions. Noting that progress in reducing ozone-depleting chemicals had been more rapid than anticipated even two years earlier, they concluded that substantial further tightening of targets was feasible, eliminating "virtually all" CFCs, halons, and carbon tetrachloride by 1995 to 1997, and methyl chloroform by 1995 to 2000. While noting that these accelerated phaseouts would carry costs from re-trofitting and early capital retirement, the panel made no quantitative cost estimate but instead relied on their broad judgment of technical and economic feasibility, an aspect of their work that attracted substantial industry criticism.[78] Achieving the accelerated phaseouts would depend on several conditions being met. They would require increased short-term use of the two HCFCs already available and full com-mercial availability of several others still in development. In addition, eliminating MC significantly before 2000 would require achieving widespread cooperation and sharing of technology among many small users. Finally, parties would have to take several decisions by 1994 to accomplish phaseouts by 1997: establishing a proce-dure to identify essential uses; making arrangements for international halon-bank management, particularly to meet developing countries' needs; and providing a small "service tail" of allowed consumption to service long-lived refrigeration equipment.

The third meeting of the parties was held at Nairobi in June 1991, as the as-sessment panels were partway through their work. Although Protocol revisions could not be officially considered until the panels reported, three new conflicts and uncertainties over the future direction of the Protocol rose to prominence at this meeting. First, suggestions for general advance of phaseouts met opposition from developing-country delegations, who were concerned that available financial assis-tance might be inadequate even to meet the phaseout schedule agreed upon in 1990. Second, the science panel cochair made an informal presentation of the panel's new results suggesting that MeBr was a serious candidate for controls. Third, conflict intensified over the appropriate treatment of HCFCs. Several European nations sought other parties' endorsement of a resolution calling for strict HCFC controls, including limiting some or all of them to essential uses, capping the maximum amount that could be used, and phasing them out as soon as technically feasible. Many delegations opposed the resolution, however—not just the United States, which was the most dependent on HCFCs for refrigeration and air-conditioning, but also several other historical activists on the issue.[79] DuPont initially argued that the only acceptable HCFC control was a fixed, sufficiently distant phaseout date, but later retreated to accepting the principle of treating long-lived HCFCs more stringently than short-lived ones.[80]

Parties responded to each of these new issues by asking for further information from TEAP. In a carefully negotiated instruction, they asked TEAP to study the implications of advancing all phaseouts, "e.g., to 1997," with specific reference to developing countries.[81] At the urging of the opponents of tighter HCFC controls, parties asked TEAP to identify specific uses where a rapid CFC phaseout required HCFCs, to identify the least ozone-depleting HCFCs suitable for those uses, and to identify a feasible timetable to eliminate HCFCs. Finally, they requested a review of uses and alternatives for MeBr. With their charge to review essential-use appli-cations for halons, these new requests left TEAP with four outstanding tasks.

The 1990 decision to phase out halons included an exemption for essential uses, but delegated the task of evaluating proposed essential uses to the technology panel and the halon TOC. Lacking specific guidance from the parties on how to define an essential use, the halon TOC developed their own criteria: that the activity using halon was essential; that no adequate alternative to halon was available; and that insufficient halon was available from existing stocks. Applying these restrictive conditions, the panel expressed a "qualified opinion" that, while some types of fire and explosion risk had no adequate halon alternative available, all these essential uses could be supplied until at least 2000 by redeploying existing banks, including international redeployment to meet developing-country needs. On this basis, they recommended that parties reject all essential-use applications, subject to periodic reassessment.

In response to the new questions about phaseouts, the panel expressed confidence about developing-country phaseouts but caution about HCFCs. They concluded that many sectors in developing countries could achieve phaseouts on the same schedule as in industrialized countries, and that with proper support, most developing countries could phase out all uses in five to eight years, rather than 10 years, after the industrialized countries. Regarding HCFCs, they noted that many options under development might reduce their use, but still judged that eliminating CFCs in some sectors—particularly refrigeration and insulating foams—was likely to require HCFCs for the near term. Consequently, rapid elimination of HCFCs would risk delaying the phaseout of CFCs.[82]

Finally, in a brief annex to their report the panel provided a first assessment of MeBr, including a rough breakdown of uses and an identification of possible alternatives for three specific uses drawn from a review of published literature. As with MC in 1989, the late addition of MeBr to the panel's agenda prevented them from conducting a detailed assessment thoroughly informed by industry-held technical expertise, as they were becoming increasingly practiced at conducting for other chemicals. As a result, their work on MeBr drew sharp criticism.

Negotiating the 1992 Revisions

After the release of the assessment panel reports in the fall of 1991, political pressure to advance phaseouts grew through the winter in both Europe and the United States. Reaction to the Science Panel report highlighted the first detection of summer ozone loss at midlatitudes and the large reduction in CFCs' contribution to climate change, which undercut U.S. plans to use CFC cuts for greenhouse-gas abatement.[83] The same day as the science panel press conference, DuPont announced it would advance its production phaseouts, for CFCs to the end of 1996 and for halons to the end of 1994. EPA officials also endorsed advancing existing bans by several years, while other industry actors—including both the Alliance and ICI—discussed further advances in vague but generally favorable terms.[84]

Environmentalists did not initially respond to the panels' new emphasis on MeBr, but instead stressed early elimination of HCFCs. Using chlorine-loading analysis and the same target measures as the science panel, activists argued for rapid elimination of both short and long-lived HCFCs to reduce the near-term peak. They also argued that the standard measure of a chemical's contribution to ozone depletion, its steady-state ODP, was inappropriately favorable to HCFCs. Steady-state ODPs compare the ozone losses caused by a unit emission of each chemical over an infinite time horizon. By this method, HCFCs had much lower ODPs than CFCs because of their short atmospheric lifetimes: equal emissions of a CFC and an HCFC would cause similar ozone losses for a decade or two, but over a century or more the CFC would remain in the troposphere and continue to deliver chlorine to the stratosphere, while the HCFC would essentially be gone. Consequently, HCFCs' contributions to ozone loss or chlorine loading relative to those of CFCs were three to five times larger over short than over long time horizons. Arguing that while steady-state comparisons were appropriate for managing long-term risks, a short time horizon was required as the near-term peak emerged as the greatest remaining risk, environmentalists used HCFCs' high short-term ODPs as the basis to call for their phaseout as early as 1995. They also rejected the technology panel's judgment that HCFCs were necessary for the rapid elimination of CFCs, identifying various nonozone-depleting technologies they contended were immediately available but being inappropriately neglected.[85]

Despite mounting calls to endorse faster phaseouts both domestically and internationally, the U.S. administration made no change in its position through the fall of 1991. Growing increasingly impatient with the lack of administration movement, three environmental organizations petitioned the EPA on December 3 to enact accelerated phaseouts under the precautionary obligations of the Clean Air Act, threatening a lawsuit if the EPA did not grant the request. The petition asked that existing phaseouts be advanced to 1992–1995, that long-lived HCFCs be eliminated between 2000 and 2005, and that MeBr be eliminated by the start of 1993. It cited the Science Panel report to provide the scientific basis for further concern and to support the high leverage available from MeBr cuts, but made no reference to TEAP; rather, it based its claim that the proposed cuts were feasible on the examples of a few leading companies such as Nortel, whose uses were among the easiest to eliminate.[86] While there was no immediate administration response, EPA officials said they expected the coming Protocol revisions to advance the phaseouts substantially, although less than the petition requested.[87]

The NASA and European Arctic expeditions in the winter of 1991–1992 brought more disturbing news. January observations found the northern stratosphere chemically primed for acute ozone losses like those in the Antarctic. Ozone losses as large as 20 to 30 percent were projected if the Arctic stratosphere remained cold and stable until the sun rose in March.[88] Even more worrisome, these extreme conditions were not confined to high latitudes, but were found in air parcels as far south as Cuba. While these may simply have spun off the Arctic vortex, they may also have indicated that Antarctic-like processes to enhance ClO were taking place on aerosols at the tropical tropopause, the coldest part of the earth's atmosphere outside the Antarctic stratosphere.

Scientists participating in the U.S. Arctic expedition were so disturbed by these observations that they held a press conference on February 3, at which they warned of the risk of extreme ozone losses later in the spring and suggested that large northern-hemisphere losses might be generally more likely than had been believed.[89] The announcement provoked immediate and strong reactions, greatly increasing pressure on the Bush administration. In a speech the next day, Senator Gore warned of an "ozone hole over Kennebunkport" (the site of President Bush's vacation home in Maine) and charged the administration with irresponsibility. On February 6, a Senate resolution calling for phaseout of all ozone-depleting substances "as fast as possible" passed by 96 votes to none. A similar resolution three months earlier had been blocked by Senate Republicans at the administration's request, but no attempt was made to block this one—in part due to the departure of Chief of Staff John Sununu, previously the major source of administration resistance. Rather, the White House immediately announced its general support for advancing phaseouts, and followed up with specific proposals on February 11. These asked U.S. producers to cut CFC production to half of 1986 levels immediately (only slightly below current production), endorsed advancing existing phaseouts to the end of 1995, and stated that the administration would consider advancing the present U.S. phaseout of HCFCs and enacting one for MeBr.[90]

European governments also advanced their phaseout proposals through early 1992. Environment ministers agreed informally on February 22 to advance phaseouts to year-end 1995, and one month later formalized this agreement and made it the EC negotiating position for the coming Protocol revisions.[91] Germany continued to move ahead of the EC position. With the 1991 law banning production of CFCs by the start of 1995 and halons by the start of 1992, all political parties had by this time agreed to advance the domestic bans further, to 1993.[92]

The Alliance had privately urged the change in U.S. policy, and filed a petition with the EPA on the same day as the White House announcement, requesting a nearly identical accelerated phaseout schedule. The Alliance petition stated that U.S. producers were "likely to respond favorably" to the request for immediate cuts, and supported production phaseouts at the end of 1995, but requested a small "service tail" of consumption through the end of 1999. In addition, the petition proposed advancing the phaseout of the three longest-lived HCFCs (22, 141b, and 142b) to the end of 2009 for new equipment, with all production ceasing by the end of 2019.[93] While this request for tighter controls on the chemicals the Alliance was dedicated to defending might appear surprising, the petition was in fact a defensive response to the even stronger controls being sought by NGOs. By this time, the combination of anticipated shortages and rapid innovation was reducing use of CFCs faster than was projected even in 1990. Users in the most difficult sectors were bearing significant costs, having to make large capital expenditures during a recession for HCFC-based equipment whose usable lifetime was uncertain, and often facing long waits for equipment availability because all firms in each sector tended to move together.[94] But because these reductions were proceeding far ahead of regulatory targets, the Alliance had little interest in opposing the proposed advances, as long as regulatory phaseouts did not come so soon as to force even faster reductions than were occurring through market forces. Having decided against expanding their membership and mandate to include MeBr, the Alliance was not concerned with opposing NGO demands for a rapid phaseout.

Rather, the Alliance had two primary goals: maintaining a small service tail of CFCs to service existing equipment with no drop-in alternatives, and ensuring adequate lifetimes for HCFCs and associated capital equipment. In accepting earlier phaseouts of the longest-lived HCFCs, they sought to protect full product lives for the shorter-lived ones. They presented chlorine-loading calculations to show that their schedule made only small changes in peak chlorine and the rate of decline from that proposed by the NGOs.[95] More broadly, the Alliance criticized the general approach of tightening controls each time more ozone loss was observed. Since it was known well before 1990 that stratospheric chlorine would increase for a decade or two, whatever controls were adopted, increased losses should have been anticipated and should not necessarily justify further tightening of controls.

The risk that scientists had identified on the basis of extreme chemical conditions, which prompted their February statement and these subsequent political events, failed to materialize because of extremely benign weather conditions. The Arctic stratosphere warmed sharply at the end of January, breaking up the vortex much earlier than usual and dispersing the extreme chemical conditions before the arrival of sunlight needed for the final ozone-loss step. The maximum ozone loss was 10 to 20 percent, about half that projected. Scientists regarded the outcome as an extreme piece of good luck, which one described as "dodging a bullet."[96]

Negotiations to revise the Protocol resumed in April 1992, and continued with working group meetings in July and October, leading to a diplomatic conference at Copenhagen in November. As in 1990, the negotiating agenda was strongly shaped by the reports of the science and technology panels and negotiations opened with most delegations broadly agreeing to advance existing phaseouts by a few years. With initial proposals mostly differing by only one or two years, agreement was easily reached by July to advance phaseouts of CFCs, CT, and MC to the start of 1996, and of halons to the start of 1995.[97] By October, a consensus was near that halon-bank management would allow the halon phaseout to be advanced one more year, to the start of 1994.[98] All delegations also agreed from the outset that phase-outs should include essential-use exemptions, and most agreed to extend bans to hydrobromofluorocarbons (HBFCs), a group of potential halon substitutes.[99] As in 1990, there were differences over interim cuts: the Europeans sought steep interim cuts to match the rapid reductions they could achieve in aerosols, while the United States opposed any interim cuts because it needed longer lead times to commercialize alternatives for the most difficult sectors, particularly refrigeration.[100]

The major areas of negotiation were the treatment of HCFCs and MeBr, and the financial mechanism. Conflict over HCFCs followed the pattern of the prior two years. Environmental groups sought immediate limits and early phaseouts, arguing that HCFCs contributed to ozone loss in the near term when the risk was greatest, and that zero-ODP alternatives were available in all sectors. This second claim was sharply disputed despite the success of the hydrocarbon-based "Greenfreeze" refrigerator, which both the chemical and the refrigeration industries had called impossible. Evidence that similar nonfluorocarbon alternatives were immediately available for all sectors, particularly other refrigeration uses and insulating foams, was much weaker.[101]

In negotiations over HCFCs, the United States and the EC reversed their roles from 1990. The United States sought controls similar to those it advocated in 1990—phased reductions in consumption of the highest-ODP chemicals first, to a near-phaseout by 2020 with a small service tail to 2030. Reflecting recent changes in their domestic policy and industry strategy, the EC now sought stronger controls, combining an immediate production cap, restrictions on specific applications, and a "virtual phaseout" by 2015–2020. After ICI's decision to invest predominantly in HFCs, the United Kingdom was a particularly vigorous proponent of strict HCFC controls. In July, Tolba proposed a compromise close to the European proposal, which combined an immediate cap at 3 percent of baseline CFC production with a ban on use in new equipment after 2000.[102] These positions made little atmospheric difference, but they were intensely fought. While ICI and German movement away from HCFCs made the European stance less costly, Atochem's position as the world's largest HCFC producer generated substantial conflict within the EC.[103]

By September, U.S. industry and officials had accepted the principle that maximum allowable HCFC consumption would be capped between 2 to 4 percent of baseline CFC consumption. Led by the United Kingdom, the Europeans continued to seek a phaseout in 2015 to 2020.[104] Tentative agreement was reached in October to set the cap at 3.1 percent, with reductions reaching 99.5 percent in 2020 and a phaseout in 2030.

MeBr raised more difficult conflicts. Like MC in 1990, MeBr came on the agenda after being highlighted by the science panel, bringing a new group of firms into the issue.[105] There was substantial uncertainty in its atmospheric budget, with estimates of the anthropogenic fraction of emissions ranging from 5 to 50 percent, but its estimated ODP was high—0.7 in steady state, and 5 to 7 over 10 years. The Netherlands had eliminated MeBr in 1991, while in the United States it clearly fell under the Clean Air Act requirement to ban any substance with an ODP greater than 0.2. The U.S. delegation initially proposed a 2000 ban to match the regulation being developed under the Clean Air Act, but received little support. Moreover, the proposed U.S. regulation was also facing strong opposition from the U.S. Department of Agriculture (USDA) and agricultural producers.[106] Tolba's July proposal included an MeBr freeze in 1995 and a 50 percent cut in 2000, with an exemption for use in conjunction with quarantine of international trade. [107] The EC was divided, with the Dutch and British advocating cuts of 70 and 25 percent, respectively, while France and the Mediterranean states opposed any restrictions. The Council of Ministers agreed to a compromise on October 20, freezing production beginning in 1995.[108] Tolba proposed to add a 25 percent cut in 2000. But under the leadership of an official of the Dead Sea Bromide Corporation, Israel emerged as a forceful opponent of MeBr control. Israel, which had just joined the Protocol, opposed even listing MeBr, and gained substantial developing-country support led by Kenya and South Africa.[109]

As in 1990, discussions of the financial mechanism were the most difficult. Because the London decisions had established the fund only on an interim basis with three years of funding, it had to be reauthorized and its subsequent size agreed upon. These issues unexpectedly became contentious in July, when several European delegations proposed to abolish the fund and merge it with the Global Environment Facility (GEF), a newly established environmental fund affiliated with the World Bank. At the June 1992 "Earth Summit," the GEF had been named the interim funding mechanism for the new Climate Change Convention, subject to a set of developing-country conditions to make GEF's operations more transparent and its North-South division of control more equal than other Bank programs. But despite these conditions, the GEF was still a creation of the industrialized countries associated with the World Bank. Most developing countries regarded it skeptically and viewed the new European proposal as a direct attack on the 1990 bargain. Although the United States had promoted the World Bank as a funding vehicle in 1990, it had since become reconciled to the Multilateral Fund, and also opposed the new proposal.[110] Although several factors prompted the new position, its proximate cause was that many longtime ozone delegates were moving to other duties. Because the agreement now contained financial obligations, new delegates included Finance Ministry officials, who knew little of the history of the issue or the nature of past agreements. The newcomers also brought interests in maintaining donor control over financing and in holding down costs, which were beginning to grow as national reduction programs became operational. Estimated costs for 1994–1996 were $500 million, substantially larger than for 1991–1993, while the suggestion continued to circulate that the total cost of eliminating CFCs in India alone would exceed $2 billion.[111]

Developing countries had other reasons for suspicion that industrialized countries were trying to back out of the 1990 bargain as well. Several chemical firms were beginning to argue that developing countries should be discouraged from building up an independent capacity in chemical alternatives, but rather should buy from larger-scale, lower-cost plants in the industrialized countries that had already borne the up-front development costs.[112] Moreover, many countries were in arrears on their promised donations to the fund. In November 1992 the fund was short more than 20 percent of contributions for 1991 and more than 30 percent for 1992.[113] Although much of the arrears was from Russia, which had fallen into political turmoil soon after undertaking financial commitments in London, others not making their promised payments included the United Kingdom, France, and Italy, leading advocates of abolishing the fund. Finally, it became clear in July that several delegations were assuming that a decision to advance industrialized-country phaseouts would automatically advance developing-country deadlines by the same amount. Several developing-country delegations sharply denied this, and the government of China announced on August 4 that it was proceeding with planned phaseouts in 2010—ten years after the London phaseout, not the changes being discussed for 1992.[114]

Fourth Meeting of the Parties, Copenhagen, November 1992

Parties met in Copenhagen, in the context of further disturbing scientific reports. The 1992 Antarctic ozone hole set another record for both spatial extent and depth, and for the first time extended for a few days over populated areas of South America, where it reduced total ozone to half its normal value.[115] In addition, a few days before the meeting, the WMO announced winter 1991–1992 seasonal-average ozone readings 12 percent below normal from ground-based instruments over northern Europe, Russia, and Canada, lower than had been observed in the 35 years of continuous ozone monitoring.[116]

The prior working-group agreements to advance the phaseout of halons to the first of 1994 and of other London chemicals to the first of 1996 were quickly confirmed, resisting a Nordic call to advance them one more year.[117] As both producers and users were growing confident they could achieve rapid phaseouts, the industry representatives present also supported these advances.[118] Although the EC proposed that HBFCs be treated on the delayed schedule of HCFCs, it was decided after brief discussion to eliminate them on the same schedule as CFCs.[119]

TEAP was asked to review essential use nominations for halons in 1993, and for all controlled chemicals in 1994. Since essential-use exemptions were available for all chemicals being phased out, it was clear that much would ride on the stringency of the essential-use process. Although environmentalists initially denounced the provision as a dangerous loophole, the U.S. delegation signaled its intention to press for a narrow interpretation of essential uses by announcing that the United States expected to propose only one essential use: propellants for metered-dose medical aerosols.[120]

The EC split on HCFCs, with France seeking to raise the cap to 4 percent while others sought to lower it to 2.5 percent. It was agreed to retain the tentative working-group agreement to set the cap at 3.1 percent, plus baseline HCFC use. Commenting on the European division, *The Economist* noted acidly that ICI, which had stopped producing HCFCs in favor of HFCs, "complained virtuously that Ministers were not being tough enough."[121] Unlike all other controlled substances, controls on HCFCs applied only to consumption, not production. Phased reductions beginning in 2004 would cut consumption to 0.5 percent of the initial cap by 2020, and to zero by 2030.[122] This final 10 years was the service tail for long-lived equipment on which the United States had insisted. Despite environmentalists' continued arguments that shorter-term ODPs were the appropriate measure for calibrating controls on individual HCFCs, the Protocol adopted a set of ODPs approximately equal to steady-state values.

On MeBr, the United States advocated a phaseout by 2000 in the industrialized countries and by 2010 in the developing countries, despite continuing division within the administration,[123] while many other delegations supported Tolba's proposal for a 25 percent cut. With the southern European countries, Israel, and several developing countries seeking to delay any consideration of cuts until 1995, the EC's compromise proposal for a 1995 freeze of production and consumption in industrialized countries was adopted. Although disputes remained over MeBr's budget and ODP, delegates provisionally set its ODP at 0.7, as recommended by the science panel.

Parties also reviewed a report from the Implementation Committee, which had been charged in 1991 with elaborating a compliance regime for the Protocol by preparing two "indicative lists": a list of measures parties might take in case of noncompliance, and one of situations that would constitute noncompliance. They fulfilled the first task with an exhaustive list of possible responses, but could not accomplish the second because it required reengaging the sharpest conflict of 1990: whether failing to make the agreed contribution to the fund constituted noncompliance with the Protocol. Divisive as this question was in 1990, it had since become even more so, since several large countries were in fact failing to make their agreed contributions.[124]

Moreover, the broader conflict over the fund that had surfaced in July continued in Copenhagen with greater intensity. With several European delegations still pressing to fold the fund into the GEF and threatening otherwise to renege on their commitments, developing countries protested by leaving the preparatory meeting for a full day.[125] North-south tension rose still further when a group of Eastern European and former Soviet states announced explicitly what was widely known: that they could not meet their financial commitments. They sought a new, intermediate status that would give them the same control obligations as other industrialized countries but not the financial obligations, provoking a sharp confrontation between these two groups of states. Although the situation of these states was clearly both extreme and unforeseen in 1990, granting their request would risk other petitions to escape the financial commitments based on special circumstances and, more seriously, would reduce the size of the fund by one third—a shortfall that no other parties were offering to offset by increasing their contribution. The petitioners were given vague reassurances but no explicit release from their financial obligations. Facing strong support for the fund from all other delegations, the Europeans relented and agreed to funding of $113.3 million for 1993 (the same annual level as 1992), after which a permanent fund would be established.[126] Parties could not agree on the funding level after 1993, however, so they simply set a range and deferred the decision to the next meeting.[127]

As a gesture to the Europeans, parties also agreed to commission an independent review of the Fund's effectiveness in 1994, which would allow one more chance to argue for merging it with the GEF.[128] For their part, developing countries objected to proposals to tighten or broaden controls without increasing the size of the fund, since the initial funding level was negotiated on the basis of the estimated costs of meeting the 1990 obligations. Further tightening or broadening of controls could reasonably be expected to increase compliance costs, just as continued technological progress should reduce them. But while such cost reductions would be reflected in project assessments conducted under the fund's procedures, the fund was not adequately financed to meet even the full costs of the 1990 agreements, and augmenting the fund—whether to fund existing obligations more adequately or to meet the additional costs of more stringent ones—would require explicit diplomatic agreement. Seeing no increase offered, and spotty compliance with even the original financial targets, the developing countries could reasonably view themselves as being asked to take on tighter controls and bear the additional cost themselves. Rejecting this, they successfully demanded an amendment to Article 5 of the treaty (the article that defined their special status) suspending any further tightening of their commitments pending a review of their situation by the 1995 Meeting of the Parties.[129]

The changes adopted in Copenhagen illustrate both the great progress of the regime, and the obstacles to further action that grew more serious as the regime advanced. The rapid advance of controls on the original chemicals was powerful testimony to the rapid process of innovation under way, and to the influence of the technology and science assessment bodies. On the other hand, the weak action on MeBr showed the limits of the ability of scientific assessment to move decisions and, conversely, the extent to which regime progress relied on linked processes of technological change and its political recognition—which were not available this time. This revision also illustrated how the increasing progress in solving the problem as originally defined—controlling ozone depletion by limiting emissions of CFCs and halons—brought a progressive shift toward harder problems. Even attempts to broaden the regime to new chemicals with different structures of actors and interests posed powerful challenges to the regime's norms and institutions. The progressive shift of the decision agenda toward issues of finance, technological assistance, control of institutions, and overseeing implementation—a natural progression in a maturing regime, particularly to be expected in the first major environmental negotiation after the June 1992 Earth Summit,[130] posed even harder challenges.

Nations continued to advance their implementation of ozone-depleting substance regulations domestically immediately after Copenhagen. The fight over MeBr shifted to the national and EC level, where it remained as contentious as it had been in the international negotiations. In the United States, new regulations proposed by the EPA were delayed through 1992 by the White House Office of Management and Budget (OMB) in response to objections by MeBr supporters, issued as proposed regulations in January 1993 in the final days of the Bush administration, frozen again by incoming OMB officials in the Clinton administration, then finally reproposed in March 1993 and enacted in final form in December. These regulations advanced existing phaseouts to match the Copenhagen agreements, although this made no difference to most CFC users, who were racing to avoid a May 1993 CFC product-labeling requirement. The regulations also phased out MeBr at the start of 2001 and enacted selective early phaseouts for the most ozone-depleting HCFCs, for instance, by 2003 for HCFC-141b.[131]

In the EC, ministers agreed in December 1992 to advance existing phaseouts to the start of 1995, one year ahead of the Copenhagen target, but failed to agree for nearly two years on whether controls on HCFCs and MeBr should remain at Copenhagen levels or be substantially tightened.[132] After positions began to converge in mid-1993, ministers agreed in December on a compromise that reduced MeBr 25 percent in 1998, reduced the HCFC cap to 2.6 percent, and advanced the HCFC phaseout to 2015 with a 35 percent cut in 2004.[133] Activist states, environmental groups, and the European Parliament—seeking to assert its expanded powers after the Masstricht Treaty—all sought tighter controls, but did not succeed. Ministers adopted their agreement as a regulation in June 1994 with only minor changes.[134]

8.3 Third Round of Protocol Revision, 1992–1995

A third, similar round of Protocol revision followed the 1992 Copenhagen amendments, now on a three-year cycle. While meetings of the parties continued annually, the next revision was planned for the sixth meeting, to be held in Vienna at the end of 1995, with the assessment panels beginning their work in late 1993 and reporting in late 1994. With all initially controlled chemicals now scheduled for early elimination in the industrialized countries, debate continued to focus on MeBr, the HCFCs, and the linked issues of aid and chemical controls in developing-country parties. As foreshadowed in the 1992 negotiations, these issues were deeply divisive.

The broad lines of conflict over these issues, which were to intensify over the next few years, were mapped out at the next meeting of the parties, at Bangkok in November 1993. The size of the fund for 1994 to 1996, deferred in Copenhagen, had to be decided here. Parties approved a three-year budget of $510 million despite developing-country objections that this was inadequate in view of the increasing pace of national reductions, the shortfalls in payments received, and the reclassification of a few parties as developing countries.[135] In addition, two nonbinding phaseout resolutions continued the 1992 debates over MeBr and HCFCs. A Danish resolution to phase out MeBr "as soon as technically possible" gained 16 supporters, including the United States, Canada, the United Kingdom, and Germany. A European-led resolution to limit HCFCs to "absolutely necessary" uses and advance their phaseout to 2015 gained 22 supporters but was opposed by the United States.

Atmospheric Science: New Results and Scientific Assessment

New results after Copenhagen broadly continued the trends of the previous few years. Observed disruptions of the stratosphere continued to grow more alarming, and models remained unable to account for observed ozone losses, especially at temperate latitudes. More hopefully, this period also saw the first evidence that the ozone regime was having an environmental effect. Although atmospheric concentrations of the major ozone-depleting chemicals were still growing, their growth rate in the troposphere had slowed by about half since 1990.[136] Carbon tetrachloride had already peaked in the troposphere and begun to decline, the first ozone-depleting chemical to do so, while concentrations of several CFC alternatives were increasing as expected. These were hopeful signs, but early ones. The trends described the concentration of chemicals in the troposphere, not the stratosphere, and growth had only slowed. If these trends continued, stratospheric concentrations would increase for several more years before starting to decline. Peak chlorine and bromine in the troposphere was predicted for about 1994, with the peak in the stratosphere—and, by inference, maximum ozone depletion—three to five years later.

In the meantime, observed depletion at both polar and temperate latitudes continued to increase. The Antarctic ozone holes continued to set records for severity and extent in 1992 and 1993, while global ozone losses accelerated 1 to 2 percent in the same years, even relative to the prior trend. In the northern hemisphere, the winters of 1991–1992 and 1992–1993 had ozone 15 to 20 percent below normal at high latitudes, with large shorter-term losses covering the whole hemisphere.[137] In 1994, however, ozone levels returned closer to the long-term declining trend, suggesting that the extreme values of 1992 and 1993 in part reflected the contribution of stratospheric sulfate aerosols from Mt. Pinatubo. Estimation of both regional and global trends were still complicated by discrepancies between ozone measurements from different sources.[138]

Despite some improvements, model calculations remained unable to account for the observed pattern of losses. The one area of improvement was the upper stratosphere, where after years of embarrassing discrepancies, 2-D models were finally able to replicate observed trends. Despite improved representations of surface chemistry, however, models still could not replicate observed decreases in the lower stratosphere, underpredicting them as much as threefold in high-latitude winter.[139] Several processes involving halogen chemistry on sulfate-particle surfaces were known to occur, and some observations in the lower stratosphere suggested these processes could represent important ozone sinks there, but these processes were extremely hard to model.[140] It was widely assumed that these processes accounted for the losses that models could not capture, but this remained a matter of speculation.[141]

The Protocol's science assessment repeated its now long-tested and progressively refined model. Its report this time was larger in scale than the last two under the Protocol, making it the most ambitious and comprehensive review of atmospheric science relevant to stratospheric ozone since the assessment of 1986. In addition to its review of stratospheric science and the effects of halocarbons, it included a thorough review of stratospheric effects of aviation. The panel report highlighted continued observations of ozone losses and the early evidence that the Protocol's controls were beginning to work. Although there were some preliminary indications

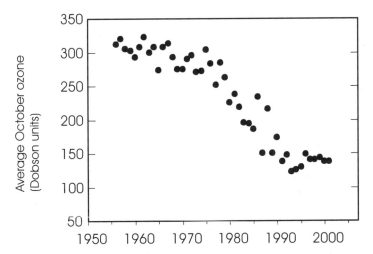

Figure 8.1 The Ozone Hole through the 1990s. After sharp increases through the 1980s, the extent, depth, and duration of the ozone hole increased more gradually through the 1990s. Source: data from J. D. Shanklin, British Antarctic Survey.

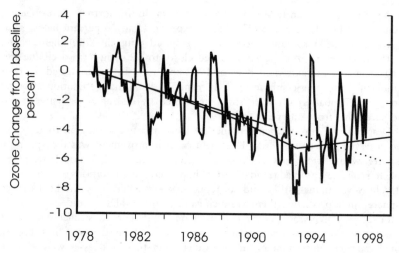

Figure 8.2 Trends in Northern Hemisphere Ozone. The heavy line shows deviations in monthly-average total ozone from their historical baseline, averaged over north temperate latitudes (25 to 60 degrees). At these latitudes, ozone losses increased through about 1995 and have remained roughly constant since. Known natural oscillations (from the seasons, the 11-year solar cycle, and the quasi-biennial oscillation) are removed. The upper trend line shows the fitted trend through mid-1991, a decline of 2.9 percent per decade. The lower line fits separate trends over subintervals. Source: UNEP 1999a.

of a trend in total-ozone in the tropics, the panel cautiously reported this observation as "little or no trend." They were also cautious in characterizing northern hemisphere losses, which they stressed did not constitute an ozone hole; and in attributing observed Arctic and midlatitude depletion to halogen chemistry. In the Arctic, they stated only that calculations "suggested" the reductions were caused by halogen chemistry,[142] while for midlatitude losses they concluded only that ozone losses in the midlatitude lower stratosphere were more strongly influenced by HOx and halogen chemistry than by NOx chemistry.[143]

The panel also reported significant continuing uncertainties over the atmospheric chemistry and budget of MeBr.[144] Recent results included some changes in reaction kinetics, and increased estimates of emissions from biomass burning and of natural sinks in the ocean and soils. While calling the aggregate effect of these results "questionable," the Panel still reduced MeBr's ODP, estimated as 0.7 in 1992, to 0.6.[145]

As in prior assessments, the center of the panel's discussion of policy implications was a comparison of future halogen concentrations resulting from alternative control scenarios. Under the baseline, now taken as commitments adopted in Copenhagen and London, the Panel projected stratospheric chlorine would peak at 3.8 ppb around 2000 and recover to 2 ppb around 2045. Northern midlatitude ozone loss would peak around 12–13 percent in winter and spring (versus about 10 percent already realized), and about 6–7 percent in summer and fall (versus 5 percent already realized). Losses could be larger if there were another major volcanic eruption or an unusually cold Arctic winter with a stable stratospheric vortex. .

With long-lived ozone-depleting chemicals already being rapidly eliminated, the available ways to lower atmospheric chlorine or bromine further were quite limited. Based on consultation with partics and the secretariat, the panel considered four ways to tighten controls: eliminating all MeBr emissions by 2001; eliminating HCFC emissions by 2004; and destroying the existing banks of halons and of CFCs, so none of these banked chemicals would ever be emitted. For each alternative, they reported by how much it would reduce integrated atmospheric chlorine-equivalent above 2 ppb. Of these four measures, eliminating MeBr gave the largest effect, reducing integrated excess chlorine by 13 percent, while the others gave reductions of 10 percent (destroying the halon bank), 5 percent (eliminating HCFCs), and 3 percent (destroying the CFC bank). Conversely, 20 percent noncompliance with existing controls until 2002, dropping to zero by 2005, would increase integrated excess chlorine by 9 percent.[146]

The UV Effects Panel again provided modest further elaboration of previously identified effects, including more detailed characterization of impacts on marine ecosystems under the Antarctic ozone hole. They noted increasing concern about potential immune-system effects leading to possible increases in the incidence of some infectious diseases. Such effects were most likely for infections that have a phase in the skin, such as malaria. Experiments had shown that papilloma virus and HIV-1 could be activated by UV, an effect that would not increase the rate of infection, but could alter its course. In addition, animal experiments had shown that UV exposure could increase the progression of leishmaniasis and the incidence of herpes. Finally, it was newly suggested that UV could alter decomposition in terrestrial ecosystems, with possible subsequent effects on biogeochemical cycles.[147]

In contrast to previous scientific assessments of ozone, this one was conducted while a revisionist backlash was under way against the Protocol and the scientific consensus supporting it, principally in the United States. The backlash began in 1992, partly in reaction to the February press conference where scientists called attention to a heightened risk of severe northern hemisphere ozone loss that was not subsequently realized. Through 1993 the movement gained support from several conservative political figures and a few scientists with no prior expertise in the issue. In some cases the arguments identified real uncertainties or weaknesses in scientific understanding, such as the inability of models to explain observed midlatitude depletion. More frequently, they featured claims long demonstrated to be false: CFCs cannot reach the stratosphere because they are heavier than air; most chlorine in the stratosphere comes from natural sources like sea spray or volcanoes, not CFCs; and the Antarctic ozone hole is a natural phenomenon that was observed in the 1950s.[148] The backlash appealed to some members of the 104th Congress, who sponsored hearings to debunk supposedly alarmist science supporting unsound policy decisions, and introduced bills to weaken or abolish controls on ozone-depleting chemicals.[149] In response, the Science Panel put substantial effort into making their report widely accessible and addressing some of the erroneous claims being advanced. In a new publication, "Common Questions about Ozone," they rebutted the major spurious claims being advanced without referring explicitly to the backlash. The backlash gained little support, even from industry, as proposals to roll back controls failed, and the movement was in decline by 1996.[150]

Technology, Economics, and Negotiations, 1993–1995

The tasks delegated to TEAP this time were substantially expanded, and its work more closely integrated with parties' negotiations, than on previous occasions. TEAP's initial tasks, assigned at Copenhagen, included developing essential-use recommendations; recommending strategies for managing the halon bank; reporting on the feasibility of implementing trade restrictions for products "produced with, but not containing" CFCs;[151] evaluating technologies for recovery and recycling of controlled substances; evaluating alternatives to HCFCs in the sectors presently most dependent on them; and conducting the first full assessment of MeBr uses, emission-reduction measures, and possible alternatives.[152] In addition, after parties resumed negotiations for the 1995 Protocol revisions, they repeatedly posed additional questions to TEAP, asking them to revisit and elaborate on points they had already addressed, and to review new questions to help resolve points of dispute. TEAP consequently issued several reports through 1994 and 1995.

As they did for halons in 1991, TEAP reviewed parties' requests for essential-use exemptions from the coming bans. They reported their second set of halon recommendations in July 1993, and parties reviewed them in November.[153] Unlike the first round, these recommendations were controversial. In the first round, the Panel had judged that supplies from the halon bank were adequate to supply all essential uses, so they did not have to engage the question of which uses were truly essential and had no acceptable alternatives. Consequently, it had remained uncertain how stringently these determinations would be made and thus how large a loophole the essential-use process would represent. In this round, 15 countries initially proposed essential uses of halons; nine of whom withdrew their applications after panel members pointed out alternatives that they had not adequately considered.[154] On reviewing the remaining applications, the panel recommended—and the parties decided—that all be rejected, after which Parties were more restrained in subsequent requests for essential-use exemptions.[155] The next essential-use review, completed in March 1994, was a much larger-scale activity because for the first time it considered all controlled chemicals, not just halons. Many applications were submitted, but the panel recommended only three, all for CFCs: aerosol propellants for medical metered-dose inhalants (MDIs); cleaning and bonding applications for the solid-fuel rocket boosters on the U.S. space shuttle (for which the panel urged manufacturers to join ICOLP to evaluate alternatives); and laboratory and analytic uses.[156]

As in previous rounds, TEAP's largest responsibility was to assess progress in reducing ozone-depleting chemicals and prospects for future reductions in each sector. Their major report on these matters was prepared through 1994, discussed with parties while still in preparation at meetings in July and October, then officially released in November 1994.[157] They noted great progress in reductions achieved to date, with a few areas of concern. The halon phaseout had taken effect at the start of 1994 with little disruption, and most industrialized nations were on track to making the other phaseouts at the start of 1996, those in Europe one year earlier by using stockpiled CFCs for near-term needs. Remaining concerns included servicing existing refrigeration equipment, and two issues related to solvent use: some users that had quickly switched to HCFC-141b faced the challenge of making a second early transition; and many firms, especially small MC users and European firms that had mistakenly assumed their uses would qualify as essential, were starting their transition planning much too late and so might face serious disruptions or be unable to meet the looming EC phaseout.

Looking forward, the panel discussed the need for HCFCs and the possibility of advancing phaseouts in developing countries. For HCFCs, they cautioned that while many sectors did not need them, they remained essential for several sectors. Although nonfluorocarbon options (especially hydrocarbons) were proven for some refrigeration uses, many refrigeration applications needed HCFC-22 at least for the few years until HFC blends were available. For a few percent of solvent uses, HCFCs 141b and 225 also still appeared to be essential as transitional substances. Finally, for insulating foams and fire-protection applications with tight space restrictions, acceptable zero-ODP alternatives were unlikely to be available until at least 1996. In the developing countries, the panel reported great progress in some sectors but major remaining problems, including the marketing of obsolete, CFC-dependent capital equipment and the risk of both inadequate CFC supply and flagging commitment to technical cooperation after the industrialized-country phaseouts were completed. In aggregate, they judged developing-country phaseouts by 2000 to be feasible, but only if projects under the multilateral fund were implemented without delay.

Parties' negotiations for further Protocol revisions began at an OEWG meeting in July 1994. These negotiations resumed the now long-standing debates between the United States and the EU over whether further tightening of controls should target HCFCs or MeBr, and between the industrialized and developing countries over the linked issues of developing-country controls and financial assistance. European delegations pressed to tighten HCFC controls, but met a significant setback in TEAP's report that HCFCs remained essential in some applications and necessary to meet energy efficiency standards in others. Germany and the Nordic states attacked these conclusions, principally on the basis of the European experience with hydrocarbon refrigerators, and charged TEAP with bias against nonfluorocarbon alternatives. Several European delegations asked TEAP to reexamine the need for HCFCs and the feasibility of advancing their phaseout before 2030. Developing-country delegations unanimously opposed this advance, and many also opposed asking TEAP to study it, so the request was tabled.[158]

Conflict over HCFCs and MeBr persisted at the November 1994 meeting of the parties in Nairobi. After acrimonious debate, it was finally agreed to ask TEAP to assess the feasibility of stricter HCFC and MeBr controls, but developing countries demanded the qualification that TEAP's responses were "for discussions only, in no way construed as recommendations for action."[159] With the MeBr freeze in industrialized countries to come into force in three months, parties agreed only with difficulty on a relatively narrow working definition of the "quarantine and pre-shipment" exemption to the freeze, to be reviewed the following year.[160] At this meeting, parties also approved TEAP's recommended essential-use exemptions with two exceptions: they limited all exemptions to two years, and narrowed the set of drugs for which metered-dose aerosol inhalants were exempted.

A full-scale assessment of MeBr, the most important new project of TEAP, was released in November 1994. The assessment was conducted by a 65-member TOC composed principally of industry experts from various usage sectors and from firms developing alternatives. In contrast to the first technical assessment of CFC alternatives, which had excluded the CFC manufacturers, this TOC included MeBr producers who, unlike CFC producers, had no plans to market alternatives to their current product. Predictably, this was a highly contentious group: manufacturers and many users fought to have the report conclude that there were no alternatives to MeBr, charges of bad faith were widespread, and manufacturers and many users attacked the report on its release. Leaders of the process reported that they experienced here the kind of obstruction they had expected from CFC producers in 1989 (and for this reason excluded them), but had never experienced from them once they were included.[161]

The TOC reported that replacement of MeBr was virtually complete in a few industrialized countries and under way in several others, and that research was under way to tell to what extent alternatives used in leading countries would work elsewhere. They concluded that no single alternative could replace MeBr in all its uses, but that alternatives existed for "a number" of current applications and "significant reductions" were feasible. TOC members differed markedly in their quantitative estimates of what reduction was feasible, however, from 50 percent by 1998 to "a few percent" by 2001. In their crucial concluding statement, the TOC and TEAP leadership wrote that "while alternatives are available for the majority of current uses, the MBTOC did not identify technically feasible alternatives, either currently available or at an advanced stage of development, for less than 10 percent of 1991 MeBr use."[162] Finally, the panel declined to give advice on a precise definition of the "preshipment and quarantine" exemption included in the Copenhagen language, recommending instead that parties consider relying on an essential-use process rather than such a broad exemption by usage classes.

TEAP delivered additional information on HCFC and MeBr controls in March 1995, which parties discussed at OEWG meetings in May and August. Regarding the feasibility of MeBr controls, TEAP elaborated on the earlier TOC report by explicitly stating majority and minority views: a majority of the TOC judged 50 percent cuts feasible by 2001, while separate minorities believed that either a full phaseout, or emission reductions of only a few percent, were feasible by that date.[163] Regarding HCFCs, TEAP sought to dissuade parties from overzealous restrictions, noting two factors that parties "may wish to take into account": that non-HCFC alternatives have their own toxicity and environmental problems, and that some companies demonstrated environmental leadership by commercializing HCFC technologies long before they were proven or practical. In view of these considerations, "parties may wish to consider the advantages of halting final HCFC production after a time that allows reasonable recovery of investment costs."[164] The May working group meeting also received a briefing from the Science Panel stressing that the recovery of the ozone layer would require controls on HCFCs and MeBr in both the industrialized and the developing countries, and a synthesis report that assessed proposed scenarios of further HCFC and MeBr controls in terms of effects on integrated chlorine-loading and feasibility. The most stringent HCFC reductions being proposed were described as "technically feasible, but possibly not economically feasible."[165]

The most serious crisis of these negotiations, once again, was finance. In an open letter circulated in September 1994, the Indian environment minister had denounced late fund contributions and threatened to withdraw from the Protocol if industrialized countries did not honor their funding commitments. Of the $393 million pledged for 1991–1994, only $216 million had been received and $25 million released to projects. Much of the delay was in disbursements from implementing agencies, which had been allocated $153 million but had disbursed less than one-sixth of that.[166] Conflict over finance and fund management was still simmering in

July 1995 when delegates considered the consultant's report on fund operations that had been agreed upon in 1992, which criticized long delays in implementing projects, particularly by the World Bank.[167] By the summer of 1995 the backlog for funding approval was also growing rapidly, in part because many countries had adopted voluntary national phaseout targets ahead of those in the Protocol. While these targets advanced the Protocol's goals, they required more funding than was budgeted—even if all promised contributions were being made, which they were not—and the resultant shortage prompted the Executive Committee to adopt restrictive criteria for project funding. Many developing country officials began supporting the view expressed by the Indian environment minister the previous year, stating that without adequate funding they might not be able or obliged to meet their phaseout commitments.[168] In this increasingly tense environment, China and the G-77 announced in August that they would accept no further controls until industrialized countries stated how much they would increase the fund. Awkwardly, the Protocol required review of control measures in 1995, but deferred fund replenishment until 1996. Parties responded to the deadlock by asking TEAP to assess economic and financial implications of several scenarios for tighter controls for developing counties.

Adding to the north-south conflict, the Indian delegation infuriated many industrialized country representatives by stating that they considered the Protocol's meaning of "basic domestic needs" to include exports to other developing countries, and proposing an amendment to the Protocol's trade provisions that would remove any restrictions on exports from developing-country parties.[169] A group of OECD nations counterproposed an amendment that would allow Article 5 producers to export products containing controlled substances to other Article 5 parties under tightly drawn restrictions. As they were increasingly doing with contentious matters, parties deferred consideration of both proposed amendments and asked TEAP to assess them.[170]

As on prior occasions, several alarming new scientific observations were widely publicized as the meeting of the parties approached. Northern hemisphere depletion set another record in the unusually cold winter of 1994–1995, reaching seasonal averages 9 percent below the long-term average in midlatitudes and 20 percent below in high latitudes, despite the stratosphere's recovery from the aerosols injected by Mount Pinatubo in 1993.[171] The WMO announced in September that the 1995 Antarctic ozone hole was the deepest and largest ever, with ozone levels already down 25 to 30 percent by the end of August.[172] In addition, a new study suggested significant ecological harm from a breakdown product of several HCFCs and HFCs.[173]

TEAP delivered a final update in November, responding to parties' August questions on trade measures and tighter controls for developing countries. They avoided the specific request to assess the proposed trade amendments entirely, instead providing a general discussion of the benefits of trade and of competitive markets. To the request for cost estimates of various accelerated control schedules for developing countries, they simply referred parties to a recent consultant's report and stated it was the best source on the matter.[174] For HCFCs and MeBr they declined to provide quantitative cost estimates, instead providing their usual judgments of technical and economic feasibility. For HCFCs, TEAP estimated that a 2010 freeze and a 2040 phaseout would be both feasible and cost-effective, based on the consultant's cost estimates and data from existing multilateral fund projects.[175] For MeBr, they estimated that a 25 percent cut in 2005 with a 2011 phaseout would be feasible, and would cost between $86.6 and $326.7 million, while a 2001 phaseout was technically infeasible.

Vienna Revisions, December 1995

The controversies of the past two years persisted at the seventh meeting of the parties, held at Vienna in December 1995, on the tenth anniversary of the Vienna Convention. The United States and the EU sought substantially tighter controls on MeBr and HCFCs, respectively, eventually agreeing to substantial tightening of MeBr and slight tightening of HCFCs. Financial issues also persisted, as well as several newly prominent implementation problems, of which the most serious was noncompliance in the former Soviet bloc. Led by Greenpeace, environmental groups used the meeting to mount a forceful public attack on continuing production of all ozone-depleting chemicals, denouncing the producers and publishing estimates of each company's production and profits from ozone-depleting chemicals since 1985.[176]

For HCFCs, the EU proposed the compromise that had emerged from tough internal negotiations through the fall: reducing the cap to 2 percent, and advancing the phaseout to 2015.[177] The U.S. delegation and industry opposed the proposal, arguing that the assessment panels showed this to be both infeasible and environmentally insignificant, since it would only achieve only 2 percent reduction in integrated chlorine loading.[178] The two sides finally agreed to leave the phasedown schedule unchanged, but lower the cap from 3.1 percent to 2.8 percent, which consumption was not expected to exceed in any case.[179] In a more significant advance, the developing countries for the first time agreed to accept a phasedown schedule for HCFCs, albeit a far-distant one: a freeze at 2015 levels beginning in 2016, and a phaseout by 2040, with these measures to be reviewed in 2000. Environmentalists attacked this slow schedule, arguing that OECD producers could export HCFCs aggressively to boost developing countries' baselines, since the OECD countries had only consumption limits, not production limits.[180]

On MeBr, the U.S. delegation pushed for a 2001 phaseout, citing the science panel's conclusion that this would bring the largest environmental benefit of any measures considered. In contrast to its stance in 1992, Israel expressed its support for any MeBr controls, including a phaseout.[181] Internally divided, the EC proposed only a 50 percent cut by 2005, although several member-states wanted stronger controls. A compromise was negotiated in which the present cap on production and consumption at 1991 levels would be cut 25 percent in 2001 and 50 percent in 2005, with a phaseout by 2010.[182] Even this relaxed phaseout schedule included a 10–15 percent allowed production increase to meet developing countries' basic domestic needs, an exemption for quarantine and preshipment use (which was, however, defined fairly narrowly), and—at U.S. insistence—a provisional exemption for "critical agricultural uses," a response to U.S. domestic opposition to MeBr controls from USDA and other agricultural interests. EC delegations attacked this potentially broad exemption as vague and dishonest, and parties asked TEAP to examine how the exemption might be implemented and to review possible critical uses.[183] Against substantial opposition led by Kenya, developing countries agreed to freeze MeBr production and consumption from 2002 at a baseline defined by their four-year average over 1995–1998, with TEAP to review possible phasedown schedules in 1997. Environmental groups attacked this three-year baseline extending into the future, concerned that it could be inflated by a rapid increase in consumption, although several major developing countries indicated that they planned to cut MeBr much more aggressively than this.[184]

By this time funding was running so short, due to the combination of slow payments and accelerated national phaseout plans, that some countries were being asked to slow their phaseout plans.[185] Despite the weak state of the financial mechanism, a few industrialized-country delegations sought to advance the developing-country phaseouts of basic substances to the start of 2006, or 10 years after the Copenhagen dates.[186] A group led by India blocked the proposal, so the phaseout remained at 2010, 10 years after the London dates, despite the widespread adoption of voluntary early phaseout targets by many developing countries.[187] Developing countries agreed to undertake new controls on HCFCs and MeBr in conjunction with informal agreements, affirmed in a decision of the meeting, that funding would be increased in the next round and that developing countries would not be expected to take on additional controls without adequate assistance.[188] Parties dealt with the conflict over developing-country exports by affirming that developing countries may export if no Protocol limit is violated, but also set up a system to monitor exports, viewed by some as a first step toward export controls.[189]

A final difficult issue confronted here was the treatment of the former Soviet and Warsaw Pact states. The Protocol treated these just like OECD states, requiring both early reductions and financial contributions. None had made their required financial contributions, and several had informed the parties in 1994 that they were unlikely to be able to meet their control commitments without assistance. Parties had begun to craft a response in mid-1994, led by the Protocol's Implementation Committee, with technical consultation provided by TEAP and additional funds to be provided through the GEF.[190] At Vienna, parties proposed a program of monitoring and further assistance, but out of concern for the CFC black market also imposed restrictions on exports of controlled substances from parties receiving such assistance: Russia and other former Soviet states would be allowed to export only to each other. The Russian delegation lodged a protest and walked out of the meeting room as the decision was adopted.[191]

With the 1995 assessment completed, parties also discussed the charges of the Assessment Panels for the next round. All panels were officially requested to give the parties year-by-year updates, as they had been doing informally, and—in response to the increasingly recognized linkages between ozone depletion and climate change—were invited to offer assistance to the subsidiary bodies of the Framework Convention on Climate Change. In addition, the science panel was asked to examine the potential consequences of noncompliance, and of potential sources of stratospheric changes other than halocarbons, in particular aviation.[192]

TEAP needed substantial reorganization, as the nature and magnitude of demands on it had changed. As the industrialized-country phaseouts approached, firms were becoming less willing to bear the substantial costs of participating, even as developing-country needs for technical assessment were increasing. Moreover, there was little further need for evaluating alternatives to the original chemicals: TEAP's major jobs were increasingly evaluating alternatives for HCFCs and MeBr, and addressing specific questions from the parties (e.g., CFC destruction technologies, appropriate treatment of process agents) that required specialized technical expertise not represented on the panel. In addition, as TEAP's stature and influence in the regime had grown, its work had attracted increasing governmental attention, and some governments who saw their preferred decisions rejected on the basis of contrary advice from TEAP sought to restrain its independence. At this meeting the delegations of Sweden and Switzerland, who suspected that TEAP's continued defense of the necessity of HCFCs reflected a bias toward HCFC-dependent industry rather than impartial technical judgment, sought to increase governmental control over TEAP's membership and operations. With support of many other delegations, TEAP leadership resisted this attempt to bring it under greater political control, but also proposed various changes in its operations to address shifting needs, consolidating its operations by merging or eliminating subbodies whose workload had decreased, and further increasing participation of developing-country experts.

The MeBr TOC was a particular problem. With the TOC increasingly polarized into two factions, one asserting that no alternatives existed and one seeking to develop and demonstrate the availability of alternatives, TEAP proposed to reorganize the body as a smaller group "focusing on the technical and economic viability of MeBr alternatives," and to reevaluate "participation by organizations that market ozone-depleting substances but do not offer alternatives"—a clear reference to the obstructive behavior of MeBr producers. Parties deflected the attempt to increase political control over TEAP by charging an advisory group to review TEAP's operations and report in one year. For their part, MeBr promoters after the Vienna meeting increasingly circumvented the assessment process and attempted to make their case directly to delegations and national officials that significant reductions were infeasible and that TEAP and the TOC were politically driven and biased.[193]

8.4 Further Assessment and Review of Decisions, 1995–1999

Negotiating the Montreal Amendments, 1996–1997

Two more rounds of assessment and revision of Protocol measures followed the 1995 Vienna adjustments. The Assessment Panels continued to provide frequent informal reports to parties, and completed another full assessment round in 1998. Parties further amended the Protocol at their 1997 meeting in Montreal, and in 1999 at Beijing.

With controls off the agenda for the moment, the parties' 1996 meeting, in San Jose, Costa Rica, was principally devoted to reauthorizing the fund for 1997–1999.[194] In Vienna, funding had figured centrally in negotiations over developing-country controls, but the level for the next period had not been decided. Some developing-country delegations had proposed renewal at levels as high as $800 million to $1 billion, but instead TEAP was asked to estimate the fund's expenses through a detailed examination of national phaseout programs and projects. Reporting to parties in August 1996, TEAP estimated that only $436 million was needed to meet developing-country targets—$20 million less than the previous three years' budget—plus an additional $40–$60 million to continue funding countries' existing accelerated-phaseout programs.[195] Developing countries initially objected to the low figure, particularly in view of suggestions of substantial increases made in Vienna, but parties nevertheless reached agreement relatively easily. Benedick noted that several favorable factors smoothed these negotiations, including a sharp improvement in donor countries' payments to meet their existing fund obligations, their clear willingness to support moderately higher funding than TEAP suggested, and the return to negotiations of former UNEP director Mustafa Tolba, now representing Egypt.[196] Parties agreed to three-year replenishment of $540 million, including $74 million carried over from 1994–1996.[197]

In an important item of business left from Vienna, parties also considered here the report of the advisory group on TEAP. In a strong show of support, parties rejected the proposal to impose more political control on TEAP, and formally adopted terms of reference that closely followed TEAP's own suggestions. An explicit code of conduct for members of TEAP and its committees was also adopted, which formalized long-standing informal expectations regarding members' responsibilities to avoid conflicts of interest.[198] In an indication of continuing tension between TEAP and the European delegations that had led the effort to restrain its independence, EU delegates sharply criticized a report of the halon TOC which recommended that a substantial fraction of existing stocks of banked halon could be destroyed, stating that they had not provided enough data to support their recommendation.[199]

At the ninth meeting of the parties, held at Montreal in September 1997, European nations continued their attempts to advance the HCFC phaseout and restrict specific uses. The EU had already lowered its own HCFC cap to 2 percent in early 1996, and proposed here that the Protocol's HCFC phaseout be advanced to 2015 to match their own requirement, with interim cuts starting in 2004.[200] Meeting widespread opposition led by the United States and Canada, the proposal was once again defeated. Debate over MeBr also continued to be characterized by uncertainty and conflict. New scientific results, including increased estimates of MeBr's removal by oceans and soils and some changes in chemical kinetics, had continued to decrease its estimated ODP. Parties had reduced its official ODP from 0.7 to 0.6 in 1995, but the new results made its true value appear closer to 0.4—down by nearly half from 1992, but still double the level that triggered a fast phaseout under the U.S. Clean Air Act and high enough that MeBr remained highest priority for further international controls.[201] The methyl bromide TOC and industry continued to disagree sharply over how much its use could be reduced. When a new report of the TOC concluded that 75 percent reductions were feasible in both industrialized and developing countries by 2001, industry countered that the estimate was too optimistic and simply reflected political pressure from the United States. The TOC responded with an uncommonly blunt statement that the industry was obstructing progress by working to preserve the status quo rather than attempting to develop and implement alternatives.

Against this backdrop, several industrialized countries proposed advancing the MeBr phaseout from the 2010 agreed in Vienna. The United States continued to seek the earliest phaseout, in 2001, but with a potentially broad exemption for "critical uses." The EU proposed a 2005 phaseout with more tightly drawn exemptions. Parties agreed on 2005, with a somewhat limited critical-use exemption.[202] Developing countries also agreed to move beyond their 2002 freeze, adding a 20 percent cut in 2005 and a phaseout in 2015, with additional funding to be provided for MeBr demonstration and test projects.[203] Criteria to define "critical uses" were agreed upon, and TEAP was asked to publish a list of approved critical uses. A use could be judged critical if "significant market disruption" would result from its elimination, as long as proven alternatives were not available and best practices were used to reduce emissions. Emergency use of up to 20 tonnes per year was also allowed for rapid response to dangerous pest outbreaks.

Efforts continued at the 1997 meeting to reduce the illegal trade in CFCs that had undermined chemical alternative markets for several years. This black market first came to prominence in 1994 and 1995, when production cuts and the U.S. excise tax drove the retail price of CFCs to nearly $15 per pound (up from $1 a pound in 1989).[204] Although many alternatives were in fact cheaper than CFCs by this time, the expense of capital conversion in some refrigeration and fire-fighting uses provided the incentives to delay conversion that drove the black market. At its peak around 1995, the market was estimated as 10–20 Kte per year of illegal imports to the United States and 6–15 Kte into Europe.[205] Most black-market CFC originated as illegal production in Russia, whose 100 Kte production capacity was half the world's total. With increased customs enforcement effort and initial progress on illegal Russian production, black-market imports to the United States appeared to drop sharply in 1996 and 1997, to about 5 Kte.[206] To further restrict illegal trade, parties agreed in Montreal to adopt an import licensing system for all ozone-depleting chemicals in 2000, although a more restrictive proposal to ban exports of stockpiled CFCs was not enacted because of U.S. opposition.

Plans also began in 1996 and 1997 to phase out the largest essential-use exemption, the 13 Kte of CFCs used annually as propellants in metered-dose inhalants (MDIs), specialized medical aerosols used to treat asthma and other respiratory conditions. With the first CFC-free MDI marketed in 1995 and many others planned for marketing by 2000, TEAP recommended in 1996 that parties phase out the exemption by 2005. After discussing various ways to gradually increase pressure on firms to move away from CFCs without risking disruptions to patient care, parties agreed to prepare national transition strategies to phase out the exemption.[207]

The 1998 Assessments

The 1997 Montreal amendments were enacted with only interim reports and briefings from the panels, the first time the Protocol had been revised without a full assessment. In the next full assessment, completed in late 1998, the science panel reported that tropospheric concentrations of all ozone-depleting chemicals, and of chlorine, had peaked as expected in 1994 and had begun to decline. This early decline came from sharp cuts in methyl chloroform, the controlled substance with the shortest atmospheric lifetime. Tropospheric concentrations of chlorine from longer-lived CFCs, and of bromine, were still increasing. The combined concentration of chlorine and bromine in the stratosphere was predicted to peak before 2000.

Despite slowing growth of source gases, ozone loss continued unabated in both the Antarctic and the Arctic. After sharp growth through the early 1990s, the depth and extent of the Antarctic ozone hole grew more slowly, with column losses usually exceeding 50 percent through September and October and reaching 70 percent for a week or so. In the Arctic, six of the prior nine winter-spring seasons had shown ozone losses in some months that were 25 to 30 percent below the average levels of the 1960s, associated with unusually cold winter temperatures in the Arctic stratosphere. Ozone loss had slowed over midlatitudes, however, with little further reduction after the stratosphere's recovery from the 1991 Mount Pinatubo eruption. Column ozone over 1994–1997 at midlatitudes averaged about 4 percent below 1979 levels year-round, and 5.4 percent in winter. Based on new indications of larger MeBr removal in oceans and soils, the panel recommended revising its ODP from 0.6 to 0.4.[208]

The panel once again assessed the consequences of several further changes to

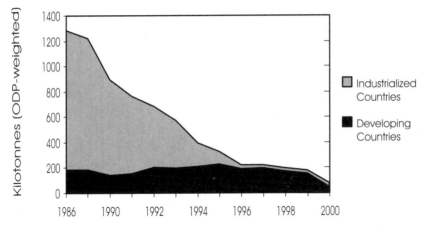

Figure 8.3 Consumption of CFCs and Halons under the Protocol. World consumption of CFCs and halons has declined by about 95 percent since 1986. While most of this reduction occurred in the industrialized countries (upper area), consumption in developing countries (lower area) began to decline in the late 1990s. Source: UNEP 2002.

control measures on integrated future stratospheric halogen. Relative to a baseline of the 1997 controls (though those had yet gained few ratifications), only small further reductions were possible. The largest remaining opportunity was to eliminate halon emissions worldwide by 2000, by stopping production and destroying the stock in existing equipment: this would reduce integrated halogen loading by 9 percent for halon-1211 and by 7 percent for halon-1301. Accelerating global phaseouts of HCFCs, CFCs, and MeBr gave smaller reductions, of which the largest was a 5 percent reduction from eliminating global HCFC production by 2004.[209] The panel also voiced the new concern, however, that returning stratospheric chlorine to 2 ppb might not restore the ozone layer. Rather, they noted that stratospheric temperatures and concentrations of several other species—including methane, nitrous oxide, water, and aerosols—would all influence by how much a specified halogen abundance would destroy ozone. Changes in these conditions, as well as natural variability, could preclude detecting any recovery of ozone for 20 years or more after halogen concentrations peaked.

The effects panel once again reported modest elaboration in understanding of UV effects on human health, terrestrial and aquatic ecosystems, biogeochemical cycles, air quality, and materials. The panel report also followed the lead of the science panel and added a section answering "frequently asked questions" about UV effects. One significant change was that some recent research had suggested that UV might be responsible for some of the widely observed decline in amphibian populations worldwide, but the panel noted that many other factors likely contributed to the decline and that the experimental evidence on UV effects on frogs, toads, and salamanders showed conflicting results.[210]

TEAP reported on prospects for further chemical controls, essential uses, and implementation. They reported that "virtually all" developing countries would meet the approaching 1999 freeze on the original chemicals. With developing countries

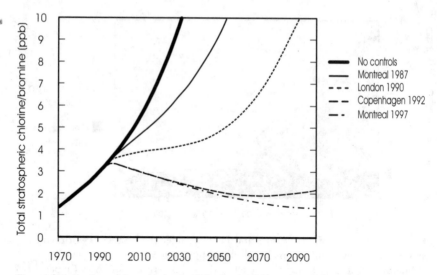

Figure 8.4 Projected Stratospheric Chlorine and Bromine: Effects of Protocol Adaptation. Under present controls, total stratospheric chlorine and bromine (expressed as chlorine equivalent) is projected to decline slowly toward its natural level of about 0.7 ppb, passing through 2 ppb—the level at which the Antarctic ozone hole appeared—around 2040. Source: UNEP 1999b.

not required to eliminate CFC consumption until 2010, however, TEAP noted that some countries could face CFC shortages after 2004 or 2005 unless some export production continued in industrialized countries—which the Protocol permitted, but which was not planned. As they had done for several years, the panel reported that HCFCs and HFCs remained essential for some refrigeration uses, in both industrialized and developing countries. The methyl bromide TOC increased its estimate of the fraction of use for which alternatives were already available to more than 95 percent, and noted that they could not find a single crop that needed MeBr. Finally, TEAP recommended reducing the number of essential-use exemptions, and expressed concern about new ozone-depleting chemicals, particularly two new substances being marketed as solvent alternatives, n-propyl bromide (NPB) and chlorobromomethane (CBM).

Negotiating the Beijing Revisions, 1997–1999

With the new assessments under way and their reports scheduled for late 1998, parties underwent one more series of negotiations to amend the Protocol, beginning in mid-1998, proceeding through their tenth meeting in November 1998 in Cairo and concluding with amendments adopted at their eleventh meeting, at Beijing in November 1999. These negotiations focused on the by now thoroughly familiar issues of further tightening of HCFC and MeBr controls and finance, and on the new issue of linkages, both geophysical and institutional, between stratospheric ozone and global climate change.

The EU continued to pursue tightening of existing chemical controls. In new regulations proposed in late 1997 and agreed to by ministers in December 1998, the EU decided to ban all CFC use, even stockpiled or recycled, by 2000; to tighten HCFC controls to include a ban on exports to nonparties from the first of 2004, on consumption from 2009, and on production, which had not previously been controlled, from 2025; and to ban MeBr consumption from 2005, with interim cuts that slightly favored the southern member-states that used the most.[211] The ban on HCFC production was enacted over strong objections from European producers, who argued that as long as other parties were allowed to consume and import HCFCs, European production cuts would simply be replaced by U.S. production.

In discussions through 1999 and in a formal proposal at the Beijing meeting, the EU pressed similar advances in the Protocol's controls on HCFCs, MeBr, and newly developed ozone-depleting chemicals, which they argued were needed to strengthen ozone-layer protection and to reflect progress in developing alternatives. For HCFCs, they proposed to add production controls to existing consumption controls and to ban trade with nonparties. For industrialized countries, production would be frozen from 2005 and phased out from 2025; for developing countries, production controls would parallel existing consumption controls, with a phaseout in 2040. Many delegations opposed the proposals, led by Canada and China, while environmental groups attacked the opponents for seeking to increase HCFC exports to developing countries—whose HCFC consumption was allowed to grow unconstrained through 2015. Delegations agreed only to freeze production, from 2004 in the industrialized countries—with an extra production allowance to export for developing countries' "basic domestic needs"—and from 2016 in the developing countries.[212] In addition, trade in HCFCs was banned from 2004 with nonparties to Copenhagen—more than 40 countries, including China, India, and Russia.[213] An accompanying decision asked TEAP to assess whether the industrialized countries' 2004 production freeze risked creating shortages of HCFCs in developing countries.

For MeBr, the point of conflict was the exemption for "quarantine and preshipment" uses, which was nearly one-quarter of consumption and increasing in many countries. An EU proposal to freeze the quantity exempted at present levels was opposed by many delegations, including the United States, Canada, China, and many developing countries. Unable to reach agreement on a quantitative limit, the conference instead agreed on a definition of preshipment use, in the hope of limiting arbitrary growth under this exemption; required parties to report data on these uses; and asked TEAP to report again on alternatives to MeBr in these uses by 2003.[214] After this meeting, having failed for several years to advance the Protocol's MeBr phaseout to match the 2001 domestic phaseout, the EPA responded to congressional direction by issuing a revised MeBr rule that set back the U.S. phaseout to match the 2005 phaseout in the Protocol.[215]

The prospect of new ozone-depleting chemicals had been a concern since 1997, when parties adopted a standard procedure for deciding whether additional chemicals should be controlled by the Protocol, although in a weaker form than the EU had proposed it. Parties had also asked TEAP to assess the first two such chemicals proposed for controls, bromochloromethane (BCM) and n-propyl bromide (NPB).[216] The EU had enacted an immediate ban on BCM in their December 1998 regulations,[217] and argued in Beijing for the Protocol to follow this ban—and, more broadly, for an expedited procedure to let new chemicals be controlled without time-consuming Protocol amendment and ratification. Parties agreed to ban BCM in 2002 with a possible essential-use exemption, but many opposed allowing control of new chemicals without amending the Protocol. Instead, a weaker decision asked the assessment panels to monitor such substances and ask parties to consider ways to expedite revisions.[218]

Work continued through this period to bring Russia into compliance and eliminate its CFC production as a potential source of illegal international trade. In 1996, the World Bank had begun a special initiative to raise funds to close Russia's production capacity. By 1998 Russia was producing 9 percent and consuming 6 percent of world CFCs, but still had nearly half of world production capacity. Agreement was reached in October 1998, when 10 donor countries committed $19 million to close Russian CFC and halon production by 2000, in addition to the $10 million for this purpose and the $60 million to convert Russian usage sectors already provided by the GEF.[219]

After a long, slow start, the multilateral fund was finally achieving a substantial record of successful project development and reduction of ozone-depleting chemical usage. From its establishment to 1999, the fund approved $900 million of project funding to eliminate a projected 117 Kte of annual ODS use, about 60 percent of total baseline consumption in developing countries. Landmark negotiations concluded in 1999 to grant $150 million to China and $82 million to India to close their CFC and halon production facilities over 10 years.[220] These agreements were particularly important because halon and CFC production in these countries had grown much more rapidly than expected, even while consumption declined.[221] The fund continued to operate on a three-year cycle, however, with its renewal level negotiated by the parties each time, so another replenishment had to be negotiated for 2000–2002.

At their 1998 meeting in Cairo, parties asked TEAP to establish a task force to assess the appropriate renewal level. The task force reported to the June 1999 OEWG meeting that because growth in developing-country consumption had grown less than expected during their 1995–1997 baseline period, these countries could achieve their required freeze during the next period with funding of only $306 million, rather than the $465 million earlier projected. TEAP also suggested, however, that replenishing at $500 million would help accelerate the momentum toward a developing-country phaseout.[222] Parties failed to agree on the renewal level at two OEWG meetings, however, so the matter was referred to the Beijing conference. Discussions here focused on the linked questions of the level of replenishment and whether funds should be provided on concessional-loan terms (rather than grants) and counted toward the total. Bargaining followed a clear trade-off between the funding level and the counting of concessional loans, with the outcome that loans would not be included, but the fund would be renewed at less than the previous level, $440 million of new funding plus $35.7 carried over from 1997–1999. Although some developing countries charged that they were being required to do more with less, the actual burden of the declining funding was not clear: required cuts were growing deeper and broader, but rapid development of alternatives was continuing to reduce their cost.[223]

Linkages between stratospheric ozone depletion and global climate change first came onto the Protocol's agenda through discussion about HFCs in mid-1998. HFCs were not controlled or even listed under the Protocol because they did not contribute to ozone depletion, but had periodically been proposed for control because of their large global warming contribution. At a working group meeting in July 1998, a group of European and developing countries called for study of alternatives for HFCs and for treatment of HFCs to be coordinated by parties to the Montreal Protocol and the Kyoto Protocol on climate change. The United States, with industry support, objected to the proposal on the procedural grounds that HFCs were not controlled by either Protocol—a fine point, since HFCs were included in the basket of multiple gases that the Kyoto Protocol jointly controls. The question was deferred by asking TEAP to consult with its counterpart climate bodies to harmonize technical assessments.[224] After meeting in October, these bodies proposed that TEAP members help expand the treatment of HFCs in the 2000 assessment of climate change. Parties to both the Montreal Protocol and the Framework Convention on Climate Change (FCCC) approved this cooperation at their November 1998 meetings, asking that the bodies work jointly to identify options to limit emissions of HFCs and perfluorocarbons (PFCs), and to evaluate how these options could affect implementation of the ozone regime.[225] At the prompting of the Alliance, the U.S. delegation initially opposed the proposal, but after receiving no support they instead worked to modify the decision to keep HFCs as much as possible within the FCCC—which was substantively more appropriate, but also promised slower and less effective treatment—and to ensure that no other assessment of HFCs was conducted under the Montreal Protocol.

The TEAP task force on HFCs and PFCs reported in Beijing that while HFCs were critical to phasing out ozone-depleting chemicals, the implementation of the Montreal and Kyoto Protocols need not conflict. In particular, since alternatives to HFCs were available for some applications, developing and transitional economies could, with financial assistance, leapfrog HFC use in those applications. In subsequent discussion, Switzerland attacked the report on both substantive and procedural grounds, arguing that its analysis of national HFC regulations was too general and did not reflect the complexity of the issue. Greenpeace argued for binding controls on HFCs and argued that TEAP was not impartial on the matter because its membership included employees of chemical manufacturers.[226]

HFCs linked the ozone and climate issues because they were part of the ozone solution but part of the climate problem—alternatives to ozone-depleting chemicals that contributed to greenhouse warming. A second type of ozone-climate linkage became increasingly prominent in 1998 and 1999, as parties heard in briefings from the science panel.[227] Evidence was growing that processes associated with global climate change were exacerbating ozone loss, risking the delay of ozone layer recovery by up to several decades even as stratospheric chlorine and bromine declined; or, worse still, that climate-related processes could extend the period of maximum risk of ozone loss, or trigger extreme losses at middle and high northern latitudes similar to those in the Antarctic.

The potential importance of a temperature-related feedback from climate to ozone had long been recognized. Greenhouse warming, by cooling the stratosphere as it warms the surface and troposphere, would increase ozone loss by increasing formation of the polar stratospheric clouds (PSCs) on whose surfaces the crucial loss chemistry takes place. But two results published in 1999 suggested additional, stronger linkages that could sharply increase the risk of extreme northern hemisphere losses. One suggested that the Arctic stratosphere was poised on a threshold of extreme denitrification, so slight additional cooling could substantially increase ozone loss. The other proposed an even larger feedback—that by warming the tropical tropopause, climate change could allow more water to enter the stratosphere, raising the threshold temperature below which rapid heterogeneous ozone-loss processes occur and possibly bringing spring Arctic ozone losses of 50 percent, rather than the roughly 15 percent presently realized. These results shed light on two puzzles, one long-standing and one emerging at the end of the 1990s: that

Table 8.1 The Montreal Protocol and Its Amendments: Summary

	Montreal 1987	London 1990	Copenhagen 1992	Vienna 1995 (adjustment only)	Montreal 1997	Beijing 1999
CFCs	Cut 5 major CFCs 50% 7/98	Phase out 15 CFCs 2000	Phase out 1996			
Halons	Freeze 3 major halons 1992	Phase out 3 halons 2000	Phase out 1994			
Carbon tetrachloride	Not controlled	Phase out 2000	Phase out 1996			
Methyl chloroform	Not controlled	Phase out 2005	Phase out 1996			
Special provisions for developing (Article 5) countries, above chemicals	Controls delayed 10 yrs; ICs may overproduce 10% to export for "basic domestic needs"	Controls delayed 10 yrs; production for "basic needs" exports up to 15% post-IC phaseout		Phase out CFCs, halons, CT by 2010; MC by 2015		
HCFCs	Not controlled	Report data; voluntary resolution to phase out by 2040, 2020 "if possible"	Freeze 40 HCFCs 1995 (base 1989 HCFC + 3.1% CFC); cut 99.5% 2020; phase out 2030 (consumption only)	Baseline reduced to 2.8%, Art. 5 countries freeze 2016, phase out 2040		Freeze production 2004, Art. 5, 2016

Methyl bromide	Not controlled	Not controlled	Freeze 1995 (base 1991)	Phase out 2010; Art. 5 freeze 2002 (base 1995–98)	Phase out 2005; Art. 5 phase out 2015	Definition of "pre-shipment use" exemption
Other chemicals	None	None	Phase out 34 HBFCs 1996	Art. 5 phase out HBFCs by 1996		Phase out BCM 2002
Data reporting required	Production, imports, exports, by chemical, base year, then annually	EC may combine import, export data for all member-states				
Trade restrictions with nonParties	Chemicals: ban exports 1993 (Art. 5 countries only); imports 1990: "products containing"; ban imports 1993	Ban chemical exports from all parties; new chemicals, ban export and import 1993, products containing, ban imports 1996	Trade restrictions extended to HBFCs, but not HCFCs or MeBr		Ban export and import of MeBr; national import/export license systems; ban all exports from parties violating production phaseout	Ban imports, exports of HCFCs, BCM
Size of multilateral fund		$160–$240 million (1991–93)	$510 million (1994–96)	$540 million (1997–99)		$475 million (2000–2002)

1. Interim controls are omitted; only final control levels shown. Except as otherwise noted, controls are effective January 1 of the year shown.
2. "Article 5" countries are developing countries consuming less than 0.3 kg of controlled chemicals per person per year.
3. ICs = industrialized countries.

stratospheric models still could not quantitatively reproduce the observed magnitude and spatial distribution of observed ozone losses; and that observed ozone losses were continuing to grow despite stratospheric chlorine's beginning to decline.[228]

In sum, negotiations from 1997 to 1999 made modest further progress, although with substantial conflicts and procedural difficulties. Delegations complained about ineffective premeeting communication regarding proposed revisions and significant procedural confusion during the Beijing meeting. As an indication of this confusion, the Beijing amendment was adopted in such haste that it included a drafting error that set the excess MeBr production allowance for basic domestic needs at 15 percent for one period rather than the intended 10 percent.[229] These problems reflected the combined effects of the increasingly complex network of obligations associated with the regime, and the decreasing marginal importance of further action for major industrialized countries, since nearly all feasible actions to eliminate the major ozone-depleting chemicals had been enacted or implemented.[230] After gaining its twentieth ratification, the Beijing amendment entered into force in February 2002.

8.5 Progress and Status of the Ozone Regime

By adapting repeatedly to changing circumstances, the ozone regime has made rapid, sustained progress over the decade and a half since its formation. Although the pace of ratification has slowed for the most recent Protocol amendments, the Protocol and its London and Copenhagen amendments enjoy nearly universal worldwide participation.[231] The consistent increase in stringency and breadth of controls on ozone-depleting chemicals has resulted in a reduction in worldwide production and consumption of more than 95 percent.[232] The environmental result of these reductions can be observed in the slowing and reversal of growth in atmospheric concentrations of ozone-depleting chemicals.

The success of the ozone regime derives not from the control measures adopted in the original Protocol, but from its subsequent repeated adaptation. The Protocol's core control measures have been repeatedly tightened in accordance with the institutional provisions for assessment, review, and revision drafted in the original 1987 Protocol. But the adaptation of the regime has been more substantial than simple numerical adjustments of its original controls. In addition to broadening to incorporate additional ozone-depleting chemicals not initially considered, the regime has adapted to address a series of novel and unexpected challenges. Meeting these new challenges has in some cases required developing new institutions or changing the mandates, participation, and procedures of existing institutions.

For example, the science assessment panel has retained the basic strategy that has allowed it to make authoritative summaries and syntheses of scientific knowledge, while continually adjusting the service it provides the parties from establishing the basic seriousness of the ozone-depletion risk to providing simple, scientifically defensible heuristics to help parties evaluate alternative management decisions and monitor their progress, and serving a broad educational function to help address a backlash against the current scientific understanding.[233] The adaptation of TEAP has been even more striking. In addition to repeating its initial job of advising on technological feasibility four times, it has increasingly served to bring diverse forms of technological expertise to bear on any problems the parties identify, and to support the parties in specific aspects of implementation that require focused technical expertise. When necessary, new institutions have been established or the mandate of existing ones has been expanded to perform new jobs and address new problems as they arise.

But as the ozone regime has approached completion of its initial core tasks, it has faced increasing complications and new challenges. The tightening of existing controls to phaseouts harnessed technological and market forces to make cuts progressively easier and simplified negotiations on some key dimensions, but also complicated negotiations on other dimensions. Moving to complete phaseouts made participation of all major developing countries essential; required negotiation and implementation of a process to evaluate essential uses; turned alternatives with nonzero ozone depletion, such as HCFCs, into transitional solutions only; and turned regulatory attention to new classes of chemicals. Broadening controls to new chemicals posed additional scientific, technical, and political challenges. Chemicals added after the initial regime formation were initially given lower priority, either because their effects were smaller, more uncertain, or more contested, or because opportunities to reduce them were initially less evident. Since each new chemical also brought new groups of stakeholders who lacked the accumulated experience, trust, and socialization of those already engaged, such additions often required revisiting old scientific arguments, policy controversies, and negotiation details. However many chemicals were already controlled, adding new ones never became easy—as the controversy over the attempt in the late 1990s to accelerate the process reveals.

Moreover, as maximum feasible chemical controls were enacted and implemented, the regime had to focus on qualitatively different issues, such as implementation and finance. National implementation in the western industrialized countries was diverse in character but largely trouble-free, as these nations had both the intention and the capacity to fulfill their obligations, but national implementation elsewhere posed more serious challenges. The failure of implementation in Russia and other post-Soviet states following collapse of their economies in the early 1990s allowed an international black market to develop, imposed a substantial additional financial burden on the regime that required an ad hoc solution involving the GEF, and exacerbated already serious tensions between the OECD and developing-country parties. The delayed onset of developing-country controls posed different implementation problems, particularly concerned with reducing consumption in sectors of many small users without serious disruption. These problems have become more prominent even as international attention has shifted away from the issue because phaseouts in OECD nations are nearly complete. The shift of focus to developing-country implementation has also directed negotiations toward questions—such as how much money will be transferred, who will control it, what conditions recipients must meet, and how much discretion they have in spending it—that are inherently and unavoidably divisive. The need to replenish the fund every three years, while appropriate in view of changing knowledge of needs, has kept these battles active.

Finally, as the phasedown of halogenated ozone-depleting chemicals approaches completion, other potential sources of ozone loss and linkages with other issues, such as climate change, have become more important, but cannot be managed adequately within the ozone regime. Linkages can be of several types. Solutions to one problem can contribute to another, as HFCs reduce ozone loss but contribute to climate change. Or one problem can exacerbate another, as greenhouse warming now appears to be hindering recovery of the ozone layer. In each case, the sharp focus and limited mandate of each regime impairs its ability to respond adequately. The ozone regime has made little progress on HFCs, because as nonozone-depleting chemicals they lie outside its mandate. Neither the ozone regime, nor the weaker regimes being developed for other global environmental issues such as climate, yet appear able to coordinate learning or action to manage linked problems, in which multiple linked human activities and emissions affect multiple linked dimensions of the global environment. Learning how to manage such linkages will be the crucial next step in protecting the global environment.

9

The Theoretical and Practical Significance of the Ozone Regime

9.1 The Ozone Regime and Its Explanatory Challenges

By the late 1990s, the achievements of the ozone regime were so remarkable that it was becoming easy to forget the years of effort that were required to establish the regime. Within 10 years of its formation, the regime had gained nearly universal participation in a program of large-scale policy and behavior change. It had repeatedly adapted its central provisions in response to new scientific information and technological capabilities, and had significantly promoted the technological developments that made this adaptation possible. It had expanded its scope three times to control additional ozone-depleting chemicals, and had surmounted several major challenges. Moreover, relative to the typical pace of diplomacy and large-scale policy change, this was all accomplished with extraordinary rapidity.

The book thus far has principally developed an empirical account of the ozone issue, although with points of explanation and interpretation woven through the historical account. This final chapter pulls these threads of explanation and interpretation together to develop a series of more abstract and general arguments. It identifies causal factors that shaped the history of this issue, traces the processes and conditions that mediated their influence, and probes the extent of their generalizability and applicability to other issues.

The history of the ozone issue falls into three broad periods. First, a decade of deadlock persisted from 1975 to 1985 despite repeated attempts, diverse in character and institutional setting and led by multiple actors, to develop international cooperation. A rapid burst of regime formation followed from 1986 to 1988, which produced the first agreement on concrete actions and most elements of the subsequent regime. Finally, the period since 1989 has been one of sustained progress, in the elaboration and increasing density and universality of the regime, and in the repeated adaptation of the regime's core control measures in response to new information and capabilities. These three distinct phases of the issue pose three explanatory challenges: explaining the dramatic transition from sustained deadlock by which the regime was established; explaining the persistent deadlock that preceded this transition; and explaining the sustained progress in building and adapting the regime that followed it.

The following sections explore these explanatory challenges. The next section examines determinants of the transition to regime formation, through a static comparison of factors that distinguish the period of regime formation from the long period of prior deadlock. Sections 9.3 through 9.5 examine a set of dynamic processes that contributed to sustaining both the preregime deadlock and the subsequent adaptation and progress. Two types of processes are stressed. The first concerns interactions between scientific assessments and the rhetorical use of scientific claims about environmental risk to support policy action or delay. The second concerns interactions between the setting and revision of regulatory targets, the assessment of technologies available to pursue the targets by reducing chemical usage, and the strategic responses of private industry. Section 9.6 recasts some of the preceding theoretical arguments in practical terms, to propose a set of lessons the ozone issue may hold for other issues. The validity of such generalization will depend on the issues' similarity on important dimensions such as global scope, prominence of scientific and of technological uncertainty, and absence of a dominant national actor. The final section discusses limitations of the ozone regime, which have become evident as it has increasingly approached its objectives, and which now pose the most serious challenges for future management of global environmental change.

9.2 Breaking the Deadlock: Explaining the Transition to Regime Formation

Much recent research has sought to understand the formation and subsequent evolution of international regimes for the management of specific issues.[1] One line of work has examined the processes by which regimes are formed, with different scholars stressing negotiation, imposition by dominant actors, and spontaneous formation.[2] Others have stressed the relative contributions of power, interests, and knowledge as factors explaining regime formation.[3] Recent empirically grounded work has stressed the role of interactions between knowledge and the material factors of power and interests, particularly over issues, such as managing environmental risks, where uncertainty is prominent and policy change occurs over years to decades.[4]

The history of the ozone issue sheds important new light on these questions. The rapid formation of the ozone regime over two years, after nearly 10 years of prior deadlock, provides an uncommonly crisp record to test hypotheses about regime formation. Moreover, this transition did not represent either the new appearance of the issue on policy agendas, or a significant change in the way the issue was framed. Ozone had been repeatedly on international policy agendas during the prior 10 years of deadlock, framed in essentially the same terms: preventing ozone loss and resultant impacts through international regulatory cooperation to reduce emissions of halogenated chemicals. Consequently, the formation of the ozone regime does not fit the prominent theory of policy change that argues such large and rapid policy changes are typically caused by major changes in the policy agenda or the framing of an issue.[5]

This section examines factors potentially contributing to the transition to regime formation. It presents a comparative static analysis, which evaluates proposed factors according to whether they changed significantly between the transition and deadlock periods. A factor that did not change over the period in question cannot have exercised significant influence on the transition. A factor that did change significantly may have contributed to the transition, but the comparison alone cannot verify its importance or identify the mechanism of effect. For some factors, however, the record allows detailed tracing of causal processes that, together with plausibility on theoretical grounds, provides further evidence to support their influence.

Table 9.1 summarizes the analysis, which is discussed in detail below. In addition to the transition period of 1986–1987, the comparison subdivides the initial deadlock into two periods, 1978–1982 and 1982–1985, in order to distinguish outcomes more precisely. While only the third period achieved any concrete agreements to manage the issue, the two prior periods failed in different ways. The first period yielded no action at all despite several attempts, while the second yielded the Vienna Convention, a formal international treaty of predominantly symbolic significance containing no concrete commitments.

The results of the comparison are significantly at odds with prevailing explanations of the formation of the ozone regime. The two factors most widely cited as important causes can be regarded as shifts in the supply of opportunities to reduce ozone-depleting chemicals, and in the demand for such reductions: advances in technological options to reduce the chemicals; and advances in scientific knowl-

Table 9.1 Explaining the Formation of the Ozone Regime

	1977–1982	1982–1985	1986–1987
Action Outcome	Nothing	Symbolic	Strong
Potential Explanatory Factors			
Availability of CFC alternative technologies	Baseline: early research on HCFCs and HFCs	~ Unchanged	~ Unchanged
Scientific basis for concern about ozone-loss tisk	Baseline: large projected losses, not verified	Less	~ Baseline
Institutional context for negotiations	~ None	Little (+)	More (++)
Coalition of activist delegations	Yes	Yes	Yes
U.S. leadership	Yes	Yes (−)	Yes (+)
Authoritative assessment	No	No (+)	Yes

Note: Plus and minus signs denote approximate quantitative comparisons between the strength of the factor operating in the later periods and the initial period (for example, attempted U.S. leadership was present through all periods but was weaker in 1982–1985 and stronger in 1986–1987 than it was in 1977–1982).

edge or concern regarding the risk of ozone loss and its consequences. The comparison shows these were unlikely to have been important, because neither underwent any significant change between the relevant periods. But in contrast to substantive scientific advances, the availability of an authoritative scientific assessment does distinguish the transition from the previous periods. Detailed tracing of processes of influence also supports the suggestion that scientific assessment was of decisive importance. A third widely proposed factor is powerful leadership, either by a single hegemony or by a coalition of sufficient size and influence to force the outcome. The comparison rejects this claim in its strong form, but does support a specific, narrow effect of vigorous activism by a small coalition of U.S. officials as influential in shaping the specific terms on which the regime was formed, but not in the fact of its formation. The comparison also supports the possible importance of an institutional context for international negotiations.

The Myth of a Technological Breakthrough

The first factor often proposed to account for formation of the ozone regime, a breakthrough in developing CFC alternatives, simply reflects an erroneous understanding of the record. No breakthrough occurred over the relevant period to make CFC alternatives substantially more available, either to one firm or to all. The major alternatives were known to be feasible and partly developed by 1980, when firms abandoned their initial research programs, and were no further advanced by 1986. The principal advances in both halocarbon and nonhalocarbon CFC alternatives occurred in 1988 or later, when the regime was largely formed and many actors were pursuing CFC phaseouts. Moreover, the details of initial position changes by U.S. industry—a modest retreat in 1986 led by the Alliance, and an attempt to reverse the strong U.S. negotiating position in 1987—are inconsistent with the changes being driven by CFC producers hoping to profit from alternatives markets.

The record does support two related claims about the influence of CFC alternatives or industry strategy on regime formation, but these are much weaker and more limited. First, although knowledge of alternatives within the industry was unchanged from 1980 to 1986, knowledge outside the industry did change after the 1986 revelations about the earlier development programs and the reasons for their termination. These revelations, probably inadvertent, made the activists in the EPA more confident that alternatives could be commercialized, and strengthened their position in internal policy debates. Their attempt to run an independent expert assessment of barriers to applying chemical alternatives, while not fully successful, gave them somewhat more confidence and support. These hints that replacing CFCs would be easier than believed contributed to the activists' willingness to take an extreme negotiating position, although even a highly optimistic view of the state of alternatives in 1987 gave no confidence that the United States could achieve the Protocol's 50 percent cuts at reasonable cost.

Second, the second shift in industry position two years later, led by DuPont's March 1988 announcement, requires a more complex mix of explanations. Several factors motivated this shift, all related to the development of the regime that was under way: for example, increasing evidence of risk, the specific challenge posed by the Trends Panel, the momentum of negotiations, and progress in resumed alternatives development. But a dawning, perhaps belated, realization by managers at DuPont and other major CFC producers that alternatives markets might offer advantages potentially offsetting their risks surely made this change of position easier.

Only several years later, after producers had made large commitments to developing alternatives, did their interests undergo a more substantial shift. Beyond a certain point in alternatives development, continued availability of CFCs became a barrier to marketing the alternatives, and manufacturers' interest shifted to completing the transition by killing off CFCs as quickly as could be done without disruption. But the rejection of the primary claim, that a technological breakthrough accounted for regime formation by causing a shift in the interests of major industry actors, is clear: there was no breakthrough at the time in question; industry preferences and positions shifted only modestly, in line with those of the most reluctant states; and the controls in the 1987 Protocol were adopted against the strongly expressed preferences of major industry actors in both the United States and Europe.

The Myth of a Decisive Advance in Scientific Knowledge

The second factor widely proposed to account for the transition is a major advance in scientific knowledge of the stratosphere and the risks of ozone loss. This claim is at best highly misleading and incomplete, and more likely wrong. Several related forms of the claim have been made, which differ in the mechanisms of influence proposed and their relative orientation toward explanation or normative justification of the outcome. An informal view widely expressed by practitioners is that the 1987 treaty was "science-driven," in the sense that some decisive advance in scientific understanding was achieved and effectively conveyed to the policy debate. By this account, new knowledge caused the subsequent negotiated outcome, presumably by altering the salience of potential risks of inaction, or their perceived balance with the costs of action, for a sufficiently numerous and powerful group of policy actors. The causal mechanisms by which scientific knowledge achieves this policy influence are left implicit. Other, more nuanced accounts have proposed specific mechanisms of influence. These include persuasive advice offered by eminent scientific advisers to powerful political leaders; individual intermediaries conversant with both science and policy domains who operate as "knowledge brokers," conveying advances in knowledge with appropriate framing and characterization of policy implications; and "epistemic communities," transnational networks of scientists and scientifically knowledgeable officials and advocates, who develop consensually shared interpretations of relevant scientific knowledge and give consistent advice to political leaders.[6] But if they are to provide a persuasive explanation for the changed policy outcome, all these mechanisms require a change in substantive scientific knowledge big enough to matter. In the formation of the ozone regime this condition was not met.

At the time of the transition to regime formation, from late 1986 through 1987, there had been no scientific advance that significantly increased the confidence that CFCs had caused or would cause ozone loss. There had been no decisive new result that settled the reality or likelihood of ozone depletion, or even any aggregate diminution of recognized uncertainties, controversies, and discrepancies. This is not to argue that no progress had been achieved in understanding the stratosphere over the prior ten years. A great deal of progress had been made, but research had generated new questions, uncertainties, and discrepancies between theory and observations at least as fast as it had resolved old ones. Quantitative model projections of future ozone loss had fluctuated widely with new results, but steady-state projections always remained within the uncertainty range initially stated in 1976. Even if these short-term fluctuations in projected depletion are accorded policy significance, they cannot account for this transition, because they reached their highest levels in 1979, their lowest in 1983, and in 1986 stood near the low end of their historical range.[7] Although general confidence was growing that the largest problems in understanding the stratosphere were close to resolution, this confidence was not yet supported by improved agreement between theory and observations, and was, moreover, undercut by the unresolved claims of large observed depletion in Antarctica and the global satellite record.

More broadly, primary scientific claims—meaning new observations, theories, or model results advanced by individual scientists, published in the scientific literature, or informally verified through discussion and consensus in the relevant research community—had little direct policy influence over the entire history of the ozone issue. Primary scientific claims put issues onto policy agendas, but never resolved policy debates or significantly influenced outcomes. Rather, as argued below, scientific claims exercised significant influence on policy outcomes only through the intermediation of official scientific assessments.

But what about the ozone hole? Was it not the primary factor driving negotiation of the 1987 Protocol, and does it not therefore represent a decisive direct influence of a primary scientific result on a policy outcome, independent of assessment? The lurking presence of the ozone hole was clearly prominent in negotiators' minds through the 1987 negotiations. It effected a widespread elevation of concern and vividly dramatized the limitations of available scientific knowledge—especially highlighting the often overlooked truth that uncertainty cuts both ways, so true risks can be either more or less than present consensus estimates. But several factors militate against its having had a large causal influence over the negotiated outcome.

Although shocking, the hole was both less unique and less prominent in the debate than it has come to appear in retrospect. The ozone hole was not the first or only cataclysmic image related to ozone loss, nor the only claim of large, unexpected loss circulating at the time. Claims of extreme future losses and potentially catastrophic impacts had been advanced periodically over the prior 10 years.[8] The contemporaneous claim of large global losses in the satellite record was as prominent in policy debates and news coverage as the ozone hole, despite lacking a similarly vivid visual image. Moreover, over the crucial period CFCs were only suspected, not known, to cause the hole. Policy makers agreed widely that this strong suspicion was not enough to support stringent controls—indeed, U.S. activists worried that regulations adopted on that basis would not survive legal challenge—so negotiators agreed not to found their decisions on the hole. In this regard, all negotiators' accounts and the scientific briefings they received concur. Rather, both activists and opponents grounded their positions in the consensus results of the 1986 assessment. Some actors (admittedly few and marginal) even argued that environmental changes in Antarctica did not matter, or that the mystery posed by the hole called for several more years' delay of CFC controls while further research investigated its cause. The very rapidity with which controls were tightened after CFCs were authoritatively verified to cause the hole underscores the limited role the hole played in the initial 1987 decision.

The strongest influence that can be attributed to the hole in 1987 is that it weakened the confidence of those advocating only a freeze based on a narrow interpretation of the consensus in the 1986 assessment, and bolstered the confidence of those seeking strong cuts as a precautionary response to the same consensus. By giving a bargaining advantage to the advocates of strong cuts, it is very likely that the ozone hole contributed, with other factors, to the specific outcome of 50 percent cuts rather than a smaller cut or a freeze. But the regime would very likely have formed at the same time, in a similar form, had there been no ozone hole.[9]

Not (Just) Scientific Knowledge, but Authoritative Scientific Assessment

In contrast with the limited policy influence of primary scientific claims, statements in official scientific assessments did move major actors to change positions, resolve debates over contested policy-relevant claims, and stimulate the formation of policy coalitions. In some cases, the statement in question simply repeated a scientific claim that was already widely known and verified, but that exercised policy influence only when restated in an assessment. In these cases, the assessment simply validated the scientific claim to policy makers, giving effect to its preexisting potential to influence policy decisions. In other cases, the crucial assessment statements were consensually supported syntheses that had not been, or could not be, made outside an assessment. In these cases, the assessment contributed an independent influence on policy, which was not reducible to the mere validation of a prior primary scientific claim.

The strongest instances of assessments exerting such policy influence are the two major atmospheric-science assessments released at the beginning and end of the transition period, the 1986 NASA/WMO assessment and the 1988 Ozone Trends Panel. Evidence for their effects comes from detailed tracing of processes of influence. The 1986 assessment presented an authoritative review of all topics relevant to atmospheric ozone, unprecedented in its scale, participation, and ambition. This assessment was widely perceived to have one primary message, based on its comparison of stratospheric models: if other pollutants continued their recent growth trends, there would be little or no ozone loss if CFC emissions remained roughly constant, but large losses if they grew by more than about 3 percent annually. In addition, this assessment forcefully confirmed the recent resumption of CFC growth.

These two messages—that CFCs were growing and that growth would bring large ozone losses—decisively influenced the 1986–1987 negotiations. Major U.S. industry actors reviewed their long-standing positions in response to this assessment and changed them to support international controls on emissions growth. The major European governments changed their position to support CFC controls at a level near current production. Formerly the strongest opponents of action, these actors all took new positions consistent with the implications of the assessment and cited the assessment as the reason for their change. Whatever the shares of sincere persuasion and tactical response in these changes—being persuaded of the assessment's conclusions, or calculating that enough others would be so persuaded that controls were now likely, so continuing to oppose them would risk being isolated and discredited—they can clearly be attributed to the assessment. The effect of the assessment, initially operating through these position changes, was to bound the range of feasible negotiated CFC controls on the upper end, at a level ranging roughly from a freeze at present levels to a modest, perhaps two-fold, increase.

The assessment did not imply any particular policy outcome. Rather, it limited the bargaining range by limiting the range of proposals that could be defended, in two steps. First, by making authoritative statements linking policy positions with environmental consequences, the assessment deprived certain positions of the scientific claims formerly advanced to support them: that is, it was no longer possible to assert that unrestricted CFC growth was unlikely to cause significant ozone loss. Second, given a preexisting policy consensus that certain environmental outcomes were unacceptable, removing the scientific basis for these positions, made it impossible to advocate them and retain standing as a responsible participant in policy debates: that is, making it impossible to claim that unrestricted growth would not cause large ozone loss made it unacceptable to advocate unrestricted growth. Faced with these changes in what was rhetorically defensible, the strongest opponents all took new positions roughly where the assessment drew the limit of defensibility. By excluding bargaining positions beyond this point, the assessment imposed a lower bound on feasible agreements. This represented a decisive influence on the agreement actually reached, even though it was substantially stronger than this lower bound.

The Ozone Trends Panel appeared in March 1988, at the end of the transition period. This unusual assessment combined an authoritative review and restatement of results that were already well known and verified—in particular, the consensus formed in late 1987 that CFCs were the primary cause of the ozone hole—with new primary analysis of ozone data to resolve contested claims over the existence of a significant global trend. The new analysis concluded that such a trend did exist, in winter and year-round at middle and high latitudes, and moreover was substantially larger than models predicted. As with the 1986 assessment, evidence for the policy influence of this assessment is strong. Its release prompted DuPont to rapidly review its position and decide within one week to stop making CFCs and endorse a global phaseout. Many other government and industry actors followed with similar announcements, after awaiting independent confirmation of the assessment's new trend analysis a few months later. This delayed response might seem to weaken the argument that it is authoritative assessment statements, not verified scientific claims, that influence policy. But this was a highly anomalous assessment statement, based on new primary analysis done as part of the assessment. This statement was consequently less well validated than normal assessment statements, and policy actors' awaiting its independent verification does not rebut the general argument.

This assessment's influence derived from a somewhat different mechanism than that of 1986. It authoritatively resolved a contested scientific question that had previously become widely accepted as a decisive condition for policy action—the existence of a statistically significant global ozone trend. Since 1975 DuPont had stated that they would stop making CFCs if "creditable scientific evidence" proved them harmful, and since the early 1980s the detection of a significant global trend had become widely accepted—despite its obvious problems—as a proxy for such harmfulness. Responses to the panel's conclusion that CFCs caused the ozone hole provide even stronger evidence of the influence of assessment. This finding was widely known and thoroughly verified four months before the release of the assessment, but had provoked no changes of position. After the assessment release, however, actors cited this among the reasons the assessment led them to reconsider and change their positions, although the assessment did scarcely more than reprint what was already a well-known conclusion.

Variants of Leadership, Hegemony, and Forcing Coalitions

A third set of factors proposed to explain the 1986–1987 transition concerns the exercise of leadership or hegemony, by the United States alone or in conjunction with a forcing coalition of like-minded activist nations. Although superficially attractive, these claims in their strongest forms fail to distinguish the transition from pre-1980 deadlock. Officials from the same coalition of activist nations pursued international cooperation in the late 1970s with similar energy and commitment to what they brought in 1986–1987. In the United States, activist officials had similar levels of senior political support in the two periods, in some respects more in the earlier than the later period. After leading these early efforts, American officials withdrew in the early 1980s but rejoined the activists by 1984. In all three periods there was vigorous support for action by officials from Canada, Sweden, and Norway, plus more restrained support from Switzerland, Austria, the Netherlands, Denmark, and Germany.[10] There were only two changes in the coalition pressing for action between the transition period and the earlier periods: increased senior political commitment in Germany, making it more willing to challenge the EC no-action consensus; and the addition of New Zealand and Australia to the group. Similarly, the coalition most strongly opposed to controls (initially the United Kingdom, France, Italy, Japan, and the Soviet Union, with Spain and Greece added in the transition period) underwent little change. Consequently, any strong claim attributing the regime formation of 1986–1987 to U.S. hegemonic leadership, or to the formation of a forcing coalition favoring action, can be rejected.

There are, however, two narrower claims for influence of U.S. leadership that can be persuasively supported, each focusing on a particular group of U.S. officials. First, it was through the institutional and financial commitment of officials in one U.S. institution, NASA, that the influential 1986 scientific assessment came about. It did not have to be NASA, or even a U.S. institution that did this: another body with enough resources and scientific standing could have done it. Moreover, part of the strategy that made this assessment so influential was to broaden its international sponsorship and participation so far that U.S. participants could not control the results. Still, it was scientific and bureaucratic leadership by U.S. officials that made it possible.

Second, the coalition of activist officials who had gained control of the U.S. domestic agenda in 1986 exercised a blunter form of U.S. leadership. To the shock of both their domestic opponents and their international counterparts, this coalition advanced a negotiating position so extreme that even the other long-time activists believed it insupportable. With informal allies in Congress, they succeeded in establishing unilateral U.S. action with sanctions against nations not following as the salient alternative to a strong negotiated agreement. The credibility of this threat was maintained with great delicacy through early1987, despite efforts by domestic opponents to restrain the delegation and revoke the threat. The combination of the extreme U.S. bargaining position with this threat was primarily responsible for the agreement to cut CFCs by half, rather than following the assessment-based consensus for a freeze or a smaller cut.

An Institutional Context for International Negotiations

An international institutional context for negotiations can provide several functions to advance the achievement and implementation of policy agreements. An institutional context increases parties' ability to contract agreements and their confidence that costly measures or policy adjustments they undertake will be reciprocated for mutual benefit.[11] An institutional structure also allows the continuity of staff effort, even at a tiny scale, that is necessary to ensure that meetings and discussions continue. Internationally authorized meetings in turn require national submissions and impose deadlines. These requirements, supported by international staff, can provide resources and political support to national officials responsible for the issue and, moreover, require senior national officials to attend actively to the issue, prepare positions and supporting arguments, and search for ways of demonstrating progress.[12] The aggregate effect is to establish a default state of gradual forward movement.

The degree of institutionalization of international ozone negotiations increased consistently over the three periods. The lack of any institutional context for the ad hoc process of the late 1970s imposed uncertainty and disorganization on the proceedings and extra administrative burden on the activists, and provided many opportunities for opponents to delay and obstruct progress with no cost or risk of embarrassment. The convention negotiations from 1982 had official status under UNEP with associated secretarial support. Even this modest increase in institutional structure granted continuity of proceedings, allowed the possibility of intersessional work to prepare negotiating agendas, and created an expectation that major nations would participate, present positions, be prepared to defend them substantively, and at least pretend to work toward agreement in good faith. The Protocol negotiations that began in late 1986 retained these advantages, with added momentum conferred by the convention itself and the accompanying agreements on technical workshops and resumption of Protocol negotiations. These steps further elevated the level of national engagement required, the expectations of continued work toward a concrete agreement, and the political need for a success. This increasingly dense web of meetings, requirements, and expectations maintained slight forward pressure at all times, and provided opportunities for large leaps forward when other conditions became favorably aligned, as they did in late 1986.

The increasing institutionalization of negotiations after 1985 also increased the procedural power of UNEP director Tolba, which he exploited by taking much stronger stances than are normal or safe for an international bureaucrat, advancing strong substantive positions, engaging in backroom maneuvering, and at times browbeating delegations. Although officially only the secretariat to these negotiations, he exercised the prerogatives of a chair, using tactics that would be risky even for a real chair. He was instrumental in helping the strong control position held by only a few delegations, albeit strong ones, to achieve as much as it did.[13]

The functions provided by international institutional support to the ozone negotiations also helped to maintain progress through varying domestic circumstances in leading nations, allowing leadership to shift as necessary among individuals and governments to whomever was willing to take it on. The international process worked like a flywheel, providing the momentum that moved consultations through the inevitable periods of reduced commitment or internal dissent that affected even the strongest national leaders. The clearest examples of this effect followed the attempted reversals within the U.S. government that threatened to scuttle agreement at late stages in 1985, 1987, and 1990. Each time, the momentum of the international process, other states' stake in a successful outcome, the banal requirement to attend meetings with a supported position, and the perceived need for the United States to participate responsibly in what it had contributed to starting, all helped domestic activists stave off the attempted reversals. The effect of growing institutional support is clear in the large-scale contrast between the deadlock and transition periods, and became even clearer after the regime was established.[14]

A Viable Negotiating Proposal

A final difference between the deadlock and transition periods, to which negotiators' accounts ascribe great influence, was a basic change in the policy proposal being advanced that made it more broadly acceptable. Throughout the deadlock period, the activists proposed that others replicate the aerosol bans that they had already enacted. This proposal offered many points for attack and nothing to persuade the reluctant. It was useful only as a near-term, interim measure, since most growth was in nonaerosol uses. By proposing a discrete, specific change of policy, it allowed little room for negotiation or flexible implementation to respond to differing national circumstances. Although good evidence suggested the social cost of aerosol bans was low, they would abolish the largest business line of the European CFC producers, who until 1985 effectively controlled policy in several nations. Most fundamentally, because the proposal would impose no costs on its proponents, it suffered two fundamental weaknesses: it was acutely vulnerable to charges of hypocrisy, and it failed to signal a serious commitment by the activists.

In contrast, the activists in late 1986 began proposing comprehensive CFC emission cuts, having briefly considered and rejected them twice previously. These proposals had none of the flaws of the aerosol-ban proposal. They were rationally related to the environmental goal being pursued rather than needing secondary justification as an immediately available stopgap measure. They allowed room for both negotiation over the specific stringency of cuts to be adopted, and flexibility in means of national implementation. Most crucially, they would impose higher costs on the activists advancing them than on the reluctant states, whose continued use of aerosols gave them large low-cost reduction opportunities. The proposals thereby both signaled the activists' seriousness of intent and avoided any hint of hypocrisy.

The apparently strong effect of this new proposal suggests powerful advice for activists seeking to craft effective negotiating proposals. But as an explanatory variable, it poses a more serious risk of endogeneity—that changed politics made the proposal viable, rather than the proposal's viability changing the politics—than the other factors considered. Activist officials had considered similar proposals twice before, and rejected them because they were too confident that their aerosol-ban approach would succeed, because the proposals appeared too complex, and—crucially—because they would impose costs on the activists and so required a higher level of domestic commitment. Indeed, a similar proposal was considered domestically in the United States in 1980, and rejected. That the activists advanced this proposal in 1986 does not mean it was a new idea, but that they were now willing to take costly measures to secure others' agreement. They signaled their seriousness not just by how forcefully they advocated their proposal, but by their willingness to bear costs to secure agreement.

Rejecting "Easy Agreement" Claims

The comparison of the 1986–1988 transition with the two prior periods of deadlock suggests that two widely proposed factors—a decisive advance in scientific knowledge or concern, and a significant technological breakthrough in CFC alternatives—cannot account for the transition. Instead, the comparison highlights three factors as potentially important. First, authoritative scientific assessments presenting sufficiently strong policy-relevant conclusions appear to have caused both the formation of a new bargaining consensus supporting international controls in 1986, and the proliferation of calls for complete CFC phaseouts in 1988. Second, aggressive bargaining tactics by a faction of U.S. officials, supported by congressional allies who provided a credible threat of unilateral U.S. action, appear to have been primarily responsible for the 1987 agreement to cut CFCs in half rather than following the consensus for a freeze. Third, the increasingly dense international institutional context for negotiations appears to have promoted both the weak agreement of 1985 and the stronger measures of 1987.

In rejecting either a scientific or a technical breakthrough as accounting for the transition, the comparison also rejects claims that structural characteristics of the ozone issue made regime formation predetermined or easy. These claims come in theoretical and practical forms: the theoretical claim that formation of a cooperative regime was so likely a priori that it does not require explanation and holds little opportunity for generalizable inference; and the practical argument that solving the problem was so easy that the terms of its solution and associated strategies hold little guidance for managing other issues. The 10-year persistence of deadlock despite repeated attempts at international cooperation allows rejection of the broadest form of these claims, that the issue was predisposed for easy resolution throughout its history. The comparison of deadlock and transition periods rejects the more limited claim that the issue became easy in 1985 or 1986 through structural shifts that were not themselves results of actors' attempts to manage the issue. The two factors most often proposed to account for such a shift are precisely the two that the detailed comparison rejects: a decisive advance in scientific knowledge or concern, and a shift in industry interests due to progress in CFC alternatives.

9.3 Sustaining Deadlock, Sustaining Progress: The Need for Dynamic Explanations

An even more striking feature of the ozone issue's history than the rapid transition to regime formation is the extreme stability of the periods before and after. Before the transition, deadlock persisted for 10 years, during which the issue was continually on international policy agendas and repeated attempts were made to achieve concrete cooperative action. After its formation, the regime underwent consistent, repeated development, elaboration, and strengthening of its core chemical control measures, institutions, and procedures. This process of adaptation is fundamental to understanding the effectiveness of the regime. The Montreal Protocol of 1987 did not solve the problem of ozone depletion by industrial halogen chemicals. To the extent that the problem has been solved, it has been by the subsequent adaptation, refinement, and expansion of the regime. In these dynamic processes of adaptation lie the most important insights, both theoretical and practical, to be drawn from the ozone regime.

It is common enough for regimes to evolve in response to changing conditions. Indeed, similar patterns of progression toward more effective management have been identified in studies of other international environmental regimes.[15] Still, the rapidity and effectiveness of the ozone regime's adaptation, and its strategic use of the production and incorporation of new information, are without peer. The regime's formal structure of nested legal instruments and requirements for periodic assessment and review have been widely noted, and their legal foundation and significance examined.[16] But while these structures were necessary to allow the regime to adapt, they cannot account for how and why it did adapt so rapidly and effectively.

Rather, the extreme consistency of deadlock before the regime's formation and of progress afterward suggests the need to examine dynamic processes that sustained and stabilized distinct equilibrium states of deadlock and progress, respectively. Elucidating these processes poses different explanatory challenges than the static issues explored above, but is potentially of both greater theoretical interest and greater practical utility, particularly the dynamic processes that sustained the continued progress after 1988. One such process contributing to regime progress was simply the continuing effect of an increasingly dense institutional context, as discussed above, which continued to operate more strongly after adoption of the Protocol. Each new institutional layer created an increasingly full calendar of activities, a more extensive set of obligations and deadlines, further opportunities for parties to compete for leadership roles, and procedural authority for international institutions and officials with interests in continuing successes. Moreover, the creation of a growing set of specific obligations in the Protocol, both procedural and substantive—including production and consumption limits for ozone-depleting chemicals and obligations to report data, to participate in multiple subsidiary bodies created for various tasks, and to support periodic assessments and reviews of the Protocol—also increased the influence of the institutions that were the custodians of these obligations.

While this institutional effect is familiar, the stability of the early deadlock and later progress was also supported by two more powerful dynamic processes, which have not been previously recognized and are of substantial general importance. The first of these concerns the availability of knowledge about technical options to reduce ozone-depleting chemicals and the interactions between this knowledge, technological assessment, and regulatory controls. The second concerns the linked evolution of scientific knowledge, its presentation and synthesis in scientific assessments, and the use of scientific claims as warrants for policy action or delay. Each of these processes operated in distinct but related modes before and after regime formation. In each case, events associated with the regime formation shifted the process from operating in a mode that supported continued deadlock, to a different mode that supported continued progress.

9.4 Technological Change, Technology Assessment, and the Ozone Regime

Technological knowledge about the feasibility, performance, and cost of potential alternatives to CFCs, and the distribution of this knowledge among economic and policy actors, played crucial roles in sustaining both the early deadlock and the later rapid progress of the ozone regime. Before the transition, such knowledge was held predominantly by the CFC producers, who had conducted initial research on chemical alternatives in the 1970s. These firms had no interest in helping the proponents of CFC controls to compel them to do something costly, difficult, and risky. Consequently, they sought to promote a pessimistic view of the viability of alternatives. They consistently characterized alternatives in highly unfavorable terms, kept tight control over relevant technical information, and attempted to keep detailed arguments about the potential to develop alternatives out of policy debates as much as possible.

Moreover, these firms were not just able to control the availability of technical knowledge to other actors; they also had some control over how much they themselves knew, since they could choose how far to investigate alternatives. Their decisions to abandon alternatives research around 1980 in part simply cut off investment in a project unlikely to be profitable, but also limited their exposure to the risk of CFC regulation. By pursuing alternatives far enough to be reasonably confident that they could develop them if needed, they limited the harms they would face under CFC regulation. But pursuing alternatives to the point of solving all major problems would have put them in the position of holding information whose revelation could harm them, by increasing the risk of CFC restrictions. As long as they were confident of success in opposing restrictions, it was better not to know how to solve the problems of commercializing alternatives. The firms trod this line skillfully. Sustaining the widespread belief that CFC alternatives were infeasible or unacceptably costly, while making no outright false statements, was a great strategic success of industry through the early 1980s.

Other actors knew much less than the CFC manufacturers about alternatives. To the extent that anyone knew about reducing CFC use through process or product change, it was the major user firms. But since most known non-CFC alternatives had been rejected in favor of CFCs or replaced by them, this knowledge had an unfavorable bias: most alternatives had known problems that would require significant effort to overcome. Moreover, since users were not mobilized in looking for alternatives absent a salient threat to CFCs, many alternatives had simply not been identified because there was no reason to look. For their part, the activists seeking to promote controls could not demonstrate that significant reductions were feasible, because they lacked the authoritative technical knowledge necessary to rebut industry claims that alternatives were highly problematic. Unable to engage industry expertise, independent attempts to assess CFC reduction opportunities were thwarted from the start. These studies' conclusions that the maximum feasible reduction in CFC use *at any price* was 25 to 50 percent illustrate how totally the industry view prevailed. This distribution of knowledge sustained a stable equilibrium in which the level of confidence that alternatives could be developed and the level of efforts made to develop them were both low. Those who wanted controls on environmental grounds could not show that they were feasible, while those with the best knowledge would not reveal it, lest others use it to make them take costly action.

The transition to regime formation did not depend on a shift in the conditions sustaining this low-confidence equilibrium, but was driven by other forces, principally the new consensus around a freeze formed by the 1986 scientific assessment. A belief that developing alternatives would be easier than widely believed no doubt contributed to the U.S. activists' 1986–1987 campaign for strict cuts, but the evidence to support such confidence remained extremely thin and the activists were also willing to risk taking a position whose feasibility was uncertain. Once enacted, however, the initial target and other elements of the regime changed the conditions that had supported the low-confidence equilibrium, establishing a new equilibrium that generated sustained progress in identifying and implementing ways to reduce ozone-depleting chemicals. The new equilibrium was supported by interactions between the periodic reviews of Protocol controls, firms' efforts to limit their exposure to regulatory and competitive risk, and a cluster of activities to produce, evaluate, and distribute technical information about alternatives, with the Protocol's technology assessment process at its hub.

Establishing these new conditions depended on the regulatory target in the initial Protocol, and the spread of calls for stronger restrictions over the next two years. The initial target and the risk of more to come transformed the business environment for firms producing and using ozone-depleting chemicals. The Protocol triggered the beginning of consolidation in CFC production almost immediately. Over the longer term the looming targets posed grave risks for CFC producers but also held opportunities for the largest and most technically sophisticated producers, since they created the market conditions necessary to commercialize chemical alternatives. For CFC users, on the other hand, the effects of both the initial controls and the threat of more were entirely harmful. Threatened with losing technologies on which they depended to varying degrees, users were offered instead the unrealized promise of costlier chemical alternatives of uncertain availability, performance, and regulatory acceptability. This prospect set off a headlong rush to reduce dependence on the threatened chemicals. User firms collaborated with CFC makers to develop fluorocarbon alternatives, but also pursued their own research and entertained third-party proposals. Diverse routes were explored, including related and unrelated chemicals, old or rejected approaches reconsidered, changes in manufacturing processes and product characteristics, and operational changes to reduce emissions and use.

Government and industry collaborated on several initiatives to help promote this innovation drive and focus its results, by publicizing the need for reductions and promoting open exchange and critical examination of proposals. These collaborations included the annual "Alternatives" conferences on ways to reduce all uses of ozone-depleting chemicals, workshops on sector-specific reduction opportunities, and cooperative organizations to undertake extended work on specific problems. These activities made technical knowledge about alternatives and emission-reduction options much more widely available than previously, breaking the predominant control over information formerly held by the CFC producers.

The Protocol's technology assessment processes, TEAP and its sectoral committees, played critical roles in promoting this wave of innovation and linking it to the evolving Protocol negotiations. The effectiveness of these bodies turned on their recognizing and exploiting a fundamental difference between technology assessments and scientific assessments: that technology assessments have a much greater ability to change the conditions of technological feasibility on which they are reporting, by advancing present technical skill, solving problems, and identifying and removing barriers to the implementation of new processes and products.

TEAP's primary and official job was to advise parties on the extent of feasible reductions in ozone-depleting chemicals in each sector. Assessing technical options to support regulatory decisions is a difficult job, and prior attempts had usually ended in failure, not just on ozone but on all environmental issues.[17] Because the best technical information on which to base feasibility judgments is usually held by industry, the job requires engaging technical experts from industry. But since the reason for seeking this information is to inform regulatory decisions, industry experts are typically reluctant to participate and to share their knowledge freely. Indeed, why should they? The attention of a firm's best technical people is among its most valuable competitive assets. Diverting them to serve on public advisory bodies, in pursuit of public goals that promise little benefit, or even harm, to the firm is extremely costly. No firm, and no industry, has an interest in helping regulators to impose burdens on them.

Despite this obstacle, TEAP succeeded in engaging a critical mass of top technical experts from the firms, research institutes, and other organizations with the most current and relevant expertise. This success was not just a matter of persuading the right experts to attend, but of eliciting their best efforts and most honest judgments with little regard for the policy positions or immediate commercial interests of their employers. Industry experts participated in this way, and their firms let them, for several reasons that reflected a combination of skillful strategic choices by regime organizers and favorable conditions. The strongest motivation to participate was the crisis that user firms faced from the looming Protocol target and the imminent risk of stricter ones, which made them desperate to reduce their reliance on ozone-depleting chemicals as rapidly as possible. TEAP organizers effectively exploited this crisis by designing the process to help firms solve the technical problems of achieving such rapid reductions and the collective-action problems of organizing to do so. TEAP's committees brought together, with antitrust protection, critical masses of the most respected experts in highly specific subfields. Working groups included representatives of both major users needing to reduce their use rapidly and firms developing diverse alternative technologies. The panel's work was orga-nized to maximize practical effectiveness and reduce extraneous risks and costs: working sessions were intensive, focused, and short; work teams had complete con-trol over their reports; activities included site visits, acquiring specific technologies for testing and modification, and whatever was needed to assess whether, how, and how well something worked. The processes of critically examining and evaluating technical alternatives and solving application problems provided both a high order of professional challenge and satisfaction, and some of the best opportunities to solve the problems of rapidly reducing CFC use, increasing both individuals' interest in participating and their firms' willingness to send them. The crucial strategic in-sight underlying this success was that the activities of gathering data and collectively solving technical problems that served the needs of the participating firms, also served the panel's purpose of giving the parties high-quality technical advice on what further reductions were feasible.

Moreover, the same activities provided still more benefits to the regime, which were not among TEAP's official responsibilities but were among its most important contributions. Experts' work on the panels to solve problems and identify and refine alternatives repeatedly revealed opportunities to reduce usage further than previ-ously recognized, not just to the point of meeting existing targets but beyond them. With industry's responses to the environmental and regulatory challenge repeatedly moving them ahead of existing regulatory obligations and the repeated identifica-tion of further reduction opportunities, it was repeatedly possible for the regime, with advice from TEAP, to tighten regulatory requirements. Moreover, as the panel's work proceeded and further opportunities were identified, individual participants increasingly took on the role of confirming the legitimacy of the assessment process and spreading information among their industry peers about the opportunities it identified, a task that their professional stature and connections made them uniquely capable of doing. Through these processes, TEAP made three basic con-tributions to the development of the regime: providing advice on feasible reductions, as charged; advancing the margin of feasible reductions through focused technical problem-solving; and advancing the reductions actually achieved, by supporting the dissemination of knowledge about emerging reduction options.

Evidence of the success of the technology assessment process is of several kinds. TEAP and its subbodies provided a huge number of specific technical judgments that were, with few exceptions, persuasive, technically supported, consensual, and, when tested by subsequent technology development, were found to be accurate or somewhat conservative. While carefully avoiding usurping parties' authority, TEAP exercised substantial influence over their decisions. Its strong, specific, carefully delimited statements of feasible reductions repeatedly shaped the rate and direction of parties' control decisions. Moreover, its conclusions were only infrequently disputed by policy actors, about equally for being too bold and too timid, even when parties did not follow them precisely. A strong measure of parties' approval of TEAP's performance is that they repeatedly asked TEAP to take on new and expanded jobs, even to answer questions that embedded too much policy to be resolved by technical deliberations (which TEAP declined). Indeed, parties relied on TEAP increasingly frequently after 1990 to help reduce their negotiating conflicts: when they met a serious obstacle in negotiations, they would identify a related quasi-technical question to pose to TEAP and defer negotiations on the issue until receiving TEAP's response. A more recent testimony to TEAP's effectiveness was the unsuccessful attempts of some parties to assert greater control over the assessment process after its conclusions repeatedly failed to support their preferred approach to Protocol revision.

At a more general level, the explanation for TEAP's effectiveness in providing public benefits to the ozone regime—good advice on the extent of feasible reductions, and promotion of further opportunities to reduce more—is that TEAP effectively coupled the provision of these public benefits with private benefits to participating individuals and firms. These private benefits were sufficient to motivate the level of participation and effort that the assessment processes needed in order to succeed. This was an innovation of substantial importance. The strategy of motivating private actors to provide public goods voluntarily by jointly providing private benefits to them has long been recognized for conventional public goods such as interest-group political organization and charitable giving,[18] but this technology assessment process applied the model to the less studied but fundamentally important process of eliciting information from private actors to support public-policy decisions. Private participants were attracted by three kinds of private benefits: help in solving the urgent technical problems they faced in reducing their own dependence on ozone-depleting substances; ideas, information, and contacts to support commercial and professional opportunities in marketing alternative technologies and associated services and information; and the professional challenge, prestige, and peer-group recognition that the process offered the individual experts who participated.

The occasions when the Protocol's technology assessment process performed the weakest provide additional support for this explanation of the foundations of its success. The assessment process has been most contentious and least effective when the balance between public and private benefits on which its success rested has been disrupted, because private benefits at stake were either too large or too small. On some occasions, the specific competitive advantages at stake have made the potential private benefits to participating firms so large—and so rival—that firms have sought to manipulate the technical deliberations and conclusions to their own advantage. These conditions have arisen most acutely over questions of when to phase out an essential-use exemption. For example, when one firm marketed the first non-CFC MDI, that firm's representatives sought to have the essential-use exemption for

MDIs revoked immediately, while those whose alternatives were still in development sought a more gradual phaseout of the exemption. These occasions have involved TEAP bodies in bargaining that has inevitably blurred the boundary between technical and political debates, but has not called the broader technical integrity of the process into question.

On other occasions, potential private benefits have been so small relative to the expected cost of regulatory controls that private participants have tried to obstruct the assessment process entirely. These more serious conditions have arisen each time parties considered broadening controls to include new chemicals, particularly MC in 1990 and MeBr in 1995. On these occasions, the assessment process had to evaluate feasible reduction opportunities with no control yet enacted, and a reasonable prospect that none would be enacted if reductions appeared too difficult or costly. Under these conditions, firms producing and using the chemicals proposed for controls had no interest in assisting the technology assessment process. Rather, their preferred strategy was to obstruct the assessment and claim that significant reductions were impossible—the strategy followed for CFCs prior to 1986. As the political context shifted and the prospect of controls grew more likely, the interests of users and of potential marketers of alternatives began to shift to support the assessment process. But no such shift occurred for those, like the MeBr producers, who had no prospect of marketing alternatives. Obstruction remained their preferred strategy even as it grew increasingly likely that controls would be enacted, since they had no prospect of mitigating their losses by developing alternatives. In the initial attempts to assess alternatives for these newly considered chemicals, TEAP's bodies had much weaker engagement of industry expertise than in their other efforts, and their judgments were consequently more weakly founded, more contested, and at greater risk of error.

TEAP was just one component, albeit a crucial one, of a broader process of interactions between regulatory targets, technological advice, and industry activity. This process stimulated and channeled a huge wave of innovation, whose results were repeatedly integrated into decisions to advance control measures. The result was that ozone-depleting chemicals were eliminated much more rapidly than was thought possible. In certain key respects, this rapid regime development was achieved by reversing the order of actions implied by a simple image of rational policy choice. Conventionally, rational policy choice is viewed as involving the assessment of risks, impacts, and responses prior to deciding on control measures. The success of this regime appears to have been associated with proceeding in nearly the opposite direction. Assessments of atmospheric science were influential in forming the regime and adopting the first target, but this target was adopted with little confidence that it could be met at reasonable cost. It was the prior enactment of this fairly stringent target, and the risk of more to come, that set in motion the subsequent processes of technological development, assessment, and strengthening of control measures through which the regime adapted so rapidly and successfully.

9.5 The Policy Influence of Scientific Knowledge and Assessment

The questions of what relationship scientific knowledge does, can, and should have with political decision-making are old and vexed. Although the suggestion that scientific knowledge either can or should be the exclusive determinant of policy choices, intermittently popular in the early twentieth century, is now regarded as quaint,[19] issues of the extent, mechanisms, and conditions of influence of scientific claims and knowledge on policy—in either positive or normative terms—remain only weakly understood. The suggestion that public decision-making is most competent and legitimate when separate domains are delineated in which the authority of expertise and of democracy predominates has a history nearly as troubled as that of older ideas that either domain's preeminence is absolute.[20] Yet the contrary view, that expert knowledge and political power are so deeply intertwined that it is futile to try to separate them, is equally unsatisfying.[21] Recent scholarship has concentrated on positive description of how the boundaries between expert and political domains are defined and adjusted in practice, resonating with many elements of more sophisticated scholarship from the 1960s and 1970s that avoided the false dichotomies then widespread.[22] This recent work has identified various forms of intermediary bodies or social networks that define the boundary and transmit information and influence across it, in both directions or just one.[23]

Two aspects of the history of the ozone issue cast light on these questions: policy actors' competitive use of scientific arguments as warrants for their preferred positions in policy negotiations; and the design and management of institutions to mediate between the domains of science and policy, and advise international policy negotiations. Arguments about scientific evidence for ozone-depletion risk were always prominent over the history of the issue, both domestically and internationally. Nearly all policy proposals were presented as founded on available scientific evidence, and nearly all disagreements over how to proceed were characterized as arising from different interpretations of the available evidence of risk. But the actual influence of scientific knowledge on policy was much more limited than this prominent rhetorical use would suggest. As argued above, the instances of direct policy influence of primary scientific claims were infrequent and of a few specific types. In contrast, there were clear instances of strong policy influence of scientific assessments. Here, I elaborate on these claims: (a) I identify the specific ways primary scientific claims did affect policy; (b) I elaborate on conditions for scientific assessments' achieving stronger influence; (c) I identify two distinct patterns in the rhetorical use of scientific claims and their assessment over time that contributed to the initial deadlock and the subsequent progress.

The core of the argument rests on the distinction between primary scientific claims and scientific assessments. By "primary scientific claims," I mean the direct results of scientific research: new observations, theories, or model results advanced by scientists individually or in groups, published in the scientific literature, or verified through criticism, attempted replication, and consensus in the relevant research community. By "assessments," I mean collective, deliberative processes of scientific experts reviewing and evaluating the state of scientific knowledge, and synthesizing it with a view to providing information of use to policy makers or decision-makers, typically in the form of an official report issued under the authority of some organization. Assessments can include restatements or critical evaluations of preexisting primary scientific claims, as well as synthetic statements that draw conclusions from the existing body of research. Assessments do not normally include new primary research or analysis, although one important ozone assessment, the 1988 Ozone Trends Panel, did.

Over the history of the ozone issue, primary scientific claims were repeatedly responsible for placing the issue on domestic policy agendas. This was achieved through public statements and advocacy by individual scientists, and through scientific results being publicized by scientists, the press, and environmental organizations. In the United States, ozone depletion from supersonic aircraft came onto the policy agenda in 1971 due to arguments advanced publicly by McDonald and Johnston, and depletion from CFCs and chlorine came onto the agenda in 1974 due to Molina and Rowland's paper and ACS presentation, and public advocacy by Rowland, Molina, Cicerone, and a few other scientists. Thereafter, every significant new finding quickly passed from being debated in scientific circles to policy circles. Not all attempts to put the issue onto policy agendas succeeded, of course, as Hampson's futile efforts beginning in 1964 attest.

Beyond their agenda-setting role, primary scientific claims were routinely, but selectively, used by policy actors to support their existing policy positions. Activists consistently adopted and publicized new claims and results that suggested more serious risks to support their arguments for policy action; opponents did the same with new results suggesting diminished risk to support their arguments that action was not justified. Used this way, new results that significantly strengthened or weakened the case for ozone loss periodically contributed to a general elevation or diminution of policy concern, or emboldened the advocates or opponents of controls on ozone-depleting chemicals.

But while this partisan, selective use of primary scientific claims was always prominent in policy debates, primary scientific claims rarely or never exercised more direct or active influence on policy outcomes, either domestically or internationally. They did not induce any policy actor to retreat from a prior position, close any debate over either policy action or its scientific justification, or provide any focal points around which coalitions formed.[24] Even a weaker form of influence, causing partisan actors to abandon an argument they were formerly making in support of their preferred position, occurred only infrequently.[25] This is admittedly a high standard for judging policy influence, but it is a standard that scientific assessments, unlike primary scientific claims, achieved on several occasions in the history of this issue. Consequently, this high standard does not lead to a predictable, generic finding of "no effect."[26]

Scientific claims were used in an additional way that contributed to sustaining the initial deadlock, when unresolved claims were widely accepted as warrants for delaying policy action. From the earliest policy debates, actors argued over what scientific evidence would justify further chemical controls. Although the qualitative confirmation of the major points of the CFC/ozone claim achieved by 1976 was sufficient to motivate aerosol controls in a few countries, only a few activists thought this evidence justified more stringent and potentially costly restrictions. But what additional evidence would justify such stronger controls? Both activists and moderates repeatedly challenged opponents to state what evidence they would accept as verification of depletion claims and as warrant for controls. In the early debate, it briefly appeared that observing ClO in the stratosphere would be the decisive evidence, but any consensus that may have existed on this point quickly fell apart after ClO was detected, with the opponents demanding demonstration that likely depletion would be quantitatively important.

In later debate, the questions that came to be widely accepted as decisive for policy action were principally framed by the opponents of early action. This was a predictable consequence of the rhetorical structure of early arguments. Since at each point, activists argued that the evidence already available was sufficient while opponents argued it was not, it was the opponents who were repeatedly pressed to state, at least illustratively, what further evidence would change their minds. Two pieces of evidence emerged from this argument as the most important conditions to justify controls: projections of "significant" ozone depletion in 2-D stratospheric models under realistic emission scenarios including multiple pollutants; and detection of a statistically significant decline in global ozone.

These points were advantageous to opponents, in that they posed a stringent challenge of demonstration and required substantial progress in research methods, technologies, and skills that in fact took about 10 years to achieve. But both were more striking for the advance they required in tools and methods than for any strong basis either had to be a decisive warrant for action. The ozone-trend claim first entered the debate in 1975, through an industry argument that ozone loss was more readily reversible than claimed, because any loss could be quickly detected and (under extreme assumptions of rapid response) arrested. Over several years, this argument about the feasibility of arresting ozone loss by responding rapidly to an observed trend evolved into the argument that observing a significant trend should be the evidentiary standard necessary to justify a response. This point gained surprisingly widespread acceptance, despite clear shortfalls of two forms. On the one hand, seeing a significant trend would not solve the problem of attributing causation: a decline could be a natural fluctuation, or could be caused by some human perturbation other than chlorinated chemicals. On the other hand, if a significant trend was caused by chlorinated chemicals, demanding its detection as the condition for action is a stark refusal to adopt a precautionary stance. In view of the long-term character of the issue and the slow reversibility of ozone loss, taking this as the threshold for action implies a strong willingness to accept risk of environmental harm.

The condition of finding significant ozone loss in 2-D models under multiple-perturbation scenarios did admit the possibility of precautionary action before depletion was realized. Moreover, the modeling advances that were required—2-D models with full chemistry and multiple-perturbation scenarios—addressed basic weaknesses of early models, and were necessary to identify latitudinal and seasonal patterns of depletion and to put depletion projections in a reasonable context of broader atmospheric changes. But like the ozone-trends challenge, this one was more striking for the large methodological progress it required than for any decisive advance it brought in the credibility of future loss projections.

Moreover, neither of these challenges addressed the largest component of uncertainty in future depletion—how much emissions would grow—which was systematically ignored throughout the debate. The neglect of emission-growth projections was particularly striking in the modeling debate, since all arguments over modeled depletion projections were fought over steady-state results. Ignoring uncertainty over emissions growth was one way that the debate was consistently simplified into a binary fight over whether action was warranted or not. Attempts to make the debate even minimally more sophisticated by considering contingent responses against potential large emissions growth never succeeded.

Since the opponents of action had the greatest influence in framing these evidentiary challenges, it is unsurprising that the large advances required to meet them served as effective delaying devices for nearly 10 years. But this delaying effect was intrinsically self-limiting, and grew progressively weaker after 1984 as it began to appear that the challenges soon would be met. The predominant but neglected uncertainty, emissions growth, became increasingly prominent after growth resumed in 1983. Model results that met the long-standing challenge began to appear in 1984 and 1985. Various claims began to appear that global ozone depletion had occurred, although these were not yet verified or even widely accepted.

Sufficiently credible resolution of these long-standing questions was achieved between 1986 and 1988—weakly, partially, and suggestively by 1986, and decisively by late 1988.[27] But the resolution gained its policy effect through assessments, not (or not exclusively) through new primary results. The 1986 assessment conducted the model comparison that showed the long-awaited 2-D models projected somewhat larger depletion than earlier models, both in the global average and in the newly possible temperate-latitude projections. This comparison also showed that continued growth would likely cause large ozone loss, and confirmed the previously disputed resumption of CFC market growth. The 1988 assessment concluded that significant and unexpectedly large losses of global ozone had occurred, and strengthened earlier findings of substantial depletion in 2-D model projections and strong resumption of CFC growth. With these assessment findings, the long-established but arbitrary evidentiary standard was met.

When the standard was met, the actors who had used it to call for delay accepted it as justifying the policy actions they had long opposed, and neither tried to use remaining uncertainty to call for further delays nor found other grounds to continue opposing action.[28] Major industry and government actors changed their positions to accept limits on CFC growth in response to the 1986 assessment, and phaseouts in response to the 1988 Trends Panel. There was no serious attempt on either occasion to dispute the assessment's conclusion, or to claim the conclusion did not require a change in position. Calling for more research had allowed opponents to delay calls for action, but carried an implied commitment to accept action if and when the research results supported it.

This record indicates that scientific assessments can exercise important influence on policy. They can accomplish this by authoritatively resolving scientific questions that have come to be accepted as crucial determinants of the seriousness of the issue and warrants for policy action. They can also accomplish this by constraining the bargaining range for subsequent action by making it impossible to claim scientific foundation for policy proposals lying outside certain limits. But while some scientific assessments exercised such influence over the history of the ozone issue, many did not. For an assessment to achieve such influence required that the state of scientific knowledge being assessed and the policy setting into which the results were delivered meet certain conditions. It also required that the process and output of the assessment meet certain conditions necessary to make it authoritative, in the sense that no policy actor attempted to claim that its statements were false.

Achieving this authoritative status depended on several linked aspects of the assessments' design, participation, sponsorship, procedures, and outputs. The assessments that achieved it—the 1986 NASA/WMO assessment, the Trends Panel, and the subsequent science panels—followed and extended the practice of earlier NASA assessments by assembling as participants a large number of the most respected experts in the field. They organized their participation, deliberations, and written products to produce reference-quality reviews of specific subfields. They also made statements that synthesized and interpreted the current state of relevant knowledge, but whose scope was limited to matters on which the participants were indeed an authoritative group, avoiding the charges leveled against the earlier NAS and U.K. assessments that they were advancing a policy agenda.[29] The scientific interest and prestige of the activity increased its attractiveness to participants, in a mutually reinforcing process whose consequence was that the reports of the activities were held in the highest scientific respect. Despite NASA leadership and funding for the 1986 and 1988 assessments, the degree of international participation and sponsorship assuaged any concern that they might reflect a national bias, with the consequence that no outsider could credibly deny their conclusions.

These elements of assessment design comprise what I call the authoritative monopoly strategy—conducting an assessment by assembling an international near-monopoly of relevant expertise to present an authoritative statement of present knowledge. This strategy allowed assessments to make statements of scientific knowledge, including certain carefully delimited synthetic statements based on such knowledge, that were beyond reproach in the practical sense that actors who opposed the policy implications drawn from them could not rebut them. Moreover, the scale of such assessments made any attempt to mount a parallel or potentially contending assessment scientifically superfluous, if not infeasible. When other assessments were conducted at all, including those by official national bodies, they were subordinated to the international assessment, providing national evaluation and perspective to meet a legal or political need to have a national assessment. Consequently, the standing of an assessment conducted by this strategy to make authoritative statements is exclusive, channeling all scientific debate through one activity and ensuring that the research community speaks with a single voice on matters of interpretation and nuance of language that could risk conflicting interpretations. While subsequent assessments conducted by the science panel had their standing enhanced by their formal status under the Protocol, the 1986 and 1988 assessments show that authoritative status can be achieved through an assessment's strategy—its process, management, and outputs, conditioned on the knowledge and policy context—independent of such official status.

Assessments that attained such authoritative status influenced the subsequent policy debate by compelling policy actors to engage their major conclusions or risk being discredited as serious participants. They accomplished this through three linked effects, each of which blocked one of the major rhetorical strategies a policy actor might use to avoid engaging the claims. The scientific stature and scale of the assessments made their major conclusions sufficiently credible that policy actors could not arbitrarily deny them. Rather, to the extent that actors sought to dispute the implied policy consequences of assessment statements, they supported their claims using concepts, arguments, and statements drawn from the assessments. The broad international participation and sponsorship of the assessments made their claims sufficiently legitimate that policy actors could not reject them as partisan without engaging their substance. And the prominent presentation of assessments, under official auspices of governmental or intergovernmental institutions, made their claims sufficiently salient that no policy actor could ignore them. This last effect is crucial in distinguishing the effect of assessment statements from that of primary scientific statements, however widely known or well verified. The prominent public statement of assessment conclusions, endorsed by powerful institutions, makes the conclusions common knowledge, in the game-theoretic sense that all relevant actors (in policy and scientific domains and on the boundary) do not just know them, but know that all others know them, and so on.[30] It is this reciprocal knowledge that compels policy actors to address the conclusions and their policy implications, even when the same claims, as well validated, carried no such compulsion when stated only in scientific forums.

The scale and expense of conducting an assessment using this strategy implies that it is not often possible to do so. The time and effort required are costly to the scientific participants, so the willingness of enough leading people to participate, which is essential for the strategy to succeed, itself provides a strong signal of their consensual view of the seriousness of the issue. Even given this level of scientific concern, such an assessment still requires resources and leadership. These assessments were powerfully driven by financing and leadership from NASA, which as the largest upper-atmosphere research funder in the world, had the money and clout to recruit top-level participants. To achieve the required scale and participation, it also helped that these assessments provided scientific benefits to participants and sponsors in parallel with the authoritative statement of current knowledge they provided to policy makers.

Prior assessments that failed to influence policy illustrate the need for each of the major elements of the authoritative monopoly strategy. Many assessments lacked the scope and stature of participation needed to avoid errors or idiosyncrasies, make focused statements of the state of knowledge, resolve ambiguities, or render judgments on outstanding debates—that is, to develop an authoritative consensus that can attract wide scientific attention and respect, and escape attack. Assessments conducted by small expert committees in particular, even when in many respects they were of exemplary scientific quality, failed to attain this authoritative status, and were attacked by both scientists and policy actors for even minor errors, omissions, or nuances and interpretations. Assessments that lacked broad international participation and sponsorship were readily dismissed as tainted by a single national viewpoint. Those that strayed too far into explicit statements of policy implications were dismissed as political documents of no credibility. When multiple independent assessments were conducted at the same time, these could readily be combed for discrepancies to support the claim that scientific knowledge was too contested and uncertain for anything to be done but conduct more research.

The 1986 and 1988 assessments were crucial in the initial formation of the ozone regime. Scientific assessments continued to employ the same strategy to provide authoritative summaries of current knowledge after the regime was formed, now formally embedded in the regime's institutions. But the functions these assessments performed for the regime, and the character of debates over the policy implications of scientific knowledge in which they were engaged, underwent a fundamental change. After the initial threshold of credibility was passed and the regime was formed, no similarly high evidentiary threshold was applied to questions of whether new results provided grounds for further action. Indeed, policy debates never again seriously engaged basic questions of atmospheric science. Even subsequent revelations of important limitations in understanding of ozone depletion never called into question the judgment that the issue was serious enough to merit continued policy action, once this judgment had initially been made. Rather, each new observation that suggested increased risk and was reported by an assessment was accepted as establishing the desirability of tighter controls, with virtually no policy argument over the confidence or reliability of the new results. Since new results for several years provided increasingly alarming atmospheric observations and projections, scientific evidence and assessments repeatedly provoked increased public and political concern and motivated the pursuit of tighter controls.

This trend could largely have been predicted as soon as ozone loss was first confirmed, because of the issue's long time constant. Once a stratospheric effect of chlorine was first detected with high significance in the noisy ozone record, it was bound to increase as stratospheric chlorine increased, driven by the large accumulated stock of CFCs in the troposphere, regardless of near-term emission reductions (which would not affect these trends for years to decades). But, predictable or not, the new observations were widely perceived to represent new information showing that the problem was worse than had been thought. The issue's long time constant had initially hindered definitive detection of a change, and so contributed to inappropriate political complacency, but now the same characteristic generated a climate of steadily increasing concern that maintained pressure for the maximal feasible acceleration of controls. A few industry actors tried to argue that the new observations were predictable and so did not warrant repeated tightening of controls, but to no effect.

With increasing evidence of risk, and a low evidentiary threshold required to establish the desirability of further action, scientific assessments no longer needed to build an authoritative case for the seriousness of the issue as a warrant for action. Rather, the primary function of these assessments shifted toward a management-support role, giving parties guidance on what kind of further action to take. To provide this guidance in the face of the failure of the gas-phase stratospheric models formerly used, assessments used the proxy measure of halogen-loading time paths instead of depletion projections, and various policy metrics they defined on the basis of these time paths. These projections became the universally accepted metric for comparing scenarios and policies, greatly increasing the assessments' utility to policy makers—ironically, since the proxies originated as a pragmatic response to a failure of scientific explanation, and were widely recognized to have serious flaws.

The existence of the Protocol's controls made it easier to conduct and defend these projections by providing a defensible baseline against which future alternative decisions could be assessed, and thus making it unnecessary to project and defend an arbitrary emissions baseline—the object of a long-standing battle before the Protocol. In addition to providing a baseline, present controls also operated as a lower bound. While the assessments' projections did consider various scenarios of noncompliance and emission growth in nonparties, they never considered a relaxation of previously agreed controls—although it is plausible that they could have, had new evidence suggested that depletion concerns were less serious than previously thought.

Although these assessments never sought to foreclose parties' prerogative to decide future controls, they retained their authoritative status and continued to channel and delimit policy debate by the same mechanisms as operated for the 1986 and 1988 assessments. Their conclusions remained for all practical purposes beyond criticism, and they continued to displace any other ozone-assessment efforts.[31] They continued to make positive statements about the differences in atmospheric consequences of alternative decisions that, if not quite making any choices on the agenda indefensible, provided strong supporting arguments to the advocates of choices yielding markedly reduced atmospheric harms. These arguments were not decisive, but in the presence of other favorable conditions, often prevailed.

The Political Insignificance of Downstream Environmental Effects

In contrast to scientific arguments about atmospheric changes in their own right, arguments about the social, economic, or ecological effects of these changes mattered only at the earliest stage of the issue, and only in a highly simplified form: plausible identification of a sufficiently serious impact was necessary to demonstrate why the proposed atmospheric change should matter, and to put it onto policy agendas. After this was achieved in 1971, no subsequent information or argument about effects, whether presented in primary scientific literature or in assessments, influenced any policy outcome or even figured prominently in any policy debate.

The 1994 Effects Panel was impressively candid in acknowledging this limited influence, noting that a few clear, important examples of harms had sufficed to demonstrate the broad necessity of protecting the ozone layer, while persistent weaknesses in the science of impacts made them unable to contribute to subsequent debates about the adjustment of control measures, or to help identify and evaluate response options, once that initial decision had been made.[32]

The insignificance of effects in the policy debate is more explicable after 1990. Once the decision had been taken to move to the extreme point of eliminating all ozone-depleting chemicals as fast as possible, neither claims about effects nor claims about the atmospheric science of ozone depletion had any further significance for the policy debate. The pace of subsequent phaseouts and broadening was determined entirely by judgment of the maximum feasible pace of the transition, not by any arguments about the harms averted.

But the lack of consideration given to effects in policy debates before this point is surprising, calling deeply into question whether the balancing of benefits and costs played any significant role in this policy debate. Not only did impacts not figure in policy debates after initially contributing to putting the issue on policy agendas, but policy discussions of responses never considered adaptations to reduce impacts. Discussions of atmospheric science and of response technologies were consistently and rigorously kept separate through the development of the issue. They were used in different ways and prominent at different stages of the regime's development: atmospheric science claims were most prominent around the time of the regime's formation, when they crossed the threshold of seriousness judged necessary to take action; technological claims were most prominent after the regime's formation, when they defined the boundary of maximally rapid extension of controls. Moreover, these two bodies of information were assessed by distinct groups using highly dissimilar processes, linked only by the charge to assess common scenarios of future controls developed in consultation with the parties.

9.6 Practical Implications: Lessons for Management of Other Issues

This section seeks to identify the points of practical guidance that the experience of the ozone issue may hold for management of other international issues. It does not simply attempt to identify specific decisions made in the ozone regime with successful outcomes to argue that these should be imitated for other issues. Rather, it identifies ways that the regime has effectively exploited clear patterns of influence operating in scientific, technological, economic, or political domains, such that similar approaches are likely to be effective on other issues when the same conditions are replicated. Some of the practical lessons identified are simply reframings of the theoretical claims discussed above, emphasizing causally important factors that actors might seek to manipulate, to the extent they are able. Other lessons do not have a connection to the points of general theoretical interest discussed above, but are of practical importance and appear to be persuasively demonstrated in the ozone record.

Institutions and Procedures for Adaptation

In contributing to long-term effectiveness, the most important elements of the 1987 Protocol were its provisions for adapting controls in response to new information and new capabilities: the requirement for periodic review of controls, the independent expert assessment panels, and the linkages between these. Such provisions will be essential if concrete measures are to be taken on any issue under significant uncertainty about whether they are correct in their stringency, form, timing, or breadth—that is, under any attempt to take precautionary action.

Although it has been argued that initial targets for an environmental regime must be weak in their stringency but wide in their participation in order to avoid shifts in emission-intensive investment that will thwart the regime's purpose and obstruct further bargaining progress,[33] narrow participation in the initial ozone targets did not impair the subsequent effectiveness of the regime because of these adaptation provisions. Although the Protocol began with limited participation, including no major developing country, the need to broaden participation was recognized from the outset and the Protocol's trade measures gave nonparties incentives to join.[34] Together with the financial provisions negotiated in 1990, these incentives were sufficient to expand participation to include the major developing countries. Narrow initial participation did not matter because the regime was adaptable enough to expand participation faster than the shift of investment to nonparties that might have generated significant leakage of emissions or obstacles to further progress.

The successful development of an adaptive regime is to be contrasted with an alternative approach that was repeatedly proposed in ozone negotiations but never taken: advance adoption of explicit contingent agreements, which would trigger specified controls upon the occurrence of future events such as observed ozone depletion or observed growth in CFC markets. This approach would likely be much less effective at adapting to changed information than the approach actually taken here, which left future contingencies unspecified but constructed institutions and procedures to consider them. The two crucial weaknesses of the contingent agreement approach are that it presumes that all contingencies and the appropriate response to them can be specified in advance, neglecting the diversity and unexpectedness of new information that could be—and in fact was—received; and that it presumes that future parties would reliably agree on whether and when the triggering signal had been observed—which is implausible if parties make the determination directly, and risks politicizing assessment bodies if the task is delegated to them. In the ozone regime, less precise advance specification of the conditions for future action gave future parties the discretion they needed to leave future assessments independent and nonpolitical—a crucial lesson for the allocation of responsibility between political and assessment bodies in any regime.

The Power of Early Regulatory Targets, the Irrelevance of Their Level

The initial regulatory targets of the 1987 Protocol put into motion the dynamic processes that drove the subsequent adaptation of the ozone regime. These targets served three purposes. First, by bounding uncertainty over future emissions growth they provided a baseline against which to calibrate projections of future stratospheric change, thereby closing long-standing arguments over future emissions projections and eliminating the largest, most neglected uncertainty in future ozone loss. Second, they created market opportunities for CFC alternatives, making their development worthwhile even if they initially appeared to be uncompetitive. Third, by imposing vague but potentially large risks on industries producing—and especially using—CFCs, they forcefully directed attention and development effort to the problems of reducing use and developing alternatives. The second and third functions were by far the most important, even though they operated strongly only in the nations that had already controlled aerosols.

None of these effects depended on the initial target being at the correct level, either in the sense of being socially optimal (as judged in retrospect or at the time) or of being the level ultimately agreed upon after subsequent revisions. Indeed, as was widely recognized at the time, there was no substantive basis to justify the particular level of a 50 percent CFC reduction. This level was arbitrary, a round number in the middle of the bargaining range defined by the two major factions' opening proposals. If the activists' judgment of remaining uncertainties was correct, the 50 percent target was far too weak to solve the problem; if the opponents' judgments were correct, it was far stronger than necessary. It was not even confidently known to be attainable without serious cost and disruption in those countries that had already banned aerosols.

But the initial target did not have to be substantively correct in any sense to serve the purposes it did for subsequent regime development. Rather, it had to satisfy conditions defined in strategic, not substantive, terms. It had to be stringent enough to impose significant risk on its targets over whether they could meet it at reasonable cost, in order to provoke a serious reassessment of uses and a serious search for alternatives. It had to be far enough in the future to allow time to develop and implement alternatives. Yet it had to be credible, which in turn required that it not be so stringent or so distant that opponents would be tempted to gamble that they could reverse it, rather than making the efforts required to meet it. These conditions helped ensure a broad search for alternatives, not just those already identified and known to be feasible. But while the initial 50 percent target clearly did met these conditions, it is plausible that a less stringent target, perhaps in the freeze to 20 percent cut range sought by the Europeans and by industry, would also have met them—particularly since much of the response was motivated not by the target alone but also by the widely perceived risk that stricter ones might soon be enacted.

While the conditions defining a suitable target in this dynamic setting are distinct from those that define an appropriate static target, one condition applies to both cases: targets must be defined over a sufficiently broad and diverse set of activities—in this case, usage sectors for controlled substances—to diversify against the risk of some uses being difficult and costly to reduce despite best efforts. The ozone case forcefully shows the importance of defining targets broadly, since several uses (e.g., solvents and halons) were much easier to reduce than expected, while others (e.g., commercial refrigeration and insulating foams) were as hard or harder.

Many other global environmental issues, most notably global climate change, and possibly many nonenvironmental issues as well, must also be managed adaptively, by modifying specific policies in response to changes in knowledge and capabilities. In managing all such issues, initial targets are likely to serve the same strategic purposes as they did in ozone, and their appropriate level is likely to be defined by the same conditions. In the wake of the U.S. rejection of the Kyoto Protocol, it has become popular to criticize emission targets as a fundamentally inappropriate approach to reducing greenhouse gas emissions, but the arguments being advanced for this general rejection of targets fail to consider the dynamic role of targets proposed here.[35] When targets are viewed in this sense, the strongest case to be made against the Kyoto targets is that their control period is too near to allow the processes of developing and implementing emission-reduction options to yield results. Otherwise, feeble political commitments to any greenhouse-gas emission targets thus far have meant that the ability of such targets to launch adaptive processes such as occurred in ozone is simply untested.

Harnessing Feedbacks between Regulation, Technology Assessment, and Innovation

The Protocol's institutional arrangements allowed the regime to adapt in order to manage the ozone-depletion problem, but the main vehicle by which the problem was actually managed was the rapid development of technical options to reduce use of ozone-depleting chemicals, spurred by feedbacks between regulatory targets, technology assessment, and industry responses. The central driver of these feedbacks, and the most important and generalizable innovation of the ozone regime, was the Protocol's system for technology assessment. No similar system has been developed for any other issue, but many aspects of its design and operation are likely to be applicable to other issues if the required conditions are in place.

To drive the regime's adaptation as it did, this system had to solve two problems that have thwarted nearly all attempts to use technology assessment to advance a regulatory regime: motivating the required degree and character of participation by private-sector technical experts; and preventing discussions and conclusions from being biased to support the material interests of participating firms. Moreover, linkages between these problems made it impossible to solve them separately, since attempts to provide private benefits to motivate participation risked impairing the objectivity of the outputs. As discussed above, the problem of motivating participation was solved by designing the technical assessment process to jointly provide public benefits to the regime and private benefits to participants. Participants derived three kinds of benefits. First, the firms that depended on use of ozone-depleting chemicals faced acute needs to reduce their usage because of the current and anticipated regulatory targets. To solve their urgent technical and business problems, these firms needed the opportunities the assessment process provided to share relevant expertise with others facing similar problems. Offering this private benefit posed no threat to the regime or the integrity of the assessment process, since these private interests were closely aligned with the regime's goals: both were served by rapidly identifying diverse alternatives, solving problems in applying them, and conducting high-quality, objective technical assessment of them. This close alignment of interests depended, however, on the conditions that attended the formation of the ozone regime, the existing regulatory target and the threat of more stringent targets that preceded the work of the technology assessment process.

The second type of private benefits included various rewards to participating individual experts, such as professional challenge, enhanced professional stature, the respect of eminent peers, and recognition for helping to solve a serious global problem. In this case, too, the goals of the regime and the private interests of participants were fully aligned, although here the relevant private interests were those of participating individuals rather than their employers.

The third type of private benefits included diffuse commercial and professional opportunities to profit materially from the success of the regime and from participating in the assessment, principally through opportunities to market alternatives and associated services, information, and professional expertise. It was these benefits that required the most delicate management, since they were important for motivating participation but posed the risk that the assessment process would be captured: that its conclusions would fail to reflect the best judgments available from honest technical deliberations, but would instead be biased to serve the interests of particular firms or factions.

Such capture, in a particularly simple form, was the bane of all attempts to conduct technical assessment of CFC reduction opportunities before the Protocol, which were routinely captured by those seeking to preserve the status quo by arguing that large reductions would be ruinously costly. The risk of this particular form of capture was greatly reduced by the Protocol's initial regulatory targets, although assessment organizers still tried to defend against it by excluding CFC producers from the first assessment. While these firms' contributions to subsequent assessments suggest that this punitive precaution was not necessary, the same risk was realized in the behavior of methyl chloroform and methyl bromide producers when these chemicals were first proposed for inclusion in the Protocol.

The more serious risk of capture under the Protocol's assessment process was by interests favoring or opposing particular alternatives, seeking to bias the findings of the group with excessive claims for their preferred alternatives or attacks on their rivals. This risk was especially serious, because design of the process to promote effective participation by private-sector experts meant that normal methods to check bias through transparent proceedings or scientific peer review were not available. Instead, organizers sought to limit this risk principally through the mix of participants in each work group. Participation in each group was balanced to include those associated with multiple competing technological alternatives and representing a wide range of material interests, while the high professional stature and closely overlapping expertise of individual participants promoted a highly critical, nondeferential working environment in which implausible or weakly supported claims were vigorously questioned. In addition, all participants were required to agree to a set of ground rules intended to keep the proceedings technical and restrain tendentious argument.[36]

Different participants' private interests posed risks of capture to differing degrees, and not all had to be neutralized. Promoters of particular alternative technologies had to be counterbalanced by advocates of different approaches and by users. More broadly, the diffuse interests of participants who foresaw economic opportunities in rapid phasedowns, whether through consolidation of existing markets or through profitable opportunities in new ones, were balanced by the interests of users, who would bear the brunt of too-rapid phasedowns. Still, the process depended on the specific interests of individual participants—for example, competitive advantages that turned on the evaluation of a specific, proprietary technology—not overwhelming participants shared interests in solving common problems or in general opportunities from the transition to alternatives (or, failing that, being weak enough to be restrained by balanced participation and ground rules). When this condition could not be met, the process was least effective.

The structural factors that allowed this strategy to be so effective are likely to be present on many other issues. A similar strategy could effectively advance the development, knowledge, and implementation of technological options to manage other environmental issues; provided the two central conditions are met. Participating experts must derive private material benefits sufficient to motivate their serious participation, but not so strong and divisive as to preclude open technical deliberations. In addition, participants' domains of expertise must overlap enough that experts from multiple firms can collaborate effectively, benefit from collaboration, and restrain each other's tendencies to bias deliberations for their own benefit. This strategy has been applied once to technical assessment for climate change, for the high-GWP gases that are implicated in both the ozone and the climate issues, with promising results. It promises much wider applicability to specific aspects of the climate-change problem and to other issues on which technological innovation promises a significant contribution to effective management.

This strategy carries several costs and limitations, however. First, the process cannot be transparent, since it relies on closed deliberations and nonattribution to free participating experts to make independent judgments. Such secret deliberations may be unproblematic if they are overwhelmingly technical in content, but can represent significant loss of democratic accountability if they consider technical questions that imply trade-offs over political values. This charge has been made against the ozone technology assessment process in its repeated defense of the continuing need for HCFCs as interim solutions for reducing CFCs. However appropriate, this conclusion has combined technical judgments of the availability of non-halocarbon alternatives in specific sectors, with the strategic judgment that penalizing the firms whose early investment in HCFCs facilitated the initial movement from CFCs by sharply curtailing the useful life of their investments would be unwise. Second, the assessment process has no provision for independent outside review comparable to scientific peer review, but relies instead on participating experts and panel leadership to police each other's work for both technical quality and bias. Such review may be the best that is attainable if the best technical information is, as assumed, privately held and thus likely to be unknown to independent reviewers, but it also carries evident risks. Finally, the assessments have consistently declined to make quantitative cost estimates, or even to clearly specify what they mean by economic, as opposed to technical, feasibility. Although these omissions have drawn strong criticism, they have conferred clear advantages on the assessments. Panel leaders have argued that quantitative cost estimates have little value in a context of rapid technical change and moving regulatory targets. Such estimates would also be vulnerable to attack for their methods and details, as the opaque process of evaluating technical options and adding up feasible potential has not been.

The Authoritative Monopoly Strategy for Scientific Assessments of Environmental Risk

Atmospheric-science assessments of ozone beginning with the 1986 assessment, and the contrast with prior ineffective assessments, have demonstrated a strategy by which scientific assessments can effectively contribute to policy debates and the requirements for achieving it, both before and after regime formation. This strategy is likely to be applicable to other issues, provided the conditions that enabled its effectiveness in this case are present. The relevant research community must be cohesive enough that there can be wide agreement on who constitutes an authoritative group. The field must be able to make progress on policy-relevant questions over politically relevant periods (e.g., of a few years to a decade). Finally, there must exist some policy body empowered to act on the results of such questions. This strategy will not be feasible independent of the underlying conditions of scientific knowledge and concern, however. The high participation and effort required and the associated costs will preclude such an assessment strategy unless there is a strong scientific consensus that the issue is serious. A variant of this strategy has been used in atmospheric-science assessments of climate change under Working Group 1 of the Inter-governmental Panel on Climate Change (IPCC), with the crucial difference that the IPCC grants more authority to political oversight bodies and correspondingly less to the assessment's scientific leadership. Although these assessments have enjoyed fair success, they have not yet achieved a breakthrough in influence comparable to that of the 1986 ozone assessment, and their slower, more cumbersome process has risked exhausting the needed scientific participants and making them unwilling to continue participating.

Domestic-International Linkages in Negotiations

The negotiation of the first concrete international controls in 1987, and the long prior period during which the activists who had already enacted domestic aerosol bans called in vain for others to follow them, illustrate a general strategic lesson that might be called "the trap of unilateral leadership." A nation or coalition that makes an initial unilateral move, in the hope of providing leadership or demonstrating feasibility, risks becoming stuck in a situation where no further movement is possible either domestically or internationally. The early movers who banned aerosols reached a point beyond which further domestic controls appeared to grow rapidly more costly. They were unable to achieve further domestic cuts because their opponents pointed to the preferability of cheap controls abroad to expensive ones at home, and to the failure of other nations to follow the initial unilateral moves. They were unable to secure international movement because they could not provide any effective inducements, either offers or threats. Having already made the easy reductions, they could neither offer these as reciprocal concessions nor credibly threaten either to withdraw them or to impose sanctions on others for not following. The stance of calling for others to "do what we have done" was therefore not just rhetorically weak because it appeared nonserious or hypocritical, but also was strategically weak because of the difficulty of coupling it to reciprocal offers or threats.

Having thus uncoupled domestic from international action, the activists were unable to secure any further movement internationally or domestically until several external factors shifted in their favor in 1986 and 1987. It was only when the activist faction gained control of the U.S. agenda that it became possible to relink further, potentially high-cost domestic action with the first concrete international controls. Moreover, the trap of unilateral leadership did not disappear with adoption of the Protocol. On the two subsequent occasions when nations sought to use significant unilateral movement to leverage international movement—European restrictions on HCFCs and American restrictions on methyl bromide—the tactic achieved limited success or none, for reasons similar to those that thwarted the activists prior to 1986. Early movers were stuck asking others to follow them, with limited reciprocal concessions to offer.

The generalization of this lesson to other international environmental issues, as indeed to any negotiations over individual contributions to a collective good, is powerful and straightforward: activists seeking to advance the collective goal should not make their maximum feasible contributions unilaterally in advance, but at least partly exchange them for reciprocal contributions by others. The validity of this advice depends, of course, on the number and size of the actors considering an advance initiative and the size and details of the initiative. If the activists' initial step can be designed so that it makes it easier for others to follow or harder for them not to, then some level of unilateral first move may be appropriate, although not to the point that the activists can offer no further movement in response to others joining.[37]

Repeated Negotiations and Ratchets

After initial formation of the ozone regime, the bargaining range for subsequent revisions of controls was always bounded below by those previously agreed. Several factors favored this one-way character of future changes. Once major industry actors had begun investing in the rapid development of alternatives, they had a strong stake in avoiding reversals. The assessment panels, whose reports defined the negotiating agenda, never considered a return to weaker targets. With weaker targets never on the agenda, advocates of further action had the advantage in bargaining that they could propose more stringent cuts than they actually sought, while opponents could propose nothing weaker than the status quo. The consistency of subsequent changes has led some observers to describe the ozone regime as a ratchet—capable of moving in only one direction.[38] But whether this represented a general feature of the regime's design is not clear, however, since new information after the regime formation so consistently indicated greater environmental concern and easier opportunities to reduce.

As argued earlier, it is on balance likely that new information on both environmental harm and technological capacity will shift in the direction of favoring increasing, rather than decreasing, controls. On the technological side, serious pursuit of emission-reduction options is likely to identify more, better, and cheaper alternatives than previously known. On the environmental side, if initial action on a long-term issue is delayed until a high evidentiary threshold is met, progressively larger disruptions are likely to be observed for some time after the initial evidence, whether or not action is taken. These are tendencies, of course, not universal effects: despite the bias toward new information favoring stronger controls, new knowledge will sometimes moderate earlier views of the risk, or find persistently high cost and difficulty of tightening controls. The possibility that new knowledge can go in this direction implies that a true ratchet cannot be an appropriate instrument of adaptive management. Indeed, an adaptive regime must be able to retreat under strong contrary evidence, if it is to be capable of advancing with some boldness under uncertainty: otherwise, the risk of moving too far and getting stuck would promote making only minimal acceptable movements. While it appears unlikely that the specific institutions and procedures of the ozone regime were capable of moving in only one direction, the uniformity of subsequent changes in knowledge means that their capacity to move effectively in the opposite direction, or in orthogonal directions, was not tested.

This pattern—rapid accumulation of bad news after the first significant demonstration of harm—will likely be a feature of any long-term issue. If and when the evidence for global climate change becomes sufficient to induce greenhouse-gas controls, for example, a long period of ever more extreme evidence of climatic disruption can be expected thereafter. This likelihood implies advice for opponents of greenhouse-gas controls that is sharply at odds with their present tactics. The higher the standard of demonstration required to accept the first concrete measures to limit greenhouse-gas emissions, the more rapid and disruptive the subsequent pressure for increasingly strict limits is likely to be. Consequently, those who oppose such controls would better protect their long-term interests by taking measures now to reduce the likely disruption they will face from such a rapid future tightening of controls, by supporting early adoption of modest restrictions on the future growth of emissions.

The combined effects of the Protocol's adaptive mechanisms and the direction of new knowledge and capabilities were so powerful that controls were rapidly shifted to complete elimination of the major ozone-depleting chemicals. This extreme-point outcome transformed the subsequent evolution of the regime, making quantitative comparisons of burdens and past contributions moot. All subsequent debates over controls were reduced to two issues: the timing of phasedowns, which determined the dynamics of the transition from old to new markets; and the broadening of controls to new chemicals. The prominence of discrete judgments and discrete decisions—for example, judgments of feasibility or of sufficient evidence to proceed, and decisions to eliminate specified uses or specified chemicals—is a striking characteristic of the ozone issue, but its implications for other issues are unclear. It may merely indicate the importance of framing decisions and judgments in discrete terms, even when the factors shaping judgments and the actual dimensions of decisions are more continuous. If so, the lesson for other issues would be to be alert for opportunities to frame proposed judgments or decisions in categorical terms. Alternatively, this tendency may reflect unique characteristics of the ozone issue, in particular that it became possible to eliminate the offending chemicals at modest cost. If this is the case, then insights from the ozone issue that depend on its ability to exploit such discrete decisions are less likely to be applicable to other issues that admit no resolution through similarly categorical elimination of certain technologies, products, or activities, but can be managed only through partial, quantitative measures. Whether an issue does or does not admit of categorical resolution is a matter of how the issue and potential responses to it are framed, but it appears at present that climate change and other issues strongly linked to energy use are much less likely to exhibit this discrete character than did ozone. For issues that do not, the lessons from ozone that strongly depend on this—such as the irrelevance of downstream impacts, and the shift in function of scientific assessment after initial regime formation—are unlikely to apply.

9.7 The Challenge of Adaptive Management

The fundamental challenge of global environmental change is mastering adaptive management: designing institutions, procedures, and policies to help generate new information, and to adapt appropriately to new information as it appears.[39] This challenge concerns design of an entire regime in view of the state of knowledge, the structure of the issue, and the technology and strategic environment of the affected industries and activities. The success of the ozone regime relative to other environmental regimes lies principally in its having integrated principles of adaptive management into its design and effectively implemented them. It has repeatedly used new scientific and technological information to make revisions to its chemical control measures that have been at least roughly consistent with the implications of the new information. In terms of the primary target it has adopted—total stratospheric chlorine and bromine concentrations—it has already reversed the prior trend of disruption and is likely to restore the environment to a relatively undisturbed state within a few decades, assuming full implementation of controls now coming into force in developing countries, and no significant backsliding.

In accomplishing this, the ozone regime has benefited from several fortunate conditions. Short-term political conditions in 1986 and 1987 made it possible to enact an initial regulatory target strong enough to set the subsequent processes of technology development and adaptation in motion. New information that appeared after initial regime formation consistently indicated greater severity of the problem, and greater ease of reducing contributions to it, allowing subsequent adaptations to follow existing market, regulatory, and political momentum by tightening controls. The culmination of this process in agreements to eliminate all controlled chemicals profoundly simplified subsequent negotiation and decision-making, and maximally promoted technological and market processes to make the transition to new alternatives. Finally, it is easier to design such a system the first time. Undertaking such a project of designing institutions and procedures to generate strong substantive action depends in part on not facing sophisticated opponents who foresee the second step they oppose and block the crucial institutional design decisions necessary to empower the early institutions.

Moreover, the closer the regime has approached successful completion of its original core mission—eliminating the chlorinated and brominated industrial chemicals that destroy ozone—the more it has encountered resistance to further progress, on both the policy front and the environmental front. As a policy matter, it has become increasingly difficult to achieve further reductions as effort has shifted from early, easy reductions toward the last remaining uses of each chemical (many of them treated as essential uses) and toward sectors with less capacity, and as the regime's main work has shifted toward the harder and more divisive problems of enforcing compliance and administering the system of financial assistance. In environmental terms, as halocarbon emissions have been reduced, their remaining contribution to ozone depletion has diminished relative to other chemicals and other environmental processes not included in the ozone regime, and linkages between ozone depletion and other forms of environmental change have grown in practical importance. In the early days of the ozone regime, it mattered little that the total stratospheric halogen concentration adopted as the regime's primary management target was an imperfect indicator of ozone depletion. As halocarbon emissions have been reduced, the potential disconnect between these two measures has become increasingly serious.

These qualitative shifts in the primary character of the problem and the processes contributing to it pose more severe challenges than any the ozone regime has met, but must be met for truly adaptive management to be achieved. The ozone regime has faced the strongest challenges each time a qualitative enlargement of its scope was required, albeit a modest one, in expanding controls to new chemicals. Each such expansion posed the relatively manageable challenge of requiring the regime's parties and institutions to engage new technological and scientific issues, and the much harder challenge of involving new, previously uninvolved constituencies of producers and users with limited knowledge of the issue, little practice or confidence in identifying alternatives to present practices, strong preferences for continuing the status quo, and the expectation that they can resist encroachment of the regime by political force.

The same challenges are posed even more strongly when nonhalocarbon chemicals emerge as important contributors to depletion or linkages between ozone depletion and other forms of environmental change become prominent. The ozone regime benefited throughout from sharp focus—the ability to focus on a few specific, salient scientific questions, and a few sharply targeted controls. When linkages between ozone depletion and other dimensions of environmental change have become prominent, requiring integrated management of multiple processes of human environmental perturbation, the limitations of any such single-issue regime have been exposed. Such linkages are becoming increasingly important, and the ozone regime has been quite ineffective at addressing them. It has, for example, been ineffective at addressing ways that solutions to ozone depletion contribute to other problems, such as use of HFCs. It has made no move to respond to the increasingly evident fact that remaining threats to ozone arise principally from sources other than halocarbon emissions, in particular from greenhouse-gas emissions. Early efforts to cooperate with other regimes on scientific and technical assessment have shown some promise, but the path from such preliminary cooperation to any concrete action is not at all clear. Incorporating these linkages will require either an extreme expansion of the ozone regime—a highly unlikely prospect—or construction of some new form of supervening global environmental regime with a mandate to address multiple sources and multiple effects. Such a superregime would brings in new interests, actors, and distributions of preferences and knowledge with a vengeance—and would have none of the advantages of focus, precisely defined targets, and urgency that helped the early development of the ozone regime.

The ability of present policy systems to respond properly, and in time, to linked forms of global environmental change remains deeply in doubt. The ozone regime provides the best existing example, and an impressive model of a regime that can evolve dynamically in response to changing conditions. But it is limited in addressing broader linked problems by the very clarity of focus that helped power its rapid and effective response to its own targeted issue. The construction of environmental management regimes that can exploit the lessons in adaptive management offered by the ozone regime, but can also embrace increasingly salient linkages across multiple forms of environmental change, driven by multiple forms of human activity and emission and resulting in multiple impacts, is a challenge that we have not yet begun to address.

Appendix A: List of Interviews

All interviews were conducted by the author, except as noted.

Subject	Location	Date
Martin Holdgate (Marc Levy)	Gland	Aug. 27, 1990
T. Bunge, W. Garber, and H. Brackemann, UBA (Levy)	Berlin	Sept. 3, 1990
Heinrich Klaus, BMU (Marc Levy)	Bonn	Sept. 14, 1990
Laurence Musset, Min. de l'Environnement (Levy)	Paris	Jan. 9, 1991
Jean Claude Ray, Bulle Bleue (Levy)	Paris	Jan. 10, 1991
Philippe Gaillochet, Office Parlementaire d'Evaluation des Choix Scientifiques & Technologiques (Levy)	Paris	Jan. 14, 1991
Maurice Muller, Min. de l'Environnement (Levy)	Neuilly	Jan. 16, 1991
Patrice Miran, Min. de l'Agriculture (Levy)	Paris	Jan. 22, 1991
Maurice Verhille, Atochem (Levy)	Paris	Jan. 23, 1991
Gerard Megie, CNRS (Levy)	Paris	Jan. 24, 1991
Serge Langdeau, Environment Canada	Ottawa	May 2, 1991
Danielle St. Pierre, Environment Canada	Ottawa	May 2, 1991
Alex Chisholm, Environment Canada	Ottawa	May 2, 1991 Jun. 17, 1993
John D. Reid, Environment Canada	Ottawa	May 2, 1991 June 17, 1993
Robert Hornung, Friends of the Earth	Ottawa	May 3, 1991
Brian Herman, External Affairs Canada	Ottawa	May 4, 1991
F. A. Vogelsburg, Jr., DuPont	Nairobi	June 10, 1991
Paul Cammer, HSIA	Nairobi	June 10, 1991

Subject	Location	Date
Mark Sweval, Great Lakes Chemical	Nairobi	June 10, 1991
Nick Campbell, ICI	Nairobi	June 11, 1991
Carol Niemi, Dow Chemical and HSIA	Nairobi	June 11, 1991
John Whitelaw, Australia delegation	Nairobi	June 12, 1991
Philip Woolaston, former NZ Min. of Environment	Nairobi	June 12, 1991
Edward Leviton, U.S. Commerce Dept.	Nairobi	June 13, 1991
George Strongylis, DG-11, EC Commission	Nairobi	May 22, 1991
Peter Winsemius, Netherlands (former Environment Minister)	Nairobi	May 23, 1991
Willem Kakebeeke, Netherlands delegation	Nairobi	May 23, 1991
Marcia Donner-Abreu and José Bustani, Brazil delegation	Nairobi	June 12, 1991
Steve Seidel, U.S. EPA	Nairobi	June 14, 1991
Avani Vaish, India delegation	Nairobi	June 14, 1991
Steve Lee-Bapty, U.K. delegation	Nairobi	June 18, 1991
Mike Harris, ICI; and J. von Schwein-ichen, MonteFluos	Nairobi	June 19, 1991
David Robert, BAMA (Levy)	London	Aug. 1, 1991
Adam Markham, WWF (Marc Levy)	Gland	Sept. 17, 1991
Kevin Fay, Alliance for Responsible CFC Policy	Washington, DC	Jan. 16, 1992 May 17, 1996
Richard Smith, U.S. State Dept.	Washington, DC	Jan. 16, 1992
Richard Benedick, ret. U.S. State Dept.	Telephone	Jan. 17, 1992
Eileen Claussen, U.S. EPA	Washington, DC	Jan. 17, 1992
Stephen Andersen, U.S. EPA	Washington, DC	Jan. 17, 1992 Dec. 4, 1995 May 14, 1996
Robert Watson, U.S. NASA	Washington, DC	Jan. 21, 1992
Elizabeth Cook, Friends of the Earth	Washington, DC	Jan. 21, 1992
John Hoffman, U.S. EPA	Washington, DC	Jan. 21, 1992 May 14, 1996
Sue Biniaz, U.S. State Dept.	Washington, DC	Jan. 21, 1992
David Doniger, Natural Resources Defense Council	Washington, DC	Jan. 21, 1992
Scott Hajost, Environmental Defense Fund	Washington, DC	Jan. 21, 1992
John Bennett, Greenpeace Canada (Jill Lazenby)	Telephone	June 10, 1993
Jim Kerr, Environment Canada AES	Toronto	June 14 1993
Warren Godson, Environment Canada AES	Toronto	June 14, 1993
Julia Langer, Friends of the Earth	Toronto	June 14, 1993
Barney Boville, Environment Canada AES	Toronto	June 14, 1993

Subject	Location	Date
Harold Schiff, York University	Toronto	June 14, 1993
Rao Krishna Vupputuri, Environment Canada AES	Toronto	June 16, 1993
F. Kenneth Hare (Trent University)	Toronto	June 16, 1993
Art Fitzgerald, Nortel	Toronto	June 16, 1993
G. Victor Buxton, Environment Canada	Ottawa	June 17, 1993
John W. Reed, Environment Canada	Ottawa	June 17, 1993
Wayne Evans, Trent University (formerly AES)	Telephone	June 17, 1993
David MacDonald, Canadian House of Commons	Ottawa	June 17, 1993
Gary Gallon, Can. Envt. Industries Assoc. (Jill Lazenby)	Toronto	June 21, 1993
Denise Mauzerall, formerly U.S. EPA	Cambridge, MA	June 24, 1993
Michael McElroy, Harvard University	Cambridge, MA	June 29, 1993
Nay Htun, UNEP	Cambridge, MA	May 9, 1994
John Oxley, UNIDO	Vienna	June 16, 1994
Everet Kok, UNIDO	Vienna	June 16, 1994
Cahit Gurkok, UNIDO	Vienna	June 16, 1994
John Oxley, UNIDO	Vienna	Aug. 17, 1994
T. MacQuarrie and S. Gorman, Environment Canada	Ottawa	Feb. 10, 1995
Peter Thacher, ret. UNEP	Stonington, CT	Mar. 10, 1995
Tom Land, U.S. EPA (Sydney Rosen)	Washington, DC	July 13, 1995
Lou Bock, U.S. Customs Service (Sydney Rosen)	Washington, DC	July 13, 1995
Ron Lindzen, U.S. IRS (Sydney Rosen)	Washington, DC	July 13, 1995
H. J. Banks, Methyl Bromide TOC	Vienna	Dec. 4, 1995
Rodrigo Rodriguez-Kabana, Methyl Bromide TOC	Vienna	Dec. 4, 1995
Patty Clearly, Methyl Bromide TOC	Vienna	Dec. 5, 1995
Suelhy Carvalo, TEAP	Vienna	Dec. 1995
J. C. van der Leun, Effects Panel	Vienna	Dec. 5, 1995
Gary Taylor, Halons TOC	Vienna	Dec. 4, 1995
Barbara Blum, formerly U.S. EPA	Telephone	May 6, 1996
Drusilla Hufford, U.S. EPA	Washington, DC	May 13, 1996
Katie Smyth, AFEAS	Washington, DC	May 13, 1996
John Passacantado, Ozone Action	Washington, DC	May 13, 1996
Alan Miller, University of Maryland	Washington, DC	May 14, 1996
Irving Mintzer, University of Maryland	Washington, DC	May 14, 1996
John Topping, Climate Institute	Washington, DC	May 16, 1996
Jim Losey, Skadden Arps (formerly U.S. EPA)	Washington, DC	May 16, 1996
Allison Morrill, ICOLP	Washington, DC	May 16, 1996
Pam Zurer, *Chemical and Engineering News*	Washington, DC	May 17, 1996
Tom Stoel, formerly NRDC	Washington, DC	May 17, 1996

Subject	Location	Date
David Wirth, formerly State Department and NRDC	Washington, DC	May 17, 1996
Rumen D. Bojkov, WMO	Telephone	July 1, 1996
Martin G. Wagner, formerly U.S. EPA	Telephone	July 16, 1996
Annie Roncerel, UNDP	Telephone	July 16, 1996
Lee Thomas, former EPA Administrator	Telephone	July 16, 1996
Daniel Albritton, NOAA	Boulder, CO	July 28, 1997
Susan Solomon, NOAA	Boulder, CO	July 28, 1997
Adrian Tuck, NOAA	Boulder, CO	July 29, 1997
Carl Howard, NOAA	Boulder, CO	July 29, 1997
Martin Hoffert, Columbia University	Telephone	Nov. 12, 1998
Michael MacCracken, formerly Livermore Lab	Washington, DC	Nov. 17, 1998
Tom Morehouse, formerly USAF	Warrenton, VA	May 8, 1999
Mike Shapiro, formerly U.S. EPA	Telephone	July 16, 1999
Richard Barnett, former Chairman, Alliance	Telephone	July 16, 1999
Dixon Butler, NASA	Telephone	July 16, 1999
Steve Weil, formerly U.S. EPA	Telephone	July 17, 1999
Milton Russel, formerly U.S. EPA	Telephone	July 18, 1999
Joseph Cannon, formerly U.S. EPA	Telephone	July 19, 1999
Janice Longstreth, U.S. EPA	Telephone	July 19, 1999
Robert Worrest, formerly U.S. EPA	Telephone	July 20, 1999
F. Sherwood Rowland, UC Irvine	Telephone	July 20, 1999
Donald Wuebbles, University of Illinois	Telephone	July 21, 1999
Robert Hudson, U of Maryland (formerly NASA)	Telephone	Aug. 9, 1999
Michael Prather, UC Irvine, formerly NASA	Telephone	Aug. 9, 1999
Donald Clay, U.S. EPA	Telephone	Aug. 10, 1999
Mario Molina, MIT	Telephone	Aug. 24, 1999
Judy Odoulamy (Kosovich), formerly U.S. EPA	Telephone	Aug. 27, 1999
Donald Heath, formerly NASA Goddard	Telephone	Aug. 27, 1999
Carroll (Pegler) Bastian, formerly NSF and EPA	Telephone	Aug. 28, 1999
James Brydon, formerly Environment Canada	Telephone	Aug. 30, 1999
Warren Muir, formerly U.S. EPA	Telephone	Aug. 31, 1999
Shelby Tilford, formerly NASA	Telephone	Sept. 2, 1999
James Anderson, Harvard University	Cambridge, MA	Jan. 7, 2000
Steven Wofsy, Harvard University	Cambridge, MA	Jan. 10, 2000
Marilyn Bracken, formerly U.S. EPA	Telephone	Sept. 21, 2000
Walter Komhyr, NOAA	Telephone	Sept. 20, 2001

Appendix B: Archival Sources

74-4-1 Statement to Subcommittee on Public Health and Environment, T. Stoel, Jr., and K. Ahmed, NRDC, Dec. 11–12, 1974.

76-1-3 Note by the United States, Doc. no. 27, 19th session, Environment Committee, OECD, Paris, Dec. 13, 1976.

76-2-1 Press Release and Background Statement on NAS ozone report, DuPont, Sept. 20, 1976.

76-4-2 Statement at public meeting on CFCs, R. E. Train, EPA, Dec 3, 1976.

76-4-4 Memo to file: History of NAS Study on the CFC Issue, C. L. Bastian, NSF, Washington, DC, Aug. 2, 1976.

76-4-5 Press release, "New Information Being Evaluated by Ozone Panel," NAS, Washington, DC, May 3, 1976.

77-1-2 Report on the UNEP Meeting of Experts on the Ozone Layer, UNEP, Washington, DC, Mar. 8, 1977.

77-1-3 UNEP/CCOL.1/11, Report of the First Session, CCOL, Geneva, Nov. 1–3, 1977.

77-1-4 Work Plan: Outer Limits, UNEP, Nairobi, Sept. 26, 1977.

77-2-4 Cover letter from H. W. Crawford to J. Brydon, July 15, 1977.

78-1-1 UNEP/CCOL.11/4, Report of the Second Session, CCOL, FRG, Nov. 28–30, 1978.

78-1-2 Presentation to International Meeting on CFCs, T. B. Stoel, Munich, Dec. 6–8, 1978.

78-4-2 B. Blum, Munich speech and press release, EPA, Dec. 7, 1978.

79-1-2 UNEP/CCOL.III/5, Report of the Third Session, CCOL, Paris, Nov. 20–23, 1979.

79-2-1 *Briefing Book, Chlorofluorocarbon/Ozone Depletion Issue*, DuPont, Aug. 1979.

79-4-1 EPA-560/12-79-003, Report on the Progress of Regulations to Protect Stratospheric Ozone, Aug. 1979.

80-1-2 Agreed Conclusions: Meeting on International Regulation of CFCs, Oslo, Apr. 15, 1980.

80-2-1 *Fluorocarbon/Ozone Update*, DuPont, June 1980.

80-2-2 Information paper, Alliance, Oct. 1, 1980.

80-2-3 *Fluorocarbon/Ozone Update*, DuPont, Jan. 1980.

80-2-4 *Fluorocarbon/Ozone Update*, DuPont, Apr. 1980.

80-2-5 *Fluorocarbon/Ozone Update*, DuPont, July 1980.

80-2-6 *Fluorocarbon/Ozone Update*, DuPont, Oct. 1980.

80-2-7 "An Assessment of Chlorofluorocarbon-Ozone Science," response to ANPR, CMA, Dec. 1980.

80-2-8 Letter to S. D. Jellinek, EPA, from C. N. Masten, Director, Freon Products Division, DuPont, June 16, 1980.

81-1-1 Executive Summary, CCOL, Copenhagen, Oct. 12–16, 1981.

81-1-2 Report on OECD Experts Meeting on Modelling of CFCs, R. K. R. Vupputuri, Paris, Sept. 15 and 16, 1981.

81-1-3 Press release, UNEP, Oct. 12–16, 1981.

81-1-4 Memorandum to the Cabinet, J. Roberts, Minister of the Environment, 1981.

81-1-6 UNEP/WG.69/6, An Assessment of Ozone Layer Depletion and Its Impact, CCOL, Stockholm, Jan. 20–29, 1982.

81-1-8 UNEP/WG.69/3, Towards an Ozone Convention, UNEP Secretariat, Stockholm, Jan. 20–29, 1982.

81-1-9 UNEP/WG.69/9, Letter to all governments from F. Sella, Bilthoven, Netherlands, Nov. 11–14, 1980.

81-1-10 Letter to the Minister from the Deputy Minister, J. B. Seaborn, Canada, Dec. 22, 1981.

81-1-11 Memorandum from ARPD to ADMA, A. J. Chisholm, Canada, Dec. 31, 1981.

81-1-12 Third WMO Statement on Modification of the Ozone Layer Due to Human Activities, Nov. 25, 1981.

81-1-13 Agenda for Stratospheric Advisory Committee Meeting, Nov. 20, 1981.

81-1-19 UNEP/CCOL/V: Report of the 5th Session, CCOL, Copenhagen, Oct. 12–16, 1981.

81-1-20 Report on CFCs, OECD, Environment Committee, Mar. 6, 1981.

81-2-2 "Summary of Critical Points," Alliance for Responsible CFC Policy, 1981.

81-2-3 Comments on draft OECD fluorocarbon paper, Mar. 1981.

81-2-4 *Fluorocarbon/Ozone Update*, DuPont, Apr. 1981.

81-2-5 *Fluorocarbon/Ozone Update*, DuPont, Nov. 1981.

81-2-6 Effect of CFCs on the Atmopshere: Summary of Research Program, CMA, June 15, 1981.

81-4-2 Press release, "USA-WMO Meeting of Experts on the Stratosphere," Hampton VA, May 18–22, 1981.

81-4-3 Proposal for a council decision on CFCs, Commission of the European Communities, Oct. 8, 1981.

81-4-4 AERE Report R10268, "The Impact of CFCs on Stratospheric Ozone: An Update," R. J. Murgatroyd and R. G. Derwent, Aug. 10, 1981.

82-1-1 Technical Annex on CFCs, notes by A. Chisholm, Jan. 20–29, 1982.

82-1-4 A Review of Government Responses to Decision 8/7B and 9/13B Concerning CFCs and Risks to the Ozone Layer, UNEP Secretariat, Stockholm, Jan. 20–29, 1982.

82-1-5 Measures under Consideration by EEC, EEC, Stockholm, Jan. 20–29, 1982.

82-1-9 Handwritten notes from Day 1 UNEP meeting, A. Chisholm, Stockholm, Jan. 20, 1982.

82-1-10 Handwritten notes from Day 2 UNEP meeting, A. Chisholm, Stockholm, Jan. 21, 1982.

82-1-15 UNEP/WG.78/5, Assessing Socio-economic Impact of Alternative Strategies for Protection of Health and the Environment from Depletion of the Ozone Layer, UNEP Secretariat, Geneva, Nov. 2–11, 1982.

82-1-16 Report of meeting with Canadian industry representatives, A. Chisholm, Apr. 6, 1982.

82-1-20 Opening statement, P. S. Thacher, Geneva, Dec. 10, 1982.

82-1-27 Draft Convention for the Protection of the Ozone Layer with commentary, UNEP Secretariat, July 21, 1982.

82-1-32 Letter to Mr. Usher and Mr. Bohte from G. Diprose, Oct. 15, 1982.

82-1-34 Memo to H. de Lemos, DAED, Office of the Environment Programme, from B. Bohte, Nov. 10, 1982.

82-1-55 Collected correspondence of A. Miller re State and EPA domestic consultations for Convention negotiations, 1982.

82-1-56 Informal discussion papers, Sweden, EEC, Canada, Netherlands, Jan. 1, 1982.

82-1-57 "Note to Past Reviewers," with paper "Assessing Socio-economic Impact of Alternative Strategies for Protection of Health and the Environment from Depletion of the Ozone Layer," J. Kosovich, OECD, July 9, 1982.

82-2-1 Strobach letter to T. Wilson, OES, Alliance, Dec. 6, 1982.

82-2-2 *Fluorocarbon/Ozone Update*, DuPont, June 1982.

82-4-3 Report to Congress, Progress of Regulations to Protect Stratospheric Ozone, EPA, Feb. 1982.

83-1-2 Meeting report, UNEP Meeting of Legal/Technical Experts, A. Chisholm, Geneva, Apr. 11–15, 1983.

83-1-3 UNEP/WG.78/10, Revised Draft Ozone-Layer Convention, UNEP Secretariat, Geneva, Apr. 11–13, 1983.

83-1-11 UNEP/WG.94/L.1, Draft Report of the Working Group, J. Sola, Geneva, Oct. 17–21, 1983.

83-1-18 UNEP/WG.94/L.1/Add.4, Recommendations for Future Work, Working Group, Geneva, Oct. 17–21, 1983.

83-1-26 Collected correspondence to/from Miller with State and EPA, re continuing Convention negotiations, 1983.

83-2-1 Comments on Clean Air Coalition position paper, "CFCs and Ozone Depletion," Alliance, June 10, 1983.

83-2-2 Letter to T. Wilson, Dept. of State, from D. R. Strobach, Alliance, Oct. 3, 1983.

83-2-3 Legislative Alert no. 2, DuPont Freon Products Division, Mar. 3, 1983.

83-3-1 Collected letters re Clean Air Act, A. S. Miller and D. R. Strobach, 1983.

83-4-2 Report to Congress, "The Progress of Regulation to Protect Stratospheric Ozone," EPA, Apr. 1983.

84-1-1 Handwritten note re UNEP—sponsored Ozone Layer Protection Convention (written by U.S.), Spring 1984.

84-1-2 Cable from R. Lederman, Canadian delegation in Nairobi, May 21, 1984.

84-1-3 Cable from R. Lederman, Canadian delegation in Nairobi, May 23, 1984.

84-1-6 Briefing note update, A. Chisholm, 1984.

84-1-7 Meeting report, UNEP Global Ozone Framework Convention, A. Chisholm, Vienna, Jan. 16–20, 1984.

84-1-11 Introduction of draft Protocol, address by Canadian delegate, A. Chisholm, Geneva, Oct. 22, 1984.

84-1-13 Draft telegram, Canada, Sept. 5–7, 1984.

84-1-14, 84-1-15 Telegrams, UNEP, UK, 1984.

84-1-18 Appendix 1, Executive Summary, UNEP, Geneva, Oct. 15–19, 1984.

84-1-19 Briefing note, UNEP Global Framework Convention/Control Protocol, Toronto, Sept. 5–7, 1984.

84-1-20 Briefing note, re stratospheric ozone layer, A. Chisholm, 1984.

84-1-21 Series of telexes, Geneva, Oct. 1984.

84-1-24 "Production, Sales, and Calculated Release Data for CFC-11 and CFC-12," Fluorocarbon Program Panel, 1984.

84-1-25 UNEP/WG.110/L.1/Add.4, draft report, Group of Legal and Technical Experts, Geneva, Oct. 22–26, 1984.

84-1-26 Draft report of the Canadian delegation, A. Chisholm, Geneva, Oct. 22–26, 1984.

84-1-28 UNEP/WG.110/4, Report of the Ad Hoc Group of Legal and Technical Experts, Geneva, Oct. 22–26, 1984.

84-1-29 UNEP/WG.110/4, Annex IV, Third Revised Draft Protocol on CFCs, Group of Legal and Technical Experts, Geneva, Oct. 22–26, 1984.

84-1-31 Meeting notes from State Dept. briefing, A. S. Miller, Jan. 4, 1984.

84-2-2 Comments on draft Rand report, "Projected Use, Emissions, and Banks of Potential Ozone Depleting Substances," DuPont, Freon Products Division, Dec. 1984.

84-2-3 Letter to M. R. Hughes from G. Cox and R. Ward, CMA, Apr. 4, 1984.

84-2-4 Comments on the U.S. position paper, Fluorocarbon Program Panel of the CMA, Apr. 1984.

84-3-1 Letter from A. Miller to R. Watson, Jan. 30, 1984.

84-4-1 Materials for NRDC lawsuit, 1984.

84-4-8 "Preliminary Results and Conclusions", transparencies from press conference, Antarctic ozone, Sept. 30, 1987.

84-4-10 Collected slides, D. Albritton, 1984.

85-1-1 Protecting the Ozone Layer, R. E. Benedick, Geneva, Jan. 21, 1985.

85-1-4 Protecting the Ozone Layer, U.S. Position, Secretary of State, Vienna, Mar. 18–22, 1985.

85-1-21 Final Act, Conference of Plenipotentiaries on the Ozone Layer, UNEP, Vienna, Mar. 18–22, 1985.

85-1-24 Vienna Convention for the Protection of the Ozone Layer, The Final Act, UNEP, 1985.

85-1-26 Convention for Protection of the Ozone Layer and Protocol on CFCs, and attached U.K. note, Plenipotentiaries Conference, Vienna, Mar. 18–22, 1985.

85-1-27 UNEP/IG.53/5, Declarations made at the time of adoption of the Final Act of the Conference of Plenipotentiaries on the Protection of the Ozone Layer, Vienna, 1985.

85-1-28 Press release, "European Community Is Holding Up Negotiations on UNEP/World Framework Convention for the Protection of the Ozone Layer," European Environment, Brussels, Feb. 13, 1985.

85-2-1 Economic Consequences of the U.S. Ban on CFCs as Aerosol Propellants, Alliance, 1985.

85-2-2 Letter from Strobach to A. Miller, Sept. 19, 1985.

85-2-3 Letter to Benedick, CMA, Geneva, May 3, 1985.

85-2-4 Letter to Richard Benedick, from Cox and Ward on behalf of FPP, May 30, 1985.

85-3-4 Papers for research staff meeting, A. Miller, Jan. 10, 1985.

85-4-10 Slides, Daniel Albritton, 1985.

86-1-1 UNEP Workshop on Control of CFCs, A. Chisholm and G. V. Buxton notes, Rome, May 26–30, 1986.

86-1-2 UNEP/WG.148/2, Report of the Workshop on Control of CFCs, Rome, May 26–30, 1986.

86-1-3 Press release, "Chafee Opens Hearings on Ozone Depletion and 'Greenhouse Effect,' " J. H. Chafee, 1986.

86-1-4 Witness list, Hearing on Ozone Depletion, the Greenhouse Effect, and Climate Change, June 10, 1986.

86-1-5 L. Thomas testimony, Senate Subcommittee on Environmental Pollution, June 11, 1986.

86-1-14 Telex, meeting of Working Group of Legal and Technical Experts on CFC Protocol, Geneva, Dec. 4, 1986.

86-1-16 Limiting the Buildup: Policies to Control Chlorine in the Stratosphere, I. Mintzer, WRI, Sept. 1986.

86-1-18 Preparing for a CFC Phaseout, D. A. Wirth, and D. D. Doniger, NRDC, Leesburg, VA, Sept. 8–12, 1986.

86-1-19 Trade Issues Related to International CFC Control, S. O. Andersen, EPA, Leesburg, VA, Sept. 8–12, 1986.

86-1-20 Analysis of Global Application of a Production Capacity Cap, S. Seidel, EPA, Leesburg, VA, Sept. 8–12, 1986.

86-1-21 Control Strategies to Achieve Ozone Depletion Limits, J. Hoffman, EPA, Leesburg, VA, Sept. 8–12, 1986.

86-1-22 Potential Health and Environmental Effects of Ozone Depletion and Climate Change, S. Seidel, D. Tirpak, and J. S. Hoffman, EPA, Leesburg, VA, Sept. 8–12, 1986.

86-1-23 Overview paper on topic 6B, G. Brasseur and C. Bevington, Leesburg, VA, Sept. 8–12, 1986.

86-1-24 Analysis of the Importance of Various Design Factors in Determining the Effectiveness of Control Strategy Options, M. J. Gibbs, ICF, Inc., EPA, Leesburg, VA, Sept. 8–12, 1986.

86-1-25 The Potential Impact on Atmospheric Ozone and Temperature of Increasing Trace Gas Concentrations, M. J. Gibbs, ICF Inc., EPA, Leesburg, VA, Sept. 8–12, 1986.

86-1-26 Equity of Ozone Protection Strategies, S. O. Anderson, EPA, Leesburg, VA, Sept. 8–12, 1986.

86-1-27 Ozone: Response to Emissions of CFCs, N2O, CH4, and CO2, FPP/CMA, Leesburg, VA, Sept. 8–12, 1986.

86-1-28 Agenda, UNEP Workshop on Protecting the Ozone Layer, Leesburg, VA, Sept. 8–12, 1986.

86-1-29 Notes by A. Chisholm, UNEP Workshop on Protecting the Ozone Layer, Leesburg, VA, Sept. 8–12, 1986.

86-1-30 Global Atmospheric Problems, A. Chisholm, Leesburg, VA, Sept. 8–12, 1986.

86-1-31 Notes to press and participants, UNEP Workshop on the Ozone Layer, Leesburg, VA, Sept. 8–12, 1986.

86-1-32 Letter to J. Hoffman from R. Oijerholm, Sept. 8–12, 1986.

86-1-33 Control Strategy Alternatives in Meeting Future CFC Demands, J. S. Hoffman, Leesburg, VA, Sept. 8–12, 1986.

86-1-35 An Assessment of Costs of Alternative Regulatory Strategies, D. M. Ambler, Leesburg, VA, Sept. 8–12, 1986.

86-1-36 Thoughts on a Possible CFC Protocol: A Use Limit Scenario, Leesburg, VA, Sept. 8–12, 1986.

86-1-37 Approach to the Control of CFCs, B. Gidaspov, Leesburg, VA, Sept. 8–12, 1986.

86-1-38 Austrian contribution to UNEP Workshop on CFCs Part II, Leesburg, VA, Sept. 8–12, 1986.

86-1-39 A Possible Regulatory Strategy for the control of CFCs, Japan, Leesburg, VA, Sept. 8–12, 1986.

86-1-40 Regulatory Strategies for CFCs, H.-J. Nantke, FRG, Leesburg, VA, Sept. 8–12, 1986.

86-1-41 Potential Ozone Responses to Chlorofluorocarbon Control Strategies, EC, Leesburg, VA, Sept. 8–12, 1986.

86-1-42 Factors That Affect the Cost of Protecting the Stratosphere, S. O. Andersen, EPA, Leesburg, VA, Sept. 8–12, 1986.

86-1-43 Economic Instruments for the Control of CFCs, R. Valiani, Leesburg, VA, Sept. 8–12, 1986.

86-1-44 The Risk of Alternative Strategies to Protect the Ozone Layer, A. Miller, Leesburg, VA, Sept. 8–12, 1986.

86-1-45 A Global Capacity Cap: Equity, Trade, and Implementation, L. Gundling, Leesburg, VA, Sept. 8–12, 1986.

86-1-46 The World Ceiling Production System of CFCs and Its Advantages, P. M. Dupuy, Leesburg, VA, Sept. 8–12, 1986.

86-1-47 Impacts of Possible Strategies Controlling CFCs: User Industry Viewpoint, FEA, Leesburg, VA, Sept. 8–12, 1986.

86-1-48 Timing of Regulations to Prevent Stratospheric Ozone Depletion, J. K. Hammitt, Leesburg, VA, Sept. 8–12, 1986.

86-1-52 Memo on Geneva meeting, G. V. Buxton, Geneva, Dec. 1–5, 1986.

86-1-54 Handwritten note, A. Chisholm, Geneva, Dec. 8, 1986.

86-1-59 Notes from meeting with L. Thomas, G. V. Buxton and A. Sheffield, 1986.

86-1-60 Talking Points—Ozone Layer, G. V. Buxton and A. Sheffield, 1986.

86-1-61 UNEP/WG.148/2, Report of the First Part of the Workshop on the Control of CFCs, Rome, May 26–30, 1986.

86-1-62 UNEP/WG.151/2, 5th Revised Draft Protocol on CFCs, Ad Hoc Working Group of Legal and Technical Experts, Geneva, Dec. 1–5, 1986.

86-1-63 UNEP/WG.151/L.1, Draft Protocol on CFCs or Other Ozone-modifying Substances (written by Canada), Geneva, Dec. 1–5, 1986.

86-1-64 UNEP/WG.151/L.2, Revised Draft Protocol on CFCs (submitted by the United States), Geneva, Dec. 1–5, 1986.

86-1-65 UNEP/WG.151/Background 5, Assessment Statement—Effects of Ozone

Layer Modification, Ad Hoc Working Group of Legal and Technical Experts, Geneva, Dec. 1–5, 1986.

86-1-67 Canadian Draft Proposal for a Protocol to Protect the Ozone Layer: An Explanation, Dec. 1, 1986.

86-1-71 Reporting cable, national proposals, UNEP, U.S. Delegation, Geneva, Dec. 1–5, 1986.

86-1-73 UNEP/WG.151/CRP.5, Protection of Ozone Layer Protocol Negotiations, EC, Geneva, Dec. 2, 1986.

86-2-1 Policy statement supporting possible global CFC production capacity cap, R. Barnett, Alliance, 1986.

86-2-2 Speech at the National Press Club, R. Barnett, Alliance, Washington, DC, Sept. 16, 1986.

86-2-6 CFCs and the Ozone Layer statement, European CFC Producers, Oct. 1986.

86-2-11 DuPont position statement on the chlorofluorcarbon/ozone/greenhouse issues, Sept. 1986.

86-2-12 Presentation to EPA Workshop, "A Search for Alternatives to the Current Commercial CFCs," D. S. Strobach, Alliance, Washington, DC, Feb. 24, 1986.

86-2-13 Remarks to EPA Workshop, R. Barnett, Alliance, Mar. 7, 1986.

86-2-15 Letter from R. Barnett to J. D. Negroponte, Asst. Secy. of State, OES, 1986.

86-2-16 Recent Research Results and Future Directions, CMA, FPP, 1986.

86-2-17 Letter to Freon Products customers from J. Glas, DuPont, Sept. 26, 1986.

86-3-2 Press release, NRDC, Oct. 3, 1986.

86-3-3 Testimony of A. Miller, WRI, Senate Committee on Foreign Relations, Mar. 18, 1986.

86-3-4 Memo to G. Speth, WRI, from K. von Moltke, Oct. 2, 1986.

86-3-5 Miller/Mintzer working papers, A. Miller et al., WRI, 1986.

86-3-6 Draft EDF position statement on stratospheric ozone depletion, Oct. 16, 1986.

86-3-7 Statement on CFCs and related compounds, environmental group consortium, Oct. 31, 1986.

86-3-8 Draft CFC bill, D. Doniger, I. Mintzer, R. Pomerance, A. Miller, 1986.

86-4-1 The European Community Approach to Control of CFCs, D. W. Pearce, Metra Consulting Group Ltd., 1986.

86-4-3 Presentation slides, "Options for an International Protocol on Protecting the Stratosphere," Office of Air and Radiation, USEPA, Oct. 14, 1986.

86-4-4 Congressional Record, address by Sen. J. H. Chafee, Oct. 8, 1986.

86-4-7 Letter from senators to G. Schulz, Oct. 6, 1986.

86-4-10 Collected slides, D. Albritton, 1986.

86-5-4 "Ozone Policy," G. Darst, AP, 1986.

87-1-2 Sixth Revised Draft Protocol on CFCs, UNEP, 1987.

87-1-3 Ministerial briefing note, Ozone Layer Issue, G. V. Buxton, Mar. 26, 1987.

87-1-4 UNEP/WG.172/2, Report of the Ad Hoc Group on the Work of Its Third Session, Geneva, Apr. 27–30, 1987.

87-1-5 UNEP/WG.172/CRP.9, Conclusions of the Scientific Working Group, Geneva, Apr. 27–30, 1987.

87-1-6 Briefing material, ozone layer depletion and the UNEP Ozone Protocol, A. Chisholm, Apr. 22, 1987.

87-1-7 Memo on GATT legality of ozone layer trade restrictions, Canadian Dept. of Justice, May 1, 1987.

87-1-8 Press release, "Global Agreement Reached to Protect Ozone Layer," Environment Canada, May 1, 1987.

87-1-12 Briefing note on informal consultations, G. V. Buxton, Brussels, June 28–29, 1987.

87-1-13 Telex, U.S. position, UNEP CFC Protocol, 1987.

87-1-15 Telex with handwritten note, J. R. Wright, 1987.

87-1-16 7th Revised Draft Protocol on CFCs, Legal Drafting Group, The Hague, July 6–9, 1987.

87-1-19 "Science Agencies to Investigate Ozone Hole," C. Redmond, P. Waller, and L. Blum, NASA press release 87–114, July 28, 1987.

87-1-20 Meeting summary (with DuPont on Ozone Protocol), G. V. Buxton, Toronto, Aug. 20, 1987.

87-1-29 Telex, Entry into Force Provision, Ozone Protocol, J. Allen, G. V. Buxton, and A. Chisholm, Sept. 1, 1987.

87-1-30 Telex, Diplomatic Conference on Protocol to Ozone Layer Convention, EEC status, Sept. 2, 1987.

87-1-34 Canadian delegation instructions, UNEP Protocol for the Ozone Layer, J. Allen and E. G. Lee, 1987.

87-1-37 List of participants, Conference of Plenipotentiaries on the Ozone Layer, Montreal, Sept. 14–16, 1987.

87-1-45 Briefing note on Ozone Layer Meeting, P. M. Higgins, Geneva, Apr. 27–30, 1987.

87-1-47 Ministerial briefing note, Vienna, Feb. 23–27, 1987.

87-1-51 UNEP/WG.167/2, Report of the Ad Hoc Group on the Work of Its Second Session, Vienna, Feb. 23–27, 1987.

87-1-53 UNEP/WG.167/INF.1, Ad Hoc Scientific Meeting to Compare Model Assessments of Ozone Change for Strategies for CFC Control, Ad Hoc Group of Legal and Technical Experts, Geneva, Apr. 27–30, 1987.

87-1-57 Outgoing and incoming telegrams, U.S. Dept. of State, 1987.

87-1-63 U.S. position paper, Vienna, Feb. 23–27, 1987.

87-2-1 *Fluorocarbon/Ozone Update*, DuPont, Mar. 1987.

87-2-5 Congressional statement to Senate Environment and Public Works Committee, R. Barnett, Jan. 28, 1987.

87-4-7 Summaries, Airborne Antarctic Ozone Experiment, D. Albritton and R. Watson, Washington, DC, Sept. 30, 1987.

87-4-13 Hearings, U.S. House of Representatives, Committee on Energy and Commerce, Subcommittee on Health and the Environment, Mar. 9, 1987.

87-4-10 Collected slides, D. Albritton, 1987.

88-2-4 Letter from DuPont to Sens. M. Baucus and D. Durenberger, 1988.

88-2-5 *Fluorocarbon/Ozone Updates*, "Protecting the Ozone Layer, the Search for Solutions," July, Aug., and Dec. 1988.

88-2-6 Statement of support for the FSPI's CFC voluntary phaseout program, NRDC, EDF, and FOE, Apr. 12, 1988.

88-2-9 Press release announcing formation of PAFT and attached letter from J. Glas, DuPont, Jan. 5, 1988.

88-2-10 Press release, "Industry Announces Phaseout Program," FSPI, Apr. 12, 1988.

88-2-11 Letter from Sens. Baucus, Stafford, and Durenberger to R. E. Heckert, DuPont, Feb. 21, 1988.

88-2-12 "Ozone Science: Recent Findings," presentation slides, M. MacFarland, DuPont, 1988.

88-3-1 Remarks to first Alternatives Conference, D. A. Doniger, Jan. 14, 1988.

88-3-2 Comments on EPA's proposed regulations, D. Doniger, D. Wirth, and N. Dean, Feb. 8, 1988.

88-4-10 Collected slides, D. Albritton, 1988.

89-1-3 UNEP/OzL.Pro.WG.I(2)/4, Final Report by the Open-ended Working Group of the Parties to the Montreal Protocol, Nairobi, Sept. 4, 1989.

89-1-4 UNEP/OzL.Pro.WG.II(I)/3, Discussion paper on possible regulations of HCFCs and HFCs, Open-ended Working Group of the Parties to the Montreal Protocol, Geneva, Nov. 1989.

89-1-6 Report of the First Session of the Second Meeting of the Open-ended Working Group of the Parties of the Montreal Protocol, Geneva, Nov. 23, 1989.

89-1-7 UNEP/OzL.Pro.1/5, Report of the Parties to the Montreal Protocol on the Work of Their First Meeting, Conference of the Parties to the Vienna Convention for the Protection of the Ozone Layer, Helsinki, May 6, 1989.

89-1-9 Draft UNEP Integrated Report, Nairobi, July 25, 1989.

89-1-11 UNEP/OzL.Pro.WG.II(1)/4, Synthesis Report, Assessment Panels, UNEP, Dec. 6, 1989.

89-1-13 UNEP/OzL.Pro.WG.II(1)/CRP.2, "Transient Scenarios for Atmospheric Chlorine and Bromine," M. J. Prather and R. T. Watson, Geneva, Nov. 13, 1989.

89-1-16 Report of the Legal Drafting Group, Geneva, Nov. 20, 1989.

89-2-1 Production and Release of Methyl Chloroform—1989 Data, European Chlorinated Solvent Association, 1989.

89-2-5 Company Policy in the Face of Global Concerns: Ozone as a Model, J. P. Glas, Jr., DuPont, Sept. 1989.

89-2-6 *Fluorocarbon/Ozone Update*, DuPont, Aug. 1989.

89-3-3 *Atmosphere* 1, no. 4 (winter 1986).

89-3-5 U.S. EPA presentation.

89-3-6 "What Lies Ahead for the Ozone Layer: Ozone Protection Campaign," Friends of the Earth International, 1989.

89-3-7 Safeguarding the Ozone Layer and Global Climate from CFCs and Related Compounds, Citizen Symposium on CFCs, London, Mar. 4, 1989.

89-4-3 Congressional testimony, R. Watson and W. G. Rosenberg, Washington, DC, May 19, 1989.

89-4-4 Congressional testimony, W. Rosenberg, May 19, 1989.

89-4-5 Congressional testimony, C. Auer, R. Watson, E. B. Claussen, and W. G. Rosenberg, May 19, 1989.

89-4-7 Congressional testimony, R. Smith, May 19, 1989.

89-4-8 Statement by the Federal Minister for the Environment, K. Topfer, Helsinki, May 2–5, 1989.

89-4-11 Press release, "Canada to Phase Out CFCs," L. Bouchard, Feb. 20, 1989.

89-4-16 Diplomatic cable, U.S. delegation summary of OEWG meeting, Nov. 1989.

90-1-21 Report of the Second Meeting of the Parties to the Montreal Protocol, London, June 29, 1990.

90-1-27 State Dept. cables and attached statement of B. Kalvalsky, World Bank, 1990.

90-2-3 Appraisal, "The Ozone Issue and Regulation," ICI, Chemicals and Polymers, June 1990.

90-2-6 Congressional testimony (in letter), A. D. Bourland, Feb. 8, 1990.

90-2-7 Congressional testimony, A. W. Braswell and P. Cammer.

90-3-1 Briefing, National Action Plans. Who Is Banning the Ozone Destroyers?, Friends of the Earth, 1990.

90-4-9 Transient Scenarios for Atmospheric Chlorine and Bromine, M. J. Prather and R. T. Watson, Nairobi, 1990.

90-4-11 Congressional testimony, R. T. Watson, May 15, 1989.

90-4-28 Statement by the Chief of Staff, White House, June 15, 1990.

91-1-3 Proposed decision, Sweden, Finland, Norway, Switzerland, Austria, Canada, and United States, Nairobi, June 14, 1991.

91-2-4 Slides, "DuPont's Response to the Stratospheric Ozone Issue," F. A. Vogelsburg, London, Mar. 12–13, 1991.

91-2-8 Du Pont's Response to the Stratospheric Ozone Issue, speech by F. A. Vogelsburg, London, Mar. 12–13, 1991.

91-2-12 An Industry Perspective on Phaseout of CFCs, presentation by F. A. Vogelsburg, NAE, Apr. 22–25, 1991.

91-2-14 Request for letter of reference from E. Claussen to F. A. Vogelsburg, July 12, 1991.

91-2-16 Letter to author from N. C. Campbell, Chloromethanes, ICI Chlorchemicals, Nov. 20, 1991.

91-3-5 Petition to Accelerate Phase-out Schedule for Ozone-Depleting Chemicals, NRDC, FOE, EDF, Dec. 3, 1991.

91-3-8 "The Global Business Outlook for CFC Alternatives," presentation to conference "Science and Public Policy," P. Usher, London, Mar. 12–13, 1991.

91-3-9 Press release, "Environmentalists Say Grim Ozone Report Requires Urgent Response," FOE, Oct. 22, 1991.

92-1-13 UNEP/OzL.Pro.4/15, Report of the Fourth Meeting of the Parties to the Montreal Protocol on Substances That Deplete the Ozone Layer, Copenhagen, Nov. 25, 1992.

92-1-17 Tasks Assigned to TEAP by the 4th Meeting of the Parties to the Montreal Protocol, L. B. Campbell, 1992.

92-2-7 Production report of CFCs 113, 114, and 115 and attached letter, G. Thornton, Feb. 1992.

92-2-8 Petition to W. Reilly for accelerated CFC phaseout, Alliance, Feb. 11, 1993.

92-2-9 Press release, "Du Pont Able to Increase HFC-134a Capacity with New Technology," DuPont, July 1992.

92-3-2 NRDC memo, New Amendments to the International Ozone Treaty and Recent Update on EPA Ozone Protection Rules, D. Doniger, Dec. 14, 1992.

92-3-3 Opening National Proposals, FOE, OEWG meeting, Geneva, Apr. 6–15, 1992.

92-3-4 Pamphlet, FOE, 1992.

94-1-5 Report Under Paragraph 8 of Article 5 of the Montreal Protocol, ICF Inc., Dec. 23, 1994.

94-3-1 Report on methyl bromide issues at Montreal Protocol meeting, M. Miller, Nairobi, Oct. 3–7, 1994.

94-5-1 *Global Environmental Change Report* (Cutter Information Corp.) 6, no. 9, (May 13, 1994).

95-1-2 UNEP/OzL.Pro.7/12, Report of the 7th Meeting of the Parties to the Montreal Protocol on Substances that Deplete the Ozone Layer, Vienna, Dec. 27, 1995.

95-1-3 UNEP/OzL.Pro/WG.1/11/10, Report of the Eleventh Meeting of the Open-ended Working Group of the Parties to the Montreal Protocol, Nairobi, May 13, 1995.

95-1-4 Statement of the G-77 and China, 7th Meeting of the Parties to the Montreal Protocol, Vienna, Dec. 1995.

95-1-5 1991–1995 Summary Contribution Status, Trust Fund for the Multilateral Fund, UNEP, 7th Meeting of the Parties to the Montreal Protocol, Vienna, Dec. 1995.

95-1-7 UNEP/OzL.Pro.7/Prep/CRP.1, proposal for a critical agricultural use exemption for methyl bromide, submitted by the United States, 1995.

95-1-8 Study on the Financial Mechanism of the Montreal Protocol, COWI Consultants, Mar. 1995.

95-1-9 UNEP/OzL.Pro.7/Prep/CRP.4/Rev.1, Decision 7/31, Assessment Panels, 1995.

95-2-9 Memo to Alliance members, results of Vienna Montreal Protocol meeting, D. Stirpe and K. Fay, 1995.

95-3-2 "The Ozone Layer Destroyers: Whose Chlorine and Bromine Is It?," Greenpeace International, Nov. 1995.

95-3-3 "Deadly Complacency: CFC Production, the Black Market, and Ozone Loss," J. Vallette, *OzoneAction*, Sept. 1995.

95-4-1 Statement to the meeting of the parties, Montreal Protocol, Minister of the Environment, Israel, Dec. 5, 1995.

96-1-1 UNEP/OzL.Pro.8/12, Report of 8th Meeting of Parties, Montreal Protocol, San Juan, Costa Rica, Dec. 19, 1996.

96-2-2 Press release, "AlliedSignal licenses key high-efficiency, CFC-free refrigerant to Du Pont," AlliedSignal Corp., May 28, 1996.

96-2-3 Science update on atmospheric lifetime and ozone depletion potential of methyl bromide, N. D. Sze and M. K. W. Ko, AER Inc., Nov. 13, 1996.

96-2-4 Metered-Dose Inhalers, briefing packet for the 8th Meeting of the Parties, Montreal Protocol, International Pharmaceutical Aerosol Consortium, San Jose, Costa Rica, Nov. 19, 1996.

97-1-2 UNEP/OzL.Pro.9/CRP.7, Control of New Substances with Ozone-depletion Potential, revised draft decision by the EC, Sept. 11, 1997.

97-1-3 UNEP/OzL.Pro.9/12, Report of the 9th Meeting of the Parties to the Montreal Protocol, Montreal, Sept. 25, 1997.

98-4-1 *Stratospheric Update*, EPA, Oct. 1998.

Notes

Note on Sources

The empirical account of the book is based on approximately 150 interviews with participants in the ozone issue from 10 countries—scientists, national and international officials, environmental advocates, and representatives of firms and industry associations—of whom five granted me full access to their files. These are listed in Appendix A. In addition, a large body of archival material was collected through attendance at meetings and repeated contact with participating individuals and institutions over a period of ten years. This archival material, which provides a unique record of the development of this issue, has been donated to the Harvard Library's Environmental Science and Public Policy Archives, where it is held as the Edward A. Parson Stratospheric Ozone Collection. Citations of the form (87-1-25) refer to accession numbers in this collection. Archival sources used are listed in Appendix B. In addition to primary material, the account draws on a comprehensive review of existing secondary sources, including published accounts by participants, advocates, and journalists through various stages of the issue's history; scientific publications and conference reports; and trade press. Notes use shorthand notation for sources that are used repeatedly. For newspapers, these are *NYT* (*New York Times*), *WSJ* (*Wall Street Journal*), *WP* (*Washington Post*), *LAT* (*Los Angeles Times*), *CSM* (*Christian Science Monitor*). For frequently used technical and trade-press sources, the abbreviations are *IER* (*International Environment Reporter* [Bureau of National Affairs, Washington D.C.]), *ENDS* (*The ENDS Report* [Environmental Data Services, London]), *CMR* (*Chemical Marketing Reporter*), *C&EN* (*Chemical and Engineering News*), *UAPB* (*NASA Upper Atmospheric Program Bulletin*), D&CI (Drug and Cosmetic Industry).

Chapter 1

1. Benedick 1991 (enlarged edition 1998). Ambassador Richard Benedick led the U.S. delegation on the issue during the crucial 1985–1987 period and continued to observe negotiations closely thereafter. Other important participants' accounts include Dotto and Schiff 1978; Bastian 1982; Wirth, Brunner, and Bishop 1982; and Stoel, Miller, and Milroy 1980, on early policy and scientific debates in the United States and internationally.

2. Major journalistic accounts include Roan 1989; Cagin and Dray 1993; Brodeur 1975, 1986; and Gribbin 1988.

3. Previous scholarly treatments of the ozone issue include Morrisette 1989; P. M. Haas

1992b; Parson 1993; Litfin 1994; and Grundmann 2001. In addition, several broader studies of international environmental politics have examined ozone as one case, including Rowlands 1995; Soroos 1986; and Social Learning Group 2001. Finally, some scholars have examined specific subcomponents of the ozone regime, in particular its financial mechanism (e.g., Kauffman and de Sombre 1995; Biermann 2000) and its provisions for implementation (e.g., Barratt-Brown 1991; Parson and Greene 1995; Victor 1997; Greene 1998).

4. Krasner 1983; Haggard and Simmons 1987; Strange 1983; E. B. Haas 1990; Keohane 1984; Rittberger and Mayer 1993; Jacobson and Weiss 1990; Victor, Raustalia, and Skolnikoff 1997; Young 1994; Levy, Young, and Zurn 1995.

5. Schelling 1960; Raiffa 1982; Sebenius 1993; Susskind 1994; Sjostedt 1993; Lax and Sebenius 1986.

6. Waltz 1979; Krasner 1991; Oye 1986; Keohane 1984.

7. See, e.g., Jervis 1997; Parson 1993; Parson and Clark 1994; Social Learning Group 2001; Young 1989; Gruber 2000; Sebenius 1992; Mitchell 1994.

8. Putnam 1988; Evans, Jacobson, and Putnam 1993; Elster 1995.

9. See, e.g., Collingridge and Reeve 1980; Ezrahi 1980; Dahl 1989; Fischer 1990; Jenkins-Smith 1990; Jasanoff 1990; Heclo 1978; Keck and Sikkink 1998; Sabatier and Jenkins-Smith 1993, 1999; P. M. Haas 1992a; Litfin 1994; Guston, 1997; Mitchell, Clark, Cash, and Alcock, forthcoming.

10. Jaffe, Newell, and Stavins 2001.

11. Porter and van der Linde 1995; K. Palmer, Oates, and Portney 1995.

12. Chwe 2001.

13. Joint provision of private and public goods has long been recognized as a strategy for providing traditional public goods such as public recreational or cultural facilities or interest-group political organization (Olson 1965), but has not previously been recognized for eliciting privately held information needed to guide policy.

14. Stone 1993; Sebenius 1991; Skolnikoff 1993; Eckstein 1975.

Chapter 2

1. This discussion of the earliest scientific research on ozone and the stratosphere draws extensively on Schmidt 1988; H. S. Johnston 1982, 1992; London and Angell 1982; and Munn 1979.

2. A nanometer is 10^{-9} meter, or one one-billionth of a meter. Light with wavelengths between about 400 nm (violet) and 750 nm (red) is visible to the human eye. Nearby shorter wavelengths lie in the ultraviolet (UV) region, longer wavelengths in the infrared (IR).

3. The tropopause, the boundary between the troposphere and the stratosphere, occurs much lower in polar regions (around 8 km or 25,000 ft) than near the equator (around 16 km or 50,000 ft).

4. Total ozone is measured in Dobson units (DU): 100 DU equals the quantity of ozone that would form a layer 1 mm thick at sea level. Total ozone quantities were typically about 240 DU year-round near the equator, with early spring maxima at high latitudes of about 440 DU in the Arctic and 360 DU in the Antarctic.

5. Götz, Meetham, and Dobson 1934

6. The four reactions of the Chapman system of ozone creation and destruction are as follows (Chapman 1930):
Ozone creation:

$$O_2 + UV\ (< 242\ nm) \rightarrow 2\ O \tag{1}$$

$$O + O_2 + M \rightarrow O_3 + M \tag{2}$$

(where M denotes a third molecule that does not interact chemically but must be present to carry off energy released in the process to stabilize the newly formed ozone molecule.)
Ozone destruction:

$$O_3 + UV\ (< 320\ nm) \rightarrow O + O_2 \tag{3}$$

$$O + O_3 \rightarrow 2O_2 \tag{4}$$

7. Brewer 1949; Dobson 1956, summarized in NASA/WMO 1986, p. 664.

8. Godson and Hare interviews.

9. NAS 1965.

10. Dütsch 1961; G. B. Hunt 1965.

11. Revelle and Suess 1957; Keeling and Pales 1965.

12. HOx refers collectively to H, HO, and HO_2. The reactions of the proposed HOx cycle are as follows (NAS Panel, 1976; Johnston 1982):

$$HO + O_3 \rightarrow HO_2 + O_2 \tag{5}$$

$$HO_2 + O \rightarrow HO + O_2 \tag{6}$$

Net: $O_3 + O \rightarrow 2O_2$.

13. Bates and Nicolet 1950; McGrath and Norrish 1960; Cadle 1964; Hampson 1964, 1966; Leovy 1969; Harrison 1970.

14. After Nicolet 1970 suggested that reaction (5) was too slow, Langley and McGrath 1971 measured it and found it 5,000 times too slow. Later work showed this measurement to be wrong, however, and the correct rate only about ten times slower than hypothesized by Hunt (H. S. Johnston 1982).

15. H. S. Johnston 1982; Murcray et al. 1968. The NOx catalytic cycle proposed by Crutzen was as follows:

$$NO + O_3 \rightarrow NO_2 + O_2 \tag{7}$$

$$NO_2 + O \rightarrow NO + O_2 \tag{8}$$

Net: $O_3 + O \rightarrow 2O_2$

16. McElroy and McConnell 1971; Crutzen 1971; Nicolet and Vergison 1971.

17. The reactions of the chlorine catalytic cycle are as follows:

$$Cl + O_3 \rightarrow ClO + O_2 \tag{9}$$

$$ClO + O \rightarrow Cl + O_2 \tag{10}$$

Net: $O_3 + O \rightarrow 2O_2$

18. H. S. Johnston 1992, p. 25, reports declining a request from an anonymous NASA official in December 1972 to publicize the chlorine-ozone risk of the shuttle, as he had earlier publicized the NOx risk of supersonic aircraft, and speculates that the same official may have seeded interest in chlorine elsewhere.

19. Hoshizaki et al., 1973; Stolarski and Cicerone 1974; Wofsy and McElroy 1974; summaries of the controversies and suggestions of attempted suppression are in H. S. Johnston 1992; Dotto and Schiff 1978.

20. Several major accidents had shown the dangers of toxic refrigerants, particularly in large systems. The worst was a 1929 leak in a Cleveland hospital that killed more than 100 people (Hounshell and Smith 1988, p. 155).

21. The discussion of early applications of CFCs that follows draws extensively on Cogan 1988; Manzer 1990; Cagin and Dray 1993; and Hammitt and Thompson 1997.

22. DuPont developed a numerical system to name the new chemicals, with separate digits denoting the number of fluorine, hydrogen, and carbon atoms in a molecule. The rightmost digit counts fluorine atoms; the second, hydrogen atoms plus one; the leftmost, carbon atoms minus one (dropped if zero). All remaining locations in the molecule are filled with chlorine. Hence, $CFCl_3$ is denoted CFC-11; CF_2Cl_2 is denoted CFC-12; CHF_2Cl is denoted HCFC-22; $CF_2ClCFCl_2$ is denoted CFC-113. The letter prefixes used with these numerical codes have changed over time. Initially, DuPont's Freon brand name was in wide generic use, although other firms marketed the chemicals under different trade names after DuPont's patents expired in the 1950s. In early debate over the environmental impact of these chemicals, as DuPont

campaigned not to have their brand name used to denote the chemicals generically. They were described by several generic names until usage settled on "chlorofluorocarbons" (CFCs) in the 1980s. Here, the chemicals will be denoted CFC-11, CFC-12, and so on.

23. The instrument could detect substances present at tiny concentrations, as low as a few parts in 10^{11} (Lovelock and Simmonds 1980).

24. This conversation, at the 1971 Gordon Conference "Environmental Science: Air," in New Hampton, NH, has been recounted in Dotto and Schiff 1978; Gribbin 1988; and Lovelock 1988. McCarthy was research director of DuPont's Freon Products Division, which produced their fluorocarbons.

25. "Seminar on the Ecology of Fluorocarbons," Andover, MA, Oct. 29–Nov. 1, 1972 (testimony of Roy Schuyler, Subcommittee on the Upper Atmosphere, U.S. Senate, 9/18/75).

26. At this point, the CMA was named the Manufacturing Chemists Association (MCA). The Fluorocarbon Panel directed this research program and the CMA remained its institutional sponsor until 1990.

27. Soulen 1979; Fluorocarbon Panel 1979.

28. Accounts of how Rowland became interested in this problem, which was unrelated to his prior work, and how Molina and Rowland approached it, are in Rowland and Molina 1994 & Dotto and Schiff 1978, pp. 10–19.

29. E.g., "Were this trifling quantity of atmospheric ozone removed, we should all perish." Charles Abbott of the Smithsonian Institution, quoted NYT 10/30/33; cited by Cagin and Dray 1993, p. 136.

30. Berkner and Marshall 1965, 1967. Catastrophe theories proposed by Ruderman 1974; Reid et al. 1976; and Whitten et al. 1976 were widely reported (e.g., NYT 3/31/75, p. 36; 2/19/76, p. 21; 5/15/77, p. D9).

31. Lovelock 1982 (p. 241) claims the need for UV screening was exaggerated, citing evidence from the geological record and contemporary ecosystems that many organisms tolerate much stronger UV than would occur even with no ozone layer. Elsaessar 1978 argues that ozone's role in early life was being misrepresented, since the ozone estimated to be needed was less than 10 percent of the modern ozone layer.

32. Testimony of J. E. McDonald, Committee on Appropriations, U.S. House of Representatives, 3/19/71.

Chapter 3

1. Horwitch 1982, pp. 215–39.

2. Testimony of J. E. McDonald, Subcommittee on Transport, House Appropriations Committee, 3/2/71, p. 302; CTAB Scientific Subcommittee meeting, Boulder, CO, Mar. 18–19, 1971.

3. The NOx-ozone cycle was identified for the natural stratosphere by Crutzen (1970), erroneously dismissed as an SST concern by the Study of Critical Environmental Problems (SCEP) 1970, p. 69, and independently developed by Johnston at the Boulder meeting.

4. NYT 5/18/71, p. 78; 5/30/71, p. D7.

5. Though several later accounts attribute the SST cancellation to ozone concerns and identify the decision as an example of U.S. environmental alarmism, contemporary and participants' accounts are unanimous that the influence of the ozone issue was minor. See, e.g., Dotto and Schiff 1978; Broderick and Oliver 1982; Schneider 1989; NYT 3/18/71, pp. 1, 28; 3/19/71, p. 1.

6. Broderick and Oliver 1982, p. 11; NYT 5/18/71, p. 78; 5/30/71, p. D7; editorial, 6/13/71, p. D12.

7. Mormino, Sola, and Patten 1975, p. vi

8. Dotto and Schiff 1978, pp. 67–68; Glantz, Robinson, and Krenz 1985, p. 582; McElroy and Salawitch 1989.

9. Grobecker 1981.

10. CIAP Monograph I, Panel on the Natural Stratosphere 1975, pp. 3-162–3-170.

11. Johnston and Quitevis 1975. Any process that converts NO to NO_2 without consum-

ing odd-O can increase ozone because NO_2 photolyzes to NO + O and the oxygen atom forms ozone. The balance between ozone destruction and creation from NOx depends on competition for NO_2 between photolysis (which increases ozone) and reaction with O (which reduces ozone). Ozone formation dominates below about 13 km because of the scarcity of oxygen atoms.

12. Broderick and Oliver 1982, p. 16.

13. Van der Leun and Daniels 1975.

14. Glantz, Robinson, and Krenz 1982, p. 58.

15. Grobecker, Coroniti, and Cannon 1974, p. vii.

16. E.g., *NYT* 1/22/75, p. 41; *Christian Science Monitor* (subsequently *CSM*) 2/5/75; *Pittsburgh Press* editorial, quoted in Dotto and Schiff, 1978, p. 83.

17. *NYT* 2/5/75, p. 73, and 3/31/75, p. 36; T. Donahue, letters to *Science* 187 (3/28/75): 1142–43, and *EOS* 56 (1975): 210, with Grobecker responses; H. S. Johnston letter to *Chemical and Engineering News* (subsequently *C&EN*), 4/21/75, p. 5; Johnston and Donahue testimony to House Subcommittee on Government Operations, 11/13/75.

18. "Deception Charged in Presentation of SST Study," *Science*, 11/28/75, p. 861; Dotto and Schiff 1978, pp. 103–10.

19. Canada, Australia, and the Soviet Union also produced SST reports, but these were research summaries with no attempt to draw policy-relevant conclusions (Pittock 1972; Kerr and McElroy 1976; Budyko and Karol 1976).

20. The NAS committee used the Livermore model, establishing a practice followed by later NAS ozone committees of basing their assessments on one model with variable inputs rather than on multiple models. The problem of the omitted smog chemistry was widely known, but could not be corrected with available computing power. The NAS committee tried to address it by assigning a factor of 100 uncertainty bound to depletion from subsonic aircraft, but this approach still excluded the real possibility of ozone increases (NAS 1975, p. 29). When this chemistry was added to models a few years later, subsonic aircraft were calculated to slightly increase ozone in the upper troposphere and in the total column (Broderick and Oliver 1982, p. 16).

21. COMESA 1975; COVOS 1976; Grobecker 1981.

22. Hoffert and Stewart 1975.

23. COMESA 1975, pp. xv–xviii, 69–71; Hoffert and Stewart 1975, p. 53; Hughes and Wynne 1992, p. 2; Hoffert interview.

24. Wynne et al. 2001; Brickman, Jasanoff, and Ilgen 1985; Boehmer-Christiansen and Skea 1990. Limited evidence of more direct political purposes for these assessments comes from the striking similarity of a British minister's defense of the Concorde in 1971 and the COMESA conclusions four years later (Hansard [record of Parliamentary debates, 823], 228–29, 10/25/71; COMESA 1975, pp. xviii, xxii), and from a French official's later description of both studies as attempts to counter American alarmism (*IER* 3/10/83, pp. 686–87).

25. *NYT* 3/14/75, pp. 41, 77.

26. "Giscard Speaks Out," *Newsweek* 7/25/77; May 1979; Grieco 1979; Gribbin 1988; Brenton 1994, p. 135; Lovelock 1982, p. 243; interviews with ICI and U.K. officials. British officials were also incensed that open U.S. decision processes and financial support from U.S. Concorde opponents gave British opponents much more influence than they had enjoyed at home.

27. Lufthansa's chairman called the Concorde "economically impossible" (Owen 1997, p. 98); Gribbin 1988, p. 23, argued that judging the Concorde to be profitable required ignoring all development costs and taking a strongly biased view of operating costs.

28. *Times* (London) 4/5/75, p. 3; and 9/18/75, p. 5; *NYT* 10/12/75, p. 25.

29. Panel on the Natural Stratosphere 1975, pp. 1–13, 5-125–5-152; Panel on the Perturbed Stratosphere 1975, pp. 1-27–1-30.

30. The only news coverage was an article in the Orange County edition of the *Los Angeles Times* (6/28/74, p. 3). Gribbin (1988, pp. 50) notes that the article failed to attract attention even though *Nature* ran a commentary highlighting its importance.

31. Rowland and Molina, 1994; *New Scientist* 9/26/74, p. 781.

32. Dotto and Schiff (1978) provide a detailed summary of this two-year period, informed by firsthand experience.

33. *NYT* 9/26/74, p. 1; 9/27/74, p. 82. *Time* magazine also covered the story briefly in October (10/7/74, 93–94).

34. Rowland, who had advocated an aerosol ban since the ACS meeting, appeared with NRDC staff at the press conference announcing the petition (*NYT* 11/21/74, p. 29).

35. Rogers hearings, Dec. 10 and 11, 1974; H.R. 17577, sponsored by Rogers and Esch, HR 17545 sponsored by Aspin. Both proposed aerosol bans with essential-use exemptions. *NYT* 12/12/74, p. 50.

36. *NYT* 9/26/74, p. A1.

37. Senior managers and scientists from chemical companies were the most temperate, emphasizing the claims' "unproven" character or specific assumptions on which they depended, although they also engaged in some tendentious argument (see, e.g., testimony of I. Sobolev, reprinted in *Aerosol Age*, 8/75, p. 18; and statements of R. McCarthy at AEB press conference, reported in *Drug and Cosmetic Industry* (*D&CI*), 11/74, p. 71; in *NYT* 11/2/74, p. 59; and in Rogers and Esch subcommittee testimony, 12/11/74, p. 379). More extreme and colorful attacks came from unnamed spokesmen, public-affairs officials, and representatives of aerosol firms. For example, "Extremists in the areas of ecology and consumer protection today are waging a more effective war on American industry than the hosts of enemy saboteurs ever dreamt of doing during World War II" (Robert Abplanalp, acceptance speech for Packaging Man of the Year award, reprinted in *D&CI* 11/74, p. 70; similar quotations in *D&CI* 12/74, p. 41; and *Aerosol Age*, 5/75).

38. The calculation, by Crutzen (1974a), indicated that delaying an aerosol ban from 1975 to 1978 would increase peak ozone loss from 1.2 to 1.7 percent, an increase of 0.5 percent.

39. R. McCarthy, testimony to Rogers and Esch subcommittees, 12/10/74, p. 381.

40. E.g., *NYT* 6/30/75, p. 30.

41. "Until the first results of this industry research become available, the available facts do not rank as proof that fluorocarbons will lead to ozone depletion." R. McCarthy in *CMR* 206, 14 (9/30/74): 1; H. S. Johnston 1992, p. 29; F. S. Rowland, testimony to Rogers and Esch subcommittees, 12/10/74.

42. Crutzen 1974a; Rowland and Molina 1975; Cicerone, Stolarski, and Walter 1974; Wofsy, McElroy, and Sze 1975.

43. Pack et al. 1977; Murcray et al. 1975; Wilkniss et al. 1975; summaries in Interdepartmental Committee for Atmospheric Sciences 1975, Appendix A; and *NYT* 12/13/75, p. 23.

44. Schmeltekopf et al., 1975; testimony of F. S. Rowland, Senate Subcommittee on the Upper Atmosphere, 9/23/75, p. 791.

45. Lazrus et al. 1976; Farmer and Raper 1977.

46. Zander 1975; *NYT* 9/10/75, p. 1.

47. Seeing ClO does not establish the source of the chlorine, the length of the catalytic chain, or the absence of other reservoirs or sinks. But with the other pieces, "Many scientists say the results [of attempts to detect stratospheric ClO] may be widely accepted as proof or disproof of the whole theory" (*WSJ* 12/3/75, p. 1).

48. The changes were all modest. The two largest changes (slower Cl + O_3, faster Cl + CH_4) reduced calculated depletion by half to two-thirds (Watson et al. 1976, reported at 4th CIAP Conference, Feb. 1974, and ACS, 4/9/75). Other new results went the other way, so the changes in total reduced calculated losses by about 20 percent. Industry representatives selectively publicized the results reducing ozone loss, stating that earlier depletion estimates were too high "by 300 percent" (R. Orfeo testimony to Bumpers subcommittee, p. 737; Cicerone testimony to House Subcommittee on the Environment and Atmosphere, 5/20/75; *NYT*, letters, W. Hoskins, 7/3/75, p. 30; R. Cicerone, 7/9/75, p. 36; F. S. Rowland, 8/8/75, p. 26).

49. Rowland et al. 1976.

50. Eggleton, Cox, and Derwent 1976; *NYT* 5/7/76, p. 1, 5/13/76, p. 24. An incorrect set of HCl measurements that suggested most upper-stratospheric chlorine might be tied up in chlorine nitrate compounded the confusion, as industry groups claimed that they decisively refuted ozone-depletion claims (*WSJ* 6/30/76, p. 8).

51. Bastian 1982.

52. Sharp disagreement among scientists at the hearing over the policy significance of recent research results, occurring at the same time as the debacle over the CIAP executive summary and press conference, widespread comment ("Public Credibility on Ozone," *Science* 187, no. 1182 [3/28/75]; *NYT* 3/31/75, p. 36).

53. Federal Task Force on Inadvertent Modification of the Stratosphere (IMOS) 1975, pp. 5, 13; *NYT* 6/13/75, p. 1.

54. Senior DuPont officials obtained a leaked copy of the report's galley proofs and contacted White House officials to try to block publication of the report, requiring the committee to hastily advance publication by two weeks (*New Yorker* 6/9/86, p. 73; Bastian 1982, p. 177); R. A. Bantham, op-ed, *NYT* 8/28/75, p. 33.

55. T. F. Cairns, DuPont research director and NAS member, 1/76 (quoted in Dotto and Schiff 1978, pp. 264–65).

56. Dotto and Schiff 1978, pp. 262–86; C. L. Bastian memo to the file, NSF, 8/2/76 (76-4-4), p. 2. The harshest attacks on IMOS came from British sources (e.g., Lovelock and Scorer in *New Scientist* 6/19/75, p. 643; and *Fortune* 8/75, p. 192; discussion at 10th Aerosol Congress, London, 9/75, in *Aerosol Age* 11/75, p. 18).

57. *NYT* 5/3/75, p. 37; 6/17/75, p. 30; 6/22/75, p. C3; editorial, 6/24/75, p. 32.

58. Bastian 1982, p. 179; *NYT* 5/5/75, p. 23, and 7/15/75, p. 39.

59. *NYT* 12/24/75, p. 24.

60. *NYT* 9/16/75, p. 34.

61. R. Cicerone, testimony, Senate Subcommittee on the Upper Atmosphere, 9/23/75, p. 992; *NYT*, editorial, 10/21/75, p. 36.

62. Dotto and Schiff 1978, p. 175

63. Testimony of V. J. Marriott, Senate Subcommittee on the Upper Atmosphere; *NYT* 6/22/75, p. C3, and 9/19/75, p. 39.

64. COAS was announced at the February 1975 IMOS hearing by its chair, Gillette's VP for consumer affairs. COAS worked closely with the Aerosol Education Bureau, a body formed in the 1960s to combat a trend of teenagers inhaling aerosols as intoxicants (*Aerosol Age* 5/75, p. 18; Dotto and Schiff 1978, p. 150–55).

65. *NYT* 7/8/75, p. 12; *LAT* 7/28/75, p. B1.

66. DuPont's research director, T. F. Cairns, contacted the Academy to distance himself from the Scorer tour (memo circulated to panel members summarizing Cairns's call, and private comments by R. McCarthy on the difficulty of restraining PR people, quoted in Dotto and Schiff 1978, pp. 157–60).

67. DuPont press release and cover letter, 9/20/76 (76-2-1); NAS *Letter to Members* 7, no. 1 (11/76): 2–3.

68. *NYT*, 6/22/75, p. 3.

69. Denunciations of anti-aerosol advertising in *Aerosol Age* 7/75, 7–9; *D&CI* 12/74, p. 135.

70. Both the Rogers-Esch and Aspin bills made regulatory decisions depend on Academy findings. The Aspin bill was most extreme, requiring regulation to come into force unless the NAS president found "no substantial danger" to human life or the environment (74-4-1, p. 5). While it was obviously impermissible to give such direct regulatory authority to the NAS president, who was not a government official, similarly broad precautionary authority was given to the EPA administrator in the 1977 CAA amendments.

71. C. Bastian, memo to file, 8/2/76 (76-4-4).

72. An NAS draft press release, 5/3/76 (76-4-5), identified three results from the prior two months as responsible for their delay: chlorine nitrate; the resultant importance of NO_2 in controlling the amount of active chlorine in the stratosphere; and increased stratospheric water vapor from temperature feedbacks. Also *NYT* 5/7/76, p. 1.

73. Loss calculations used the Livermore model, developed by panel member Julius Chang and his colleagues. The panel considered several proposed CFC sinks and found none of them plausible, but still made this 20 percent reduction in depletion in case other sinks might be identified (NAS Panel 1976, p. 14).

74. Of this uncertainty, a factor of 1.1 was due to uncertainty in present emissions, 3 to transport, and 6 to the rates of seven key reactions—including a factor of 3 from one reaction, HO + HO$_2$ (NAS Panel 1976, p. 160). Critics of the panel's quantification of uncertainty noted that the revision that soon made the largest change in loss estimates, NO + HO$_2$, was not even one of their seven key reactions (UK STRAC 1979, pp. 165–66).

75. E.g., "Scientists Back New Aerosol Curbs to Protect Ozone in Atmosphere," *NYT* 9/14/76, p. 1; "Aerosol Ban Opposed by Science Unit," *WP* 9/14/76, p. 1.

76. NAS Committee 1976, p. 7.

77. *NYT* 9/17/76, p. 14; 9/18/76, p. 20.

78. Reports of this discussion are in Dotto and Schiff 1978, p. 284; Brodeur 1986; Schneider 1989, p. 223; and Lovelock 1982, who likens Pittle's questioning of panel member Fred Kaufman to the trial of Galileo.

79. The committee's chair subsequently expressed satisfaction with the quick regulatory response, stating that they had meant at most two years to *enactment* of regulations, which might require an earlier start given regulatory delays (testimony of J. Tukey, Senate Subcommittee on Upper Atmosphere, 12/15/76).

80. The EPA had commissioned an initial study by A. D. Little in 1975. The second study, by International Research and Technology Corporation, was completed in early 1977 (Federal Register [FR] 42, no 93 [5/13/77]: 24542–49; Stoel, Miller, and Milroy 1980, pp. 47–48).

81. Memo from Bastian and Muir to Peterson and Stever, 9/24/1976, quoted Bastian 1982, p. 189.

82. *EPA Environmental News*, 10/12/76; *NYT* 10/16/76, p. 6; *WSJ* 10/18/76, p. 12.

83. Wirth, Brunner, and Bishop 1982, p. 219; *NYT* 11/23/76, p. 21.

84. 42 FR 24535, 5/13/77; Wirth, Brunner, and Bishop 1982, p. 226–36.

85. 43 FR 11300, 3/17/78.

86. Effective 10/15/78 for CFC manufacture, 12/15/98 for filling cans, and 4/15/79 for interstate shipment.

87. 11 Kte of annual usage was exempted, slightly less than 5 percent of original use. The EPA periodically reviewed the exemptions, but made few changes to the original set (EPA Reports to Congress 82-4-3, 83-4-2).

88. EPA press release, 3/15/78 (78-4-2).

89. *NYT* 11/20/76, p. 33; 5/13/77, p. D1.

90. Section 157 (b). Both TSCA and the CAA amendments were in development through the early CFC debate, and both contained the authority to regulate CFCs. Aerosol regulations were promulgated under TSCA because it was enacted first, but subsequent regulation of ozone-depleting chemicals has been under the CAA.

91. Stoel, Miller, and Milroy 1980, pp. 58–59; summarized in statement of Lee Thomas (86-1-5), p. 5.

92. T. Stoel, testimony to Rogers and Esch subcommittees (74-4-1); *NYT* 11/21/74, p. 29.

93. John deKaney and Mario Molina, quoted in *WSJ* 9/25/78, p. 12.

94. Foley and Ruderman 1973; H. S. Johnston, Whitten, and Birks 1973; Goldsmith et al. 1973; Chang and Duewer 1973.

95. Hampson 1974. Hampson was also the first person to propose ozone losses 10 years earlier. Although he published this new suggestion first, the Livermore team was secretly working on the same problem, after stumbling on it through a data-entry error in modeling the effects of the 1960 nuclear test. In one run they inadvertently replaced one blast by 1,000, and calculated losses of more than half the ozone layer (MacCracken interview).

96. *NYT* 9/6/74, p. 1, 10/5/75, p. 8, 10/17/74, p. 7; ACDA publication 81, 1974.

97. E.g., Lovelock (1982), a relentless antagonist of ozone-loss claims until the late 1980s, cited the claim that ozone loss would be the worst environmental effect of a nuclear war as the most ridiculous instance of "ozone hysteria." (See also *NYT*, editorial, 11/12/74, p. 38; Schneider and Mesirow 1976.)

98. Crutzen and Birks 1982; Baum 1982, pp. 28–29.

99. Editorial, *Times* (London), 6/14/75, p. 13; U.K. DOE 1976, p. 6.

100. Van Eijndhoven et al. 2001.

101. Still, a 1978 study of 12 countries found that national authority to regulate CFCs existed in 9, and other forms of regulatory capacity existed in the remaining 3 (Stoel, Miller, and Milroy 1980, pp. 30–31).

102. See, e.g., R. Train speech to NATO Committee on the Challenges of Modern Society, 12/3/76 (76-4-2).

103. Parson et al. 2001, p. 239; interviews with Hare, Bastian, Bracken, and Thacher.

104. Schreurs 2001; *IER* 2, no. 12 (12/12/79): 1000.

105. Sokolov et al. 2001.

106. Cavender-Bares, Jäger, and Ell 2001; Stoel, Miller, and Milroy 1980, pp. 45, 123.

107. *IER* 1, no. 1 (1/10/78); Schiff, Stoel, Watson, and Tuck interviews.

108. U.K. DOE 1976.

109. Bastian 1982, p. 185.

110. *IER* 1, no. (12/10/78): 427; *IER* 5, no. 10 (10/2/82): 308.

111. White 1982.

112. Stoel, Miller, and Milroy 1980, p. 261; DeReeder 1976, p. 6–1.

113. "Note by the United States" (76-1-3).

114. U.S. Department of State, 1975; Stoel, Miller, and Milroy 1980, p. 46; Bastian 1982, p. 184.

115. OECD Chemicals Group, "The Economic Impact of Restrictions on the Use of Fluorocarbons, Final Report," ENV/Chem/77.2 (rev.), Paris, 1977; Stoel, Miller, and Milroy 1980, p. 84; Stoel 1983, p. 66.

116. UNEP "Work Plan: Outer Limits," 9/26/77 (77-1-4).

117. Stoel, Miller, and Milroy 1980, p. 273. (UNEP funding of this meeting also indicates WMO's limited interest in the issue.)

118. Decision 29 of the 3rd Governing Council, 5/2/75.

119. Caldwell 1990, pp. 72–82.

120. Proceedings in UNEP (1979b); background papers, Plan of Action (77-1-2), including recommendation, p. 12.

121. Thacher 1993, pp. 15–16; Stoel, Miller, and Milroy 1980, p. 81; Bastian 1982, pp. 184, 191–92; *NYT*, 3/10/77, p. 16; interviews with Boville, Hare, Schiff, Stoel, Thacher.

122. Stoel, Miller, and Milroy 1980, p. 275.

123. Stoel 1983, p. 52; Wirth, Brunner, and Bishop 1982, p. 230.

124. The environment minister also exercised existing statutory authority in March 1977 to require all firms to report CFC production and imports—an advantage in developing regulations that many governments lacked (Stoel, Miller, and Milroy 1980, p. 52; Parson et al. 2001, p. 240; *IER* 2, no. 5 [1979]: 666).

125. The study, by the Battelle Institute of Frankfurt, appeared in May 1976. It drew on the EPA consultant's study (IRTC) to project losses of 11,000 jobs and DM 500–600 million of capital (Stoel, Miller, and Milroy 1980, pp. 49, 128–47).

126. The German government lacked authority to demand production data, however, and thus had to trust firms' statements that they were complying (Enquete Commission of the 11th German Bundestag 1991).

127. EPA Report (79-4-1), p. 45.

128. Metra Consultants 1976, cited in Stoel, Miller, and Milroy 1980, pp. 49, 73, 154.

129. Van Eijndhoven et al. 2001.

130. Resolution C133, OJ C 133, 6/7/78; Haas and McCabe 2001; Cavender-Bares, Jäger, & Ell 2001; *IER* 1, no. 1 (1/10/78).

131. Stoel, Miller, and Milroy 1980, p. 44; *NYT* 1/30/78, p. 14.

132. Interview with R. Lonngren, 7/13/78, cited in Stoel, Miller, and Milroy 1980, p. 171.

133. The CCOL meeting was Nov. 28–30 in Bonn, the regulatory meeting, Dec. 6–8 in Munich.

134. Stoel's presentation (78-1-2); study reprinted as Stoel, Miller, and Milroy 1980.

135. *IER* 2, no. 1 (1/10/79): 477–78.

136. The CCOL *Ozone Layer Bulletin* (UNEP 1979a) includes a remarkably blunt assessment of the meeting.

137. Enquete Commission of German Bundestag 1989, p. 192; *IER* 2, no. 1 (1/10/79): 461; discussion of Germany interagency division in *IER* 3, no. 4 (9/4/80).

138. *New Scientist* 80, no. 1124 (10/12/78): 94.

139. Decision 80/372, EEC OJ L 90, 4/3/80, p. 45; *IER* 1, no. 3 (3/10/78): 61; *IER* 2, no. 6 (6/13/79): 726; Haigh 1990, p. 268; van Eijndhoven et al. 2001; Winsemius interview.

140. Oslo report, Apr. 14–16, 1980 (80-1-2); *IER* 3, no. 3 (3/12/80), pp. 87–88; Bastian 1982, p. 194.

141. *IER* 3, no. 5 (5/14/80): 169–70. (The critic failed to note that Dutch CFC use was low because of several years of intentional reductions.)

142. *IER* 3, no. 6 (6/11/80): 237–38; Blum, Bracken, Odoulamy, and Weil interviews.

143. Bastian, Blum, Brydon, Odoulamy, and Reid interviews.

144. Stoel and Ahmed testimony, EPA regulatory hearings (74-4-1).

145. NAS Panel 1976, p. 42; *IER* 2, no. 1 (1/10/79): 478.

146. *IER* 1, no. 2 (2/10/78); *IER* 2, no. 4 (4/11/79): 597; *IER* 1, no. 5 (5/10/78): 130.

147. *IER* 1, no. 4 (4/10/78): 93; *IER* 1, no. 5 (5/10/78): 121; Brickman, Jasanoff, and Ilgen 1985, pp. 276–85.

148. Stoel, Miller, and Milroy 1980, p. 52.

149. John Mills of ICI, quoted in *ENDS* 139, no. 8 (1986): 16; Howard interview.

150. DuPont 1979 briefing book (79-2-1) reports initial evidence of toxicity in an Ames test, and recommends a two-year inhalation study.

151. McCarthy and Bower 1980, p. 138.

152. Manzer 1990.

153. E.g., McCarthy testimony, Senate Subcommittee on the Upper Atmosphere, 9/23/1975; Cicerone, letter to Bumpers, appended to hearing.

154. DuPont 1979 briefing book (79-2-1); J. Steed, *C&EN* 11/24/86, cites expenditure of $15 million over the life of the program, matching the total reported in June 1980.

155. Three firms filed patents for catalytic processes to produce HFC-134a: ICI in 1977 (*Chem. Abs.* 97:91719 [1982], 91:19875 [1979], 90:137242 [1978]), Daikin Kogyo in 1978 (*Chem. Abs.* 93:167600 [1980], 95:135053 [1981]), and DuPont in 1980 (*Chem.Abs.* 96: 180749 [1982]). DuPont also published research on lubricant compatibility of 134a in 1978 (*Chem.Abs.* 90:74121 [1978]). Other patent filings from alternatives research included several HCFC refrigerant blends (DuPont for several HCFC blends in 1976 and 1977; Allied for a blend of HCFCs 22 and 124 in 1980); an HCFC-123 blend to reduce use of CFC-11 in blowing rigid polyurethane foam (Allied, U.S. Pat. 4,624,970); and a process to blow polystyrene foams with HCFC-22 (DuPont 1975, *Chem.Abs.* 89:44703 [1978]). DuPont researchers also published discussions of many new HCFC and HFC uses in the journal *Research Disclosures* in 1976 and 1977 (146, pp. 13–14; 154, p. 4). (This journal provides quick publication of corporate research findings for which patents are not being sought, to establish "prior art" and prevent others from patenting them.)

156. The reasons included toxicity, inadequate performance, and lack of a viable manufacturing process. Work continued on six: HCFCs 22, 141b, and 142b, and HFCs 134a, 143a, and 152a (DuPont *Fluorocarbon/Ozone Update*, June 1980 (80-2-1); J. Steed, *C&EN* 11/24/86; D. Strobach, paper presented to EPA workshop (86-2-12), 2/24/86, shows work on HFC-134a also stopped because of no commercial synthesis route.

157. Palmer et al. 1980, p. 14. Updates of this analysis in 1982 and 1986 modestly increased the estimates of maximum feasible reductions, to about one-third (Mooz et al. 1982; Camm et al. 1986).

158. The three CFC sinks the panel judged plausible in 1976 had been rejected but a new one had been proposed: photolysis while adsorbed on desert sand. The panel noted that its uncertainty range should be broadened to account for omitted chemistry, but did not attempt to quantify this (NAS Panel 1979, pp. 6, 10, 18).

159. This range extended from the panel's steady-state point estimate to the lowest emission-growth estimate (NAS Committee 1979, pp. 6, 10, 74, 105).

160. NAS Committee 1979, pp. 23, 25–27, 196, consultations listed in Appendix L. While the report stresses that CARCE did not rely solely on Rand and DuPont reports but formed

its own judgments, it is clear from the text that its reliance on these sources was substantial.

161. The regulations applied to CFCs 11, 12, 113, 114, and 115; HCFC-22; and halon 1301, the first time a halon was proposed for control as an ozone depleter (EPA Regulatory Alert and Press Release 2742, 4/15/1980).

162. DuPont *Fluorocarbon/Ozone Update* 6/80 (80-2-1); *IER* 3, no. 5 (5/14/80): 170–71.

163. 45 FR 66726, 10/7/80; Bastian, Odoulamy, Shapiro, and Weil interviews.

164. *IER* 3, no. 8 (8/13/80): 337; Shapiro interview.

165. Bastian 1982, pp. 190–91; Miller interview.

166. DuPont, *Fluorocarbon/Ozone Briefing Book* (79-2-1); *Fluorocarbon/Ozone Update*, 1980 through 1982 (80-2-1, 80-2-3, 80-2-4, 80-2-5, 80-2-6, 81-2-4, 81-2-5, 82-2-2).

167. E.g., Fluorocarbon Panel 1980. While the panel was sponsoring good unbiased research, much of which went against industry's interests, they sacrificed their credibility among scientists—even those accepting their money—by also publishing highly partisan synthesis reports such as this one.

168. By using assumptions highly unfavorable to alternatives, this study by the Battelle Institute of Columbus found a huge energy penalty, 30–50 billion gallons of fuel over ten years (*IER* 3, no. 9 [9/10/80]: 401).

169. DuPont briefing book (79-2-1), p. 12; *Fluorocarbon/Ozone Update*, 4/80 (80-2-4); *Fluorocarbon/Ozone Update*, 6/80 (80-2-1), says 7 to 10 years, but is quite optimistic in tone regarding technical feasibility.

170. To make this case, the Alliance made prominent use of small insulating-foam contractors, who faced large CFC costs and no adequate substitutes. See, e.g., interview with L. L. Cockrell by R. Scheer, 2/13/93 (transcript appended to Scheer senior thesis, "Reporters on the Loose," Harvard College, 1993), discussing his congressional testimony: "I'm not an idiot. The only reason I was there was because I was a small businessman with 40–45 employees, and the lobbyists from the plastics industry felt like maybe I gave them a perspective . . . that was more effective than having DuPont or Allied testify."

171. By late 1980, the EPA was receiving regular congressional threats that if it tried to exercise its authority to enact further CFC regulations, it risked having it legislatively revoked (Muir interview).

172. E.g., Alliance, "Information Paper," 10/1/80 (80-2-2) criticized the approach as "untested" and incorrectly claimed that the EPA sought to use the approach to reduce usage 50–70 percent in the United States, rather than worldwide (where most of it could be easily achieved through aerosol cuts). The proposal was also attacked for transfers and transitional effects, which the ANPR had not adequately considered (Shapiro and Warhit 1983).

173. Bastian 1982, p. 195.

174. 146 FR 23629, 4/27/81; Bastian 1982, p. 195; Wirth, Brunner, and Bishop 1982, p. 238.

175. Kraft 1984.

176. *ENDS* 80, no. 9 (1981): 23.

177. DuPont briefing book (79-2-1), p. 13.

178. Bastian, Odoulamy, Shapiro, and Weil interviews.

179. A. Miller reports only being able to staff an environmental panel for a 1981 congressional hearing only by writing all testimony and fielding all questions for himself and two representatives of other nongovernmental organizations (Miller interview; *IER* 4, no. 9 [12/9/81]: 1127).

Chapter 4

1. Fluorocarbon Panel 1976, p. 1; *NYT* 10/1/75, p. 1.

2. McCarthy, Bower, and Jesson 1977.

3. This presumption, while reasonable, can of course be wrong if significant nonlinearities or thresholds are present, in which case a small anthropogenic increase on a large natural baseline may have major impacts.

4. Sobolev testimony to Rogers and Esch hearings; COAS study of 1975 Alaskan eruption, *NYT* 10/1/75, p. 48.

5. Lovelock and Rasmussen both observed concentrations of 100–150 ppt. Historical release was initially estimated as equivalent to 4 ppt, then revised to 160 ppt, of which 120 ppt would remain if the only sink were the stratosphere (NAS Panel 1976, p. 43–44).

6. McCarthy testimony, Senate Subcommittee on the Upper Atmosphere, 9/18/75; Singh et al. 1979; Rasmussen et al. 1980.

7. NAS Panel 1976, p. 9. Atmospheric lifetimes are stated as the time for an initial concentration to be reduced by a factor of e (about 2.718)—a standard convention for describing exponential decay processes.

8. NAS Panel 1976, pp. 179–201.

9. The existence of this sink was suggested by Lovelock's observation that CCl_4 concentrations were lower downwind of the Sahara, and the fact that CFCs photolyze faster in quartz vessels. Although this sink was invoked by opponents of CFC controls through 1980, it was clear from the earliest measurements that even if real, it was too small to matter (U.K. STRAC 1979, p. 177; von Schweinichen 1980, p. 152; DuPont briefing book 1979 (79-2-1), p. 23; McCarthy and Bower 1980, p. 128; NAS Panel 1976, p. 66–67).

10. "Evaluation of Methodology for Analysis of Halocarbons," U.S. Dept. of Commerce report NBSIR 78-1480, June 1978; NAS Panel 1976, pp. 36–38; Lovelock and Simmonds 1980, pp. 47–50; Jesson 1982, pp. 43–44.

11. The results to be provided by the ALE network were often stated in the misleading form "It will detect a 10 year lifetime after three years of operation, and a 20-year lifetime after five" (see, e.g., DuPont briefing book 1979, 79-2-1). In fact, at any time, the network could exclude lifetimes far enough above or below the true value. For example, with a true lifetime of 10 years, the network could exclude (with 95 percent confidence) lifetimes outside 6–20 years after 3 years, or 8–13 years after 5 years; with a true 50-year lifetime, the network could exclude lifetimes below 20 years after 3 years (McCarthy and Bower 1980, fig. 8.3).

12. The uncertainty bounds were 58–117 years for CFC-11 and 64–400 years for CFC-12. Cunnold et al. 1986.

13. The new rates for reactions of chlorine with ozone and with methane ($Cl + O_3 \rightarrow ClO + O_2$, and $Cl + CH_4 \rightarrow HCl + CH_3$), figured prominently in Sept. 1975 Senate hearings (Watson et al. 1976; Watson 1977).

14. Early proposals included reaction of chlorine with ammonia and with sulfuric acid aerosols (I. Sobolev testimony, Rogers and Esch subcommittee, Dec. 1974); Rowland, Spencer, and Molina 1976; NAS Panel 1976, pp. 206–216.

15. These couplings also made ozone loss due to HOx depend on how much NOx and Cl are present. As Crutzen 1979 pointed out, this coupling made HOx injections, which had been neglected since 1970, important again in a stratosphere perturbed by Cl.

16. Summaries of progress in stratospheric kinetics through this period are in Broderick and Oliver 1982; Sze 1982; H. S. Johnston 1982, 1984, 1992; Rowland 1991.

17. Duewer et al. 1977; Widhopf, Clatt, and Kramer 1977; Turco et al. 1978; *NYT* 7/17/77, p. 1. For responses that highlight the reduced NOx effect from these changes while ignoring the increased Cl effect, see Fluorocarbon Panel letter to H. Gutowsky, 6/22/77 (77-2-4); Sir B. Mason (UKMO), speech to World Climate Conference, WMO, Geneva, Feb. 12–23, 1979; *WSJ*, editorial, 5/15/79, p. 26.

18. The new measurement of $NO + HO_2 \rightarrow HO + NO_2$ (Howard and Evenson 1977) was discussed at the AGU meeting in June 1977, and in NASA 1977, pp. 155–57.

19. The prior rate for this reaction ($HO_2 + O_3 \rightarrow HO + 2O_2$) had been estimated to fit the observed distribution of ozone given 1970 chemistry, and not revised since (Zahniser and Howard 1978; Crutzen and Howard 1978).

20. The new ozone-loss pathway was particularly effective in the lower stratosphere because it did not require the presence of oxygen atoms (Ravishankara et al. 1977; Birks et al. 1977; Crutzen 1979; NASA 1979, p. 336).

21. Molina and Molina, 1978; NASA 1979, p. 336; Sze 1982, p. 147; NASA/WMO 1982, fig. 1-120.

22. Fluorocarbon Panel letter to H. Gutowsky, chair of 1976 NAS Panel (77-2-4).

23. Peroxynitric acid formed by the reaction, $HO_2 + NO_2 \rightarrow HO_2NO_2$, which was initially found to be fast (Cox, Derwent, and Hutton 1977; Howard and Evenson 1977). UV absorption measurements suggested it photolyzed slowly enough that it would mostly reduce ozone loss from chlorine, but by 1981 its effect was known to be small: about half reacts with HO, reducing ozone loss to chlorine, while the rest photolyzes to HO and NO_3, nearly offsetting this reduction (Molina and Molina 1981; Graham, Winer, and Pitts 1978; NASA 1979, p. 337; Sze 1982, pp. 147–48).

24. Lazrus et al. 1976; Yung et al, 1980; NASA 1979, p. 337.

25. NAS Panel 1979, pp. 9–10; NASA/WMO 1982 reports no major change in Cl or N reactions for three years.

26. H. S. Johnston 1982; McElroy 1982.

27. The significant changes were a tripling in the rate of HO with HNO_3, reducing calculated ozone loss by a third (Wine et al. 1981); a fivefold increase in the rate for HO with peroxynitric acid (Littlejohn and Johnston 1980); and smaller increases in the rates of HO with HO_2, NO_2, and H_2O_2 (Keyser 1980; Sridihan, Reimann, and Kaufman 1980; DeMore et al. 1981; NASA/WMO 1982, pp. A14–A17).

28. NASA/WMO 1982, pp. A-15, 3–4.

29. UNEP CCOL 1983; NAS 1984; NASA/WMO 1986, p. 725, table 13–2.

30. Rowland testimony, House Subcommittee on Health and the Environment, 3/9/87, pp. 21–22; Rowland interview (June 1984), quoted in the *New Yorker* 6/9/86, p. 82.

31. NAS Panel 1976, pp. 149–150.

32. See, e.g., NASA 1977, pp. 226–27, criticizing treatment of uncertainty in NAS Panel 1976.

33. Early simple 2-D results were reported by Cunnold et al. 1975 and Vupputuri 1976, summarized in NAS Panel 1976, p. 13, and U.K. STRAC 1979.

34. In 1976, stratospheric models typically represented 60 to 100 chemical and photochemical reactions. The standard reaction set had grown to more than 150 by 1980, and more than 250 by 1985, still considering only homogeneous gas-phase chemistry (NAS Panel 1976, p. 108; NASA/WMO 1986).

35. See, e.g., DuPont *briefing book* (79-2-1, pp. 25–26), U.K. DOE 1976; U.K. STRAC 1979. It was occasionally argued that even 2-D models were inadequate, and that only 3-D models could responsibly be used for policy. One support for this claim was that even zonal averaging in 2-D models can give nonsensical results: regions where zonally averaged diffusion moves against the zonally averaged concentration gradient so the diffusion parameter should be negative (Pyle and Houghton 1980; Mahlman 1975).

36. NASA 1977, pp. 139–152; NASA 1979, p. 338.

37. Dickinson, Liu, and Donahue, 1978.

38. Atmospheric N_2O concentration increased 0.2 percent annually from 1960 to 1980 (R. F. (Weiss 1981). Although this increase occurred at the same time as a large increase in human nitrogen fixation (from a few percent of the natural rate to about one third), the source of the atmospheric increase, the human share of current N_2O emissions, and the projected human contribution to future emissions growth all remained controversial (Crutzen 1974b; McElroy, Wofsy, and Young 1977; H. S. Johnston 1977; Liu et al. 1976).

39. Two percent annual increases in methane's atmospheric concentration in the early 1980s made it increasingly important for understanding stratospheric chemistry. Methane's lifetime is long enough that a significant fraction reaches the stratosphere, where it reacts with energetic oxygen atoms to release HO. It is thus both a major source of stratospheric HOx and the major sink for chlorine (Sze 1982; Khalil and Rasmussen 1983).

40. Groves, Mattingly, and Tuck 1978; Groves and Tuck 1979.

41. NASA 1979, p. 354; U.K. STRAC 1979, pp. 184–89.

42. Penner 1981; Wuebbles, Luther, and Penner 1983; NASA/WMO 1982.

43. Miller et al. 1981. First results of this model by DuPont scientists were reported in NASA/WHO/1982; see also Cicerone 1982, p. 151.

44. Dickinson 1982, p. 161.

45. Cicerone, Walters, and Liu 1983.

46. Prather, Mc Elroy, and Wofsy 1984.

47. NASA/WMO 1986, pp. 764–71; Prather, Wuebbles, and Isaksen interviews.

48. NASA/WMO 1986, p. 728.

49. Testimony of F. S. Rowland, Senate Subcommittee on Environmental Pollution, 6/10/86.

50. NASA/WMO 1986, p. 745.

51. NASA/WMO 1986, p. 658. Estimated uncertainty from future emissions gave a range of 0.3 to 11.1 percent.

52. Isaksen and Stordal 1986; ENDS 139 (8/86): 1.

53. Wuebbles and Connell 1984 separately introduced heterogeneous reactions of chlorine nitrate with H_2O and HCl in the Livermore model (F. S. Rowland testimony, Senate Subcommittee on Environmental Pollution, 6/10/86).

54. NASA/WMO 1986, p. 781

55. Crutzen, Iasksen, and Reid 1975; Heath, Kraeger, and Crutzen 1977.

56. NASA 1979, p. 352; London and Angell 1982, p. 35.

57. Foley and Ruderman 1973; Chang and Duewer 1973; Goldsmith et al. 1973; Angell and Korshover 1976.

58. London and Angell 1982, p. 22; H. S. Johnston, Whitten, and Birks 1973; Elsaessar 1978.

59. Ruderman and Chamberlain 1973; Angell and Korshover 1976; NAS Panel 1976.

60. Chang Duewer, and Wuebbles 1979; NASA 1979, pp. 350–51; H. S. Johnston 1982, p. 126.

61. Anderson 1982, p. 271

62. NASA 1979, p. 339.

63. Anderson 1982, pp. 244–48.

64. Noxon 1979 (observations first presented at the Conference on Ozone in Logan, UT, Sept. 1976).

65. NASA/WMO 1986, pp. 618–32; Albritton slides for chairman's workshop, Les Diablerets, 7/11/85 (85-4-10).

66. Flights of 7/28/76 and 7/14/77, reported in Anderson, Margitan, and Stedman 1977.

67. DuPont briefing book 1980 (79-2-1); Broderick and Oliver 1982, p. 21.

68. Orfeo testimony, Senate Subcommittee on the Upper Atmosphere, 9/1975, p. 732.

69. C&EN 10/15/79, p. 27; UAPB 80-6, p. 5; Anderson quotes in C&EN 9/13/82, p. 23.

70. Berg et al. 1980; NASA/WMO 1982, pp. 1–184.

71. NASA/WMO 1982; Anderson 1982, p. 221.

72. Anderson 1975; Anderson et al. 1981; Burnett and Burnett 1980.

73. Anderson 1982, p. 282.

74. Wennberg et al. 1999.

75. Wofsy and Logan 1982, p. 200; NASA/WMO 1986.

76. NYT 2/26/80, p. C1; 6/2/81, p. C1; 9/17/82, p. A16.

77. NAS 1984, p. 34; NASA/WMO 1986, p. 609, fig. 11-2.

78. NAS 1984, pp. 36, 82; Solomon and Garcia 1983; Roscoe 1982.

79. NASA 1979, p. 336.

80. Anderson 1982, p. 279; Wofsy and Logan 1982, p. 199.

81. A related claim was that consumer rejection of aerosols in the 1970s showed that market responses were adequate to manage the environmental risk of CFCs, so regulation was not necessary (e.g., John W. Dickinson, chair of COAS, quoted in NYT 10/1/75, p. 3).

82. Production for nonaerosol uses averaged 7 percent growth through the late 1970s. Based on its industry surveys, Rand claimed that continued 7 percent growth was most likely, while EPA claimed that FPP data suggested 9 percent growth abroad (Stoel 1983, p. 84; Rabin 1981, p. 34; Palmer et al. 1980).

83. DuPont letter to Jellinek, 6/16/80 (80-2-8); Fluorocarbon Panel 1980, p. 10; DuPont comments on 1979 NAS Panel, January 1980 (80-2-3 and 81-2-4).

84. The NAS 1982 and 1984 assessments reported only steady-state estimates except for

one table in NAS 1984, reproduced from Wuebbles 1983; NASA/WMO 1982 noted that steady-state depletion was varying linearly with steady-state emissions, but made no other mention of emission growth (pp. 3-12-3-14).

85. Wolf 1980; UNEP CCOL 1981 (81-1-8); EPA Report to Congress (82-4-3); OECD 1983; NASA/WMO 1986, p. 77.

86. Rowland et al. 1982 (initially presented to American Chemical Society, Apr. 1982).

87. UNEP CCOL 1983, p. 11; UAPB 81-3, p. 7.

88. E. P. Borisenkov and Y. E. Kazakov, *Tr. Gl. Geofiz. Obs.* 438 (1980): 62–74.

89. *ENDS* 93 (10/82): 5; UNEP CCOL 1983, p. 11; Gamlen et al., 1986.

90. NAS 1984, p. 72; NASA/WMO 1986, p. 7 confirms that explaining the CFC-12 concentration observed by ALE requires a substantial additional source relative to the Panel figures.

91. UAPB 84-2, p. 7.

92. Fluorocarbon Panel letter of 6/19/84, appended to (84-1-24).

93. NASA/WMO 1986, p. 69.

94. Seidel and Keyes 1983; NAS 1984, p. 102; Wuebbles 1983.

95. Gibbs 1986; Quinn et al. 1986; Nordhaus and Yohe 1986, p. 27. Nordhaus and Yohe's probabilistic projection method gave median growth of 4 percent, which would reach 6 ppb by the mid-twenty-first century.

96. UNEP CCOL, 1983. The panel used the tactic of stating only near-term results through 1986, when it reported new runs of the DuPont 2-D model showing ozone loss only through 2005. (Fluorocarbon Panel, "Recent Research Results and Future Directions," Feb. 1986, 86-2-16.)

97. DuPont comments on Dec. 1984 Rand draft (84-2-2); CEFIC comments to Rand summarized in Quinn et al. 1986, p. 129. Similar industry arguments that use rhetorical tactics more typical of environmentalists to oppose regulation, invoking fixed resource constraints and ignoring the possibilities of substitution and technological change, are in von Schweinichen, 1980.

98. CEFIC, cited in NASA/WMO 1986, p. 77 (noting that two years of 8 percent growth undermined the projection).

99. This is still 13 percent below the 1974 peak for reporting companies, but omits the areas of fastest growth, CFC-113 and Chinese production. Including CFC-113, 1984 production exceeded the prior peak (NASA/WMO 1986, p. 69).

100. NAS Panel 1976, pp. 47, 216.

101. Methyl chloroform was replacing perchloroethylene and trichloroethylene, which were increasingly restricted on account of toxicity and smog formation. Through the 1960s and 1970s its production grew 16 percent annually in the United States and 30 percent worldwide, reaching 580 Kte in 1979 (Crutzen, Isaksen and McAfee 1978; McConnell and Schiff 1978; NAS 1977; NASA/WMO 1986, p. 72; Lovelock 1977; EPA 1982 Report to Congress [82-4-3], p. 6).

102. Quinn et al. 1986, citing DuPont response to their earlier estimate; NASA/WMO 1986, p. 70, citing DuPont private communication; CCOL 1984 (84-1-18), para. 4; Rowland testimony to House Subcommittee on Health and Environment, 3/9/87, p. 14 (atmospheric concentration of CFC-113 quadrupled in nine years).

103. NASA/WMO 1986, p. 71; Khalil and Rasmussen 1981. Both DuPont and ICI suggested global halon production was growing very slowly, but one series of atmospheric observations in the early 1980s found concentration increasing 20 percent annually (Lal et al. 1985).

104. UNEP CCOL 1981, paras. 22, 33.

105. For example, DuPont criticized Rand's estimate that CFC-113 production doubled from 1976 to 1979, stating that the actual increase was "much less" and that CFC-113 should not even be considered because its use is so small and nearly all for "critical applications" (DuPont comments, 84-2-2, p. 28). But DuPont's Donald Strobach, acting as science adviser to the Alliance, refused a request from the World Resources Institute for CFC-113 production data, giving the obviously fallacious reason that releasing even aggregated world production

data would compromise commercial confidentiality (letter to A. Miller, WRI, Sept. 1985, 85-2-2).

106. Dickinson 1982, p. 163.

107. Wuebbles 1981, 1983.

108. *NYT* 5/16/75, p. 38; Angell and Korshover 1973, 1976, 1978; London and Oltman 1979; H. S. Johnston, Whitten, and Birks, 1973; Komhyr et al. 1971.

109. London and Angell 1982, p. 22; Angell and Korshover 1978; London and Kelly 1974.

110. Testimony of Orfeo and Engel, Senate Subcommittee on the Upper Atmosphere, pp. 739–40; H. S. Johnston 1984, pp. 486–87.

111. Christie 1973; Angell and Korshover 1973; Ruderman and Chamberlain 1973; NASA/WMO 1982, p. 3–31.

112. NASA 1979, p. 354; Cicerone 1982, p. 149.

113. Hill and Sheldon 1975.

114. Hill, Sheldon, and Tiede 1977; NASA 1977, p. 85.

115. NAS Panel 1976, p. 281. The possibility of slow natural oscillations was a particularly serious limitation, since only two or three ozone stations had records long enough to identify even the hypothesized 11-year cycle.

116. U.K. STRAC 1979; NASA 1979; NAS Panel, 1979, p. 16; NASA/WMO 1982; CCOL 1981, para 18; Hill 1982, p. 80; Fluorocarbon Panel 1982 (81-2-3).

117. WMO 1981, para. 22.

118. The 75 ozone stations are all on land, 80 percent of them in the northern hemisphere. A few new stations were established in the 1970s, but more stations and many years of data were needed to reduce the locational bias.

119. The ozone network had severe operational problems. The WMO's program to upgrade instruments found frequent calibration errors of 7 percent, and some as large as 20 percent. Many stations made observations only on sunny days, adding further bias because high pressure is associated with low ozone. Only three stations had records that covered more than two 11-year solar cycles. Finally, even of stations reporting data regularly, almost none corrected their archived data when they found calibration errors (NASA 1977, p. 53; CCOL 1983).

120. UNEP CCOL 1981, para. 1; Fluorocarbon Panel, "Recent Research Results and Future Directions," 2/1986 (86-2-16).

121. Estimated 1970–1979 trends were 0.2 percent (Bloomfield et al. 1983), 0.28 percent (Reinsel et al. 1981), and 1.5 percent (St. John et al. 1981, the only statistically significant result) (NASA/WMO 1982, pp. 3-35–3-40).

122. Reinsel et al. 1981, 1984; NASA/WMO 1986, p. 789; Angell 1988; "Has Stratospheric Ozone Started to Disappear?" *Science* 237 (7/10/87): 131–132.

123. NASA/WMO 1986, pp. 789–790.

124. Liu et al. 1980; NASA/WMO 1982, pp. 2–7; Angell & Korshover 1983; Albritton slides from NOAA seminar, 7/20/84 (84-4-10).

125. The correction was critical because without it, the observed decline at 40 km was not statistically significant (Fluorocarbon Panel critique of April 1984 U.S. negotiating position [84-2-4], p. 8).

126. Reinsel et al. 1984 found a statistically significant decline of 0.2–0.3 percent per year at 34–43 km from 1970 to 1980 in the 13 stations with 10–24 year records, but little trend at lower altitudes. Rowland, (1984 interview with Paul Brodeur) cites this publication as confirming Heath's BUV claim (*New Yorker* 6/9/86, p. 82). Albritton slides, 1/20/87 (87-4-10) cite Angell 1988 for Umkehr decline, mostly in 1982–83; see also UAPB 84-2, p. 10; NASA/WMO 1986, p. 800; UNEP CCOL 1984 (84-1-18), para. 24; *C&EN* 11/24/86, p. 20.

127. NASA/WMO 1986, p. 20; Watson et al. 1986; "Has Stratospheric Ozone Started to Disappear?" *Science* 237 (7/10/87): p. 131–132, reporting Reinsel and Tiao analysis of 13 ozone sonde stations showing 0.5 percent per year decrease in lower stratosphere (10–25 km) and 1 percent per year increase at 0–10 km.

128. See, e.g., the Fluorocarbon Panel's critique of balloon data because they are incon-

sistent with model predictions (Fluorocarbon Panel, "Recent Research Results and Future Directions," 2/86, 86-2-16).

129. *NYT* 4/2/75, p. 9; 9/17/75, p. 48.

130. *NYT* 9/14/75; NAS 1982, p. 307.

131. Reported at NASA/WMO Stratospheric Workshop, Hampton, VA, May 18–22, 1981, and in NASA/WMO 1982, p. 3-49; controversy reported in "Satellite Date Indicate Ozone Depletion," *Science* 213 (9/4/81): 1088–89.

132. The claim was cited in parliamentary questions in Germany (Cavender-Bares, Jäger, and Ell 2001) and attacked by industry representatives for being circulated in public before peer review (e.g., Orfeo congressional testimony, 11/5/81, in *IER* 4, no. 12 [12/5/81]); *NYT* 10/20/81, p. C1; Cicerone 1982, p. 149; NAS 1984, p. 91; NASA November 1983 discussion of requirements for validating ozone measurements, UAPB 85-1, p. 8.

133. Hughes and Wynne 1992, p. 14.

134. Tuck interview. Despite continuing attempts to improve current data, few or no retrospective corrections were applied to stored data and the BAS's low opinion of the quality of the collection was amply justified.

135. Farman, Gardiner and Shanklin 1985.

136. *NYT*, 11/7/85, p. B21.

137. Stolarski et al. 1986.

138. McCormick et al. 1982; Hamill, Toon, and Turco 1986.

139. Chubachi 1984 from Quadrennial Ozone Symposium (1984), Halkidiki, Greece.

140. Komhyr, Grass, and Leonard, 1986; Komhyr interview.

141. *NYT* 11/7/85, p. B21, quotes Heath for SBUV confirmation of Antarctic depletion.

142. *LAT* 2/2/86, p. 1; G. Darst, AP 9/16/86; *NYT* 10/21/86, p. C3.

143. For example, since SBUV observations did not quite reach the earth's surface, part of the drift could arise from an increase in tropospheric ozone that Dobsons detected but SBUV did not (Fleig, Bhartia, and Silberstein 1986; "Evidence of Arctic Ozone Destruction," *Science* 240 [5/27/88]: p. 1144–45).

144. G. Darst, AP 9/16/86; UAPB SB-87-1, p. 1; Albritton slides 6/6/86 (86-4-10); "Evidence of Arctic Ozone Destruction," *Science* 240 (5/27/88): 1144–45; Watson answer to question in March 1987 testimony.

145. E.g., Canadian Parliamentary question, in (86-1-54); Buxton, note to Minister, Nov. 1986 (86-1-59); "Europe Agrees," *Nature*, 3/26/87, p. 321; testimony of D. Heath and R. Watson, House Subcommittee on Health and the Environment, 3/9/87; Heath, Watson, Rowland, Butler, and Prather interviews.

146. *C&EN* 11/24/86, pp. 20, 26; Watson testimony.

147. Watson and Rowland interviews.

148. Albritton slides, January 1987; OMB, 4/3/87; WRI, Oct. 1987; NAS symposium, 3/23/88 (87-4-10, 88-4-10).

149. Bastian 1982, pp. 180–81; NAS 1977, p. 46, notes a small program started in 1976 in the EPA, not enough allocated, then immediately proposed for cutting.

150. Testimony of J. Hoffman, Subcommittee on Natural Resources, Committee on Science, Space, and Technology, 3/10/87, p. 136; Longstreth and Worrest interviews.

151. U.K. STRAC 1979; UNEP CCOL 1983.

152. *Resources* no. 72 (Feb. 1983) (Resources for the Future, Washington, DC.), pp. 8–9.

153. J. E. McDonald, testimony to Subcommittee on Transport, House Appropriations Committee, March 2, 1971.

154. The increase is dominated by mobility and lifestyle, and began long before it could be attributed to ozone loss (Rundel and Nachtwey 1983).

155. NAS Panel 1979, p. 99; UNEP CCOL 1979, 1986; EPA 1988; *NYT* 2/7/75, p. 11; van der Leun interview.

156. Skolnick 1991.

157. D. Rigel testimony, House Subcommittee, 3/9/87 (printed testimony reproduces his editorial from *J. Derm. Surg. Oncol.* 8, no. 9, (Sept. 1982).

158. Technology and Economic Assessment Panel 1991, p. 15.

159. M. Kripke, ACS Seminar, 4/2/76, reported *NYT* 3/2/76, p. 30; no mention in assessments until NAS 1982 ("effects" section of NAS 1979 was exclusively skin cancer); Brozek et al. 1992; Jeevan, Heard, and Kripke, 1992.

160. NAS Panel 1979, p. 20; UNEP CCOL 1981, para. 51.

161. NAS Committee 1977 p. 51; NAS Panel 1979, p. 21.

162. Worrest, Van Dylce, and Thomson, 1976.

163. Thacher 1993, p. 30.

164. Scotto et al. 1988.

165. See, e.g., Crutzen 1979, p. 1827.

166. In fact, climatic effects were consistently highlighted most strongly when total depletion estimates and UV effects declined, suggesting that control advocates were motivated by a general precautionary stance, looking for reason to control (see, e.g., Canadian summary of UNEP CCOL 1983, 83-1-2).

167. E.g., Cicerone 1982, p. 147.

168. Ramanathan 1975; *NYT* 9/14/75, p. 1.

169. Ramanathan, Callis, and Boughner 1976.

170. NAS Committee 1977, p. 40.

171. UNEP/EPA June 1986 conference; *C&EN* special issue, 11/24/86.

172. Hoffman and Seidel interviews.

173. Existing studies of assessment processes and strategies include the comparison of several early climate assessments in Glantz, Robinson, and Krenz, 1985, and the sketch of possible evaluative criteria in Clark and Majone 1985. The most ambitious study of scientific assessments of global environmental issues was completed in Mitchell, Clark, Cash, and Alcock (fortcoming 2003); Farrell and Jäger (forthcoming, 2003).

174. PL 95-95, Title I, Part B, 42 USC 7450 et seq.

175. NAS Panel 1976, p. 8.

176. The chair later stated that the committee was aware of regulatory lags, so starting rule-making immediately was consistent with its recommendation, but the report is silent on this matter (Tukey testimony, 12/15/77).

177. Canadian Harold Schiff now chaired the panel, but the two Europeans were only corresponding members.

178. NAS Panel 1979, pp. 9, 10, 42.

179. The committee judged the maximum feasible reduction at any price as 50 percent, and recommended that CFC controls might have to be limited to banning new uses (NAS Committee 1979, pp. 25–27, 196, 259).

180. NAS Committee 1979, p. 24.

181. "Ozone Depletion Would Have Dire Effects," *Science* 207 (1/25/80): 394–395; NAS Committee 1979, p. 126.

182. Concerning a discrepancy between calculated and observed ClO in the upper stratosphere, the committee reported, "Some members think it unlikely to change depletion estimates by more than a factor of two, while others think this cannot be quantified." This was one of three important discrepancies highlighted in the report, but no attempt was made to quantify uncertainty associated with the other two (NAS 1982, p. 29).

183. NAS 1982, pp. 5–6, 48.

184. NAS 1982, pp. 75–76.

185. NAS 1982, pp. 104–13.

186. *C&EN* 11/24/86; *NYT* 4/1/82, p. A21; *WSJ* 4/1/82; *WP* 4/2/82, p. A12.

187. NAS 1984, p. 78, table 6-1, pp. 96, 106–7.

188. NAS 1984, p. 139; U.S. cable summary (84-1-1).

189. *ASHRAE Journal* 7/84, p. 8.

190. U.S. cable summary (84-1-1).

191. A. F. Tuck book review, "Halocarbons: Effects on Stratospheric Ozone," *QJRMS* 104 (1978): 1010.

192. Letter from H. Wiser, EPA, to M. Uman, staff director, NRC Committee, 6/16/81

(81-1-32), exhorting the committee to retain quantitative uncertainty analysis but do it better. Instead, it abandoned the effort.

193. Members of the authoring committee are identified in the COMESA and 1979 assessments, and are largely unchanged between the two, including the chair (R. Murgatroyd of UKMO). Authors are not identified in the 1976 assessment, but acknowledgments suggest they were largely the same group (also Tuck interview).

194. U.K. DOE 1976; UK Royal Commission on Environmental Pollution 1974.

195. Derwent, Eggleton, and Curtis 1976 (AERE Harwell 1-D model). Because the U.K. and NAS assessments used different models, differences in their depletion projections cannot be fully attributed to specific assumptions, but one difference predominates. Like the Crutzen and Wofsy studies, these U.K. model runs used the faster of two recommended rates for the key reaction $HO + HO_2$, while NAS (1976) used the low rate, 10 times slower. The fast rate for this reaction reduced calculated depletion by half, and was the main reason this assessment, like IMOS, calculated nearly the same loss, without considering chlorine nitrate, as the NAS Panel found including chlorine nitrate (NAS Panel 1976, p. 215; IMOS 1975, pp. 26–28).

196. Relative to COMESA, the notable additions in STRAC are James Lovelock and two ICI scientists. One member, Brian Thrush, served both on STRAC and as a corresponding member on the 1979 U.S. NAS Panel.

197. U.K. STRAC 1979, pp. 166, 194.

198. U.K. STRAC 1979, p. 4.

199. R. J. Murgatroyd and R. G. Derwent, The Impact of Chlorofluorocarbons and Other Man-made Pollutants on Stratospheric Ozone," AERE Report R10268, Aug. 1981 (81-4-4).

200. Stoel, Miller, and Milroy 1980; Thacher 1993; Hare, Schiff, and Chisholm, interviews.

201. UNEP CCOL 1977

202. A nation could join simply by stating it had an ozone research program. By 1979, members included Australia, Canada, Denmark, Germany, France, India, Italy, Japan, Kenya, the Netherlands, Norway, Sweden, the Soviet Union, the United Kingdom, the United States, and Venezuela. By 1983, Egypt, Poland, Brazil, and Argentina had also joined (83-1-2).

203. UNEP, "An Assessment of Ozone Depletion and Its Impacts," Dec. 1978, annex to CCOL 1978.

204. UNEP CCOL 1979; Schiff, Chisholm, and Watson interviews.

205. Since this was the last meeting chaired by UNEP's Ramses Mikhail before his retirement, the tone may have reflected his ability to be more forceful than normally permitted (Mikhail interview; Stoel, Miller, and Milroy 1980, p. 276).

206. UNEP CCOL 1981, Executive Summary (81-1-1); press release (81-1-3); CCOL 1983, para. 9 (82-1-20).

207. UNEP CCOL 1980, para. 10, and successive reports through 1983.

208. UNEP CCOL 1983, para. 89.

209. UNEP CCOL "Assessment: Effects of Ozone Layer Modification," 11/25/86, UNEP/ WG.151/ 5, 86-1-65.

210. E.g., UNEP CCOL 1978.

211. By 1975, Fluorocarbon Panel research contracts allowed scientists to publish with no prior review or control. (R. Dickinson, quoted in NYT 10/1/75, p. 3; Howard, Molina, and Watson interviews).

212. Fluorocarbon Panel 1980; Fluorocarbon Panel 1981, "Summary of Research Program" (81-2-6); Fluorocarbon Panel 1984, comments on EPA negotiation submissions (84-2-3, 84-2-4).

213. NASA 1977, 1979; DeMore et al. 1979, 1981; see UAPB 81-1, p. 3.

214. UAPB 81-3, p. 1.

215. NASA 1977, pp. 190–92. Most models roughly doubled their depletion estimates with the new rate.

216. NASA 1979, p. 338.

217. This calculation was explicitly presented as a 95 percent confidence interval. NASA 1979, pp. 355–58.

218. NASA 1979, p. 344, table 7-3.

219. NASA 1979, p. 339.

220. D. Ehhalt of Germany chaired the Trace Species Group. UAPB 81-3; NASA press release (81-4-2).

221. R. Watson of NASA and D. Albritton of NOAA, who chaired the assessment and one chapter, respectively (UAPB 85-1, p. 2).

222. NASA/WMO 1986, p. 64.

223. NASA/WMO 1986, pp. 76–77.

224. NASA/WMO 1986, p. 18.

225. NASA/WMO 1986, p. 14.

226. See, e.g., "The Sky Is Falling, Maybe for Real," C&EN 11/24/86, p. 3.

227. Anderson, Butler, Hudson, Rowland, Prather, Tilford, and Watson interviews.

228. J. van Horn, Fluorocarbon Panel Manager, letter to M. Uman, NRC, 8/21/81, with report attached (81-4-4).

229. The NAS assessments of 1977, 1979, and 1982 relied extensively on the parallel NASA assessments, as did all CCOL reports. U.K. DOE stopped doing assessments after 1979, but the informal 1981 update principally used the 1979 NASA assessment as its authority to criticize the 1979 NAS assessment.

230. The EPA did not commission an NAS assessment in 1985–1986, but relied on this one for its required report to Congress. The EC Commission announced in Nov. 1985 that it could not provide the required assessment in late 1985, but would wait for the NASA/WMO report to issue its own, and suggested delaying consideration of new controls to late 1986. The CCOL canceled its 1985 assessment, and simply summarized this one for its 1986 report. (COM [85] 644 Final; IER 12/11/85, p. 408; ENDS 132, no. 1 [1986]: 22–23).

231. Watson et al. 1986.

232. This was tried. See, e.g., I. Sobolev comments at EPA Mar. 1986 workshop: clarifying the cause of the ozone hole required 10 more years' delay of regulatory action to conduct the necessary research.

233. E.g., note the assessment's weak attempt to establish a place in the policy process: "Since the Convention states measures should be based on relevant scientific and technical considerations, and that nations should collaborate on scientific assessments, the need for a comprehensive evaluation by the international scientific community of all facets of the ozone issue is clear" (NASA/WMO 1986, p. 4).

Chapter 5

1. Commission communication to the Council, COM (80) 339, 6/16/80; IER 3, no. 9 (9/10/80): 435–440.

2. IER 4, no. 2 (11/2/81): 622.

3. COM (81) 261 Final 2, 10/7/81; ENDS 76 (7/81): 4–5.

4. In case stronger measures should become needed, the Commission also began drafting directives to limit specific CFC uses, to submit in 1983 (IER4, no. 8 [8/7/81]: 919–20; Haas and McCabe 2001).

5. Actual EC production was 326 Kte at the 1976 peak, declining to 304 Kte in 1979 (COM [81] 558, 10/8/81; OJ C269, vol. 24 [10/21/81].

6. Germany proposed a 50 percent aerosol cut in December, and the Netherlands and Denmark sought a complete ban in June (IER, 9/12/81, p. 1111; IER, 7/14/82, p. 277).

7. Decision 82/795 of 11/16/82; OJ L329, 11/25/82; David Pearce, "The European Community Approach to the Control of Chlorofluorocarbons," 11/86 (86-4-1).

8. Bastian, Blum, Bracken, Odoulamy, Wuebbles, Chisholm, and Reid interviews.

9. EPA Report (82-4-3); revised OECD draft of 3/17/81 (81-1-20).

10. J. Kosovich, "Note to Past Reviewers," with attached paper (82-1-57).

11. Vupputuri meeting report, AES (Canada) (81-1-2, 81-1-13).

12. Report of 6th Governing Council, May 9–25, 1978, UNEP/GC.6/L.8, pp. 4–5 (5/19/78); Bastian interview.

13. Decision 8/7B, 8th Governing Council, Apr. 16–29, 1980 (*IER* 3, no. 6, [6/11/80]: 241–42).

14. The decision was sponsored by 19 delegations, including 8 of 10 EC countries. Sweden "pulled a fast one by wording the resolution so innocuously that even the UK had to support it" (Bastian interview). Only two nations (Japan and the Soviet Union) opposed it, preferring "guidelines" to convention discussions (Decision 9/13B, 1981 GC, May 26; Tolba letter to governments, July 21, 1981 (81-1-9); government responses (82-1-4).

15. Bastian 1982, p. 194; Bastian and Thacher interviews.

16. The others were marine pollution from land-based sources, and international transport of hazardous wastes.

17. UNEP, working group of experts in international environmental law, Oct. 28–Nov. 6, 1981, Montevideo. By authorizing a separate UNEP program in international environmental law, this resolution strengthened the mandate of the ozone negotiating body and greatly increased UNEP's flexibility relative to the innocuous Governing Council decision. The attendees, mostly lawyers and diplomats not previously familiar with the ozone issue, approved rather sweeping statements about the need for international action, the inclusion of chemicals "harmful or likely to harm" the ozone layer, and the mandate of UNEP (Thacher 1993; Sand 1985; Chisholm historical review note (87-1-6); Odoulamy, Thacher, Bastian, and Bracken interviews.

18. The Swedes hoped to have a convention ready to sign for the tenth anniversary of the 1972 Stockholm Conference on the Human Environment. All the Nordic states cosponsored the proposal, including Denmark (Memorandum to Cabinet [Canada], 7/15/81 [81-1-4]; briefing note to Assistant Deputy Minister, Atmosphere [Canada], 12/31/81 [81-1-11]; draft recommendation in UNEP/GC.10/5/Add.2.

19. Canadian delegation briefing note, 12/22/81 (81-1-10). In a curious dissonance with its conclusion, the note also confidently states that a return to even 3–7 percent CFC growth would bring rapid ozone loss.

20. Chisholm briefing to AES Committee, 11/20/81 (81-1-13).

21. Wolf 1980.

22. United States production of CFCs 11 and 12 fell from 376 to 206 Kte (1974–1980), while production capacity fell from 599 to 492 Kte (1975–1980). Capacity utilization remained well under 50 percent through 1985. DuPont modestly increased its market share through the early and mid-1980s (SRI International 1995; *CMR* 3/13/89, p. 50; Mooz et al. 1986; Cogan 1988; Reinhardt 1989; *Chemical Week* 9/30/87, pp. 6–8).

23. EFTC data reported by Commission, 5/26/81; *ENDS* 76 (7/81): 4–5; *ENDS* 102 (7/83): 22.

24. OECD report, para. 19 (81-1-20).

25. Alternative Fluorocarbons Environmental Acceptability Study (AFEAS) 1996.

26. Reinhardt 1989, p. 3; Fay and Vogelsburg interviews.

27. UNEP CCOL 1981.

28. Replies from governments to UNEP (82-1-4), para. 8.

29. A. Chisholm meeting notes (82-1-9, 82-1-10, p. 9); UNEP assessment (82-1-34).

30. Meeting notes, 1/21/82, (82-1-10), p. 2.

31. Meeting documents (82-1-5, 82-1-16, 82-1-56). Other delegations noted that while EC consumption had dropped only 20 percent from its peak and the reduction could not readily be attributed to the EC initiatives, the countries with aerosol bans had achieved 40–50 percent reductions.

32. Chisholm meeting notes (82-1-1); *IER* 6/9/82, p. 212; *IER* 7/14/82, p. 285.

33. Alliance note (81-2-2), p. 4.

34. See, e.g., P. Thacher speech to 1982 CCOL meeting (82-1-20); Fluorocarbon Panel and A. Miller comments on draft convention (82-1-32, 82-1-55); "Paper Prepared by the Secretariat," UNEP/WG.78/5 (82-1-15).

35. Thacher, opening comments (82-1-20).

36. Letter from Fluorocarbon Panel to P. Usher, UNEP (82-1-32).

37. "Second Session Continued," Apr. 11–15, 1983, Geneva (83-1-11).

38. Canadian delegation meeting report (83-1-2).

39. Canadian delegation meeting report, (83-1-2); *ASHRAE Journal*, 5/83, p. 6; W. Ruckelshaus, letter to Reps. Lukens and Madigan, published in *IER* 7/11/84, p. 219.

40. The bills, introduced by Rep. Lukens and Sen. Bentsen, died in committee due to other deadlocks over the Clean Air Act, principally concerned with automobile emissions and acid rain (*Alliance News* 2, no. 2 [3/81]: 1; DuPont Freon Products Division, "Legislative Alert No. 2," 3/3/82 (83-2-3); exchange of letters between D. Strobach and A. Miller (82-2-1); *ASHRAE Journal*, 11/82, p. 51).

41. *IER* 12/8/82, pp. 538–39, reports that senior officials in the Toxics Office and the Alliance denied knowledge of the proposal. The Alliance stated that despite this proposal, they still sought legislation to restrict the EPA's authority to regulate CFCs (Losey, Hoffman, and Fay interviews).

42. Hoffman, Losey, Seidel, Cannon, Wagner, Topping, Clay, Russel, and Hajost interviews.

43. Third session, Oct. 17–21, 1983, Geneva.

44. *IER*, 11/9/83, p. 503

45. Revised draft convention, prepared by UNEP Secretariat for Apr. 1983 meeting, art. 8 (83-1-3).

46. Comments at State Department briefing, 11/22/83 (83-2-1).

47. Summarized in DuPont critique of Rand 1984 report (84-2-2), and in correspondence of A. Miller (83-1-26).

48. Fitzhugh Green remarks at State Dept. briefing, 1/4/84 (Miller notes, 84-1-31).

49. D. Strobach comments at State Department briefing, 1/4/84 (Miller meeting notes, 84-1-31).

50. A. Miller, letter to R. Watson; transcript of Watson remarks; and subsequent correspondence (84-3-1).

51. Third session, "second part" (Jan. 16–20,1984, Vienna).

52. Canadian delegation report of meeting (84-1-7).

53. Fluorocarbon Panel, response to EPA Background Paper, Apr. 1984 (84-2-4).

54. *ENDS* 124 (5/85). 22.

55. 12th Governing Council, May 16–29 1984, Nairobi.

56. Governing Council decision 12/14 (1984), para. 3.

57. Toronto, Sept. 5–7, 1984.

58. Canadian briefing note, June 15 (84-1-20).

59. E.g., cable from M. Tolba refers to recent climate meeting in Villach, Austria, in arguing that CCOL should stress climate linkages (84-1-21).

60. Fourth Session, Oct. 22–26, 1984, Geneva.

61. The European Federation of Chemical Industries (CEFIC), the European Federation of Aerosols, and the International Chamber of Commerce attended. One representative from the International Union for the Conservation of Nature had attended the December 1982 session; otherwise, no environmental group attended any negotiating session until the summer of 1986.

62. Watson spoke in his capacity as former chair of an ad hoc technical working group (84-1-28, para. 9).

63. Canadian cable summarizing informal consultations with P. Szell, U.K. DOE (84-1-14).

64. Canadian delegation cable, 10/29/84 (84-1-26); note from meeting (84-1-19).

65. Options discussed in Canadian cable, 9/21/84 (84-1-15); Chisholm note (84-1-19); U.S. cable, 2/16/85 (85-1-4); *ENDS* 121 (2/85): 12. The United Kingdom denounced the "superficially attractive concession with one option, aimed at the EC" in a note circulated in Mar. 1985 (85-1-26).

66. Remarks of R. Benedick, head of U.S. delegation (85-1-1).

67. European Community, "Protection of the Ozone Layer: Discussion Paper," Geneva, 12/2/86 (UNEP/WG.151/CRP.5) (86-1-73); "The European Community Approach to the

Control of Chlorofluorocarbons" (86-4-1); statement by European CFC producers (CEFIC) (86-4-1).

68. US cable (85-1-4); N. Haigh, Institute for European Environmental Policy, in *ENDS* 139 (8/86): 16.

69. Canadian delegation presentation and cable (84-1-11, 84-1-13); Benedick comments (85-1-1).

70. Although both European and U.S. producers had substantial excess capacity (perhaps 30 percent in Europe, more than 50 percent in the United States), more than half of European production was still going to aerosols (Cogan 1988; Reinhardt 1989). Consequently, rapid aerosol cuts could lock in a large advantage for European producers in export markets. (Note: several observers have erroneously stated that U.S. producers were near full capacity, e.g., Buxton note [87-1-3)]; Thacher 1993.)

71. The single exception was A. Miller of the NRDC, who worked on the issue from 1980 to 1985 alone and part-time, with limited support from his organization (Hoffman, Miller, and Stoel interviews).

72. Conference of plenipotentiaries, Mar. 18–22, 1985, Geneva.

73. *ENDS* 121 (2/85): 11; notes from Vienna meeting (85-1-21); Sand 1985.

74. "The Central Issue. Note by the United Kingdom," Conference of Plenipotentiaries, Mar. 18–22 1985, Vienna, and attached paper, "Convention for the Protection of the Ozone Layer and Protocol on CFCs" (85-1-26).

75. The advocates did not expect these workshops to be important (Weil interview).

76. Although negotiations were convened under the UNEP, the question of who would provide the secretariat remained open until 1984, with the United States and Canada favoring WMO over UNEP because they thought it more scientifically competent. The WMO only expressed lukewarm interest, however, and angered the activists at the 1984 Governing Council by stating that current scientific evidence did not support early negotiation of a control Protocol, so UNEP was named the secretariat (U.S. cable of instructions to delegation, 10/12/83 [83-1-26, 83-1-18]; Canadian delegation note [84-1-7]; also cables and meeting notes [84-1-2, 84-1-3, 84-1-6]).

77. International Court of Justice, "Military and Paramilitary Activities in and Against Nicaragua" (*Nicaragua* v. *United States of America*), 1984 ICJ Reports 392, 11/26/84.

78. "Declarations," Conference of Plenipotentiaries, Mar. 18–22, 1985, UNEP/IG.53/5.21 (85-1-24), p. 35.

79. Communication from the Commission to the Council, COM (85) 8, 1/16/85.

80. US diplomatic cable, 10/16/84 (84-1-21); European Environment Bureau press release PR C/18/85, "Commission Holding Up Negotiations on Convention for the Ozone Layer," Brussels, 2/13/85 (85-1-28).

81. Thacher 1993, p. 23; Victor Raustiala, and Skolnikoff 1997.

82. Benedick 1991, p. 46.

83. *NRDC v. Ruckelshaus*, No. 84-3587, U.S. District Court for DC Circuit; *NYT* 11/28/84, p. A20; *BNA Environment Reporter* 15 (12/14/84): 1384; *ASHRAE Journal* 8/83 p. 6, 3/85, p. 10, and 2/86, p. 7; Miller letters to D. Clay, 10/13/81 and W. Ruckelshaus, 5/31/83 (84-4-1); Hoffman, Miller, Seidel, and Stoel interviews.

84. L. Thomas, congressional testimony (86-1-5), pp. 5–7.

85. Remarks by R. Barnett, Barnett letter to Thomas, 3/86 (86-2-13); Fay and Barnett interviews.

86. D. Strobach, "A Search for Alternatives to the Current Commercial Chlorofluorocarbons," presentation to EPA workshop, 2/24/86 (86-2-12); Brodeur 1986, p. 87; Hoffman, Seidel, and Miller interviews.

87. J. Steed, *C&EN* 11/24/86; E. Blanchard congressional testimony, 5/13/87; *IER* 6/10/87, p. 274; *ENDS* 139 (8/86): 16; *ENDS* 141 (10/86): 6–7; *ENDS* 158 (3/88): 18.

88. J. Steed, quoted in *WP* 4/10/88, p. A1.

89. *C&EN* 11/24/86, p. 49

90. One person from the Environmental Defense Fund served on the U.S. delegation.

Previously, one person from the IUCN attended as an observer in December 1982. Otherwise, no environmental groups had attended.

91. Meeting report (86-1-2), p. 3.

92. Brodeur 1986, p. 87; Hoffman and Buxton interviews.

93. *IER*, 8/13/86, pp. 286–87; *Inside EPA* 7, no. 41 (10/10/86); Benedick 1991, pp. 48–50.

94. Papers submitted by Germany (Lothar Gündling, 86-1-45), France (P. M. Dupuy, 86-1-46), and the European Aerosol Association (86-1-47) all support the capacity cap, while the commission's paper (D. Pearce, 86-4-1) calls a cap the second-best policy, given that CFC taxes must be "punitive" to significantly affect output.

95. In fact, EPA administrator Thomas had said EPA no longer wanted an aerosol ban as early as March 1986 (*ENDS* 139 [8/86]: 16; *ENDS* 140 [9/86]: 23; UNEP 1986; Leesburg papers, 86-1-18 to 86-1-48).

96. World Resources Institute papers (85-3-4, 86-3-5, 86-1-16); A. S. Miller & Mintzer 1986.

97. J. Hoffman, presentation to Leesburg workshop (86-1-33).

98. Isaksen and Stordal 1986. The qualitative result was typical of 2-D models, in which near-zero global depletion might still show a decrease at temperate and high latitudes, but this model showed larger depletion than most.

99. CEFIC paper (86-2-6).

100. Fox 1979, p. 35; Miller, Rowland, and Wuebbles interviews.

101. D. Strobach, letter to A. Miller, WRI, Sept. 1985 (85-2-2), makes this explicit, stating that DuPont and the Alliance's criticisms of the 1980 ANPR remain "largely valid" five years later, while also using the 1980 Rand study to support his claim that alternatives are unavailable.

102. Strobach and Miller exchange of letters (83-2-1, 83-3-1); Fluorocarbon Panel critique of U.S. position paper (84-2-4); letter to R. Benedick, 5/30/85 (85-2-3, 85-2-4); Barnett, Fay, Hoffman, and Seidel interviews.

103. H. C. Mandell testimony to Senate subcommittee 5/14/87; Cogan, 1988, pp. 46–47; *WSJ* 12/2/86, p. 4; *ENDS* 139 (8/86), p. 13; AFEAS, 1996; SRI International 1995.

104. Barnett and Fay interviews.

105. The review was led by Barnett and by J. Steed, who had succeeded D. Strobach as environmental manager for DuPont's Freon Products Division and as the Alliance's science adviser a few months before.

106. "Statement of Richard Barnett," National Press Club, Washington, DC, 9/16/86, and accompanying press release (86-2-1, 86-2-2); Barnett and Fay interviews.

107. J. Glas, letter to Freon Products customers, 9/26/86 (86-2-17); DuPont Position Statement on the Chlorofluorocarbon/Ozone/Greenhouse Issues, 1986 (86-2-11); *C&EN* 11/24/86, p. 48.

108. K. von Moltke memo to J. Gustave Speth, WRI (86-3-4).

109. European Fluorocarbon Technical Committee, "Chlorofluorocarbons and the Ozone Layer," 10/86 (86-2-6); *WSJ* 11/14/86; *ENDS* 141 (10/86): 6–7.

110. Buxton briefing note 3/26/87 (87-1-3); ministerial briefing note (87-1-47); quotes from EC and Canadian officials in *New Scientist* 4/23/87, p. 22; *The Nation* 245 (10/10/87): 376; Brenton 1994, pp. 139–140.

111. The key progress in commercial HFC-134a synthesis occurred in 1991, when DuPont and ICI announced similar advances in catalysts.

112. Quotes from D. Doniger, NRDC, and R. Orfeo, Allied Signal, *C&EN* 11/24/86, p. 51; *New Scientist* 4/23/87, p. 22.

113. Pennwalt sold a blend of HCFCs 142b and 22 as a CFC-12 replacement in aerosols. They proposed to market this blend as a refrigerant, and HCFC-141b as a CFC-11 substitute (*Chemical Week* 9/30/87, pp. 6–8).

114. Krol 1992; *IER* 7/15/92, p. 471.

115. J. Glas, presentation to National Academy of Engineering, 9/29/88 (89-2-5); *C&EN* 11/24/86, pp. 48–49.

116. R. Orfeo, *C&EN* 11/24/86, p. 48.

117. Brodeur 1986; Barnett, Fay, Miller, and Doniger interviews.

118. *Chemical Week* 9/30/87, p. 7; *C&EN* 11/24/86, p. 49; *National Journal* 11/11/86, p. 2638.

119. Haas and McCabe 2001; Thacher 1993.

120. Levy 1990; *Europe Environment* 273, (1987): p. 8.

121. *ENDS* 142 (11/86): 24

122. *ENDS* 142 (11/86): 24; Konrad von Moltke memo (86-3-4); U.S. delegation cable (86-1-71).

123. EPA Air and Radiation Office viewgraphs, "Options for an International Protocol on Protecting the Stratosphere," 10/14/86 (86-4-3).

124. EPA Office of Air and Radiation viewgraphs (86-4-3); briefing note for Canadian Minister's meeting with L. Thomas, 10/21/86 (86-1-59); Barnett letter to J. Negroponte, 10/30/86 (86-2-15).

125. D. Doniger and D. Wirth (NRDC) workshop papers, June and Sept., called for 80 percent global cuts in 5 years and phaseout in 10 (86-1-18); "Draft CFC Bill," circulated by WRI, proposed 50 percent U.S. cuts in 10 years with import restrictions (86-3-8); EDF position statement, Oct. 16, proposed 50 percent cuts with tradable permits (86-3-6); *C&EN* 11/24/86, pp. 51–53.

126. L. Thomas testimony, 6/11/86 (86-1-3, 86-1-4). Thomas first proposed a CFC phaseout at a staff briefing in the summer of 1986, and decided to pursue it in internal EPA discussions developing the U.S. response to the Sept. 1986 Canadian proposal (Hoffman, Thomas, and Seidel interviews).

127. Senators' letter to Shultz, 10/6/86 (86-4-7); Chafee statement, 10/8/86 (86-4-4); Sens. Baucus and Chafee submitted bills on 2/19/87, and similar House bills were introduced in the fall of 1986 by Rep. Richardson and on 4/19/87 by Reps. Bates and Waxman (*C&EN* 11/24/86, p. 58; Shimberg 1991, p. 34).

128. Testimony of J. Negroponte (Asst. Sec. of State, Oceans and Environment), hearings of Senate Subcommittees on Environmental Protection and Hazardous Wastes, 1/28/87; discussion with Sen. Chafee reported in *IER* 2/11/87, pp. 62–63.

129. R. Barnett, editorial, "The US Can't Do the Job Alone," *NYT* 11/16/86, p. C2.

130. Dec. 1–5, 1986, Geneva. Participation was about half what UNEP predicted. Those absent included China, India, and three EC members, while incoming EC president Belgium sent only a member of its Geneva mission (U.S. delegation cable, 86-1-71).

131. Canadian proposal (86-1-63), background paper (86-1-67), meeting note (86-1-52); U.S. cable (86-1-71).

132. U.S. proposal (86-1-64), and position paper (87-1-63).

133. Conference-room paper attached to U.S. cable (86-1-71).

134. European Community, "Discussion Paper" (UNEP/WG.151/CRP.5) (86-1-73); *ENDS* 143 (12/86):24.

135. *WSJ* 12/2/86, p. 4.

136. Canadian delegation cable (86-1-14); U.S. delegation cable (86-1-71); Benedick 1998, p. 70, reports that the early departures were protests against the plenary's rejection of the request for postponement.

137. Meeting report, Feb. 23–27, 1987, Vienna, UNEP/WG.167/2 (87-1-51); "Ozone Depletion and Climate Change: A Statement by the International Environmental Community," 10/31/86 (86-3-7).

138. Buxton briefing note, 3/26/87 (87-1-3); briefing note to Minister (87-1-47).

139. *ENDS* 146 (3/87): 22.

140. *ENDS* 156 (1/88): 11; *IER* 7/8/87, p. 318.

141. Declaration of 3/18/86, summarized in Enquete Commission of the German Bundestag 1989, p. 202; *Europe Environment* 273 (1987): 8.

142. Sixth revised draft Protocol, UNEP-WG.167/2/Annex I (87-1-2).

143. R. Benedick, Feb. 27 press statement; *NYT* 2/28/87, p. 2.

144. *IER* 3/11/87, p. 100; *Environmental Policy and Law* 17 (1987): 51.

145. U.S. cable (87-1-57); McConnell 1991.

146. "Europe Agrees to Act for Protection of the Ozone Layer," *Nature* 326 (3/26/87): 321.

147. Feb. 23–27 meeting report (87-1-51); U.S. position paper (87-1-63); Wurzburg report (87-1-53), p. 78; P. Usher speech, "The Global Business Outlook for CFC Alternatives," London, 3/12/1991 (91-3-8).

148. Report of ad hoc scientific meeting, Wurzburg, Apr. 9–10, 1987, UNEP/WG.167/INF.1 (87-1-53).

149. Report of the third session, Apr. 27–30 1987, Geneva, UNEP/WG.172/2 (87-1-4), p. 3; UNEP press release, "Nowhere to Hide," 4/27/87. Participation was up to 33 countries, including 11 developing countries.

150. Conclusions of the scientific working group, UNEP/WG.172/CRP.9, 4/29/87 (87-1-5).

151. Although no developing countries were included, Benedick (1998, p. 72) states that Tolba sought to represent their interests himself, and that they were not much interested in the details of control measures at this point.

152. Testimony of R. Benedick and L. Thomas, Waxman subcommittee, 3/9/87, including responses to questions.

153. Meeting report, UNEP/WG.172/2, p. 15 (87-1-4). The Soviets were correct, in that the Governing Council decision authorizing the negotiations mentioned only CFCs (Dec. 13/18 [1985], UNEP/GC.13/16/Ann. I, p. 47).

154. Canadian delegation briefing note, 5/3/87 (87-1-45); Environment Canada press release 5/7/87 (87-1-8); *NYT* 5/1/87, p. 1.

155. *Financial Times* 4/28/87, p. 2; *Science* 5/29/87, pp. 1052–1053, quotes NRDC's Doniger that the public was being misled, and the April meeting was still far from an agreement and further still from an effective one. *Chemical Week* 5/13/87, p. 8, quotes MITI's Ichiro Araki: "Only the Canadian and Scandinavian delegates think the compromise is good enough" (but also, "Japan cannot oppose a UN decision").

156. Canadian delegation briefing note (87-1-45).

157. *IER* 6/10/87, p. 276; *IER* 7/8/87, p. 317.

158. Both the Bundestag and the Bundesrat adopted aerosol-ban resolutions in May, which were implemented in July by voluntary agreements to cut use 90 percent by the end of 1989 (Cavender-Bares, Jäger, and Ell 2001); "West Germany Strides Towards CFC Elimination by 2000," *Nature* 327, (5/14/87): 93.

159. *IER* 6/10/87, p. 276; *IER*, 7/8/87, p. 316. The European Environmental Bureau accused the United Kingdom of blocking movement beyond 20 percent on behalf of ICI, threatening to return to only a freeze if others demanded more.

160. J. Steed speech, 3/24/87, reported in *IER* 4/8/87, p. 164; Senate joint subcommittee hearings, 5/13/87; *C&EN* 11/24/86, p. 24; remarks of Pennwalt Chair E. Tuttle at 1987 annual meeting, Pennwalt press release, 4/22/87.

161. See, e.g., quotes from Robert Watson, *C&EN* 11/24/86, p. 53; Canadian briefing note (87-1-3), "95 percent cannot be justified on current science."

162. Joint hearings, House Subcommittee on Health and Environment of the Energy and Commerce Committee, and Subcommittee on Environment of Science, Space, and Technology Committee, Mar. 9–10, 1987, correspondence appended (87-4-13); Thomas interview.

163. Joint hearings, House subcommittees; Senate hearings, Subcommittees on Environmental Protection and Hazardous Wastes and Toxic Substances, May 12–14 (*IER* 4/8/87, pp. 162, 163; *IER* 5/13/87, p. 196).

164. House subcommittee hearings, testimony of Mintzer, Dudek, Wirth, Negroponte, Benedick, and Thomas (*IER* 4/8/87, p. 163; 5/13/87, p. 197).

165. J. Steed, remarks at Álliance/CEEM meeting, Mar. 24. 1987.

166. House joint subcommittee hearing, Mar. 9, p. 7.

167. *IER* 6/10/87, p. 273. (In fact, the trade measures were included in the position cleared in November.)

168. Apr. 23 letter, summarized in *IER* 5/13/87, pp. 196–197.

169. Both sides criticized the same silence—as tacitly consenting to reductions that Dingell and industry regarded as too strong; Chafee, Gore, and NRDC, as too weak.

170. NRDC press conference, May 4, in *IER* 5/13/87, pp. 196–197; "Ozone Plan Splits Administration," *Science* 236 (5/29/87): 1052–1053; *Atlanta Journal-Constitution* 5/15/87, p. 1A, 5/12/87, p. 14A; *IER* 5/13/87, p. 196.

171. *WP* 5/29/87, p. 1; *WSJ*, 5/29/87, p. 8, quotes Hodel criticizing cancer projections: "People who don't stand out in the sun—it doesn't affect them."

172. *Atlanta Journal-Constitution* 6/6/87, p. 2; *NYT* editorial, 5/31/87, p. D28.

173. Shimberg 1991, p. 48.

174. Thomas and Claussen interviews.

175. Benedick 1998, p. 63; also Negroponte comments that discord has been exaggerated, June 18, from Japan (*IER* 7/8/87, p. 316).

176. Benedick 1998, p. 73.

177. Buxton briefing note, 7/2/87, p. 4 (87-1-12).

178. Meeting of legal experts, July 6–8, The Hague; seventh revised draft Protocol (7/15/87) (87-1-16); Benedick 1998, pp. 73–74.

179. *IER* 8/12/87, pp. 375–376.

180. The Consumer Federation had campaigned for Swiss aerosol controls in 1980, but relented when the government promised to integrate CFC controls into a broad revision of environmental legislation. With this revision still not completed in 1987, it resumed its campaign (*IER* 10/14/87, p. 492).

181. The critics included J. Farman of the British Antarctic Survey, whose team had reported the ozone hole and who served on the SORG. Farman was so outraged by the published report that he issued an opposing paper calling on the U.K. government to support CFC reductions of at least 85 percent in both aerosols and foams (U.K. DDE 1987, released 8/6/87; *ENDS* 151 [8/87]. 451; Tuck interview).

182. Fifty-five countries sent delegations, including China for the first time, with sixteen represented at Cabinet or sub-Cabinet level, as well as the largest number yet of industry and environmental groups (87-1-37).

183. The GATT status of the proposed trade restrictions was unclear, although this provision made them far less objectionable. Canadian legal opinion reported in (87-1-7); draft delegation instructions, Sept. 8 (87-1-34).

184. The Canadian delegation grew so incensed with EC demands that it briefly attempted to revoke the commission's status as a party altogether, and restrict participation to sovereign states (delegation cable, Sept. 2, (87-1-30); draft delegation instructions, Sept. 8 (87-1-34).

185. Article 2(8); interviews with Woolaston, Lee-Bapty, and Thomas. Whether the EC's obligation to report consumption was separate or joint remained ambiguous, and became a point of conflict later.

186. U.S. cable to Tolba, Sept. 1 (87-1-29); Canadian delegation draft instructions, Sept. 8 (87-1-34).

187. E.g., R. Watson testimony, joint House subcommittee hearings, 3/9/87, p. 90; *C&EN* 11/24/86, p. 53; interviews with Cicerone, Sze, and Watson, cited by Litfin 1994, pp. 103–104.

188. *NYT* 10/21/86, p. C3.

189. Interviews with Hoffman, Thomas, Seidel, and Buxton.

190. D. Albritton, "The Changing Atmosphere: Possible Implications," slides, 3/23/87 (87-4-10). Claims that important new scientific information became available during negotiations—citing either Wurzburg results (P. M Haas 1992) or new BAS data available to the U.K. delegation (Brenton 1994, pp. 140–141)—are mistaken.

191. Although these models agreed well with each other, in retrospect they were all wrong, since none considered heterogeneous processes.

192. D. Albritton slides, "Summary of the Science," presented at NOAA, 4/24/87, and Senate hearings, 5/21/87 (87-4-10); slides for joint presentation by Albritton, Hoffman, and Margitan, DPC working group, May 22 (87-4-10); McElroy testimony to joint congressional

subcommittees, 10/27/87; "The Great Ozone Controversy," editorial, *Nature* 329 (9/10/87): 101; Thacher 1993, p. 25.

193. See, e.g., Canadian cable of 6/3/87 (87-1-13).

194. Other delegations were not always pleased with this U.S. role. See, e.g., McConnell 1991; Environment Canada 1988; Buxton 1988.

195. See, e.g., Barnett, Jan. 1987 testimony (87-2-5), p. 299.

196. Doniger 1988; Brenton 1994, pp. 139–140; Shimberg 1991; R. Benedick, op-ed, *WP* 1/4/88, p. A13.

197. Brenton 1994; Shimberg 1991; Jachtenfuchs, 1990 (SEA: supposed to enter into force 1/1/87, delayed to 7/1/87 by Irish court case, referendum).

198. F. Sella, UNEP, letter to governments (81-1-9), p. 11.

199. Tolba quote, in *New Scientist* 120, no. 1636 10/28/88, p. 25.

200. Buxton briefing note (87-1-20); briefing note on Environment Canada meeting 8/20/87 with DuPont Canada.

201. N. Campbell, M. Harris (ICI), J. von Schweinichen (Montefluos) interviews.

Chapter 6

1. Callis and Natarajan 1986

2. Mahlman and Fels 1986; Tung et al. 1986; Albritton slides 9/30/87 (87-4-10); Dahlem Workshop 1988, p. 235.

3. Ozone Trends Panel 1988, ch11; Dahlem Workshop 1988; Stolarski 1988.

4. Solomon et al. 1986 proposed the following cycle:

$$Cl + O_3 \rightarrow ClO + O_2$$
$$ClO + HO_2 \rightarrow HOCl + O_2$$
$$HOCl + hv \rightarrow HO + Cl$$
$$HO + O_3 \rightarrow HO_2 + O_2$$
$$\text{net: } 2O_3 \rightarrow 3O_2.$$

5. McElroy et al. 1986 proposed the following cycle:

$$Cl + O_3 \rightarrow ClO + O_2$$
$$Br + O_3 \rightarrow BrO + O_2$$
$$ClO + BrO \rightarrow Cl + Br + O_2$$
$$\text{net: } 2O_3 \rightarrow 3O_2.$$

6. Rowland et al. 1986; Wuebbles and Connell, 1984.

7. "Depletion of Antarctic Ozone," *Nature* 321 (6/19/86): 729–30; Dahlem Workshop 1988; Albritton presentations through 1986 and 1987: "Data-Short Situation" (86-4-10, 87-4-10)

8. Mount et al. 1987; Farmer et al. 1987, p. 126.

9. Solomon et al. 1987; deZafra et al. 1987; results discussed in UAPB 8/87, p. 1, and Rowland 1988.

10. Hofmann et al. 1987.

11. Solomon press conference, Oct. 20, *IER* 11/12/86, pp. 392–393; expedition papers abstracted in *UAPB* 8/87, pp. 2–5; R. Watson testimony, House joint subcommittee hearings, Mar. 9–10, 1987, pp. 46–55; *IER* 4/8/87, pp. 162, 163; Albritton slides, 12/12/86 (86-4-10); NASA news release 87-114 (87-1-19), p. 2.

12. *GRL* Nov. supp. (1986), compiled 45 papers on explanations of the hole. McFarland (R. Scheer interview) cites this volume as showing the origin of the hole was not resolved by this time, though press coverage suggested it was. Contrast Rowland testimony, Senate subcommittee, 5/12/87 (S. Rpt. 100-201, p. 18): 80–100 percent likely that CFCs cause the hole.

13. Crutzen and Arnold 1986.

14. Molina et al. 1987; Tolbert et al., 1987; *NYT* 11/27/87, p. A30.

15. Toon et al. 1986; Crutzen and Arnold 1986.

16. Molina and Molina 1987 proposed the following cycle:

$$2(Cl + O_3 \rightarrow ClO + O_2)$$
$$ClO + ClO \rightarrow Cl_2O_2$$
$$Cl_2O_2 + hv \rightarrow Cl + ClOO$$
$$ClOO \rightarrow Cl + O_2$$
$$\text{net: } 2O_3 \rightarrow 3O_2.$$

17. Proffitt et al. 1989.

18. Fahey et al. 1989.

19. "Antarctic Ozone, Preliminary Results and Conclusions: Report from the Scientists," transparencies from press conference, 9/30/87 (87-4-7, 87-4-8).

20. Albritton slides, presentation to World Resources Institute, 10/5/87 (87-4-10).

21. Dahlem Workshop 1988, p. 245.

22. Dahlem Workshop 1988, p. 252.

23. This possibility was discussed as "highly speculative" in Dahlem Workshop 1988, p. 254.

24. Rowland 1988, p. 137.

25. Testimony of D. Heath and R. Watson, House Subcommittee on Health and Environment, 3/9/87, 87-H361-85.

26. Testimony of S. Rowland, 10/17/86, House subcommittee on Environmental Pollution; Rowland and Harris presentation, AGU, December 1986, San Francisco (L. Siegal, AP, 12/9/86); Rowland et al. 1988.

27. Ozone Trends Panel 1988, pp. 241–42; Rowland et al. 1988, pp. 38–41; Rowland Senate testimony, 3/30/88; Rowland interview.

28. Ozone Trends Panel 1988, p. 4. Albritton slides, 4/5/88 (88-4-10); Rowland interview.

29. Around 40 km, the Sage (Stratosphere Aerosols and Gases Experiment) instrument showed losses of 0–8 percent with a mean of 3 percent; Umkehr showed losses of 5–13 percent with a mean of 9 percent (Albritton slides, 4/5/88 [88-4-10]; Rowland et al. 1988, p. 46).

30. The panel had intended also to use the corrected ground data to correct the drift in satellite data from instrument degradation. Although it made some progress in this task, the relatively short satellite record prevented it from separating the effects of the solar cycle, instrument drift, and a true decline.

31. By this time, the panel's full report had been reviewed (at a January meeting with panel members and 13 reviewers) and finalized, but its publication was delayed nearly three years by production, budget, and staffing problems. The report finally appeared in 1991, with the help of industry funds.

32. Ozone Trends Panel 1988, pp. 5, 750.

33. *ENDS* 158 (3/88): 5.

34. Andelin et al. 1988. Pessimistic assumptions—no new parties, rapid CFC growth in nonparties—gave 20 percent higher world CFC use by 2009. (Benedick 1998 criticizes this analysis for excessive pessimism.)

35. Letter from Sens. Baucus, Stafford, and Durenberger to R. E. Heckert, Feb. 22, 1988 (88-2-11).

36. Heckert, Mar. 4 letter, quoted in *WP* 3/5/88, p. A15; *Chemical Week* 3/16/88, pp. 18–19.

37. E.g., Reinhardt 1989, quote from Dwight Bedsole; Glas 1989.

38. *NYT* 3/26/88, p. 41.

39. Letter from R. E. Heckert to Sens. Baucus and Durenberger (note snub to Stafford, who initiated the letter but was omitted from the response) (88-2-4); DuPont "Fluorocarbon/Ozone Update," 7/88 (88-2-5).

40. J. Steed, DuPont, quoted in *NYT* 3/25/88, p. A1; *Chemical Week* 142, no. 14, (4/6/88): 7.

41. M. McFarland, DuPont, presentation slides, "Ozone Science: Recent Findings" (88-2-12).

42. For example, DuPont's explanation exaggerated the difference between OTP and the September expedition report by highlighting remaining uncertainty in catalytic cycles and

contributions of chemistry and meteorology from September, when neither of these was changed in OTP; highlighting the OTP conclusion that CFCs were the dominant cause of the ozone hole, when this was reported in September and solidly confirmed at Dahlem; and stating that OTP represented the first "extensive use" of 2-D models "in an assessment report" projecting small ozone decreases even with the Protocol. This last point begs several questions: 2-D models had shown such results for some time, so their statement in an assessment report adds no new content; moreover, the new position responds as if small ozone decreases matter, when DuPont and other industry bodies had argued for a decade that they did not (88-2-12).

43. Albritton slides, "Scientific Assessment of Stratospheric Ozone: A Summary," 8/28/89, Nairobi (89-4-10).

44. Verhille (Atochem), *IER* 2/10/88 p.111; von Schweinichen (Montefluos), *Chemical Engineering* 1/18/88, p. 22.

45. *Financial Times* 1/16/90, p. 20; *NYT* 10/14/90, p. C1.

46. *NYT* 4/11/88, p. D4.

47. Reanalyses conducted by Reinsel and Tiao, and by Allied-Signal-led "tiger team" (Hill). Two senior Allied executives interviewed in *NYT* 4/10/88, p. F2, refused to commit to phaseout.

48. *WP* 3/30/88, p. A15.

49. E.g., industry quotes in *New Scientist* 5/26/88, p. 56, "it would bring Western economies to a halt overnight"; giving up safe household refrigerators would mean that "trucks deliver ice from a huge plant in the Midlands or people die like rats of ammonia intoxication."

50. *ENDS* 158 (3/88): p. 18.

51. Wynne et al. 2001; *ENDS* 170 3/89, 11.

52. UNEP 1989f, p. 120

53. *IER* 6/1990, p. 227.

54. EC Council, 6/16/88; *ENDS* 161 6/1988, 22.

55. Enquete Commission, German Bundestag 1989, p. 54. This commission included both scientists and political officials, so its recommendations carried substantial force in subsequent political deliberations.

56. Topfer statement to first meeting of the parties, Helsinki, May 2–5, 1989 (89-4-8).

57. The ordinance was announced by the environment minister in late 1989 and adopted by the Cabinet 5/30/90, the same day the voluntary production phaseout agreement was announced (Cavender-Bares, Jägers, and Ell 2001, pp. 45, 48).

58. *ENDS* 165 (10/88) 13.

59. No date was specified. *NYT* 9/27/88, p. A20.

60. Benedick 1998, p. 105.

61. U.K. DOE 1987; *ENDS* 161 6/88): 5–6. Executive summary released in June, full report in October.

62. *ENDS* 161 (6/88): 5–6.

63. *ENDS* 161 6/88: 17–18; *ENDS* 163 (8/88): 7; *ENDS* 165 (10/88): 12–13.

64. ICI news release, cited in Benedick 1998, p. 118 (no mention of halons).

65. Observers speculated that France was acting on behalf of Atochem, which had not yet endorsed strengthening the Protocol. *ENDS* 166 (11/88): 26; Jachtenfuchs 1990, p. 268.

66. *ENDS* 166 11/1988): 6.

67. Maxwell and Weiner 1993, pp. 33–34; Oye and Maxwell 1994, p. 200; Wynne et al. 2001.

68. *NYT* 3/20/88, p. 1.

69. Hoffman, presentation to Leesburg workshop (86-1-33); Rowland et al. 1988, pp. 36–37; *NYT* 3/20/88, p. A1.

70. Hoffman and Gibbs 1988, p. 4; emphasis in original. Ozone depletion estimates are presented, heavily qualified, only in the final appendix, from Connell's 1-D model.

71. MC contributed 0.57 ppb, and CT 0.67 ppb, to 2.68 ppb of total excess Cl. Hoffman and Gibbs 1988, p. 31, exhibit 3.

72. Letter from Sens. Baucus, Mitchell, Chafee, Stafford, and Durenberger to Tolba, 11/

12/87; Doniger 1988b; NRDC press release, "EPA Proposes Minimum Effort to Protect Ozone," 11/30/87 (cited in Cogan 1988, p. 92).

73. E.g., draft convention prepared by secretariat in the summer of 1982, circulated in the fall (82-1-27).

74. Article 6: The parties "shall convene appropriate panels of experts qualified in the field mentioned and determine the composition and terms of reference of any such panels."

75. Working group on data (second session), Oct. 24–26, 1988; report in UNEP/OzL.WG.Data/2/3/Rev.2; R. Watson, who chaired the NASA/WMO 1986 assessment and the Ozone Trends Panel, and D. Albritton of NOAA were cochairs of the Atmospheric Science Panel.

76. E.g., L. Thomas letter to M. Tolba, Mar. 1988 (*IER* 4/13/88, p. 210).

77. Oct 17–18, UNEP/OzL.Sc.1/2.

78. Andersen, Buxton, Campbell, Harris, Thomas, and Vogelsburg interviews. Some charged that the panel's work suffered as a result. E.g., letter to author from N. Campbell, 11/20/91; Van Slooten 1998, p. 150.

79. Mexico ratified first, in March 1988; the United States, Canada, Sweden, and Norway followed by June.

80. Benedick 1998, pp. 116–117; ratification website; *IER* 1/88, p. 3.

81. Environment Canada press release (89-4-11).

82. *Times* (London) 3/3/89, p. 1. Note that they did not include halons.

83. *NYT* 3/4/89, p. 6. The condition that safe alternatives be available confused some other delegates at Helsinki over how serious the United States was. (Lashof, congressional testimony, reports four delegations approached Doniger for explanation of U.S. "resistance," 5/15/89.)

84. Even the French now supported stronger controls. Litfin 1994, p. 128 attributes the change to Mitterand's sponsorship of the Hague conference on climate the next month, concerned to defend his environmental credentials. The only delegation opposing a phaseout was the Soviet one; it said they needed stronger scientific evidence (*NYT* 3/7/89, p. C13).

85. Statement of "Citizens Symposium on CFCs," London, 3/4/89 (89-3-7).

86. UNEP/OzL.Pro.1/5, paras. 40-41(89-1-7) and Decision 3 (p. 14).

87. UNEP/OzL.Pro.1/5, p. 15 (89-1-7).

88. *ENDS* 172 (5/89): 30; *NYT* 5/3/89, p. A13.

89. Benedick 1998, p. 125; Rosenberg and Claussen congressional testimony, 5/19/89 (89-4-3, 89-4-4).

90. E.g., Friends of the Earth (FOE) internal memorandum (89-3-6); *Atmosphere* 2, nos. 1 and 2 (spring–summer 1989) (89-3-3); studies commissioned by FOE to criticize assumptions of EPA analysis (*IER* 5/89, p. 244.); Lashof Senate testimony, 5/19/89; DuPont "Update," 8/89 (89-2-6).

91. *New Scientist*, 2/11/89, p. 28.

92. UNEP/OzL.Pro.1/5 (89-1-7), Decision 13, p. 20.

93. Decisions 12A and 12B, pp. 17–18, referring to Protocol art. 1(5).

94. Gehring 1994, p. 272, cites UNEP/OzL.Pro.1/5, paras. 82–85 (89-1-7).

95. UNEP/OzL.Pro.1/5, Decision 11, p. 16; Gehring 1994, p. 272; Benedick 1998, p. 127.

96. UNEP 1989e, p. xvi.

97. *NYT* 10/13/89, p. A19; *NYT* 10/24/88, p. A12.

98. Proffitt et al. 1989; *NYT* 11/16/89, p. A17.

99. "Evidence of Arctic Ozone Destruction," *Science* 240 (5/27/88): 1144–1145.

100. Albritton slides, 12/13/88 (88-4-10).

101. Watson press conference, "strongly perturbed chemistry," but "no clear evidence that loss of ozone has already occurred," *NYT* 2/18/89, p. A1; "Arctic Ozone Poised for a Fall," *Science* 243 (2/24/89): p. 1007–8; contrast Evans reports of Arctic depletion, AGU, Baltimore, May 1988, Stolarski comment in *NYT* 5/18/88, p. A25.

102. Brune et al. 1991.

103. Albritton slides, July 5–7, 1989 (Geneva) (89-4-10): vertical ozone profile comparison

stresses 40 km comparison of SAGE and Umkehr, *not* lower stratosphere—i.e., still skeptical about apparent midlatitude loss mostly in lower stratosphere.

104. Albritton and Watson interviews.

105. UNEP 1989e, p. xii; Albritton slides to chairmen's meeting, July 1989 (89-4-10)

106. UNEP 1989e, p. x.

107. UNEP 1989e, p. xiv.

108. Scotto et al. 1988; *NYT* 2/28/89, p. C4.

109. Urbach 1989.

110. Roberts 1989.

111. UNEP 1991b, p. ES-2

112. Synthesis report, UNEP/OzL.Pro.WG.II(1)/4 (89-1-11), p. 9.

113. UNEP 1989f, p. ES-2; Buxton slide presentations to OEWG, Sept. 1989, and Alternatives Conference, Oct. 1989.

114. UNEP 1989f, pp. iv, 3.

115. U.S. EPA 1988.

Chapter 7

1. Worldwide CFC revenues were about $3 billion. U.S. production was about 380 Kte, more than 1 kg per capita, generating about $700 million of revenue (Cogan 1988).

2. Hammitt et al. 1986, p. 11 figs. 1.1, 1.2.

3. European CFC exports were 127 Kte in 1985 (*ENDS* 152 [9/87]: 23).

4. UNEP 1989f.

5. *NYT* 9/17/87, p. A12; *Chemical Week* 9/30/87.

6. Glas 1989, p. 144.

7. The most recent analysis had found reduction opportunities of only 20 to 40 percent in the United States (Camm et al. 1986). Even in September 1987, when a display of alternative technologies was assembled in Montreal to show delegates the promising state of reduction opportunities, the technologies on display were highly limited and experimental (Andersen, Morehouse, and Miller, 1994; Andersen, Buxton, and Taylor interviews).

8. Manzer 1990, p. 31; *Chemical Week* 142 no. 14 (4/6/88): 7.

9. J. Steed, *C&EN* 11/24/86.

10. Congressional testimony of E. Blanchard, DuPont, 5/13/87.

11. *Chemical Week* 9/30/87, p. 6, quotes estimates of J. Glas, DuPont.

12. International Chlorofluorocarbon Substitutes Committee 1986, p. 2–2; Lagow interview.

13. DuPont press release and letter to L. Thomas, 1/5/88 (88-2-9). Four other producers joined over the following year.

14. *IER* 2/10/88, p. 110; *Chemical Engineering* 1/18/88, p. 22.

15. Von Schweinichen 1989; UNEP 1989f, p. 127.

16. AFEAS 1995.

17. In Mar. 1989, a senior DuPont researcher said that banning HCFC-22 would be "a catastrophe. Everyone is counting on it as a substitute." *NYT* 3/7/89, p. C1.

18. *ENDS* 152 (9/87): 152.

19. *ENDS* 157 (2/88): 4.

20. ENDS 156 (1/88): 9–11.

21. *ENDS* 157 (2/88): 5; *ENDS* 160 (5/88): 3–4; *ENDS* 171 (4/89): 29; European industry comments on the cost of the aerosol ban in 82-4-3, 83-1-26, 85-2-1, 86-4-1.

22. Environmentalists were divided over this substitution, because it made a large immediate improvement but HCFC-22's ODP of 5 percent was near the high end of proposed alternatives. A coalition of U.S. environmental groups supported the switch on the condition that a nonozone-depleting blowing agent be adopted as soon as possible. But since Friends of the Earth UK switched its consumer campaign to foam food packaging—including that blown with HCFC-22—after its success on aerosols, FOE was briefly caught in the awkward position

of having its U.S. organization support this use while its British organization was boycotting it (*ENDS* 158 [3/88]: 15–17; *ENDS* 163 [8/88]: 7; *New Scientist* 5/26/88, p. 56).

23. Those with no hydrogen have long atmospheric lifetimes; those with too much hydrogen are inflammable; and those with intermediate shares of H and Cl but not enough F tend to be toxic (McLinden and Didion 1987).

24. DuPont and ICI quotes in *NYT* 3/31/88, p. D1; *Chemical Engineering* 1/18/88, p. 22; Glas 1989; DuPont, *Fluorocarbon/Ozone Update* 7/88 (88-2-5).

25. Two others that initially looked promising (HCFCs 132b and 133a) were found in early 1987 to be too toxic (M. Jones, "In Search of the Safe CFCs," *New Scientist* 5/26/88, p. 56), and three more (HCFCs 22 and 142b, HFC-152a), were poorer matches for 11 because of their low boiling points but were thought to be useful in blends (Dishart, Creazzo, and Ascough 1987).

26. DuPont, *Fluorocarbon/Ozone Update* 7/88 and 12/88 (88-2-5); SRI International 1995.

27. UNEP 1989e.

28. ICI switched a pilot plant from HCFC-141b to 123 in April 1990, while DuPont canceled a 141b production plant in 1993 (*Chemical Week* 12/19/90, p. 51; Bradley 1994).

29. Draft regulations published March 1993, final regulations December 1993 (58 FR 65018, 12/10/93).

30. *Fluorocarbon/Ozone Update* 7/98, 12/98; IER 1/30/91; Cox and Lesclaux 1989; Zellner 1989. (Note, however, that Atochem announced expansion of HCFC-141b in France and the United States as late as 1991.)

31. Two others (HCFC-142b and HFC-152a) had suitable properties and were already produced at small scale for polymer production, but were inflammable. Other candidates (HCFC-124, HFCs 125 and 143a) had less suitable properties but could be used in blends (DuPont, *Fluorocarbon/Ozone Update* 3/87 [87-2-1]).

32. *NYT* 3/31/88, p. D1; *Daily Telegraph* (London) 12/23/88, p. 8.

33. *Business Week* 5/17/93, p. 78.

34. These plans doubtless included some tactical overstatement to deter others from entering. *Chemicals and Materials* 5/13/91; *Financial Times* 3/13/91, p. 4.

35. DuPont press release 7/92 (92-2-9); CMR 11/23/92, p. 5; *Chemical Week* 4/17/91, p. 8.

36. CMR 12/23/92, p. 5; *ENDS* 246 (7/95): 15–16.

37. UNEP 1991b, pp. 4-1–4-3.

38. IER 7/15/92, p. 471; *ENDS* 209 (7/92): 6.

39. *Chemical Week* 12/18/91, p. 27; *Appliance* 49, no. 2, (2/92): 42; Manzer 1990; Didion and Bivens 1990.

40. Allied Signal press release 5/28/96 (96-2-2); C&EN 71, no. 11 (3/15/93): 5–6; IER 5/4/94, p. 390.

41. *Chemical Week* 4/26/89, p. 25.

42. Kaiser and Racon made no attempt to develop substitutes, and announced plans to exit the market immediately after the Protocol. Pennwalt tried to market HCFCs 141b and 142b, both of which it already produced in small volumes for specialized applications (*Chemical Engineering* 1/18/88, p. 22; *Chemical Week* 4/26/89, p. 25).

43. SRI International 1995, p. 543.7001 I.

44. For example, after Trane redesigned its large chillers to operate on HCFC-123 and Carrier switched to HCFC-22 and HFC-134a, the firms attacked each other's choice of chemical so persistently that DuPont's vice chairman publicly called on them to stop, stating that their attacks were encouraging users to delay choosing either and to stay with CFCs (Krol 1992, reported in *Appliance* 49, no. 2 [2/92]: 42).

45. Having accepted the need to eliminate CFCs, producers repeatedly had to resist proposals to do it faster than they believed possible. For example, after the 1989 assessments reported it would be feasible to eliminate CFCs in 10 years, bills were immediately introduced to eliminate them in five (*Chemical Week* 3/1/89, p. 16).

46. *Chemical Week* 12/18/91, p. 27.

47. *CMR* 8/31/92, p. 9, quotes a refrigerant distributor: "There is no question in my mind that government bodies will aim their guns at HCFCs within a year after the CFC phaseout is complete."

48. E.g., Benedick 1998, p. 103; Faucheux and Noel 1988.

49. F. A. Vogelsburg, "If society decides it doesn't want these chemicals, we're not going to waste our time producing them" (quoted in *Financial Times*, 6/22/90, p. 22); *ENDS* 184 (6/90): 4; *IER* 7/90, pp. 313–14; *CMR* 6/25/90, p. 7; *NYT* 6/23/90, p. 31; *IER* 11/7/90, p. 459; *Chemical Week* 12/19/90, p. 51.

50. ICI Chemicals and Polymers, "The Ozone Issue and Regulation: An ICI Appraisal," June 1990 (90-2-3).

51. *ENDS* 198 (7/91): 4; *ENDS* 209 (6/92): 6, noted that ICI produced only one HCFC, and would benefit if competitors with stronger commitments to the chemicals were subjected to tight controls. See also BNA, *International Environment Daily*, 3/1/93; *GECR* 3, no. 9, (5/3/91).

52. *IER* 11/6/91, p. 591; *WSJ* 3/9/93, p. B5.

53. *CMR* 3/15/93, p. 1.

54. Three HCFCs (141b, 132b, and 225) were considered, and many firms switched to HCFC-141b, especially in Europe, before it was restricted due to its high ODP (*CMR* 3/15/93, p. 1; Andersen interview).

55. CFCs 11 and 12 were available only at a $1 to $2 premium over their posted prices of about $3/lb (*WSJ* 4/29/92, p. B1). The panic was exacerbated by President George H. W. Bush's Feb. 1992 announcement of an advance of U.S. phaseouts, in which he asked producers to cut output by half immediately.

56. *ENDS* 158 (3/88): 15, quotes an ICI official that high alternative costs will prompt users to look elsewhere, to "less effective products which may not even be made by the chemical industry"; *CMR* 7/9/90 discusses reliability problems and equipment failures associated with early use of HCFC-123 in chillers.

57. D. Pearce, "The EC Approach to Control of CFCs" (86-4-1) estimated price elasticity for CFCs 11 and 12 as 0.2; Camm et al. 1986 estimated 0.01 to 0.03 for CFC-11, 0.01 to 0.07 for CFC-12, and 0.1 to 0.16 for CFC-113.

58. American Electronics Association, "The Electronics Industry, CFCs, and Stratospheric Ozone Reduction," submitted to hearings of House Subcommittee on Natural Resources, Agriculture, Research, and Environment, 3/10/87 (pp. 303–17); Rome workshop report, Mar. 1986 (86-1-2); Canadian delegation report (86-1-1).

59. Firms with early internal programs included DEC, IBM, GM, Ford, and McDonalds (*WSJ* 12/2/86, p. 4).

60. Testimony of R. Barnett, Senate Committee on Environment and Public Works, 1/28/87; *WSJ* 12/2/86, p. 4.

61. The threat of further cuts to CFCs and other fluorocarbons was present by Jan. 1988. HCFCs were at risk for their contribution to ozone loss; HFCs, for their contribution to global climate change (e.g., Doniger 1988b).

62. For example, CFC use in auto air-conditioning was sharply reduced in the 1980s by design changes that cut the average charge from 4 to 2.5 pounds and average losses from 25 to 10 percent per year (*IER* 4/8/87, p. 164).

63. Moore 1990.

64. *Drug and Cosmetic Industry* 148, no. 6 (6/91): 18.

65. UNEP 1989c, p. 1

66. *ENDS* 165 (10/88): 4; *ENDS* 172 (5/89): 7; *ENDS* 187 (8/90): 24; Bradley 1994.

67. *WSJ* 5/13/88, p. A23.

68. Cook 1996b, p. 21.

69. *IER* 2/10/88, p. 109.

70. "Industry Announces Voluntary Phaseout Program," Foodservice and Packaging Institute press release, 4/12/88 (88-2-10); NRDC, EDF, FOE, "Statement of Support for the

FSPI's Fully Halogenated CFC Voluntary Phaseout Program," 4/12/88 (88-2-6); DuPont, *Fluorocarbon/Ozone Update* 7/88, p. 7 (88-2-5).

71. Cook 1996b, p. 26.

72. Hammitt et al. 1986, p. 31; *ENDS* 175 (8/89): 7; *Plastics World* 11/89; *Journal of Commerce* 6/25/90; *Plastics World* 3/91, pp. 93–95; *CMR* 1/18/93, p. 3; *Plastics World* (1/93): 44–50; UNEP 1994c.

73. *Supermarket Business* 6/92; *Buildings* 11/92, p. 70; Bradley 1994.

74. For example, a 1987 consultant's study projected that none of the Protocol's first reduction in 1994, and only 2 percent of the second cut in 1999, would come from this sector (Atkinson 1989: conversion problems were so severe that large-scale implementation of HFC-134a by the mid-1990s "would be a major achievement").

75. They considered HCFC-22, which was used in the 1950s but whose high vapor pressure required heavy hoses and fittings; HCFC-142b, which Pennwalt promoted vigorously but the automakers viewed with skepticism because of its inflammability; and two HCFC blends (Andersen 1989).

76. All models produced in North America, Europe, and Japan were converted by 1994, while those manufactured in developing countries continued to use CFC-12 until after 2000 (UNEP 1994c; Andersen interview).

77. *NYT* 3/31/88, p. D1; Cogan 1988, pp. 111–12; *Air Conditioning, Heating, and Refrigeration News* 11/2/87; *Appliance* 49, no. 2 (2/92): 42; *ENDS* 233 (6/94): 26.

78. *Chicago Tribune* 5/23/93, p. 1.

79. *Appliance* 49, no. 2, (2/92): 42.

80. Seidel 1996.

81. *NYT* 12/19/93, p. 30.

82. *Business Week* 9/20/93, p. 85; *NYT* 2/21/1994, p. C6.

83. Gants 1988.

84. *C&EN* 7/5/93, p. 15; Cook and Kimes 1996, pp. 55–65.

85. *IER* 12/16/92, p. 835.

86. *IER* 3/24/93, p. 210.

87. *IER* 8/12/92, p. 521; *The Guardian* 11/19/92, p. 2.

88. *IER* 2/24/93, p. 123; *ENDS* 235 (8/94): 25.

89. *ENDS* 229 (2/94): 25.

90. UNEP 1994b, pp. 95–110; *Ozone Action* 19 (7/96): 7 reports China receiving grants from the multilateral fund for 31 hydrocarbon refrigeration projects.

91. *CMR* 1/18/93, p. 3.

92. Global Environmental Change Report (94-5-1).

93. UNEP 1994b, pp. 94–107; UNEP 1995b, p. EX. 4.

94. *ENDS* 240(1/95): 29–30; *ENDS* 247 (8/95): 29–30.

95. *ENDS* 232 (5/94): 29.

96. *Appliance* 49, no. 2, (2/92): 42; Smithart 1993.

97. E.g., a new line of Trane HCFC-123 chillers announced at the 1992 Alternatives Conference reduced annual coolant losses from 25 to 0.5 percent ("Trane Announces New Ozone-Friendly Chiller," *PR Newswire* 10/8/92).

98. Andersen 1989.

99. *Building Design and Construction* 11/92.

100. *Chemical Week* 1/8/86, p. 56.

101. DuPont, *Fluorocarbon/Ozone Update* 7/88, p. 7; 12/88, p. 4 (88-2-5).

102. *Chemical Week* 4/26/89; R. N. Miller 1989; Allied Signal 1989.

103. American Electronics Association, White Paper submitted to House Subcommittee on Natural Resources, 3/10/87, pp. 303–17; *NYT* 3/31/88, p. D1; *Daily Telegraph* (London) 11/23/88, p. 8.

104. *NYT* 1/14/88, p. 15.

105. J. K. Johnson 1994.

106. *IER* 2/10/88, p. 109.

107. Hayes 1989.

108. Vice President Margaret Kerr revealed Nortel's true goal under prodding by Tolba at UNEP's Oct. 1988 meeting in The Hague (Wexler 1996, p. 93; Rose and Fitzgerald 1992; Boyhan 1992).

109. Wexler 1996 pp. 90–91.

110. In addition to Nortel and AT&T, founding members were Boeing, Digital Equipment, Ford, GE, Honeywell, Motorola, and Texas Instruments, all of which followed Nortel and AT&T's lead in announcing corporate phaseout or near-phaseout goals (ICOLP, "A New Spirit of Industry and Government Cooperation," 95-2-2).

111. *NYT* 10/11/89, p. A27; Boyhan 1992; Wexler 1996, p. 93.

112. The EPA and *Department of Defense* had formed an interagency working group in Mar. 1988, to test and evaluate alternative cleaning procedures as a first step to revising the specifications (Wexler 1996, p. 92; Keane 1995).

113. AT&T reported that eliminating 80 percent of its CFC use was easy, but that it really needed ICOLP's help for the last 5–10 percent; British Aerospace reported that one technology it gained from its archrival Boeing, a lubricant used in riveting aircraft wings, was worth much more than their cost of participating (*Machine Design* 4/23/93, p. 60; *GECR* 4, no. 5, (2/26/93); ICOLP (95-2-2).

114. E.g., 3M and Petroform announced a drop-in terpene replacement for methyl chloroform in vapor degreasing at the 1992 Alternatives Conference. *R&D* 34, no. 14, (12/92): 28.

115. *Ozone Action* 8 (9/93): 4.

116. *ENDS* 184 (5/90): 16.

117. *ENDS* 182 (3/90): 8–9.

118. By 1993, an ICI official described the former CFC-113 market as "totally fragmenting, with many of the alternatives not-in-kind" (i.e., nonfluorocarbon). *International Environment Daily* 3/1/93.

119. The Protocol defined the ODP of halon-1301 as 3 and that of halon-1211 as 10, relative to a value of 1 for CFCs 11 and 12.

120. B. Tullos, Great Lakes Chemical, testimony at EPA regulatory hearings, Jan. 1988.

121. The Halon Alternatives Research Corporation (HARC), like PAFT, AFEAS, and ICOLP, received antitrust protection under the 1984 National Cooperative Research Act (Taylor, Morehouse, and Andersen interviews).

122. *IER* 1/15/91, p. 16.

123. Andersen, Morehouse, & Miller, 1994.

124. Mauzerall interview, quoting IBM official: "We don't use vacuum tubes in our computers anymore."

125. Presentation of T. Morehouse, 1992 Alternatives Conference, summarized in *R&D* 34, no.14, (12/92): 28.

126. The new chemical, CHF_2Br, was marketed as Firemaster 100 (Mauzerall 1990, p. 30).

127. *C&EN* 73, no. 5, (1/30/95): 25; *Business Insurance*, 5/18/92, p. 23.

128. *Nihon Keizai Shimbun*, 9/2/89, in Schreurs 2001.

129. *CSM* 3/5/92, p. 6; *IER* 9/15/92.

130. *ENDS* 181 (2/90): 8.; *IER* 7/31/91, p. 422.

131. *IER* 2/26/92, p. 98.

132. *Times* (London) 12/29/92; *IER* 2/10/93, p. 110.

133. *Chemical Week* 2/17/93, p. 23; *The Economist* 1/29/94, p. 69.

134. *ENDS* 233 (6/94): 10.

135. Many participants in the 1992 Alternatives Conference said ending CFCs by 1995 was very likely feasible, a stark change from two years earlier, though not all uses had yet identified solutions (*R&D* 12/92, p. 28).

136. DuPont, *Fluorocarbon/Ozone Update*, 8/89 (88-2-5); *CMR* 7/9/90; F. A. Vogelsburg, speech, Mar. 12–13, 1991, London (91-2-4); Cook 1996a, p. 8, fig. 4; Bradley 1994; UNEP 1999b, p. 23.

137. *ENDS* 184 (5/90): 15.

138. For example, the intensity of attacks on Greenfreeze suggests concern with losing much more of the refrigeration sector to hydrocarbons (*IER* 2/24/93, p. 135).

Chapter 8

1. Report of first OEWG meeting, Nairobi, Aug. 28–Sept. 5 (89-1-3).

2. Albritton and Watson slides, "Atmospheric Science Panel: Key Findings," July 1989 (89-4-10).

3. Albritton slides, presentation to NOAA administrator, 8/25/88, Washington DC (88-4-10), p. 14.

4. Draft synthesis report, Assessment Panels (89-1-9). Although the OEWG had the authority to write the synthesis report, it adopted the chairs' draft essentially unchanged.

5. "Scientific Assessment of Stratospheric Ozone 1989: A Summary," presentation by D. Albritton and R. Watson, Nairobi, 8/28/89 (89-4-10), p. 13.

6. M. J. Prather and R. T. Watson, "Transient Scenarios for Atmospheric Chlorine and Bromine" (presented as UNEP/OzL.Pro.WG.II(1)/CRP.2 at 11/13/89 OEWG meeting [Geneva] [89-1-13], appended as annex B to panel synthesis report, then published in revised form as Prather and Watson 1990.

7. "The UNEP report makes it virtually certain that other substances will be brought within the Protocol's scope next spring" (*ENDS* 175 [8/89]: 6).

8. Watson letter to delegations (90-4-9); Prather and Watson, "Transient Scenarios for Atmospheric Chlorine and Bromine" (89-1-13); Watson 1/25/90 Senate testimony, response to questions from Sen. Gore (90-4-11).

9. "Note by the Executive Director," UNEP/OzL.Pro.Asmt.1/2/Rev.1, para. 19.

10. "Note by the Executive Director," UNEP/OzL.Pro.WG.I(2)/4, para. 2 (89-1-3). Tolba first proposed controlling any HCFC with ODP over 2 percent, but retreated in response to resistance during the meeting.

11. Report of First Session of the Bureau, UNEP/OzL.Pro.Bur.1/2, Geneva, 9/29/89.

12. The Protocol's definition of "consumption" partly met the Japanese concern, since CFCs produced before 2000 could be stockpiled to meet later servicing needs (US delegation summary [89-4-16], *ENDS* 177 [10/89]: 13).

13. U.S. Chamber of Commerce comment on HR 2699 call for 1996 CFC phaseout (90-2-6).

14. *ENDS* 177 (10/89): 13; *ENDS* 181 (2/90): 35; *ENDS* 184 (5/90): 29.

15. *ENDS* 173 (6/89): 6.

16. U.S. delegation summary (89-4-16).

17. *ENDS* 181 (2/90): 35.

18. *ENDS* 177 (10/89): 14.

19. Mertens 1989.

20. McConnell and Schiff 1978.

21. Midgley 1989; P. Cammer, letter to *C&EN*, 9/4/89, p. 2; *ENDS* 181 (2/90): 7.

22. UNEP 1989b, pp. v, 13, 20; *Health and Safety at Work* 14, no. 5 (May 1992): 25.

23. The Alliance, which had gone through a major shift to a strategy of constructive engagement with international negotiations, considered broadening its membership and mandate to include MC, but decided not to. The new group, the Halogenated Solvents Industry Alliance (HSIA), had about 250 members of the 73,000 U.S. firms that used MC (Midgley 1989; *ENDS* 181 [2/90]: 7; Cammer, Fay, and Niemi interviews).

24. Watson testimony, 5/15/89 (89-4-5).

25. *ENDS* 177 (10/89): 14; report of the OEWG meeting, Nov. 1989 (89-1-16); congressional testimony of P. Cammer, Jan. 1990 (90-2-7).

26. Report of the First Session of the Second Meeting (89-1-6), p. 4.

27. European Chlorinated Solvent Association, "Production and Release of Methyl Chloroform to the Atmosphere—1989 Data" (89-2-1); N. C. Campbell (ICI), letter to author (91-2-16).

28. Multiple interviewees and Benedick (1998) speculate the Soviet Union used none.

29. E.g., Claussen 1989.

30. *ENDS* 177 (10/89): 14; *ENDS* 178 (11/89): 7.

31. *ENDS* 177 (10/89): 15.

32. *Chemical Week* 12/6/89.

33. OEWG discussion paper (89-1-4).

34. Cavender-Bares, Jäger, and Ell 2001.

35. Claussen 1989; *Chemical Week* 12/6/89.

36. An April 1990 UNEP survey estimate $2 billion as the cost of immediately scrapping existing CFC plants and replacing them with alternatives plants (*IER*, 8/89, p. 389; 4/90, p. 171).

37. McKinsey and Company, "Protecting the Global Atmosphere: Funding Mechanisms," interim report to Steering Committee for Ministerial Conference, The Netherlands (6/27/89), Ozl.Pro.Mech.1/2, paras. 14–16.

38. *ENDS* 184 (5/90): 30.

39. In November, developing countries proposed the key clauses "obligation to comply . . . will be subject to" and access to technology shall be on a "preferential and non-commercial basis" (Benedick 1991, pp. 154–55).

40. State Department cables, May 1990 (90-1-27); "Statement of Mr. Basil Kavalsky, World Bank" (90-1-27); *WP* 5/9/90, p. A1; Claussen and Biniaz interviews.

41. *ENDS* 184 (5/90): 29.

42. Friends of the Earth memorandum, survey of national phaseout positions (90-3-1).

43. *ENDS* 184 (5/90): 29; *IER* 3/90, p. 101; *IER* 6/90, p. 227.

44. *CMR* 237, no. 26 (6/25/90): 7.

45. Friends of the Earth, "The Threat Posed by a Damaging Consensus," June 1990 (89-3-2).

46. "Statement by the Chief of Staff," 6/15/90 (90-4-28)

47. Quoted in Benedick 1991, p. 170.

48. Benedick 1991, p. 172; Enquote Commission of the German Bundestag 1991, p. 628.

49. *ENDS* 184 (6/90): 4.

50. *ENDS* 185 (6/90): 4.

51. Report of Second Meeting of the Parties (90-1-21); *ENDS* 185 (6/90): 3; Benedick 1991, pp. 173–76; Claussen, Fay, and Niemi interviews.

52. R. Stolarski, paper presented to the conference, "Tropical Ozone and Atmospheric Change," Penang Malaysia, Feb. 20–23, 1990. This result was discussed at London by Isaksen and Obasi, (90-1-21), although the reported changes in equatorial latitudes were not statistically significant.

53. Donors used a regional representation system in which the United States was an entire region, and thus held a permanent seat. Votes required two-thirds majority overall, with simple majorities in both groups (meeting report, 90-1-21).

54. The EC objected to this allocation formula, since it capped any single nation's contribution at 25 percent of the total, whereas the sum of all EC member-states' contributions added to 35 percent (Report of the Second Meeting of the Parties, 90-1-21). The most salient alternative, however, was to scale contributions by national CFC use. Since the EC was opposed to providing national-level consumption data, it could not propose this alternative.

55. Benedick 1991, pp. 198–99; Bustani, Donner-Abreu, and Lee-Bapty interviews.

56. "Report of the Mechanical Working Group," UNEP/OzL.Pro.WG.II(1)/6, 11/22/89, Geneva.

57. *ENDS* 187 (8/90): 19; *ENDS* 192 (1/91): 35.

58. Phaseout dates were 1/95 for CFCs, 1/94 for halons, 1/93 for MC and CT, and 1/2000 for HCFC-22 ("Ordinance on the Prohibition of Certain Ozone-Depleting Halogenated Hydrocarbons," 5/6/91; translation in U.S. NGO petition, Dec. 1991, 91-3-5, appendix D).

59. The correction was identified in 1989 and discussed informally in London. The corrected data were presented to the AGU in December 1990 (Herman et al. 1991; UNEP 1989e, p. xix, 168).

60. "A Deepening, Broadening Trend," *Nature* 352 (8/22/91): 668–69; "Ozone Loss Hits Us Where We Live," *Science* 254 (11/1/91): 645.

61. *NYT* 4/5/91, p. A1; *IER* 4/17/91, p. 187; 4/24/91, p. 217. Other members of the administration were more restrained in their assessment. M. Deland, chair of the U.S. Council for Environmental Quality, said the new results needed further study.

62. Albritton and Watson slides, "Ozone and Greenhouse: Recent Findings" (91-4-10); UNEP 1991c, p. xii.

63. UNEP 1989e, p. ix; R. Bojkov presentation, Conference on Tropical Ozone and Atmospheric Change, Penang, Malaysia, Feb. 20–23, 1990; Isaksen and Obasi presentations to London meeting, in UNEP/OzL.Pro.2/3, 6/29/90 (90-1-21), p. 7; *NYT* 6/24/90, p. A13.

64. Tolbert et al. 1990.

65. Rodriguez, Ko, and Sze, 1991.

66. *NYT* 10/12/90, p. A8.

67. Proffitt et al. 1990; *NYT* 9/6/90, p. A25.

68. *IER* 2/13/91, p. 68.

69. Developing country representation was still very thin: one person from each of seven countries, two and three from China and India. Albritton slides, OEWG meeting, 4/6/92 (92-4-10).

70. Albritton slides, Les Diablerets meeting, Oct. 14–18, 1991 (91-4-10), p. 4.

71. UNEP 1991c, p. xiii.

72. The eruption was calculated to increase midlatitude ozone losses by 3–8 percent (Albritton slides, panel review meeting, Les Diablerets, Switzerland [91-4-10], p. 5).

73. NASA/WMO 1986, p. 722.

74. UNEP 1991c, p. xvii; Albritton slides, Les Diablerets meeting and Oct. 30 (91-4-10).

75. UNEP 1991a, p. 36; *NYT* 11/16/91, p. A6.

76. Smith et al. 1992, p. 952.

77. Participants included 240 experts from 38 countries, among them 20 developing countries (UNEP 1991b, p. x).

78. *IER* 1/29/92, p. 31.

79. Those refusing included Canada, Australia, New Zealand, and the Netherlands (*IER* 6/19/91, pp. 324; *IER* 7/3/91, pp. 363–64).

80. F. A. Vogelsburg, speech (91-2-8), presentation at NAE Conference (91-2-12).

81. Decision proposal (91-1-3); UNEP 1991b, p. ix. Here, for the first time, parties' instructions to the TEAP reflected tough bargaining over the precise language employed. Opponents of strong measures, particularly Japan and the Soviet Union, tried to prevent the TEAP from being given strong and specific instructions to assess the measures.

82. UNEP 1991b, pp. 3–6.

83. *NYT* 10/23/91, p. A1.

84. *IER* 14, no. 22, (11/6/91): 591; 11/20/91, p. 619; W. Reilly testimony to House Subcommittee on Health and Environment, 11/14/91.

85. Friends of the Earth press release 10/22/91 (91-3-9) and pamphlet (92-3-4); Makhijani and Gurney 1995, p. 120; *IER* 2/26/92, p. 103; *CMR* 2/24/92, p. 5.

86. Petition by NRDC, Friends of the Earth, and Environmental Defense Fund 12/3/91 (91-3-5).

87. E.g., Claussen 1992; *IER* 1/29/92, p. 55.

88. "New Assaults Seen on Earth's Ozone Shield," *Science* 255 (2/14/92): 797–98; *NYT* 2/4/92, p. C4.

89. *NYT* 2/4/92, p. C4.

90. *NYT* 2/7/92, p. A1; *NYT* 2/13/92, p. B16.

91. Existing EC legislation, approved in March 1991, banned production in mid-1997. The new agreement advanced the phaseouts by 18 months (*ENDS* 203 [12/91]: 23; 205 [2/92]: 8; 206 [3/92]: 37; *IER* 3/11/92, p. 122; *NYT* 3/3/92, p. 1).

92. *Chemistry and Industry*, 3/2/92, p. 155; *IER* 2/26/92.

93. Alliance petition and press release, 2/11/92 (92-2-7, 92-2-8).

94. *CMR* 242, no. 9, (8/31/92): 9.

95. Alliance petition and press release (92-2-7, 92-2-8).

96. *Science* 256 (5/8/92): 734; *NYT* 3/24/92, p. 6.

97. *IER* 5/6/92, p. 265; 5/20/92, p. 291; Friends of the Earth, "Opening National Proposal," Apr. 1992 (92-3-3).

98. Claussen 1992; *IER* 10/7/92, p. 654.

99. Most of these were just theoretical chemicals, but Great Lakes Chemical was trying to commercialize one—with an ODP higher than CFCs but lower than halons—as a halon substitute. *ENDS* 207 (4/92): 36–37.

100. *ENDS* 210 (7/92): 37.

101. *ENDS* 213 (10/92): 12, 36.

102. These caps were defined in terms of chemicals' ODP, so a 3 percent cap is equivalent to replacing all CFCs with an HCFC whose ODP is 3 percent (*ENDS* 210 [7/92]: 37).

103. France fought in the EC to raise the cap to 4 percent, and broke ranks at Copenhagen to support a weaker interim cut (*ENDS* 214 [11/92]: 14; C. Follett, Reuters, 11/24/92).

104. *ENDS* 213 (10/92): 36; Claussen 1992; *C&EN*, 10/19/92, pp. 17–18; *CMR* 242, no. 14 (10/5/92): 3.

105. Manufacturers included two in the United States (Great Lakes Chemical and Ethyl Corporation), two in France (Rhone Poulenc and Atochem), and one in Israel (Dead Sea Bromide).

106. *IER*, 11/18/92, p. 735.

107. This use, often required by importing nations to protect against introduction of pests, accounted for about 1 percent of global use and 9 percent of developing country use (*IER*, 7/15/92, p. 455).

108. *ENDS* 213 (10/92): 36.

109. *IER*, 3/11/92, p. 127; *ENDS* 210 (7/92): 37.

110. *C&EN* 70, no. 31, (8/3/92): 23; Benedick 1998, pp. 210–11.

111. UNEP/OzL.Pro.WG.1/7/2/Rev.1.; UNEP/OzL.Pro.4/8, para. 24.

112. F. A. Vogelsburg, presentation to NAE Conference, Apr. 22–25, 1991 (91-2-12).

113. "Status of Contributions Toward the Trust Fund for the Interim Multilateral Fund," cited in Gehring 1994, p. 304.

114. *IER* 8/12/92, p. 526; 7/29/92, p. 492.

115. *ENDS* 213 (10/92): 36.

116. *ENDS* 214, (11/1992): 13.

117. D. Doniger memo, "Amendments to the Ozone Treaty and EPA Ozone Protection Rules," 12/14/92, NRDC (92-3-2).

118. Presentations at 1992 Alternatives Conference, summarized in *R&D* 34, no. 14, (12/92): 28–32.

119. *ENDS* 214 (11/92): 13.

120. *C&EN* 12/7/92, p. 5; *IER* 12/2/92, pp. 769–72.

121. *The Economist* 11/28/92, p. 50.

122. Thirty-five percent cut in 2004, 65 percent in 2010, 90 percent in 2015, 99.5 percent in 2020, 100 percent in 2030. The UNEP's proposal to ban HCFCs in new equipment after 2000 and stop production in 2005 was rejected, as was the U.S. proposal to cut out longest-lived HCFCs first (*Chemical Engineering*, 11/92, p. 63).

123. *WSJ* 11/17/92, p. B4.

124. Lang 1993, p. 277; *IER* 12/2/92, pp. 769–72.

125. Advocates of abolishing the fund were the United Kingdom, France, Germany, Italy, and the Netherlands, which had played a leading role in forming the GEF (UNEP/.OzL.Pro.4/CRP.1; Gehring 1994, p. 306; Winsemius 1989).

126. D. Doniger memo, "Amendments to Ozone Treaty and EPA Ozone Protection Rules", 12/14/92, NRDC (92-3-2).

127. The agreement was that 1994 would be no less than 1993, while the 1994–1996 total would be $340 to 500 million (the current level to the secretariat's recommendation). UNEP/OzL.Pro.4/15, Decision IV/18 (92-1-13).

128. *IER* 12/2/92, pp. 769–72.

129. This decision left widespread confusion over developing-country obligations. Benedick 1998, p. 213, identifies errors by officials and commentators who thought Article 5 phaseouts were moved to 2006, when in fact they stayed at 2010 pending the review (*IER* 12/2/92, p. 771; Rowlands 1992; Parson and Greene 1995).

130. The UN Conference on Environment and Development, Rio de Janeiro, June 1992.

131. Final regulations differed from initial proposals only in delaying the phaseout one year. They also included the first list of acceptable alternatives under the SNAP (*CMR* 10/5/92, p. 3; *WSJ* 11/27/92, p. B6; *CMR* 1/25/93; *C&EN* 71, no. 4, (1/25/93): 14, 71, no. 5, (2/1/93): 23; *IER* 3/24/93, p. 209, 12/15/93, p. 929).

132. *ENDS* 215 (12/92):34, 221 (6/93):37, 225 (10/93):35; B. Love, Reuters 6/10/93.

133. *ENDS* 227 (12/93): 37; 229 (2/94): 36.

134. EU Regulation 3093-94; *IER* 6/15/94, p. 499; *ENDS* 229 (2/94): 36, 235 (8/94): 40.

135. Of this total, $455 million was new and $55 million carry over from 1991–1993. $82 million remained outstanding from 1991–93, most from former Warsaw Pact states. Reclassifying parties as developing was not a serious drain on the fund, since the largest nation affected, Korea, announced it would not seek funding (Report of Fifth Meeting of Parties, UNEP/OzL.Pro.5/12, Dec. 5/93 [93-1-5]; Lang 1994; *IER* 12/1/93, p. 881).

136. Butler et al., 1992.

137. Northern hemisphere ozone briefly dipped to 25 percent below average in early 1993, including 18 percent loss over the continental United States (NOAA announcement of 9/23/1993, reported in *IER* 10/6/93, p. 723).

138. After the correction of TOMS data in 1990, TOMS and Dobson measurements began drifting apart again near the end of the TOMS instrument's life. (It failed in May 1993, after nearly 16 years of service.) Agreement between total ozone and vertical profile measurements also remained imperfect. SBUV and SAGE satellite instruments and Umkehr readings agreed well with each other and with total-ozone measurements, but ozone sonde measurements—the only in situ observations—agreed badly with the other sources. (Albritton slides, Les Diablerets review meeting, July 18–22, 1994 [94-4-10], p. 7.)

139. UNEP 1995a, p. xvii; Albritton slides, Les Diablerets Review meeting, July 18–22, 1994 (94-4-10).

140. Observations that losses from NOx were smaller, and those from HOx and Cl/Br higher, than gas-phase chemistry predicted, supported the importance of sulfate-aerosol surface processes, since their direct effect was to reduce gas-phase concentrations of active NOx species while increasing those of HOx, Cl, and Br.

141. Albritton slides from Les Diablerets review meeting, July 18–22, 1994 (94-4-10).

142. UNEP 1995a, p. xviii; WMO press conferences, March 5 and 23, reported in *IER* 3/10/93, p. 156; *IER* 4/7/93, p. 246.

143. UNEP 1995a, p. xvi.

144. Albritton slides 11/9/93 (93-4-10), p. 6.

145. UNEP 1995a, p. xx; Albritton slides, Les Diablerets review meeting, July 18–22, 1994 (94-4-10).

146. UNEP 1995a, p. xxii.

147. UNEP 1994d; J. van der Leun, "Avoiding Disaster," *Our Planet* 9, no. 2, (1997): 9.

148. The backlash was promoted in several popular books (e.g. Limbaugh 1992; Ray and Guzzo 1990). The source of most of the claims, Maduro and Schauerhammer 1992, was published by Twenty-first Century Science Associates, an organization dedicated to "challenging the assumptions of modern scientific dogma" and associated with the fringe U.S. political figure Lyndon LaRouche. Writings of the former atmospheric scientist S. Fred Singer (e.g., *National Review* 41, no. 12 [1989]: 34; *National Interest*, Summer 1994, pp. 73–76) also spurred the backlash. Summary in "The Ozone Backlash," *Science* 260 (6/11/93):1580–1583.

149. Congressional action began with a 1992 statement by Rep. Dannemeyer that ozone-layer concerns were fraudulent (*IER* 8/12/92, p. 530). Arizona passed symbolic legislation in early 1995 allowing CFC production in the state (which had no authority to override federal law or international treaties). Three 1995 bills proposed to delay or cancel the coming CFC phaseout and to revoke the EPA's authority to restrict MeBr except on finding by the secretary

of agriculture that acceptable alternatives were available (hearings before the Subcommittee on Energy and the Environment, House Science Committee, 9/20/95).

150. *Business Week* 7/24/95, p. 47.

151. The TEAP reported that such controls were infeasible, and parties decided in 1993 not to pursue them.

152. "Tasks Assigned to TEAP by the Fourth Meeting of the Parties to the Montreal Protocol" (92-1-17).

153. *IER* 7/28/93, p. 539.

154. The nine withdrew their requests (or denied they had made any) at the August 1993 OEWG meeting when UNEP director Dowdeswell criticized the number of requests (*IER* 9/8/93, p. 643; *OzoneAction*, 9/93, p. 2).

155. *ENDS* 224, (9/93):39–40.

156. Laboratory and analytic uses were tiny, but the Panel recommended a global exemption because they feared many parties did not understand their essential nature and thus did not apply. In the one close call, the panel judged that alternatives were available for fingerprinting, but suggested that parties consider granting the exemption if they judged courts unlikely to accept the alternatives in time for the 1996 phaseout (i.e., the Panel refrained from drawing a judgment it deemed to be non-technical and under the parties' authority).

157. UNEP 1994.

158. *ENDS* 235 (8/94):40.

159. *ENDS* 237 (10/94):40.

160. M. Miller, memo to NGOs on MeBr issues at Montreal Protocol meeting, Nairobi, Oct. 3–7, 1994 (94-3-1).

161. Buxton, Andersen, and Banks interviews.

162. UNEP 1994a, p. 3.

163. UNEP 1995b, p. 72.

164. UNEP 1995b, p. 5.

165. Report of OEWG meeting, UNEP/OzL.Pro/WG.1/11/10 (95-1-3); Albritton slides, 5/8/95 (95-4-10).

166. In July 1994, arrears were $42 million for 1991–1993, and $135 million for 1994. Late payers included the United States, Japan, France, the United Kingdom, and Italy (although 1994 arrears for the United States and Japan were misleading, since payments were due late in the year) (letter quoted in *IER* 10/19/94, p. 849).

167. COWI Consultants, "Study on the Financial Mechanism of the Montreal Protocol," Mar. 1995 (95-1-8).

168. Editorial by Mohamed El Hadi Bennadji, vice chair, Executive Committee, *OzoneAction* 15 (July 1995):1.

169. Draft Decision 7/6 bis, "Basic Domestic Needs" (95-1-5).

170. *ENDS* 249 (9/95):41–42.

171. *C&EN* 6/12/95, p. 20; *IER* 2/22/95, p. 139; 4/19/95, p. 290; 5/3/95, p. 325.

172. *ENDS* 249 (10/95):36–37.

173. Tromp et al., 1995.

174. UNEP 1995c, p. III-1; ICF, Inc., "Report under Para 8, Art 5 of the Montreal Protocol," 12/23/94 (94-1-5).

175. UNEP 1995c, p. ES-2; COWI 1995 (95-1-8).

176. Greenpeace, "The Ozone Layer Destroyers: Whose Chlorine and Bromine Is It?" (95-3-2); *Chemical Week* 157, no. 22, (12/6/95):13.

177. *Chemical Week* 157, no. 10, (9/20/95):24; *ENDS* 249 (10/95):36–37.

178. U.S. delegation statement, Alliance memo (95-2-9).

179. While the U.S. delegation expressed confidence that U.S. use would stay slightly below the new cap, the Alliance was concerned that the new cap might pinch U.S. supplies in two to three years (Alliance memo, 12/29/95, [95-2-9]).

180. *ENDS* 251 (12/95):35–37.

181. Statement of the minister of the environment, meeting of the Parties (95-4-1); After

Israel blocked MeBr controls in 1992, Greenpeace mounted a campaign there. The minister agreed in August 1993 that Israel would support any international decision reached to reduce methyl bromide (*IER* 9/22/93, p. 686).

182. Report of the Seventh Meeting of the Parties (95-1-2), annex III.

183. "Proposal for a Critical Agricultural Use Exemption for Methyl Bromide," UNEP/OzL.Pro.7/Prep/CRP.1, 11/28/95 (95-1-7); *ENDS* 249 (10/95):36–37; *Chemical Week* 157, no. 10, (9/20/95):24.

184. Alliance memo 12/29/95 (95-2-9); *ENDS* 251 (12/95):35–37.

185. *ENDS* 251 (12/95):35–37.

186. The consultant's report found that a 2006 Article 5 phaseout with a service tail was cheaper than a 2010 phaseout, suggesting a clear opportunity for mutually advantageous revision of the obligation (94-1-5).

187. Statement by delegation of India on behalf of G-77 and China (95-1-4); Alliance memo, 12/29/95 (95-2-9). The group also bid for a service tail beyond 2010, with U.S. support, but the EC refused. Announcements of planned early phaseout dates in *OzoneAction* 17 (Jan. 1996):10. Some developing countries adopted mixed approaches, since much of their use was by foreign firms. E.g., Thailand identified that 80–85 percent of its ODS use was by foreign firms, and persuaded them to phase out in their Thai operations within one year of their home-country phaseout (*Our Planet 9*, no. 2 [1997]:27).

188. Report of the Seventh Meeting of the Parties (95-1-2), decision 7/4.

189. *ENDS* 251 (12/95):35–37; Alliance memo 12/29/95 (95-2-9).

190. *IER* 8/10/94, p. 659.

191. *ENDS* 251 (12/95):35–37.

192. This assessment was to be conducted with existing international bodies concerned with aviation (ICAO) and climate change (IPCC). Decision 7/31, Assessment Panels, UNEP/OzL.Pro.7/Prep/CRP.4/Rev.1 (95-1-9).

193. Author's notes and interviews from Vienna meeting of the parties; summary in Benedick 1998, p. 230.

194. *ENDS* 263 (12/96): 36.

195. *OzoneAction* 20 (10/96):1.

196. Benedick 1998, pp. 302–3; *ENDS* 263 (12/96):36.

197. Report of the Eighth Meeting of the Parties, UNEP/OzL.Pro.8/12, 12/19/96 (96-1-1).

198. Report of the Eighth Meeting (96-1-1); UNEP 1996.

199. *ENDS* 263 (12/96):36.

200. *OzoneAction* 18 (4/96): 10.

201. N. D. Sze and M. Ko (AER, Inc., "Science Update on Atmospheric Lifetime and Ozone Depletion Potential for Methyl Bromide," prepared for Methyl Bromide Global Coalition, 11/13/96 (96-2-3); UNEP 1989e, p. xiv.

202. Interim targets were agreed as 25 percent cut in 1999, 50 percent in 2001, 70 percent in 2003.

203. Report of the Ninth Meeting of the Parties, UNEP/OzL.Pro/9/12, 9/25/97 (97-1-3).

204. Brack 1996; *NYT* 4/30/95, p. 11.

205. J. Vallette, "Deadly Complacency: US CFC Production, the Black Market, and Ozone Depletion," *OzoneAction*, 9/95 (95-3-3); *Our Planet 9*, no. 2, (1997):25; *OzoneAction* special supp. 6 (2001).

206. EPA, *Stratospheric Update*, 10/1998 (98-4-1).

207. IPAC briefing packet, 11/19/96 (96-2-4); *ENDS* 271 (8/97):42–43.

208. UNEP 1999a; *OzoneAction* 27 (July 1998).

209. UNEP 1999a, p. xxiv-xxv.

210. UNEP 1998, p. 176; *OzoneAction* 25 (Jan. 1998):7.

211. Interim cuts were 60 percent for northern states (50 percent for southern) by 2001, 75 percent for northern (50 percent for southern) by 2003. *ENDS* 273 (10/97):39–40; 280 (5/98):48–49; 283 (8/98):46.

212. The baseline for the freeze mirrored that in effect for consumption controls: 1989 HCFC production plus 2.8 percent of CFC production (*OzoneAction* 37 (1/2001):7; *Earth Negotiations Bulletin* (*ENB*) 19, no. 6, (6/12/99).

213. As of Oct. 2002, virtually all nations had ratified the Protocol and its 1990 London amendment. The 1992 Copenhagen amendment had 141 parties, but not Russia, China, or India. Ratification data at www.unep.org/ozone/ratif.shtml

214. *OzoneAction* 33 (1/2000):8.

215. *OzoneAction* 37 (1/2001):5.

216. Report of the Ninth Meeting of the Parties, Sept. 9–16, 1997 (97-1-3); *ENDS* 272 (9/97):44; revised draft decision by the European Community, UNEP/OzL.Pro.9/CRP.7, 9/11/97 (97-1-2)

217. *ENDS* 288 (1/99):45–46.

218. UNEP/OzL.Pro.11/CRP.18; (*ENB*), 19, no. 6 (12/6/99).

219. *OzoneAction* 28 (10/98):5

220. *OzoneAction* 30 (4/99):1; *ENDS* 299 (12/99): 50.

221. Montzka et al., 1999.

222. Nineteenth OEWG meeting, Geneva, June 14–18, 1999. (*ENB* 19, no. 6, [12/6/99]; *ENDS* 293 [6/1999]:47–48).

223. *ENB* 19, no. 6, (12/6/99).

224. The Inter-governmental Panel on Climate Change (IPCC) and the "subsidiary bodies" to the Framework Convention on Climate Change.

225. Decision X/16 (*ENDS* 287 [12/98]; Alliance letter, 10/8/98, www.arap.org/1998-meeting.html).

226. *ENB* 19, no. 6, (12/6/99).

227. *ENDS* 299 (12/99):50–51.

228. Waibel et al., 1999; Kirk-Davidoff et al., 1999; Staehelin et al. 2001; Wardle et al., 1997.

229. Parties corrected this error in their 2000 meeting (*OzoneAction* 37 [1/2000]:7; *ENB* 19, no. 6, [6/12/99]).

230. *ENB* 19, no. 6, (12/6/99).

231. As of Oct. 2002, the 1997 Montreal amendment had 84 parties; the 1999 Beijing amendment, 34. Ratification data at www.unep.org/ozone/ratif.shtml

232. National production and consumption data reported to UNEP at www.unep.org/ozone; 1986 and 1989 baseline chemicals combined, weighted by ODP.

233. Albritton slides, "Meeting of Chapter 3 Authors," Boulder Co, 10/18/93 (93-4-10); Albritton 1998.

Chapter 9

1. Krasner 1983; Young 1994; Rittberger and Mayer 1993; Moravcsik 1997.

2. Snidal 1985; Gilpin 1981; Keohane 1984; Young 1989; Axelrod 1984; Sebenius 1992; Ellickson 1991; Krasner 1983.

3. Strange 1983; Haggard & Simmons 1987; E. M. Haas 1990; P. M. Haas 1992, 1992b.

4. P. M. Haas, Keohane, and Levy 1993; Keohane and Levy 1996; Young 1994, 1999; Parson and Clark, 1994; Clark et al. 2001; Bennett and Howlett 1992.

5. Baumgartner and Jones 1993; True, Jones, and Baumgartner 1999; Schon and Rein 1994; Hajer 1995.

6. Tolba 1989; Benedick 1991; Brenton 1994; Rowlands 1995; Lang 1991; Brooks 1982; Litfin 1994; P. M. Haas 1992.

7. Several commentators have reproduced charts of how projected steady-state depletion varied from the mid-1970s to the mid-1980s, to illustrate the high scientific uncertainty that they imply diminished after 1985 to allow policy agreement to proceed. This argument errs on two counts. It mischaracterizes uncertainty prior to 1985, abstracting one component and greatly exaggerating its importance; and the presumption that uncertainty diminished shortly

after 1985 is simply incorrect. (The seeming reduction principally reflects the fact that subsequent assessments stopped printing these historical pictures of how estimates varied.)

8. Prominent examples included doomsday images from extreme ozone losses in the early 1970s, model projections of steady-state depletion exceeding 20 percent in 1979, and the "chlorine catastrophe" hypothesis of nonlinear depletion in 1984.

9. Note, however, that the contribution of the ozone hole to the subsequent adaptation of the regime appears to have been stronger, since it was both the hole and the unexplained global losses that prompted many actors to begin calling for CFC phaseouts in 1988.

10. Support from the Netherlands, Denmark, and Germany was more restrained due not to substantive reservations but to the need not to stray too far publicly from the EC position.

11. Keohane 1984, 1988; Oye 1986; Koelble 1995; Milner 1997; Martin 1992; Moravcsik 1997.

12. P. M. Haas, Keohane, and Levy 1993; Chayes and Chayes 1995; Keohane and Levy 1996; Mitchell, 1994.

13. E.g., Thacher 1993, p. 25: Tolba "is a very tough guy," "exerted a lot of political clout," which "cost him dearly. You get to do this a couple of times, then people don't forgive you even if you're right. . . . It led to a loss of effectiveness, and he departed the programme last year."

14. Examining the record at a finer level, this issue cannot distinguish two alternative views of the influence of institutions, based on the distinction between the 1982–1985 and 1986–1987 periods. (1) Institutional context promotes reaching agreements, but absent other driving factors these can be merely symbolic. Under this hypothesis, the limited context already present post-1982 accounts for the 1985 Convention, but the greater institutionalization post-1985 made no further contribution to the stronger 1987 outcome, which was driven by other factors. (2) The expected strength of agreements increases with increasingly dense institutional context.

15. See, e.g., List 1991, on the Baltic Sea regime, and P. M. Haas 1990b, on the Mediterranean.

16. Gehring 1994; Parson 1993; Benedick 1991.

17. Clark et al. 2001.

18. Olson 1965.

19. See, e.g., the tensions evident in Dewey 1922, 1930 and explicated in Dahl 1989, echoing debate about qualifications to govern based on knowledge or reasoning ability that date back to Plato.

20. See, e.g., Kantrowitz 1967; Ruckelshaus 1985.

21. Collingridge and Reeve 1980; Cox 1983.

22. See, e.g., Jasanoff 1990, 1996; Price 1965; Bimber 1996; Fischer 1990; Gieryn 1995; Wood 1997; Brooks 1975; Ezrahi 1980.

23. See discussions of social structures identified as advocacy coalitions, issue networks, and epistemic communities in P. M. Haas 1992a, Sabatier & Jenkins-Smith 1993 and 1999, Heclo 1978, Keck and Sikkink 1998; and of intermediaries identified as individual "knowledge brokers," "boundary organizations," or formal advisory bodies in Litfin 1994, Guston 1997, Farrell and Jäger Forthcoming; Mitchell et al., forthcoming.

24. The exception was the February 1992 announcement of extreme chemical conditions in northern latitudes and the risk of severe losses later that spring, which provoked an immediate policy response including U.S. support for earlier phaseouts. This was an exceptional event, however—a primary scientific claim presented in a high-profile press conference, with institutional support, announcing an imminent environmental risk—which illustrates how high the threshold of saliency is for scientific claims to directly influence policy.

25. The limited use of primary scientific claims discussed here must be distinguished from their more extensive use as justifications for action by authoritative bodies such as regulatory agencies (Brickman, Jasanoff, Ilgen 1985).

26. Weiss 1975.

27. It is because the required demonstrations were fully achieved only at the end of the transition period that they cannot account for its beginning.

28. The few exceptions, such as the Mar. 1986 attempt to argue that the cause of the ozone hole required another 10 years of research before action, gained little support even among industry actors.

29. Concern that these assessments reflected an activist policy agenda is undercut by occasional public statements over several years by the chair, R. Watson, who drew environmentalist attacks by stating that the scientific evidence did not yet support policy action (in 1984 and 1985), or supported only limited action (in 1986 and early 1987).

30. And all know that all know that all know, and so on. Common knowledge requires that this recursive reciprocal knowledge proceed to all degrees (Aumann 1976; Chwe 2001).

31. The only other body that continued to issue ozone assessments was the U.K. SORG, but its assessments simply validated and refined those of the Science Panel. Like the earlier U.K. assessments, these added stronger policy conclusions, but the conclusions now consistently supported maximal further reductions (see, e.g., fifth report of SORG, reported in ENDS 228 [1/194]: 13).

32. UNEP 1994d, p. i.

33. Schmalensee 1996

34. These incentives worked, even though the strongest threat came from the broadest sanctions—those on products "made with" rather than "containing" the controlled substances—whose credibility was weak from the outset and that were ultimately judged to be infeasible.

35. General arguments against targets are advanced in Cooper 1998, and repeated in Laird 2000 and Victor 2001. Toman 2001 provides a rejoinder.

36. These included confidential proceedings, no precommitted votes, providing any available data and analysis to back up claims, and collectively authored reports with no attribution—all intended to free participants from worrying about their employers' positions and let them focus entirely on technical deliberations. Andersen, Banks, Buxton, Morehouse, and Taylor interviews.

37. Bohm 1990; Parson and Zeckhauser 1995.

38. Sebenius 1991.

39. Holling 1978; Gunderson, Holling, and Light 1994; Walters 1986; Lee 1993.

References

Albritton, Daniel. 1998. "What Should Be Done in a Science Assessment." In *Protecting the Ozone Layer: Lessons, Models, and Prospects*. Ed. Philippe G. Le Prestre, John D. Reid, and E. Thomas Morehouse, Jr. Boston: Kluwer.

Allied Signal Corporation. 1989. "Remarks." International CFC and Halon Alternatives Conference. Washington, DC.

Alternative Fluorocarbons Environmental Acceptability Study (AFEAS). 1995. "1995 Update." Washington, DC: AFEAS.

Alternative Fluorocarbons Environmental Acceptability Study (AFEAS). 1996. "Production, Sales, and Atmospheric Release of Fluorocarbons Through 1995." Washington, DC: AFEAS.

Andelin, John, et al. 1988. "Analysis of the Montreal Protocol." Staff report, U.S. Congress. Office of Technology Assessment, Jan. 13.

Andersen, Stephen O. 1989. "Remarks." International CFC and Halon Alternatives Conference. Washington, DC.

Andersen, Stephen O., E. Thomas Morehouse, Jr., and Alan Miller. 1994. "The Military's Role in Protection of the Ozone Layer." *Environmental Science and Technology* 28, no. 13: 586A–589A.

Anderson, J. G., H. J. Grassl, R. E. Shetter, and J. J. Margitan. 1981. "HO_2 in the Stratophere: 3 In-situ Observations." *Geophysical Research Letters* 8, no. 3: 289–92.

Anderson, J. G., J. J. Margitan, and D. H. Stedman. 1977. "Atomic Chlorine and the Chlorine Monoxide Radical in the Stratosphere: Three in Situ Observations." *Science* 198 (4 Nov.): 501–3.

Anderson, James G. 1975. "Measurements of Atomic Oxygen and Hydroxyl in the Stratosphere." In *Proceedings, 4th Conference on CIAP. Symposium no. 175*. Washington, DC: U.S. Department of Transportation.

Anderson, James G. 1982. "The Measurement of Trace Reactive Species in the Stratosphere: An Overview." In *Causes and Effects of Stratospheric Ozone Depletion: An Update*. Washington, DC: National Academy Press.

Angell, J. K. 1988. "An Update through 1985 of the Variations in Global Total Ozone and North Temperate Layer-Mean Ozone." *Journal of Applied Meteorology* 27, no. 1 (Jan.): 91–97.

Angell, J. K., and J. Korshover. 1973. "Quasi-biennial and Long-term Fluctuations in Total Ozone." *Monthly Weather Review* 101: 426–43.

Angell, J. K., and J. Korshover. 1976. "Global Analysis of Recent Total Ozone Fluctuations." *Monthly Weather Review* 104: 63–75.

Angell, J. K., and J. Korshover. 1978. "Global Ozone Variations: An Update into 1976." *Monthly Weather Review* 106: 725–37.

Angell, J. K., and J. Korshover. 1983. "Global Variation in Total Ozone and Layer-Mean Ozone—an Update through 1981." *Journal of Climate and Applied Meteorology* 22, no. 9: 1611–27.

Atkinson, Ward (Sun Test Engineering). 1989. "Mobile Air Conditioning." International CFC and Halon Alternatives Conference. Washington, DC.

Aumann, Robert J. 1976. "Agreeing to Disagree." *Annals of Statistics* 4 (Nov): 1236–39.

Axelrod, Robert. 1984. *The Evolution of Cooperation*. New York: Basic Books.

Barratt-Brown, Elizabeth P. 1991. "Building a Monitoring and Compliance Regime Under the Montreal Protocol." *Yale Journal of International Law* 16: 519–70.

Bastian, Carroll Leslie. 1982. "The Formulation of Federal Policy." In *Stratospheric Ozone and Man*. Ed. Frank A. Bower and Richard B. Ward. Vol. 2. Boca Raton, FL: CRC.

Bates, David R., and Marcel Nicolet. 1950. "The Photochemistry of Atmospheric Water Vapor." *Journal of Geophysical Research* 55, no. 3 (Sept.): 301–27.

Baum, Rudy M. 1982. "Stratospheric Science Undergoing Change." *Chemical and Engineering News* 60, no. 37 (13 Sept.): 21–34.

Baumgartner, Frank R., and Bryan D. Jones. 1993. *Agendas and Instability in American Politics*. Chicago: University of Chicago Press.

Benedick, Richard E. 1991. "Protecting the Ozone Layer: New Directions in Diplomacy." In *Preserving the Global Environment: The Challenge of Shared Leadership*. Ed. Jessica T. Mathews. New York: Norton.

Benedick, Richard Elliot. 1998. *Ozone Diplomacy: New Directions in Safeguarding the Planet*. Enlarged ed. Cambridge, MA: Harvard University Press.

Bennett, Colin J., and Michael Howlett. 1992. "The Lessons of Learning: Reconciling Theories of Policy Learning and Policy Change." *Policy Sciences* 25, no. 3 (Aug.): 275–94.

Berg, W. W., et al. 1980. "First Measurements of Total Chlorine and Bromine in the Lower Stratosphere." *Geophysical Research Letters* 7, no. 11 (Nov.): 937–40.

Berkner, Lloyd V., and L. C. Marshall. 1965. "On the Origin and Rise of Oxygen Concentration in the Earth's Atmosphere." *Journal of Atmospheric Sciences* 22, no. 3: 225–61.

Berkner, Lloyd W., and L. C. Marshall. 1967. "The Rise of Oxygen in the Earth's Atmosphere with Notes on the Martian Atmosphere." *Advances in Geophysics* 12: 309–31.

Biermann, Frank. 2000. "The Case for a World Environment Organization." *Environment* 42, no. 9 (Nov.): 22–31.

Bimber, Bruce. 1996. *The Politics of Expertise in Congress: The Rise and Fall of the Office of Technology Assessment*. Albany: State University of New York Press.

Birks, J. W., et al. 1977. "Studies of Reactions of Importance in the Stratosphere. II. Reactions Involving Chlorine Nitrate and Chlorine Dioxide." *Journal of Chemical Physics* 66: 4591–99.

Biswas, Asit K., ed. 1979. *The Ozone Layer*. Proceedings of the UNEP Meeting of Experts, Washington, DC, Mar. 1–9, 1977. Environmental Sciences and Applications, Vol. 4. Oxford: Pergamon.

Bloomfield, P., G. Dehlert, M. L. Thompson, and S. Zeger. 1983. "A Frequency-Domain Analysis of Trends in Dobson Total Ozone Records." *Journal of Geophysical Research—Oceans and Atmospheres* 88, no. NC13: 8512–22.

Boehmer-Christiansen, Sonya, and Jim Skea. 1990. *Acid Politics: Environmental and Energy Policies in Britain and Germany*. London: Bellhaven.

Bohm, Peter. 1990. "Efficiency Issues and the Montreal Protocol on CFCs." Environment Working Paper 40. Washington, DC: World Bank, Environment Department, Sept.

Boyhan, Walter S. 1992. "Approaches to Eliminating Chlorofluorocarbon Use in Manufacturing." *Proceedings of the National Academy of Sciences* 89 (Feb.): 812–14.

Brack, Duncan. 1996. *International Trade and the Montreal Protocol*. London: Earthscan.

Bradley, Rosemary F. 1994. "Fluorocarbons: The Dynamics of a Regulation-Impacted Industry." In *Chemical Economics Handbook*. Menlo Park, CA: SRI International.

Brenton, Tony. 1994. *The Greening of Machiavelli: The Evolution of International Environmental Politics*. Royal Institute for International Affairs, Energy and Environmental Programme. London: Earthscan.

Brewer, A. W. 1949. "Evidence for a World Circulation Provided by the Measurements of Helium and Water Vapour Distributions in the Stratosphere." *Quarterly Journal of the Royal Meteorological Society* 75: 351–63.

Brickman, Ronald, Sheila Jasanoff, and Thomas Ilgen. 1985. *Controlling Chemicals: The Politics of Regulation in Europe and the United States*. Ithaca, NY: Cornell University Press.

Broderick, Anthony J., and Robert C. Oliver. 1982. "The Supersonic Transport." In *Stratospheric Ozone and Man*, vol. 2. Ed. Frank A. Bower and Richard B. Ward. Boca Raton, FL: CRC.

Brodeur, Paul. 1975. "Annals of Chemistry: Inert." *New Yorker*, Apr. 7, pp. 47–50.

Brodeur, Paul. 1986. "Annals of Chemistry: In the Face of Doubt." *New Yorker*, June 9, pp. 70–87.

Brooks, Harvey. 1975. "Expertise and Politics—Problems and Tensions." *Proceedings of the American Philosophical Society* 119, no. 4 (Aug): 257–61.

Brooks, Harvey. 1982. "Stratospheric Ozone, the Scientific Community and Public Policy." In *Stratospheric Ozone and Man*. Ed. Frank A. Bower and Richard B. Ward. Vol. 2. Boca Raton, FL: CRC.

Brozek, C. M., et al. 1992. "Exposure to Ultraviolet Radiation Enhances Pathogenic Effects of Murine Virus LP-BM5, in Murine Acquired Immunodeficiency Syndrome." *Photochemistry and Photobiology* 56, no. 3: 287–95.

Brune, W. H., et al. 1991. "The Potential for Ozone Depletion in the Arctic Polar Stratosphere." *Science* 252 (May 31): 1260–66.

Budyko, M. I., and I. L. Karol. 1976. "A Study of CIAP." *Meteorologiya i Gidrologiya (Meteorology and Hydrology)* 9, no. 18 (Feb.): 103–11.

Burnett, C. R., and E. B. Burnett. 1980. "Spectroscopic Measurements of the Vertical Column Abundance of Hydroxyl (HO) in the Earth's Atmosphere." *Journal of Geophysical Research* 86: 5185.

Butler, J. H., et al. 1992. "A Decrease in the Growth Rates of Atmospheric Halon Concentrations." *Nature* 359 no. 6394 (Oct. 1): 403–5.

Buxton, G. Victor. 1988. "The Montreal Protocol: A Canadian Perspective." *European Environment Review (EER)* 2, no. 2 (July): 46–48.

Cadle, R. D. 1964. "Daytime Atmospheric O1D." *Discussions of the Faraday Society* 37: 66.

Cagin, Seth, and Philip Dray. 1993. *Between Earth and Sky: How CFCs Changed Our World and Endangered the Ozone Layer*. New York: Pantheon.

Caldwell, Lynton Keith. 1990. *International Environmental Policy*. 2d ed. Durham, NC: Duke University Press.

Callis, L. B., and M. Natarajan. 1986. "The Antarctic Ozone Minimum: Relationship to Odd Nitrogen, Odd Chlorine, the Final Warming, and the 11-Year Solar Cycle." *Journal of Geophysical Research* 91: 10771–96.

Camm, F., et al. 1986. *Social Cost of Technical Control Options to Reduce Emissions of Potential Ozone Depleters in the United States: An Update*. Prepared for U.S. EPA, N-2440-EPA. Santa Monica, CA: Rand Corporation, May.

Cavender-Bares, Jeannine, Jill Jäger, and Renate Ell. 2001. "Developing a Precautionary Approach: Global Environmental Risk Management in Germany." In *Learning to Manage Global Environmental Risks*. Ed. William C. Clark, Jill Jäger, Josee van Eijndhoven, and Nancy M. Dickson. Vol. 1. Cambridge, MA: MIT Press.

Chang, J. S., and W. H. Duewer. 1973. "On the Possible Effect of the NOx Injection in the Stratosphere Due to Past Atmospheric Nuclear Weapons Tests." Paper presented at the AIAA/AMS meeting, Denver, June.

Chang, J. S., and J. E. Penner. 1978. "Analysis of Global Budgets of Halocarbons." *Atmospheric Environment* 12: 1867–73.

Chang, Julius S., W. H. Duewer, and D. J. Wuebbles. 1979. "The Atmospheric Nuclear Tests of the 1950s and 1960s: A Possible Test of Ozone Depletion Theories." *Journal of Geophysical Research* 84: 1755–65.

Chapman, Sidney. 1930. "A Theory of Atmospheric Ozone." *Memoranda of the Royal Meteorological Society* 3: 103–25.

Chayes, Abram. 1972. "An Inquiry into the Workings of Arms Control Agreements." *Harvard Law Review* 85, no. 5 (Mar.): 905–69.

Chayes, Abram, and Antonia Handler Chayes. 1995. *The New Sovereignty: Compliance with International Regulatory Agreements.* Cambridge, MA: Harvard University Press.

Christie, A. D. 1973. "Secular or Cyclic Change in Ozone." *Pure and Applied Geophysics* 106–8: 1000–09.

Chubachi, S. 1984. "A Special Ozone Observation at Syowa Station, Antarctica, from February 1982 to January 1983." In *Atmospheric Ozone.* Ed. C. S. Zerefos and A. Ghazi. Dordrecht: Reidel.

Chwe, Michael Suk-Young. 2001. *Rational Ritual: Culture, Coordination, and Common Knowledge.* Princeton, NJ: Princeton University Press.

Cicerone, R. J. 1982. "Perturbations of the Stratosphere and Ozone Depletion." In *Causes and Effects of Stratospheric Ozone Depletion: An Update.* Washington, DC: National Academy Press.

Cicerone, R. J., R. S. Stolarski, and S. Walter. 1974. "Stratospheric Ozone Destruction by Man-made Chlorofluoromethanes." *Science* 185, no. 4157 (Sept. 27): 1165–67.

Cicerone, R. J., S. Walters, and S. C. Liu. 1983. "Nonlinear Response of Stratospheric Ozone Column to Chlorine Injections." *Journal of Geophysical Research* 88: 3647–61.

Clark, William C., and Giandomenica Majone. 1985. "The Critical Appraisal of Scientific Inquiries with Policy Implications." *Science, Technology & Human Values* no. 3 (Summer): 6–19.

Clark, William C., Jill Jäger, and Josee van Eijndhoven, Dickson, eds. 2001. *Learning to Manage Global Environmental Risks.* 2 vols. Cambridge, MA: MIT Press.

Clark, William C., et al. 2001. "Option Assessment in the Management of Global Environmental Risks." In *Learning to Manage Global Environmental Risks.* Ed. William C. Clark, Jill Jäger, Josee van Eijndhoven, and Nancy M. Dickson. Vol. 2. Cambridge, MA: MIT Press.

Claussen, Eileen (EPA). 1989. Presentation to the International CFC and Halon Alternatives Conference. Washington, DC.

Claussen, Eileen (EPA). 1992. Presentation to the International CFC and Halon Alternatives Conference. Washington, DC.

Cogan, Douglas G. 1988. *Stones in a Glass House: CFCs and Ozone Depletion.* Washington, DC: Investor Responsibility Research Center.

Collingridge, David, and Colin Reeve. 1980. *Science Speaks to Power: The Role of Experts in Policy Making.* London: Pinter.

COMESA. 1975. *Report of the Committee on Meteorological Effects of Stratospheric Aircraft, 1972–1975.* Bracknell (Berkshire): U.K. Meteorological Office.

Committee on Impacts of Stratospheric Change and Committee on Alternatives for the Reduction of Chlorofluorocarbon Emissions. 1979. *Protection against Depletion of Stratospheric Ozone by Chlorofluorocarbons.* Washington, DC: National Academy Press.

Cook, Elizabeth. 1996a. "Overview." In *Ozone Protection in the United States: Elements of Success.* Ed. Elizabeth Cook. Washington, DC: World Resources Institute.

Cook, Elizabeth. 1996b. "Wrapping It Up." In *Ozone Protection in the United States: Elements of Success.* Ed. Elizabeth Cook. Washington, DC: World Resources Institute.

Cook, Elizabeth, and Jeffrey D Kimes. 1996. "Dangling the Carrot." In *Ozone Protection in the United States: Elements of Success.* Ed. Elizabeth Cook. Washington, DC: World Resources Institute.

Cooper, Richard N. 1998. "Toward a Real Treaty on Global Warming." *Foreign Affairs* 77, no. 2: 66–79.

COVOS. 1976. *Comité d'Etudes sur les Consequences des Vols Stratospheriques*. Boulogne: Meteorological Society of France.

Cox, R. A., R. G. Derwent, and A. J. L. Hutton. 1977. "Significance of Peroxynitric Acid in Atmospheric Chemistry of Nitrogen Oxides." *Nature* 270, no. 5635: 328–29.

Cox. R. A., and R. Lesclaux. 1989. "Degradation Mechanisms of Selected Hydrochlorofluorocarbonns in the Atmosphere: An Assessment of Current Knowledge." In *Scientific Assessment of Stratospheric Ozone: 1989. Appendix, AFEAS Report*. Report of the Scientific Assessment Panel of the Montreal Protocol, WMO Global Research and Monitoring Project, rept. 20. Geneva: WMO, Jan.

Cox, Robert W. 1983. "Gramsci, Hegemony, International Relations: An Essay in Method." *Millennium* 12: 162–75.

Crutzen, P. J. 1971. "Ozone Production Rates in an Oxygen-Hydrogen-Nitrogen Oxide Atmosphere." *Journal of Geophysical Research* 76, no. 30 (Oct. 20): 7311–27.

Crutzen, Paul J. 1970. "The Influence of Nitrogen Oxides on the Atmospheric Ozone Content." *Quarterly Journal of the Royal Meteorological Society* 96: 320–25.

Crutzen, Paul J. 1973. "A Discussion of the Chemistry of Some Minor Constitutents in the Stratosphere and Troposphere." *Pure and Applied Geophysics* 106: 1385–99

Crutzen, Paul J. 1974a. "Estimates of Possible Future Ozone Reductions from Continued Use of Fluoro-Chloromethanes CF_2Cl_2, $CFCl_3$." *Geophysical Research Letters* 1: 205–8.

Crutzen, Paul J. 1974b. "Estimates of Possible Variations in Total Ozone Due to Natural Causes and Human Activities." *Ambio* 3, no. 6: 201–10.

Crutzen, Paul J. 1979. "Chlorofluoromethanes: Threats to the Ozone Layer." *Reviews of Geophysics and Space Physics* 17, no. 7 (Oct.): 1824–32.

Crutzen, Paul J., and Frank Arnold. 1986. "Nitric Acid Cloud Formation in the Cold Antarctic Stratosphere: A Major Cause for the Springtime 'Ozone Hole.' " *Nature* 324, no. 1925 (Dec. 18): 651–55.

Crutzen, Paul J., and John W. Birks. 1982. "The Atmosphere after a Nuclear War: Twilight at Noon." *Ambio* 11, no. 2–3: 114–25.

Crutzen, Paul J., and C. J. Howard. 1978. "The Effect of the HO_2 + NO Reaction Rate Constant on One-Dimensional Model Calculations of Stratospheric Ozone Perturbations." *Pure and Applied Geophysics* 116: 497–510. .

Crutzen, Paul J., Ivar S. A. Isaksen, and John R. McAfee. 1978. "The Impact of the Chlorocarbon Industry on the Ozone Layer." *Journal of Geophysical Research* 83, no. C1 (Jan. 20): 345–63.

Crutzen, Paul J., I. S. A. Isaksen, and G. C. Reid. 1975. "Solar Proton Events: Stratospheric Sources of Nitric Oxide." *Science* 189: 457–59.

Cunnold, D., F. Alyea, N. Phillip, and R. Prinn. 1975. "A Three Dimensional Dynamical-Chemical Model of Atmospheric Ozone." *Journal of Atmospheric Science* 32: 170.

Cunnold, D. M., et al. 1986. "Atmospheric Lifetime and Annual Release Estimates for CFC_{13} and $CF2C_{12}$ from 5 Years of ALE Data." *Journal of Geophysical Research* 91: 10797–817.

Dahl, Robert. 1989. *Democracy and Its Critics*. New Haven, CT: Yale University Press.

The Dahlem Workshop. 1988. *Report of the Dahlem Workshop on the Changing Atmosphere*. Report of the Dahlem Workshop held in Berlin, Nov. 1–6, 1987. Ed. F. S. Rowland and I. S. A. Isaksen. New York: Wiley.

DeMore, W. B., et al. 1979. *Chemical, Kinetic and Photochemical Data for Use in Stratospheric Modelling*. JPL Publication 79-27. Pasadena, CA: CalTech, Apr.

DeMore, W. B., et al. 1981. *Chemical, Kinetic and Photochemical Data for Use in Stratospheric Modelling*. JPL Publication 81-3. Pasadena, CA: CalTech, Jan.

DeReeder, P. L. 1976. *OECD and the Environment*. Paris OECD.

Derwent, R. G., A. E. J. Eggleton, and A. R. Curtis. 1976. "A Computer Model of the Photochemistry of Halogen-Containing Trace Gases in the Troposphere and Stratosphere." AERE Report R-8325. London: HMSO.

Dewey, John. 1927. *The Public and Its Problems*. New York: Holt.

Dewey, John. 1930. *Human Nature and Conduct: An Introduction to Social Psychology*. New York: Modern Library. First published 1922.

DeZafra, R. L., et al. 1987. "High Concentrations of Chlorine Monoxide at Low Altitudes in the Antarctic Spring Stratosphere: 1. Diurnal Variation." *Nature* 328 (July 30): 408–11.

Dickinson, R. E., S. C. Liu, and T. M. M. Donahue. 1978. "Effect of Chlorofluoromethane Infrared Radiation on Zonal Atmospheric Temperatures." *Journal of the Atmospheric Sciences* 35, no. 11: 2142–52.

Dickinson, Robert E. 1982. "Stratospheric Perturbations: The Role of Dynamics, Transport, and Climate Change." In *Causes and Effects of Stratospheric Ozone Depletion: An Update*. Washington, DC: National Academy Press.

Didion, D. A., and D. B. Bivens. 1990. "The Role of Refrigerant Mixtures as Alternatives to CFCs."*International Journal of Refrigeration* 13, no. 3 (May): 163–75.

Dishart, K. T., J. A. Creazzo, and M. R. Ascough. 1987. "The Du Pont Program on Fluorocarbon Alternative Blowing Agents for Polyurethane Foams." Paper presented at Polyurethanes World Congress.

Dobson, G.M.B. 1956. "Origin and Distribution of Polyatomic Molecules in the Atmosphere." *Proceedings of the Royal Society* A236: 187–93.

Doniger, David. 1988a. "Politics of the Ozone Layer." *Issues in Science and Technology* 4, no. 3 (Spring): 86–92.

Doniger, David. 1988b. "Remarks." International CFC and Halon Alternatives Conference. Washington, DC.

Dotto, Lydia, and Harold Schiff. 1978. *The Ozone War*. Garden City, NY: Doubleday.

Duewer, W. H., D. J. Wuebbles, H. W. Ellsaesser, and J. W. Chang. 1977. "NOx Catalytic Ozone Destruction: Sensitivity to Rate Coefficients." *Journal of Geophysical Research* 82, no. 6: 935–42.

Dütsch, H. U. 1961. "Current Problems of the Photochemistry of Atmospheric Ozone." In *Chemical Reactions in the Lower and Upper Atmosphere*. Proceedings of an international symposium organized by Stanford Research Institute, San Francisco, Apr. 18–20, 1961. New York: Interscience.

Eckstein, Harry. 1975. "Case Study and Theory in Political Science." In *Handbook of Political Science*. Ed. F. Greenstein and N. Polsby. Vol. 1. Reading, MA: Addison-Wesley.

Eggleton, Alan, Tony Cox, and Dick Derwent. 1976. "Will Chlorofluorocarbons Really Affect the Ozone Shield?" *New Scientist* 70, no. 1001 (May 20): 402–3.

Ellickson, Robert C. 1991. *Order without Law: How Neighbors Solve Disputes*. Cambridge, MA: Harvard University Press.

Elsaessar, Hugh W. 1978. "A Reassessment of Stratospheric Ozone: Credibility of the Threat." *Climatic Change* 1: 257–66.

Elster, Jon. 1995. "Strategic Uses of Argument." In *Barriers to the Negotiated Resolution of Conflict*. Ed. K. Arrow et al. New York: Norton.

Enquete Commission of the 11th German Bundestag. (Enquete-Kommission Vorsorge zum Schutz der Erdatmosphare). 1989. *Protecting the Earth's Atmosphere: An International Challenge*. Interim Report of the Study Commission of the 11th German Bundestag. Bonn: Bonner Universitäts-Buchdruckerei.

Enquete Commission of the 11th German Bundestag. (Enquete-Kommission Vorsorge zum Schutz der Erdatmosphäre) 1991. *Protecting the Earth: A Status Report with Recommendations for a New Energy Policy*. Interim Report of the Study Commission of the 11th German Bundestag. Bonn: Bonner Universitäts-Buchdruckerei.

Environment Canada. 1988. "The Montreal Protocol on Substances That Deplete the Ozone Layer." Presentation by G. V. Buxton at annual meeting of APCA. Dallas, TX: Association Dedicated to Air Pollution Control and Hazardous Waste Management, June 19–24.

EPA, Stratospheric Protection Program. 1988. *Regulatory Impact Analysis: Protection of Stratospheric Ozone*. Washington, DC: EPA, Aug. 1.

Evans, Peter B., Harold K. Jacobson, and Robert D. Putnam. 1993. *Double-Edged Diplomacy: International Bargaining and Domestic Politics*. Berkeley: University of California Press.

Ezrahi, Yaron. 1980. "Utopian and Pragmatic Rationalism: The Political Context of Scientific Advice." *Minerva* 18, no. 1: 111–31.

Fabry, Charles, and M. Buisson. 1921. "Etude de l'extrémité Ultraviolet du Spectre Solaire." *Journal of Physics and Radiation* 2: 197.

Fahey, D.W., et al. 1989. "Measurements of Nitric Oxide and Total Reactive Nitrogen in the Antarctic Stratosphere: Observations and Chemical Implications." *Journal of Geophysical Research* 94: 16665–81.

Farman, J. C., B. G. Gardiner, and J. D. Shanklin. 1985. "Large Losses of Total Ozone in Antarctica Reveal Seasonal ClOx/NOx Interaction." *Nature* 315 (May 16): 207–10.

Farmer, C. B., and O. F. Raper. 1977. "The HF:HCl Ratio in the 14–28 Km Region of the Stratosphere." *Geophysical Research Letters* 4: 827.

Farmer, C. B., et al. 1987. "Stratospheric Trace Gases in the Spring 1986 Antarctic Atmosphere." *Nature* 329 (Sept. 10): 126–30.

Farrell, Alex, and Jill Jaeger, eds. Forthcoming, 2003. *The Design of Environmental Assessments: Choices for Effective Process*. Washington, DC: Resources for the Future Press.

Faucheux, Sylvie, and Jean-Francois Noel. 1988. *Did the Ozone War End in Montreal?* Cahiers du C.3.E. Paris: Université de Paris, Centre Economie-Espace-Environnement.

Federal Task Force on Inadvertent Modification of the Stratosphere (IMOS). 1975. *Fluorocarbons and the Environment*. Washington, DC: Council on Environmental Quality, June.

Fischer, Frank. 1990. *Technocracy and the Politics of Expertise*. Newbury Park, CA: Sage.

Fleig, A. J., P. K. Bhartia, and D. S. Silberstein. 1986. "An Assessment of the Long-term Drift in SBUV Total Ozone Data, Based on Comparison with the Dobson Network." *Geophysical Research Letters* 13, no. 12 (Nov. suppl.): 1359–62.

Fluorocarbon Panel, Chemical Manufacturers Association. 1980. "An Assessment of Chlorofluorocarbon-Ozone Science." Submitted to EPA in response to notice of proposed regulations, Dec.

Fluorocarbon Panel, Manufacturing Chemists Association. 1976. "Scientific Review: The Effect of Fluorocarbons on the Concentration of Atmospheric Ozone." Prepared for NRC Panel on Atmospheric Chemistry, Mar. 1.

Fluorocarbon Panel, Manufacturing Chemists Association. 1976. "A Summary of the Research Program on the Effect of Fluorocarbons on the Atmosphere." In *The Ozone Layer*. Proceedings of the UNEP Meeting of Experts, Washington, DC, Mar. 1–9, 1977. Ed. Asit K. Biswas. Environmental Sciences and Applications, Vol. 4. Oxford: Pergamon.

Foley, H. M., and M. A. Ruderman. 1973. "Stratospheric NO Production from Past Nuclear Explosions." *Journal of Geophysical Research* 78, no. 21 (July 20): 4441–50.

Fox, Jeffrey L. 1979. "Atmospheric Ozone Issue Looms Again." *Chemical and Engineering News*, Oct. 15.

Gamlen, P. H., B. C. Lane, P. M. Midgley, and J. M. Steed. 1986. "The Production and Release to the Atmosphere of $CC_{13}F$ and $CC_{12}F_2$ (Chlorofluorocarbons CFC 11 and CFC 12)." *Atmospheric Environment* 20, no. 6.: 1077–85.

Gants, Robert. 1988. "Remarks." International CFC and Halon Alternatives Conference. Washington, DC.

Gehring, Thomas. 1994. *Dynamic International Regimes: Institutions for International Environmental Governance*. Studies of the Environmental Law Network International. Frankfurt: Peter Lang.

Gibbs, Michael J. 1986. *Scenarios of CFC Use 1985-2075*. Washington, DC: ICF.

Gieryn, Thomas F. 1995. "Boundaries of Science." In *Handbook of Science and Technology Studies*. Ed. Sheila Jasanoff, Gerald E. Markle, James C. Petersen, and Trevor Pinch. Thousand Oaks, CA: Sage.

Gilpin, Robert. 1981. *War and Change in World Politics*. New York: Cambridge University Press.

Glantz, Michael H., Jennifer Robinson, and Maria E. Krenz. 1982. "Climate-Related Impact Studies: A Review of Past Experiences." In *Carbon Dioxide Review: 1982*. Ed. William C. Clark. New York: Cambridge University Press.

Glantz, Michael H., Jennifer Robinson, and Maria E. Krenz. 1985. "Recent Assessments." In *Climate Impact Assessment: Studies of the Interaction of Climate and Society*. SCOPE 27, International Council of Scientific Unions. Ed. Robert W. Kates, Jesse H. Ausubel, and Mimi Berberian. Chichester, UK: Wiley.

Glas, J. P. 1989. "Protecting the Ozone Layer: A Perspective from Industry." In *Technology and the Environment*. Ed. J. H. Ausubel and H. E. Sladovich. Washington, DC: National Academy Press.

Goldsmith, P., et al. 1973. "Nitrogen Oxides, Nuclear Weapon Testing, Concorde, and Stratospheric Ozone." *Nature* 244: 545–51.

Goldstein, Judith, and Robert O. Keohane, eds. 1993. *Ideas and Foreign Policy: Beliefs, Institutions, and Political Change*. Cornell Studies in Political Economy. Ithaca, NY: Cornell University Press.

Götz, F. W., A. R. Meetham, and G. M. B. Dobson. 1934. "The Vertical Distribution of Ozone in the Atmosphere." *Proceedings of the Royal Society of London*, A145: 416.

Graham, R. A., A. M. Winer, and J. N. Pitts, Jr. 1978. "Ultraviolet and Infrared Absorption Cross Sections of Gas Phase HO_2NO_2." *Geophysical Research Letters* 5: 909–11.

Greene, Owen. 1998. "The System of Implementation Review in the Ozone Regime." In *The Implementation and Effectiveness of International Environmental Commitments: Theory and Practice* Ed. David G. Victor, Kal Raustiala, and Eugene B. Skolnikoff. Cambridge, MA: MIT Press.

Gribbin, John. 1988. *The Hole in the Sky: Man's Threat to the Ozone Layer*. New York: Bantam.

Grieco, Joseph M. 1979. "The Concorde SST and Change in the British Polity." *World Politics* 31 (July): 518–38.

GRL. 1986. "Geophysical Research Letters Supplement." Special supplement issue on causes of the Antarctic ozone hole. *Geophysical Research Letters* 13, no. 12 (Nov.): 1191–362.

Grobecker, A. J., S. C. Coroniti, and R. H. Cannon, Jr. 1974. *The Report of Findings: The Effects of Stratospheric Pollution by Aircraft*. DOT-TST-75-50, U.S. Department of Transportation, Climatic Impact Assessment Program. Springfield, VA: National Technical Information Service, Dec.

Grobecker, Alan J. 1981. "The Management and Organization of the DOT Climatic Impact Assessment Study (CIAP)." Talk given at National Center for Atmospheric Research, Sept. 3.

Groves, K. S., S. R. Mattingly, and A. F. Tuck. 1978. "Increased Atmospheric Carbon Dioxide and Stratospheric Ozone." *Nature* 273: 711–15.

Groves, K. S., and A. F. Tuck. 1979. "Simultaneous Effects of CO_2 and Chlorofluoromethanes on Stratospheric Ozone." *Nature* 280: 127.

Gruber, Lloyd. 2000. *Ruling the World : Power Politics and the Rise of Supranational Institutions*. Princeton, NJ: Princeton University Press.

Grundmann, Reiner. 2001. *Transnational Environmental Policy: Reconstructing Ozone*. New York: Routledge.

Gunderson, Lance, C. S. Holling, and Stephen S. Light, eds. 1994. *Bridges and Barriers to the Renewal of Ecosystems and Institutions*. New York: Columbia University Press.

Guston, David. 1997a. "Critical Appraisal in Science and Technology Policy Analysis: The Example of Science, the Endless Frontier." *Policy Sciences* 30: 233–55.

Guston, David H. 1997b. "Stabilizing the Boundary between U.S. Politics and Science: The Role of the Office of Technology Transfer as a Boundary Organization." *Social Studies of Science* 29, no. 1 (Feb.): 87–111.

Haas, Ernst B. 1990. *When Knowledge Is Power: Three Models of Change in International Organizations*. Berkeley: University of California Press.

Haas, Peter M. 1990. *Saving the Mediterranean: The Politics of International Environmental Cooperation*. New York: Columbia University Press.

Haas, Peter M. 1992a. "Introduction: Epistemic Communities and International Policy Co-ordination." *International Organization* 46, no. 1 (Winter 1992): 1–36.

Haas, Peter M. 1992b. "Banning Chlorofluorocarbons: Epistemic Community Efforts to Pro-tect Stratospheric Ozone." *International Organization* 46, no. 1 (Winter): 187–224.

Haas, Peter M., Robert O. Keohane, and Marc A. Levy, eds. 1993. *Institutions for the Earth: Sources of Effective International Environmental Protection.* Cambridge, MA: MIT Press.

Haas, Peter M., and David McCabe. 2001. "Amplifiers or Dampeners: International Institu-tions and Social Learning in the Management of Global Environmental Risks." In *Learn-ing to Manage Global Environmental Risks.* Ed. William C. Clark, Jill Jäger, Josee van Eijndhoven, and Nancy M. Dickson. Vol. 1. Cambridge, MA: MIT Press.

Haggard, Stephan, and Beth A. Simmons. 1987. "Theories of International Regimes." *Inter-national Organization* 41, no. 3 (Summer): 491–517.

Haigh, Nigel. 1990. *EC Environmental Policy and Britain.* 2d rev. ed. Harlow, UK: Longman.

Hajer, M. A. 1995. *The Politics of Environmental Discourse, Ecological Modernization, and the Policy Process.* Oxford: Clarendon Press.

Hamill, P., O. B. Toon, and R. P. Turco. 1986. "Characteristics of Polar Stratospheric Clouds During the Formation of the Antarctic Ozone Hole." *Geophysical Research Letters* 13: 1288–91.

Hammitt, J. K., et al. 1986. *Product Uses and Market Trends for Potential Ozone-Depleting Substances, 1985–2000.* Rand Report R-3386-EPA. Santa Monica, CA: Rand Corp., May.

Hammitt, James K., and Kimberly M. Thompson. 1997. "Protecting the Ozone Layer." In *The Greening of Industry : A Risk Management Approach.* Ed. John D. Graham and Jennifer K. Hartwell. Cambridge, MA: Harvard University Press.

Hampson, John. 1964. *Photochemical Behaviour of the Ozone Layer.* CARDE Technical Note #1627, PCC D46-38-03-26. Val-Cartier, Quebec: Canadian Armament Research and De-velopment Establishment, July.

Hampson, John. 1966. *Atmospheric Energy Change by Pollution of the Upper Atmosphere.* CARDE Technical Note #1738. Val-Cartier, Quebec: Canadian Armament Research and Development Establishment.

Hampson, John. 1974. "Photochemical War on the Atmosphere." *Nature* 250 (July 19): 189–91.

Harrison, Halstead. 1970. "Stratospheric Ozone with Added Water Vapor: Influence of High-Altitude Aircraft." *Science* 170, no. 395 (Nov. 13): 734–36.

Hayes, Michael E. (Petroform Inc.). 1989. "Naturally Derived Biodegradable Cleaning Agents: Terpene-Based Substitutes for Halogenated Solvents." Paper presented at the Interna-tional CFC and Halon Alternatives Conference. Washington, DC. Oct. 10–11, 1989.

Heath, D. F., A. J. Krueger, and Paul J. Crutzen. 1977. "Solar Proton Event: Influence on Stratospheric Ozone." *Science* 197 (Aug. 26): 886–89.

Heclo, Hugh. 1978. "Issue Networks and the Executive Establishment." In *The New Amer-ican Political System.* Ed. Anthony King. Washington, DC: American Enterprise Institute.

Herman, J. R., et al. 1991a. "Global Average Ozone Change from November 1978 to May 1990." *Journal of Geophysical Research—Atmospheres* 96, no. D9 (Sept. 20): 17,297–305.

Herman, J. R., et al. 1991b. "A New Self-Calibration Method Applied to TOMS and SBUV Backscattered Ultraviolet Data to Determine Long-term Global Ozone Change." *Journal of Geophysical Research—Atmospheres* 96, no. D4 (Apr. 20): 7531–45.

Hill, W. J., and P. N. Sheldon. 1975. "Statistical Modeling of Total Ozone Measurements with an Example Using Data from Arosa, Switzerland." *Geophysical Research Letters* 2: 541–44.

Hill, W. J., P. N. Sheldon, and J. J. Tiede. 1977. "Analyzing Worldwide Total Ozone for Trends." *Geophysical Research Letters* 4: 21–24.

Hill, William J. 1982. "Ozone Trend Analysis." In *Stratospheric Ozone and Man.* Ed. Frank A. Bower and Richard B. Ward. Boca Raton, FL: CRC.

Hoffert, Martin I., and Richard W. Stewart. 1975. "Stratospheric Ozone—Fragile Shield?" *Astronautics and Aeronautics* 13 (Oct.): 42–55.

Hoffman, John S., and Michael J. Gibbs. 1988. *Future Concentrations of Stratospheric Chlorine and Bromine.* EPA 400/1-88/005. Washington, DC: U.S. EPA, Office of Air and Radiation, July.

Hofmann, D. J., J. W. Harder, S. R. Rolf, and J. M. Rosen. 1987. "Balloon-Borne Observations of the Development and Vertical Structure of the Antarctic Ozone Hole in 1986." *Nature* 326, no. 6108 (Mar. 5): 59–62.

Holling, C. S., ed. 1978. *Adaptive Environmental Assessment and Management.* Chichester, UK: Wiley.

Horwitch, Mel. 1982. *Clipped Wings: The American SST Conflict.* Cambridge, MA: MIT Press.

Hoshizaki, H., J. W. Myers, and K. O. Redler. 1973. "Potential Destruction of Ozone by HCl in Rocket Exhausts." LMC-D-354202. Palo Alto, CA: Lockheed Palo Alto Research Lab, Sept.

Hounshell, David A., and John Kenly Smith, Jr. 1988. *Science and Corporate Strategy: Du Pont R&D, 1902–1980.* New York: Cambridge University Press.

Howard, Carlton J., and K. M. Evenson. 1977. "Kinetics of the Reaction of HO_2 with NO." *Geophysical Research Letters* 4, no. 10: 437.

Hughes, Peter, and Brian Wynne. 1992. "The Issue of Climate Change in the United Kingdom." Working draft, Apr.

Hunt, B. G. 1966. "Photochemistry of Ozone in a Moist Atmosphere." *Journal of Geophysical Research* 71, no. 5 (Mar. 1): 1385–98.

Hunt, B. G. 1965. "The Need for a Modified Photochemical Theory of the Ozonosphere." *Journal of Atmospheric Science* 23: 88–95.

Interdepartmental Committee for Atmospheric Sciences. 1975. *The Possible Impact of Fluorocarbons and Halocarbons on Ozone.* ICAS 18a-FY 75. Washington, DC: Federal Council for Science and Technology, May.

International Chlorofluorocarbon Substitutes Committee. 1986. *Draft Report.* EPA Contract no. 68-02-3994. Ed. R. Lagow. Washington, DC: EPA.

Isaksen, I. S. A., and F. Stordal. 1986. "Ozone Perturbations by Enhanced Levels of CFCs, N_2O, and CH_4: A Two-Dimensional Diabatic Circulation Study Including Uncertainty Estimates." *Journal of Geophysical Research—Atmosphere* 91, no. D4 (Apr. 20): 5249–63.

Isaksen, Ivar S. A. 1990. "The Stratospheric Ozone Problem: An Update of Recent Findings." Institute Report Series. Oslo: University of Oslo, Institute of Geophysics, June.

Jachtenfuchs, Markus. 1990. "The European Community and the Protection of the Ozone Layer." *Journal of Common Market Studies* 28 (Mar.): 261–77.

Jacobson, Harold K., and Edith Brown Weiss. 1990. "Implementing and Complying with International Environmental Accords: A Framework for Research." Paper presented at the annual meeting of the American Political Science Association. San Francisco, Sept.

Jaffe, Adam B., Richard G. Newell, and Robert N. Stavins. 2001. "Technological Change and the Environment." In *The Handbook of Environmental Economics.* Ed. K. G. Maler and J. Vincent. Amsterdam: North-Holland/Elsevier Science.

Jasanoff, Sheila. 1990. *The Fifth Branch: Science Advisors as Policymakers.* Cambridge, MA: Harvard University Press.

Jasanoff, Sheila S. 1996. "Science and Norms in Global Environmental Regimes." In *Earthly Goods: Environmental Change and Social Justice.* Ed. F. Hampson and J. Reppy, 173–97. Ithaca, NY: Cornell University Press.

Jäger, Jill, et al. 2001. "Monitoring in the Management of Global Environmental Risks." In *Learning to Manage Global Environmental Risks.* Ed. William C. Clark, Jill Jäger, Josee van Eijndhoven, and Nancy M. Dickson. Vol. 2. Cambridge, MA: MIT Press.

Jeevan, A., K. Gilliam, H. Heard, and M. L. Kripke. 1992. "Effects of Ultraviolet Radiation on the Pathogenesis of Mycobacterium lepraemurium Infection in Mice." *Experimental Dermatology* 1: 152–60.

Jenkins-Smith, Hank C. 1990. *Democratic Politics and Policy Analysis*. Pacific Grove, CA: Brooks/Cole.

Jervis, Robert. 1997. *System Effects: Complexity in Political and Social Life*. Princeton, NJ: Princeton University Press.

Jesson, J. Peter. 1982. "Halocarbons." In *Stratospheric Ozone and Man*. Ed. Frank A. Bower and Richard B. Ward. Vol. 2. Boca Raton, FL: CRC.

Johnson, J. K. 1994. "Moving toward Design for Environment in Manufacturing: Lessons Learned from the CFC Phaseout at AT&T." Master's thesis, Kennedy School of Government, Harvard University.

Johnston, Harold S. 1971. "Reduction of Stratospheric Ozone by Nitrogen Oxide Catalysts from Supersonic Transport Exhaust." *Science* 173 (Aug. 6): 517–22.

Johnston, H. S. 1977. "Analysis of Independent Variables in Perturbation of Stratospheric Ozone by Nitrogen Fertilizers." *Journal of Geophysical Research—Oceans and Atmospheres* 82, no. 12: 1767–72.

Johnston, Harold S. 1982. "Odd Nitrogen Processes." In *Stratospheric Ozone and Man*. Ed. Frank A. Bower and Richard B. Ward. Vol. 1. Boca Raton, FL: CRC.

Johnston, Harold S. 1984. "Human Effects on the Global Atmosphere." *Annual Review of Physical Chemistry* 35: 481–505.

Johnston, Harold S. 1992. "Atmospheric Ozone." *Annual Review of Physical Chemistry* 43: 1–32.

Johnston, Harold S., and E. Quitevis. 1975. "The Oxides of Nitrogen with Respect to Urban Smog, Supersonic Transports, and Global Methane." In *Proceedings of the Fifth International Conference on Radiation Research*. Ed. O. F. Nygaard, H. I. Adler, and W. K. Sinclair. New York: Academic.

Johnston, Harold S., G. Whitten, and J. Birks. 1973. "Effect of Nuclear Explosions on Stratospheric Nitric Oxide and Ozone." *Journal of Geophysical Research* 78: 6107–35.

Kantrowitz, Arthur. 1967. "Proposal for an Institution for Scientific Judgment." *Science* 156, no. 3776 (May 12): 763–64.

Kauffman, Joanne, and Elizabeth de Sombre. 1995. "The Montreal Protocol Multilateral Fund: Partial Success Story." In *Institutions for Environmental Aid*. Ed. Robert O. Keohane and Marc A. Levy. Cambridge, MA: MIT Press.

Keane, Therese. 1995. *The International Cooperative for Ozone Layer Protection (ICOLP), 1990–1995: A New Spirit of Industry and Government Cooperation*. Washington, DC: ICOLP, June.

Keck, Margaret E., and Kathryn Sikkink. 1998. *Activists beyond Borders: Advocacy Networks in International Politics*. Ithaca, NY: Cornell University Press.

Keeling, C. D., and J. C. Pales. 1965. "The Concentration of Atmospheric Carbon Dioxide in Hawaii." *Journal of Geophysical Research* 70: 6053–76.

Keohane, Robert O. 1984. *After Hegemony: Cooperation and Discord in the World Political Economy*. Princeton, NJ: Princeton University Press.

Keohane, Robert O. 1988. "International Institutions: Two Approaches." *International Studies Quarterly* 32, no. 4 (Dec.): 379–96.

Keohane, Robert O., and Marc A. Levy. 1996. *Institutions for Environmental Aid*. Cambridge, MA: MIT Press.

Kerr, J. B., and C. T. McElroy. 1976. "Measurement of Stratospheric Nitrogen Dioxide from the AES Stratospheric Balloon Program." *Atmosphere* 14, no. 3: 166–171.

Keyser, L. F. 1980. "Absolute Rate Constant of the Reaction $OH + H_2O_2 \rightarrow HO_2 + H_2O$ from 245 to 423 K." *Journal of Physical Chemistry* 84: 1659–63.

Khalil, M. A. K., and R. A. Rasmussen. 1981. "Increase of $CHClF_2$ (F-22) in the Earth's Atmosphere." *Nature* 292 (27 August): 823–24.

Khalil, M.A.K., and R. A. Rasmussen. 1983. "Sources, Sinks, and Seasonal Cycles of Atmospheric Methane." *Journal of Geophysical Research* 88: 5131–44.

Kirk-Davidoff, D. B., E. J. Hintsa, J. G. Anderson, and D. W. Keith. 1999. "The Effect of Climate Change on Ozone Depletion Through Changes in Stratospheric Water Vapour." *Nature* 402 (Nov. 25): 399–401.

Koelble, Thomas A. 1995. "The New Institutionalism in Political Science and Sociology." *Comparative Politics* 27, no. 2 (Jan.): 231–43.

Komhyr, W. D., E. W. Barrett, G. Slocum, and H. K. Weickmann. 1971. "Atmospheric Total Ozone Increase during the 1960s." *Nature* 232: 390–91.

Komhyr, W. D., R. D. Grass, and R. K. Leonard. 1986. "Total Ozone Decrease at South Pole, Antarctica, 1964–1985." *Geophysical Research Letters* 13: 1248–51.

Kraft, Michael E. 1984. *Environmental Policy in the 1980s: Reagan's New Agenda*. Ed. Norman J. Vig and Michael E. Kraft. Washington, DC: Congressional Quarterly Press.

Krasner, Stephen. 1991. "Global Communications and National Power: Life on the Pareto Frontier." *World Politics* 43: 336–66.

Krasner, Stephen D., ed. 1983. *International Regimes*. Ithaca, NY: Cornell University Press.

Krol, John A. 1992. "Remarks." International CFC and Halon Alternatives Conference. Baltimore.

Laird, Frank N. 2000. "Just Say No to Greenhouse Gas Emissions Targets." *Issues in Science and Technology* 17, no. 2 (Winter): 45–52.

Lal, S., R. Borchers, P. Fabian, and B. C. Kruger. 1985. "Increasing Abundance of CF_2 BrCl in the Atmosphere." *Nature* 316: 135–36.

Lang, Winfried. 1991. "Negotiations on the Environment." In *International Negotiations: Analysis, Approaches, Issues*. Ed. Victor Kremenyuk. San Francisco: Jossey-Bass.

Lang, Winfried. 1993. "Ozone Layer." In *Yearbook of International Environmental Law*. Vol. 3, *1992*. Ed. Gunther Handl. London: Graham and Trottman.

Lang, Winfried. 1994. "Ozone Layer." In *Yearbook of International Environmental Law*. Vol. 4, *1993*. Ed. Gunther Handl. Oxford: Clarendon Press.

Langley, K. F., and W. D. McGrath. 1971. "The Ultraviolet Photolysis of Ozone in the Presence of Water Vapor." *Planetary and Space Science* 19: 413.

Lax, David A., and James K. Sebenius. 1986. *The Manager as Negotiator*. New York: Free Press.

Lazrus, A. L., B. W. Gandrud, R. N. Woodward, and W. A. Sedlacek. 1976. "Direct Measurements of Stratospheric Chlorine and Bromine." *Journal of Geophysical Research* 81: 1067.

Lee, Kai N. 1993. *Compass and Gyroscope: Integrating Science and Politics for the Environment*. Washington DC: Island.

Leovy, C. B. 1969. "Atmospheric Ozone: An Analytic Model for Photochemistry in the Presence of Water Vapor." *Journal of Geophysical Research* 74, no. 2: 417–26.

Levy, Marc. 1990. "Protecting the Ozone Layer to Save Our (Electoral) Skins." Paper presented to Harvard Global Environmental Policy Seminar, Dec. 2, 1990.

Levy, Marc A., Oran R. Young, and Michael Zurn. 1995. "The Study of International Regimes." *European Journal of International Relations* 1: 267–330.

Limbaugh, Rush. 1992. *The Way Things Ought to Be*. New York: Pocket Books.

List, Martin. 1991. *Umweltschutz in Zwei Meeren*. Munich: Tuduv.

Litfin, Karen T. 1994. *Ozone Discourses: Science and Politics in Global Environmental Cooperation*. New York: Columbia University Press.

Littlejohn, D., and H. S. Johnston. 1980. "Rate Constant for the Reaction of Hydroxyl Radicals and Peroxynitric Acid." Paper presented at the AGU meeting. San Francisco, Dec.

Liu, S. C., R. J. Cicerone, T. M. Donahue, et al. 1976. "Limitations of Fertilizer Induced Ozone Reduction by the Long Lifetime of the Reservoir of Fixed Nitrogen." *Geophysical Research Letters* 3: 157–60.

Liu, S. C., et al. 1980. "On the Origin of Tropospheric Ozone." *Journal of Geophysical Research* 85, no. C12 (Dec. 20): 7546–52.

London, Julius, and James K. Angell. 1982. "The Observed Distribution of Ozone and Its Variations." In *Stratospheric Ozone and Man*. Ed. Frank A. Bower and Richard B. War. Vol. 1. Boca Raton, FL: CRC.

London, Julius, and Jean Kelly. 1974. "Global Trends in Atmospheric Ozone." *Science* 184: 987–89.

London, Julius, and S. J. Oltman. 1979. "The Global Distribution of Long-Term Total Ozone

Variations during the Period 1957–1975." *Pure and Applied Geophysics* 117: 345–54.

Lovelock, J. E., and P. Simmonds. 1980. "Halocarbons in the Atmosphere." In *Chlorofluorocarbons in the Environment: The Aerosol Controversy*. Proceedings of the Fluorstrat 78 Symposium of the Society of Chemical Industry, Brighton, Oct. 5–6, 1978. Ed. T. M. Sugden and T. F. West. Chichester, UK: Ellis Horwood.

Lovelock, James. 1988. *The Ages of Gaia: A Biography of Our Living Earth*. The Commonwealth Fund Book Program. New York: Norton.

Lovelock, James E. 1977. "Methyl Chloroform in the Troposphere as an Indicator of OH Radical Abundance." *Nature* 267 (May 5): 32.

Lovelock, James E. 1982. "Epilogue." In *Stratospheric Ozone and Man*. Ed. Frank A. Bower and Richard B. Ward. Vol. 2. Boca Raton, FL: CRC.

Maduro, Roger, and Dale Schauerhammer. 1992. *The Holes in the Ozone Scare: The Scientific Evidence That the Sky Isn't Falling*. Washington DC: Twenty First Century Science Associates.

Mahlman, J. D. 1975. "Some Fundamental Limitations of Simplified Transport Models as Implied by Results from a Three-Dimensional General-Circulation Tracer Model." In *Proceedings of the Fourth Conference on CIAP*. Ed. T. M. Hard and A. J. Broderick. Washington, DC: U.S. Department of Transportation.

Mahlman, J. D., and S. B. Fels. 1986. "Antarctic Ozone Decreases: A Dynamical Cause?" *Geophysical Research Letters* 13: 1316–19.

Makhijani, Arjun, and Kevin Gurney. 1995. *Mending the Ozone Hole: Science, Technology, and Policy*. Cambridge, MA: MIT Press.

Manzer, L. E. 1990. "The CFC-Ozone Issue: Progress on the Development of Alternatives to CFCs." *Science* 249, no. 4964 (July 6): 31–35.

Martin, Lisa L. 1992. "Interests, Power, and Multilateralism." *International Organization* 46, no. 4 (Autumn): 765–92.

Mauzerall, Denise L. 1990. "Protecting the Ozone Layer: Phasing Out Halon by 2000." *Fire Journal* 85, no. 5 (Sept./Oct.): 22–31.

Maxwell, James H., and Sanford Weiner. 1993. "Green Consciousness, or Dollar Diplomacy? An Analysis of the British Response to the Threat of Stratospheric Ozone Depletion." Copy held is Apr. 1992 mimeo, Technology Business and Environment Program, MIT. *International Environmental Affairs* 5: 19–41.

May, Annabelle. 1979. "Concorde—Bird of Harmony or Political Albatross: An Examination in the Context of British Foreign Policy." *International Organization* 33, no. 4 (Autumn): 481–508.

McCarthy, R. L., and F. A. Bower. 1980. "The Fluorocarbon/Ozone Issue: An Industrial View." In *Chlorofluorocarbons in the Environment: The Aerosol Controversy*. Proceedings of the Fluorstrat 78 Symposium of the Society of Chemical Industry, Brighton, Oct. 5–6, 1978. Ed. T. M. Sugden and T. F. West. Chichester, UK: Ellis Horwood.

McCarthy, R. L., F. A. Bower, and J. P. Jesson. 1977. "The Fluorocarbon-Ozone Theory—I. Production and Release—World Production and Release of CCl_3F and CCl_2F_2 (Fluorocarbons 11 and 12) Through 1975." *Atmospheric Environment* 11: 491–97.

McConnell, Fiona. 1991. "Review of Benedick's Ozone Diplomacy." *International Environmental Affairs* 3, no. 4 (Autumn): 318–20.

McConnell, J. C., and Harold I. Schiff. 1978. "Methyl Chloroform: Impact on Stratospheric Ozone." *Science* 199 (Jan. 13): 174–77.

McCormick, M. P., et al. 1982. "Polar Stratospheric Cloud Sightings by SAM II, Sage and Lidar." *Journal of Atmospheric Sciences* 39: 1387–97.

McElroy, M. B., S. C. Wofsy, and Y. L. Young. 1977. "The Nitrogen Cycle: Perturbations Due to Man and Their Impact on Atmospheric N_2O and O_3." *Philosophical Transactions of the Royal Society of London* B277: 159–81.

McElroy, Michael B. 1982. "Chemistry and Modeling of the Stratosphere." In *Stratospheric Ozone and Man*. Ed. Frank A. Bower and Richard B. Ward. Vol. 1. Boca Raton, FL: CRC.

McElroy, Michael B., and J. McConnell. 1971. "Nitrous Oxide: A Natural Source of Stratospheric NO." *Journal of Atmospheric Science* 28 (Sept.): 1095–98.

McElroy, Michael B., and Ross J. Salawitch. 1989. "Changing Composition of the Global Stratosphere." *Science* 243, no. 4892 (Feb. 10): 763–70.

McElroy, Michael B., Ross J. Salawitch, Steven C. Wofsy, and Jennifer A. Logan. 1986. "Reductions of Antarctic Ozone Due to Synergistic Interactions of Chlorine and Bromine." *Nature* 321 (June 19): 759–62.

McGrath, W. D., and R. G. W. Norrish. 1960. "Studies of the Reactions of Excited Oxygen Atoms and Molecules Produced in the Flash Photolysis of Ozone." *Proceedings of the Royal Society of London* A254: 317–26.

McLinden, Mark O., and David A. Didion. 1987. "Quest for Alternatives." *ASHRAE Journal* 29, no. 12 (Dec.): 32–42.

Mertens, J. A. (Dow Chemical Co.). 1989. "Methyl Chloroform: A Valuable and Controllable Alternative." Paper presented at the International CFC and Halon Alternatives Conference. Washington, DC. Oct. 10–11, 1989.

Midgley, P. M. 1989. "The Production and Release to the Atmosphere of 1,1,1-Trichloroethane (Methyl Chloroform)." *Atmospheric Environment* 23, no. 12: 2663–65.

Miller, Alan S., and Irving M. Mintzer. 1986. *The Sky Is the Limit: Strategies for Protecting the Ozone Layer.* World Resources Institute Research Report 3. Washington, DC: World Resources Institute, Nov.

Miller, C., et al. 1981. "A Two-Dimensional Model of Stratospheric Chemistry and Transport." *Journal of Geophysical Research* 86, no. C12 (Dec. 20): 12039–65.

Miller, R. N. (Solvent Engineering, Inc.). 1989. "Remarks." Presented at International CFC and Halon Alternatives Conference. Washington, DC.

Milner, Helen V. 1997. *Interests, Institutions, and Information: Domestic Politics and International Relations.* Princeton, NJ: Princeton University Press.

Mitchell, Ronald B. 1994. *Intentional Oil Pollution at Sea: Environmental Policy and Treaty Compliance.* Cambridge, MA: MIT Press.

Mitchell, Ronald B., W. C. Clark, D. W. Cash, and F. Alcock, eds. Forthcoming, 2003. *Global Environmental Assessments: Information, Institutions, and Influence.* Cambridge, MA: MIT Press.

Molina, L. T., and M. J. Molina. 1978. "Ultraviolet Spectrum of HOCl." *Journal of Physical Chemistry* 82: 2410.

Molina, L. T., and M. J. Molina. 1981. "UV Absorption Cross Sections of HO_2NO_2 Vapor." *Journal of Photochemistry* 15: 97–108.

Molina, L. T., and M. J. Molina. 1987. "Production of Cl_2O_2 from the Self Reaction of the ClO Radical." *Journal of Physical Chemistry* 91: 433–36.

Molina, M. J., T. L. Tso, L. T. Molina, and F. C.-Y. Wang. 1987. "Antarctic Stratospheric Chemistry of Chlorine Nitrate, Hydrogen Chloride and Ice: Release of Active Chlorine." *Science* 238: 1253–57.

Montzka, S. A., et al. 1999. "Present and Future Trends in the Atmospheric Burden of Ozone-Depleting Halogens." *Nature* 398, no. 6729 (Apr. 22): 690–94.

Moore, Curtis A. 1990. "Industry Responses to the Montreal Protocol." *Ambio* 19, nos. 6–7 (Oct.): 320–23.

Mooz, W. E., et al. 1982. *Technical Options for Reducing Chlorofluorocarbon Emissions.* Prepared for U.S. EPA, R-2879-EPA. Santa Monica, CA: Rand Corp., Mar.

Mooz, William E., Kathleen A. Wolf, Frank Camm, and Rand Corporation. 1986. *Potential Constraints on Cumulative Global Production of Chlorofluorocarbons.* RAND R-3400-EPA. Santa Monica, CA: Rand Corp., May.

Moravcsik, Andrew. 1997. "Taking Preferences Seriously: A Liberal Theory of International Politics." *International Organization* 51, no. 4 (Autumn): 513–53.

Mormino, J., D. Sola, and C. Patten. 1975. *Climatic Impacts Assessment Program: Developmant and Accomplishments, 1971–1975.* DOT-TST-76-41, U.S. Department of Transportation, Climatic Impact Assessment Program. Springfield, VA: National Technical Information Service, Dec.

Morrisette, Peter M. 1989. "The Evolution of Policy Responses to Stratospheric Ozone Depletion." *Natural Resources Journal* 29, no. 3 (Summer): 793–820.

Mount, G. H., R. W. Sanders, A. L. Schmeltekopf, and S. Solomon. 1987. "Visible Spectroscopy at McMurdo Station, Antarctica, 1: Overview and Daily Variations of NO_2 and O_3, Austral Spring 1986." *Journal of Geophysical Research* 92, no. D7 (July 20): 8320–28.

Munn, R. E. 1979. "Stratospheric Ozone Depletion—An Environmental Impact Assessment." In *The Ozone Layer.* Proceedings of the UNEP Meeting of Experts, Washington DC, Mar. 1–9, 1977. Ed. Asit K. Biswas. Environmental Sciences and Applications, Vol. 4. Oxford: Pergamon.

Murcray, D. G., T. G. Kyle, F. H. Murcray, and W. J. Williams. 1968. "Nitric Acid and Nitric Oxide in the Lower Stratosphere." *Nature* 218, no. 5136: 78–79.

Murcray, D. G., et al. 1975. "Detection of Fluorocarbons in the Stratosphere." *Geophysical Research Letters* 2: 109–12.

NASA. 1977. *Chlorofluoromethanes and the Stratosphere.* NASA Reference Publication 1010. Ed. Robert D. Hudson. Washington, DC: NASA, Aug.

NASA. 1979. *The Stratosphere: Present and Future.* NASA Reference Publication 1049. Washington, DC: NASA.

NASA/WMO. 1982. *The Stratosphere 1981: Theory and Measurements.* WMO Global Ozone Research and Monitoring Project, rept. 11. Geneva: WMO.

NASA/WMO. 1986. *Atmospheric Ozone: 1985.* WMO Global Ozone Research and Montering Project, rept. 16. Geneva: WMO.

National Academy of Sciences, Committee on Atmospheric Sciences, Panel on Ozone. 1965. *Atmospheric Ozone Studies: An Outline for an International Observation Program.* Washington, DC: National Academy Press.

National Academy of Sciences, Climate Impacts Committee. 1975. *Environmental Impacts of Stratospheric Flight: Biological and Climatic Effects of Aircraft Emissions in the Stratosphere.* Washington, DC: National Academy Press

National Academy of Sciences, Committee on the Impacts of Stratospheric Change. 1976. *Environmental Effects of Chlorofluoromethane Release.* Washington, DC: National Academy Press.

National Academy of Sciences, Panel on Atmospheric Chemistry. 1976. *Halocarbons: Effects on Stratospheric Ozone.* Washington, DC: National Academy Press.

National Academy of Sciences, Committee on the Impacts of Stratospheric Change. 1977. *Response to the Ozone Protection Sections of the Clean Air Act Amendments of 1977: An Interim Report.* Washington, DC: National Academy Press.

National Academy of Sciences, Committee on Impacts of Stratospheric Change and Committee on Alternatives for the Reduction of Chlorofluorocarbon Emissions. 1979. *Protection Against Depletion of Stratospheric Ozone by Chlorofluorocarbons.* Washington, DC: National Academy Press.

National Academy of Sciences, Panel on Atmospheric Chemistry. 1979. *Stratospheric Ozone Depletion by Halocarbons: Chemistry and Transport.* Washington, DC: National Academy Press.

National Academy of Sciences. 1982. *Causes and Effects of Stratospheric Ozone Reduction: An Update.* Report prepared by the Committee on Chemistry and Physics of Ozone Depletion and the Committee on Biological Effects of Increased Solar Ultraviolet Radiation. Washington, DC: National Academy Press.

National Academy of Sciences, Committee on Causes and Effects of Changes in Stratospheric Ozone. 1984. *Causes and Effects of Changes in Stratospheric Ozone: Update 1983.* Washington, DC: National Academy Press.

Nicolet, Marcel. 1970. "Ozone and Hydrogen Reactions." *Annals of Geophysics* 26: 531–46.

Nicolet, Marcel, and E. Vergison. 1971. "L'Oxyde Azoteux dans la Stratosphere." *Acta Aeronautica* 90: 1–16.

Nordhaus, William D., and Gary W. Yohe. 1987. "Forecasting the Growth of Chlorofluorocarbon Emissions: An Econometric Approach." Paper prepared for UNEP workshop on the control of ChloroFluorocarbons, Rome. May 26–30, 1986.

Noxon, J. F. 1979. "Stratospheric NO_2, 2. Global Behavior." *Journal of Geophysical Research—Oceans and Atmospheres* 84, no. NC8: 5067–76.

Olson, Mancur. 1965. *The Logic of Collective Action: Public Goods and the Theory of Groups.* Cambridge, MA: Harvard University Press.

Organization for Economic Cooperation and Development (OECD). 1983. "Economic Aspects of CFC Emissions Scenarios." OECD Chemicals Group draft report, Paris, Sept 15, 1983.

Owen, Kenneth. 1997. *Concorde and the Americans: International Politics of the Supersonic Transport.* Washington, DC: Smithsonian Institution Press.

Oye, Kenneth, ed. 1986. *Cooperation under Anarchy.* Princeton, NJ: Princeton University Press.

Oye, Kenneth A., and James H. Maxwell. 1994. "Self-Interest and Environmental Management." *Journal of Theoretical Politics* 6, no. 4: 593–624.

Ozone Trends Panel. 1988. *Report of the International Ozone Trends Panel 1988.* Global Ozone Research and Monitoring Project, rept. 18. Geneva: WMO.

Pack, D. H., J. E. Lovelock, G. Cotton, and C. Curthoys. 1977. "Halocarbon Behavior from a Long Time Series." *Atmospheric Environment* 11: 329.

Palmer, Adele R., et al. 1980. *Economic Implications of Regulating Chlorofluorocarbon Emissions from Nonaerosol Applications.* Prepared for U.S. EPA, R-2524-EPA. Santa Monica, CA: Rand Corp., June.

Palmer, Karen, Wallace E. Oates, and Paul R. Portney. 1995. "Tightening Environmental Standards: The Benefit-Cost or the No-Cost Paradigm?" *Journal of Economic Perspectives* 9, no. 4 (Fall): 119–32.

Panel on Ozone, Committee on Atmospheric Sciences, National Academy of Sciences. 1965. *Atmospheric Ozone Studies: An Outline for an International Observation Program.* Publication 1348. Washington, DC: National Academy Press.

Panel on the Natural Stratosphere, CIAP. 1975. *The Natural Stratosphere of 1974.* CIAP Monograph 1, DOT-TST-75-51. Washington, DC: Institute for Defense Analysis, Sept.

Panel on the Perturbed Stratosphere, CIAP. 1975. *The Stratosphere Perturbed by Propulsion Effluents.* CIAP Monograph 3, DOT-TST-75-53. Washington, DC: Institute for Defense Analysis, Sept.

Parson, Edward A. 1993. "Protecting the Ozone Layer: The Evolution and Impact of International Institutions." In *Institutions for the Earth: Sources of Effective International Environmental Protection*, Ed. Peter M. Haas, Robert O. Keohane, and Marc A. Levy. Cambridge, MA: MIT Press.

Parson, Edward A., and William C. Clark. 1994. "Sustainable Development as Social Learning: Theoretical Perspectives and Practical Challenges for the Design of a Research Program." In *Bridges and Barriers to the Renewal of Ecosystems and Institutions.* Ed. Lance Gunderson, C. S. Holling, and Stephen S. Light. New York. Columbia University Press.

Parson, Edward A., and Owen Greene. 1995. "The Complex Chemistry of the International Ozone Agreements." *Environment* 37, no. 2 (Mar.): 16–43.

Parson, Edward A., and Richard Zeckhauser. 1995. "The Unbalanced Commons: Climate Change and Other International Environmental Problems of Collective Action." In *Barriers to the Negotiated Resolution of Conflict.* Ed. K. Arrow et al. New York: Norton.

Parson, Edward A., et al. 2001. "Leading While Keeping in Step: Management of Global Atmospheric Issues in Canada." In *Learning to Manage Global Environmental Risks.* Ed. William C. Clark, Jill Jäger, Josee van Eijndhoven, and Nancy M. Dickson. Vol. 1. Cambridge, Massachusetts: The MIT Press.

Penner, J. E. 1981. "Trend Prediction for O_3: An Analysis of Model Uncertainty with Comparison to Detection Thresholds." *Atmospheric Environment* 16: 1109–15.

Pittock, A. Barrie. 1972. "Evaluating the Risk to Society from the SST: Some Thoughts Occasioned by the AAS Report." *Search* 3, no. 8 (Aug.): 285–89.

Porter, Michael E., and Claas van der Linde. 1995. "Toward a New Conception of the Environment-Competitiveness Relationship." *Journal of Economic Perspectives* 9, no. 4 (Fall): 97–118.

Prather, M. J., and R. T. Watson. 1990. "Stratospheric Ozone Depletion and Future Levels of Atmospheric Chlorine and Bromine." *Nature* 344, no. 6268 (Apr. 19): 729–34.

Prather, Michael J., Michael B. McElroy, and Steven C. Wofsy. 1984. "Reductions in Ozone at High Concentrations of Stratospheric Halogens." *Nature* 312, no. 5991 (Nov. 15): 227–31.

Price, Don. 1965. *The Scientific Estate.* New York: Oxford University Press.

Proffitt, M. H., D. W. Fahey, K. K. Kelly, and A. F. Tuck. 1989. "High-Latitude Ozone Loss Outside the Antarctic Ozone Hole." *Nature* 342 (Nov. 16): 233–37.

Proffitt, M. H., et al. 1989. "In Situ Ozone Measurements within the 1987 Antarctic Ozone Hole from a High-Altitude ER-2 Aircraft." *Journal of Geophysical Research* 94: 16547–55.

Proffitt, M. H., et al. 1990. "Ozone Loss in the Arctic Polar Vortex Inferred from High-Altitude Aircraft Measurements." *Nature* 347, no. 6288 (Sept. 6): 31–36.

Putnam, Robert D. 1988. "Diplomacy and Domestic Politics: The Logic of Two-Level Games." *International Organization* 42, no. 3 (Summer): 427–60.

Pyle, J. A., and J. T. Houghton. 1980. "Modelling Stratospheric Motions and Their Influence on Ozone." In *Chlorofluorocarbons in the Environment: The Aerosol Controversy.* Proceedings of the Fluorstrat 78 Symposium of the Society of Chemical Industry, Brighton, Oct. 5–6, 1978. Ed. T. M. Sugden and T. F. West. Chichester, UK: Ellis Horwood.

Quinn, Timothy H., et al. 1986. *Projected Use, Emissions, and Banks of Potential Ozone-Depleting Substances.* RAND Note no. N-2282-EPA. Santa Monica, CA: Rand Corp. Jan.

Rabin, Robert L. 1981. "Ozone Depletion Revisited: EPA Regulation of Chlorofluorocarbons." *Regulation: AEI Journal on Government and Society* 5 (Mar./Apr.): 32–38.

Raiffa, Howard. 1982. *The Art and Science of Negotiation.* Cambridge, MA: Harvard University Press.

Ramanathan, V. 1975. "Greenhouse Effect Due to Chlorofluorocarbons: Climatic Implications." *Science* 190 (Oct. 3): 50–52.

Ramanathan, V., L. B. Callis, and R. E. Boughner. 1976. "Sensitivity of Surface Temperature and Atmospheric Temperature to Perturbations of Stratospheric Concentration of Ozone and Nitrogen Dioxide." *Journal of the Atmospheric Sciences* 33, no. 6: 1092–112.

Rasmussen, R. A., L. E. Rasmussen, A. K. Khalil, and R. W. Dalluge. 1980. "Concentration Distribution of Methyl Chloride in the Atmosphere." *Journal of Geophysical Research* 85: 7350–56.

Ravishankara, A. R., et al. 1977. "A Study of the Chemical Degradation of $ClONO_2$ in the Stratosphere." *Geophysical Research Letters* 4: 7–9.

Ray, Dixy Lee, and Louis R. Guzzo. 1990. *Trashing the Planet: How Science Can Help Us Deal with Acid Rain, Depletion of the Ozone, & Nuclear Waste (Among Other Things).* Washington, DC: Regnery Gateway.

Reid, G. C., I. S. A. Isaksen, T. E. Holzer, and P. J. Crutzen. 1976. "Influence of Ancient Solar-Proton Flares on Evolution of Life." *Nature* 259, no. 5540: 177–79.

Reinhardt, Forest. 1989. *Dupont Freon Products Division.* Case number 8-389–112. Boston, MA: Harvard Business School.

Reinsel, G., et al. 1981. "Statistical Analysis of Stratospheric Ozone Data for Detection of Trends." *Atmospheric Environment* 15: 1569–77.

Reinsel, G. C., et al. 1984. "Analysis of Upper Stratospheric Umkehr Ozone Profile Data for Trends and the Effects of Stratospheric Aerosols." *Journal of Geophysical Research* 89, no. D3 (June 20): 4833–40.

Reinsel, G. C., et al. 1989. "Trend Analysis of Aerosol-Corrected Umkehr Ozone Profile Data through 1987." *Journal of Geophysical Research—Atmospheres* 94, no. D3 (Nov. 20): 16373–86.

Revelle, Roger, and Hans Suess. 1957. "Carbon Dioxide Exchange between Atmosphere and Ocean and the Question of an Increase in Atmospheric CO_2 during the Past Decade." *Tellus* 9: 18–27.

Rittberger, Volker, and Peter Mayer, eds. 1993. *Regime Theory and International Relations.* Oxford: Clarendon.

Roan, Sharon L. 1989. *Ozone Crisis: The 15 Year Evolution of a Sudden Global Emergency.* New York: John Wiley.

Roberts, Leslie. 1989. "Does the Ozone Hole Threaten Antarctic Life?" *Science* 244 (April 21): 288–89.

Rodriguez, J. M., M. K. W. Ko, and N. D. Sze. 1991. "Role of Heterogeneous Conversion of N_2O_5 on Sulfate Aerosols in Global Ozone Losses." *Nature* 352, no. 6331 (June 11): 134–37.

Roscoe, H. K. 1982. "Tentative Observation of Stratospheric N_2O_5." *Geophysical Research Letters* 9: 901–2.

Rose, Elizabeth H., and Arthur D. Fitzgerald. 1992. "Free in Three: How Northern Telecom Eliminated CFC-113 Solvents from Its Global Operations." *Pollution Prevention Review* 2, no. 2 (Summer): 297–301.

Rowland, F. S. 1988. "Some Aspects of Chemistry in the Springtime Antarctic Stratosphere." In *The Changing Atmosphere.* Report of the Dahlem Workshop on the Changing Atmosphere, Berlin, Nov. 1–6 1987. Ed. F. S. Rowland and I. S. A. Isaksen. Chichester, UK: Wiley-Interscience.

Rowland, F. S. 1991. "Stratospheric Ozone Depletion." *Annual Review of Physical Chemistry* 42: 731–68.

Rowland, F. S., N. R. P. Harris, R. D. Bojkov, and P. Bloomfield. 1988. "Statistical Error Analysis of Ozone Trends: Winter Depletion in the Northern Hemisphere." In *IOC/IAMAP Quadrennial Ozone Symposium and Tropospheric Workshop.* Gottingen, Aug.

Rowland, F. S., and M. J. Molina. 1975. "Chlorofluoromethanes in the Environment." *Reviews of Geophysics and Space Physics* 13 (February): 1–35.

Rowland, F. S., and M. J. Molina. 1994. "Ozone Depletion: 20 Years after the Alarm." *Chemical and Engineering News* 72 (Aug. 15): 8–13.

Rowland, F. S., H. Sato, H. Khwaja, and S. M. Elliott. 1986. "The Hydrolysis of Chlorine Nitrate and Its Possible Atmospheric Significance." *Journal of Physical Chemistry* 90, no. 10 (May 8): 1985–88.

Rowland, F. S., J. E. Spencer, and M. J. Molina. 1976. "Stratospheric Formation and Photolysis of Chlorine Nitrate, $ClONO_2$." *Journal of Physical Chemistry* 80: 2711–13.

Rowland, F. S., S. C. Tyler, D. C. Montague, and Y. Makide. 1982. "Dichlorodifluoromethane, CCl_2F_2, in the Earth's Atmosphere." *Geophysical Research Letters* 9, no. 4 (Apr): 481–84.

Rowlands, Ian H. 1992. "The International Politics of Environment and Development: The Post-UNCED Agenda." *Millennium* 21, no. 2 (Summer): 209–24.

Rowlands, Ian H. 1995. *The Politics of Global Atmospheric Change.* Manchester, UK: Manchester University Press.

Royal Commission on Environmental Pollution (UK). 1974. *Pollution Control: Progress and Problems.* London: Her Majesty's Printing Office.

Ruckelshaus, William D. 1985. "Risk, Science, and Democracy." *Issues in Science and Technology,* 1, no. 3 (Spring): 19–38.

Ruderman, M. A. 1974. "Possible Consequences of Nearby Supernova Explosions for Atmospheric Ozone and Terrestial Life." *Science* 184, no. 4141 (June 7): 1079–81.

Ruderman, M. A., and J. W. Chamberlain. 1973. "Origin of the Sunspot Modulation of Ozone: Its Implications for Stratospheric NO Injection." Doc. no. JSS-73-18-3. Arlington, VA: Institute for Defense Analysis.

Rundel, R. D., and D. S. Nachtwey. 1983. "Projections of Increased Nonmelanoma Skin Cancer Incidence Due to Ozone Depletion." *Photochemistry and Photobiology* 38, no. 5: 577–91.

Sabatier, Paul A., and Hank C. Jenkins-Smith. 1993. *Policy Change and Learning: An Advocacy Coalition Approach*. Boulder, CO: Westview.

Sabatier, Paul A., and Hank C. Jenkins-Smith. 1999. "The Advocacy Coalition Framework: An Assessment." In *Theories of the Policy Process*. Ed. Paul A. Sabatier. Boulder, CO: Westview.

Sand, Peter H. 1985. "Environmental Law in the United Nations Environment Programme." In *The Future of the International Law of the Environment*. Ed. R.-J. Dupuy. Dordrecht: Martinus Nijhoff.

Schelling, Thomas C. 1960. *The Strategy of Conflict*. Cambridge, MA: Harvard University Press.

Schiff, Harold. 1982. "Report on Reports: Stratospheric Ozone Depletion." *Environment* 24, no. 7 (Sept.): 25–28.

Schmalensee, Richard. 1996. "Greenhouse Policy Architecture and Institutions." Paper prepared for National Bureau of Economic Research conference, "Economics and Policy Issues in Global Warming: An Assessment of the Intergovernmental Panel Report." Snowmass, CO: July 23–24.

Schmeltekopf, A., P. D. Goldan, W. R. Henderson, et al. 1975. "Measurements of Stratospheric $CFCl_3$, CF_2Cl_2, and N_2O." *Geophysical Research Letters* 2, no. 9: 393–96.

Schmidt, Manfred. 1988. "Pioneers of Ozone Research: A Historical Survey." Presented to Quadrennial Ozone Seminar, Gottingen, Aug.

Schneider, Stephen H. 1989. *Global Warming: Are We Entering the Greenhouse Century?* San Francisco: Sierra Club.

Schneider, Stephen H., and Lynne E. Mesirow. 1976. *The Genesis Strategy: Climate and Global Survival*. New York: Plenum.

Schon, Donald A., and Martin Rein. 1994. *Frame Reflection: Toward the Resolution of Intractable Policy Controversies*. New York: Basic Books.

Schreurs, Miranda A. 2001. "Shifting Priorities and the Internationalization of Environmental Risk Management in Japan." In *Learning to Manage Global Environmental Risks*. Ed. William C. Clark, Jill Jäger, Josee van Eijndhoven, and Nancy M. Dickson. Vol. 1. Cambridge, MA: MIT Press.

Scotto, J., et al. 1988. "Biologically Effective Ultraviolet Radiation: Surface Measurements in the United States, 1974 to 1985." *Science* 239: 762–64.

Sebenius, James K. 1991. "Negotiating a Regime to Control Global Warming." In *Greenhouse Warming: Negotiating a Global Regime*. Ed. Jessica T. Mathews. Washington, DC: World Resources Institute.

Sebenius, James K. 1992. "Challenging Conventional Explanations of International Cooperation: Negotiation Analysis and the Case of Epistemic Communities." *International Organization* 46, no. 1 (Winter): 323–66.

Sebenius, James K. 1993. "Towards a Winning Climate Coalition." Mimeo, Harvard Business School, Jan.

Seidel, Stephen R. 1996. "Keeping Cars Cool." In *Ozone Protection in the United States: Elements of Success*. Ed. Elizabeth Cook. Washington, DC: World Resources Institute.

Seidel, Stephen, and Dale Keyes. 1983. *Can we Delay a Greenhouse Warming?* Washington, DC: U.S. Environmental Protection Agency, Sept.

Shapiro, Michael, and Ellen Warhit. 1983. "Marketable Permits: The Case of Chlorofluorocarbons." *Natural Resources Journal* 23, no. 3 (July): 577–91.

Shimberg, Steven J. 1991. "Stratospheric Ozone and Climate Protection: Domestic Legislation and the International Process." *Environmental Law*, (Northwestern School of Law) 21 (Summer): 2175–82.

Singh, H. B., L. J. Salas, H. Shigeishi, and E. Scribner. 1979. "Atmospheric Halocarbons, Hydrocarbons, and Sulfur Hexafluoride: Global Distributions, Sources and Sinks." *Science* 203, no. 4383 (Mar. 2): 899–903.

Sjöstedt, Gunnar, ed. 1993. *International Environmental Negotiation*. International Institute for Applied Systems Analysis. Newbury Park, CA: Sage.

Skolnick, A. A. 1991. "Is Ozone Loss to Blame for Melanoma Upsurge?" *Journal of the American Medical Association* 265, no. 24 (June 26): 3216–18.

Skolnikoff, Eugene B. 1993. *The Elusive Transformation: Science, Technology, and the Evolution of International Politics.* Princeton, NJ: Princeton University Press.

Smith, R. C., et al. 1992. "Ozone Depletion: Ultraviolet Radiation and Phytoplankton Biology in Antarctic Waters." *Science* 255, no. 5047: 951–59.

Smithart, E. (The Trane Co.). 1993. "Choosing a Building Chiller." Paper presented at the International CFC and Halon Alternatives Conference. Washington, DC. Oct. 20–22, 1993.

Snidal, Duncan. 1985. "The Game Theory of International Politics." *World Politics* 38, no. 1 (Oct.). 1–24.

Social Learning Group. 2001. *Learning to Manage Global Environmental Risks.* Ed. William C. Clark, Jill Jäger, Josee van Eijndhoven, and Nancy M. Dickson. Cambridge, MA: MIT Press.

Sokolov, Vassily, et al. 2001. "Turning Points: The Management of Global Environmental Risks in the Former Soviet Union." In *Learning to Manage Global Environmental Risks.* Ed. William C. Clark, Jill Jäger, Josee van Eijndhoven, and Nancy M. Dickson. Vol. 1 Cambridge, MA: MIT Press.

Solomon, S., and R. R. Garcia. 1983. "On the Distribution of NO_2 in the High Latitude Stratosphere." *Journal of Geophysical Research* 88: 5229–39.

Solomon, S., G. H. Mount, R. W. Sanders, and A. L. Schmeltekopf. 1987. "Visible Spectroscopy at McMurdo Station, Antarctica, 2: Observation of OClO." *Journal of Geophysical Research* 92, no. D7 (July 20): 8329–38.

Solomon, Susan, Rolando R. Garcia, F. Sherwood Rowland, and Donald J. Wuebbles. 1986. "On the Depletion of Antarctic Ozone." *Nature* 321 (June 19): 755–58.

Soroos, Marvin S. 1986. *Beyond Sovereignty: The Challenge of Global Policy.* Columbia: University of South Carolina Press.

Soulen, J. R. 1979. "A Review of the Industry-Sponsored Research into the Effect of Chlorofluorocarbons on the Concentration of Atmospheric Ozone." In *The Ozone Layer.* Proceedings of the UNEP Meeting of Experts, Washington, DC, Mar. 1–9, 1977. Ed. Asit K. Biswas. Environmental Sciences and Applications, Vol. 4. Oxford: Pergamon.

SRI International. 1995. *Chemical Economics Handbook.* Menlo Park, CA: SRI International.

Sridihan, U. C., B. Reimann, and F. Kaufman. 1980. "Kinetics of the Reaction OH + H_2O_2 → HO_2 + H_2O." *Journal of Chemical Physics* 73: 1286.

St. John, D. S., et al. 1981. "Time Series Search for Trend in Total Ozone Measurements." *Journal of Geophysical Research* 86, no. C8 (Aug. 20): 7299–311.

Staehelin, J., N. R. P. Harris, C. Appenzeller, and J. Eberhard. 2001. "Ozone Trends: A Review." *Reviews of Geophysics* 39, no. 2 (May): 231–90.

Stein, Janice Gross, ed. 1989. *Getting to the Table: The Processes of International Prenegotiation.* Perspectives on Security. Baltimore: Johns Hopkins University Press.

Stoel, Thomas B., Alan S. Miller, and Breck Milroy. 1980. *Fluorocarbon Regulation.* Lexington, MA: Lexington Books (D. C. Heath).

Stoel, Thomas B., Jr. 1983. "Fluorocarbons: Mobilizing Concern and Action." In *Environmental Protection: The International Dimension.* Ed. David A. Kay and Harold K. Jacobson. Totowa, NJ: Allenheld, Osmun.

Stolarski, Richard S. 1988. "The Antarctic Ozone Hole." *Scientific American* 258, no. 1 (Jan.): 30–36.

Stolarski, Richard S., and R. J. Cicerone. 1974. "Stratospheric Chlorine: A Possible Sink for Ozone." *Canadian Journal of Chemistry* 52: 1610–15.

Stolarski, Richard S., et al. 1986. "Nimbus 7 Satellite Measurements of the Springtime Antarctic Ozone Decrease." *Nature* 322 (August 28): 808–11.

Stone, Christopher D. 1993. *The Gnat Is Older Than Man: Global Environment and Human Agenda.* Princeton, NJ: Princeton University Press.

Strange, Susan. 1983. "Cave! Hic Dragones: A Critique of Regime Analysis." In *International Regimes.* Ed. Stephen D. Krasner. Ithaca, NY: Cornell University Press.

Study of Critical Environmental Problems (SCEP). 1970. *Man's Impact on the Global Environment: Assessment and Recommendations for Action.* Cambridge MA: MIT Press.

Sugden, T. M., and T. F. West, eds. 1980. *Chlorofluorocarbons in the Environment: The Aerosol Controversy.* Proceedings of the Fluorstrat 78 Symposium of the Society of Chemical Industry, Brighton, Oct. 5–6, 1978. Chichester, UK: Ellis Horwood.

Susskind, Lawrence. 1994. *Environmental Diplomacy: Negotiating More Effective International Environmental Agreements.* New York: Oxford University Press.

Sze, Nien Dak. 1982. "Odd Hydrogen Processes." In *Stratospheric Ozone and Man.* Ed. Frank A. Bower and Richard B. Ward. Vol. 1. Boca Raton, FL: CRC.

Thacher, Peter S. 1993. "The Early Signs of the Depleting Ozone Layer." Fourth annual Walter Orr Roberts memorial lecture. Aspen, CO: Aspen Global Change Institute, Aug. 5.

Tolba, Mostafa. 1989. "A Step-by-Step Approach to Protection of the Atmosphere." *International Environmental Affairs* 1, no. 4 (Fall): 304–08.

Tolbert, M. A., D. M. Golden, C. M. Reighs, and A. M. Middlebrook. 1990. "Heterogeneous Chemistry on Ice and Acidic Surface—Implications for Stratospheric Ozone Depletion." *Abstracts of Papers of the American Chemical Society* 200 (August 26): 73.

Tolbert, M. A., M. J. Rossi, R. Malhotra, and D. M. Golden. 1987. "Reaction of Chlorine Nitrate with Hydrogen Chloride and Water at Antarctic Stratospheric Temperatures." *Science* 238: 1258–60.

Toman, Michael. 2001. "Moving Forward with Climate Policy." Speech. Rice University, Houston, TX, May 10.

Toon, O. B., P. Hamill, R. P. Turco, and J. Pinto. 1986. "Condensation of HNO_3 and HCl in the Winter Polar Stratospheres." *Geophysical Research Letters* 13: 1284–87.

Tromp, T. K., M. K. W. Ko, J. M. Rodriguez, and N. D. Sze. 1995. "Potential Accumulation of a CFC Replacement Degradation Product in Seasonal Wetlands." *Nature* 376 (July 27).

True, James L., Bryan D. Jones, and Frank R. Baumgartner. 1999. "Punctuated-Equilibrium Theory: Explaining Stability and Change in American Policymaking." In *Theories of the Policy Process.* Ed. Paul A. Sabatier. Boulder, CO: Westview.

Tung, K. K., M. W. K. Ko, J. M. Rodriguez, and N. D. Sze. 1986. "Are Antarctic Ozone Variations a Manifestation of Dynamics or Chemistry?" *Nature* 333: 811–14.

Turco, R. P., R. C. Whitten, I. G. Poppoff, and L. A. Capone. 1978. "SSTs, Nitrogen Fertilizer, and Stratospheric Ozone." *Nature* 276: 805–7.

U.K. Department of the Environment. 1987. *Report of the Stratospheric Ozone Review Group.* London: U.K. Department of the Environment.

U.K. Department of the Environment, Central Unit on Environmental Pollution. 1976. *Chlorofluorocarbons and Their Effect on Stratospheric Ozone.* Pollution Paper no. 5. London: Her Majesty's Stationery Office.

U.K. Department of the Environment, Stratospheric Research Advisory Committee. 1979. *Chlorofluorocarbons and Their Effect on Stratospheric Ozone (Second Report).* Pollution Paper no. 15. London: Her Majesty's Stationery Office.

U.N. Environment Programme (UNEP). 1979a. *Ozone Layer Bulletin* 2, no 1 (Jan.).

UNEP. 1979b. "World Plan of Action." In *The Ozone Layer.* Proceedings of the UNEP Meeting of Experts, Washington DC, Mar. 1–9, 1977. Ed. Asit K. Biswas. Environmental Sciences and Applications, Vol. 4. Oxford: Pergamon.

UNEP. 1986. "Report of the Second Part of the Workshop on the Control of Chlorofluorocarbons." UNEP/WG.148/3. Leesburg, VA.

UNEP. 1989a. *Economic Panel Report.* Report of the Economic Panel of the Montreal Protocol. Nairobi: UNEP, July.

UNEP. 1989b. *Electronics, Degreasing, and Dry Cleaning Solvents Technical Options Report.* Report of the Solvents Technical Options Committee under the 1989 Technology Assessment of the Montreal Protocol. Nairobi: UNEP, June 30.

UNEP. 1989c. *Flexible and Rigid Foams Technical Options Report.* Report of the Foams Technical Options Committee under the 1989 Technology Assessment of the Montreal Protocol. Nairobi: UNEP, June 30.

UNEP. 1989d. *Halons Technical Options Report.* Report of the Halons Technical Options Committee under the 1989 Technology Assessment of the Montreal Protocol. Nairobi: UNEP, Aug. 11.

UNEP. 1989e. *Scientific Assessment of Stratospheric Ozone: 1989.* Report of the Scientific Assessment Panel of the Montreal Protocol. WMO Global Research and Monitoring Project, rept. 20. Geneva: World Meteorological Organization (WMO), Jan.

UNEP. 1989f. *Technical Progress on Protecting the Ozone Layer.* Report of the Technology Review Panel of the Montreal Protocol. Nairobi: UNEP, June 30.

UNEP. 1991a. *Environmental Effects Panel Report.* Report of the Environmental Effects Panel of the Montreal Protocol. Nairobi: UNEP, Nov.

UNEP. 1991b. *Montreal Protocol: 1991 Assessment.* Report of the Technology and Economics Assessment Panel of the Montreal Protocol. Nairobi: UNEP, Dec.

UNEP. 1991c. *Scientific Assessment of Stratospheric Ozone.* Report of the Scientific Assessment Panel of the Montreal Protocol. WMO Global Research and Monitoring Project. Geneva: WMO, Nov.

UNEP. 1994a. *1994 Report of the Methyl Bromide Technical Options Committee, 1995 Assessment.* Nairobi: UNEP.

UNEP. 1994b. *1994 Report of the Technology and Economics Assessment Panel.* Nairobi: UNEP, Mar.

UNEP. 1994c. *1994 Report of the Technology and Economics Assessment Panel, 1995 Assessment.* Nairobi: UNEP, Nov.

UNEP. 1994d. *Environmental Effects Panel Report.* Report of the Environmental Effects Panel of the Montreal Protocol. Nairobi: UNEP, Nov.

UNEP. 1995a. *Scientific Assessment of Ozone Depletion: 1994.* Report of the Scientific Assessment Panel of the Montreal Protocol. WMO Global Research and Monitoring Project, rept. 37. Geneva: WMO, Jan.

UNEP. 1995b. *Supplement to the 1994 Assessments: Report of the Technology and Economics Assessment Panel.* Nairobi: UNEP, Mar.

UNEP. 1995c. *Technology and Economics Assessment Panel, Report to the Parties.* Nairobi: UNEP, Nov.

UNEP. 1996. *1996 Report of the Technology and Economics Assessment Panel.* Nairobi: UNEP, June.

UNEP. 1998. *Environmental Effects of Ozone Depletion: 1998 Assessment.* Report of the Environmental Effects Panel of the Montreal Protocol. Nairobi: UNEP, Nov.

UNEP. 1999a. *Scientific Assessment of Ozone Depletion: 1998.* Report of the Scientific Assessment Panel of the Montreal Protocol. WMO Global Research and Monitoring Project, rept. 44. Geneva: WMO, Jan.

UNEP. 1999b. *Synthesis of the Reports of the Scientific, Environmental Effects, and Technology and Economics Assessment Panels of the Montreal Protocol.* Nairobi: UNEP, Feb.

UNEP. 2002. Production and Consumption of Ozone Depleting Substances under the Montreal Protocol, 1986–2000. Ozone Secretariat, Nairobi (April). Available at http://www.unep.ch/ozone/publications.shtml

UNEP, CCOL. 1977. "Report of the First Session." Nov. 1977. Geneva.

UNEP, CCOL. 1978. "Report of the Second Session." Nov. 1978. Bonn.

UNEP, CCOL. 1979. "Report of the Third Session." Nov. 20, 1979. Paris.

UNEP, CCOL. 1980. "Report of the Fourth Session." Nov. 14, 1980. Bilthoven.

UNEP, CCOL. 1981. "An Environmental Assessment of Ozone Layer Depletion and Its Impact." Oct. 16, 1981, Stockholm.

UNEP, CCOL. 1981. "Report of the Fifth Session." Oct. 16, 1981. Copenhagen.

UNEP, CCOL. 1983. "Report of the Sixth Session." Aug. 15, 1983. Geneva.

UNEP, CCOL. 1986. "UNEP Policy Support Document," Dec. 1, 1986. Geneva.

Urbach, F. 1989. "The Biological Effects of Increased Ultraviolet Radiation: An Update." *Photochemistry and Photobiology* 50, no. 4: 439–41.

U.S. Department of State. 1995. "Fluorocarbons: An Assessment of Worldwide Production,

Use and Environmental Issues." Paper prepared for the Nineteenth Meeting of the Environment Committed, OECD. Washington DC.

U.S. EPA. 1988. *Regulatory Impact Analysis: Protection of Stratospheric Ozone*. Washington, DC: EPA.

Van der Leun, Jan C., and Farrington Daniels, Jr. 1975. "Biological Effects of Stratospheric Ozone Decrease: A Critical Review of Assessments." In *Impacts of Climatic Change on the Biosphere*. CIAP Monograph 5, Part 1. *Ultraviolet Radiation Effects*. Ed. D. S. Nachtwey et al. Washington, DC: U.S. Department of Transportation.

Van Eijndhoven, Josee, et al. 2001. "Finding Your Place: A History of the Management of Global Environmental Risks in the Netherlands." In *Learning to Manage Global Environmental Risks*. Ed. William C. Clark, Jill Jäger, Josee van Eijndhoven, and Nancy M. Dickson. Vol. 1. Cambridge, MA: MIT Press.

Van Slooten, Robert. 1998. "TEAP Terms of Reference." In *Protecting the Ozone Layer: Lessons, Models, and Prospects*. Ed. Philippe G. Le Prestre, John D. Reid, and E. Thomas Morehouse, Jr. Boston: Kluwer.

Victor, David G. 1997. "The Operation and Effectiveness of the Montreal Protocol's Non-Compliance Procedure." In *The Implementation and Effectiveness of International Environmental Commitments*. Ed. David G. Victor, Kal Raustiala, and Eugene B. Skolnikoff. Cambridge, MA: MIT Press.

Victor, David G., 2001. *The Collapse of the Kyoto Protocol and the Struggle to Slow Global Warming*. Princeton, NJ: Princeton University Press.

Victor, David G., Kal Raustiala, and Eugene B. Skolnikoff, eds. 1997. *The Implementation and Effectiveness of International Environmental Commitments*. Cambridge, MA: MIT Press.

Von Schweinichen, Joachim. 1980. "A Continental European Industry Viewpoint." In *Chlorofluorocarbons in the Environment: The Aerosol Controversy*. Proceedings of the Fluorstrat 78 Symposium of the Society of Chemical Industry, Brighton, Oct. 5–6, 1978. Ed. T. M. Sugden and T. F. West. Chichester, UK: Ellis Horwood.

Von Schweinichen, Joachim. 1989. "Remarks." International CFC and Halon Alternatives Conference. Washington, DC.

Vupputuri, R. K. R. 1976. "The Steady-State Structure of the Natural Stratosphere and Ozone Distribution in a 2D Model Incorporating Radiation and O-H-N Photochemistry and the Effects of Stratospheric Pollutants." *Atmosphere* 14, no. 3: 214–36.

Waibel, A. E., et al. 1999. "Arctic Ozone Loss Due to Denitrification." *Science* 283, no. 5410 (Mar. 26): 2064–69.

Walters, Carl. 1986. *Adaptive Management of Renewable Resources*. New York: Macmillan.

Waltz, Kenneth. 1979. *Theory of International Politics*. Reading, MA: Addison-Wesley.

Wardle, D. I., J. B. Kerr, C. T. McElroy, and D. R. Francis. 1997. *Ozone Science: A Canadian Perspective on the Changing Ozone Layer*. Ottawa: Environment Canada.

Watson, R. T. 1977. "Rate Constants for Reactions of ClOx of Atmospheric Interest." *Journal of Physical and Chemical Reference Data* 6, no. 3: 871–917.

Watson, R. T., M. A. Geller, R. S. Stolarski, and R. F. Hampson. 1986. *Present State of Knowledge of the Upper Atmosphere: An Assessment Report*. NASA Reference Publication 1162. Washington, DC: NASA.

Watson, R. T., G. Machado, S. Fischer, and D. D. David. 1976. "A Temperature Dependence Kinetics Study of the Reaction of Cl with O_3, CH_4, and H_2O_2." *Journal of Chemical Physics* 65: 2126–38.

Weiss, Carol. 1975. "Evaluation Research in the Political Context." In *Handbook of Evaluation Research*. Ed. E. L. Struening and M. Guttentag. Vol. 2. London: Sage.

Weiss, R. F. 1981. "The Temporal and Spatial Distribution of Tropospheric Nitrous Oxide." *Journal of Geophysical Research* 86: 7185–95.

Wennberg, P. O., et al. 1994. "Removal of Stratospheric O_3 by Radicals: In Eitu Measurements of OH, HO_2 NO, NO_2, ClO, BrO." *Science* 266: 398–404.

Wexler, Pamela. 1996. "Saying 'Yes' to No-Clean." In *Ozone Protection in the United States: Elements of Success*, Ed. Elizabeth Cook. Washington, DC: World Resources Institute.

White, Robert M. 1982. "Science, Politics and International Atmospheric and Oceanic Programs." 5th Donald L. McKernan Lecture in Marine Affairs, University of Washington. *Bulletin of the American Meteorological Society* 63, no. 8 (Aug.): 924–33.

Whitten, R. C., J. Cuzzi, W. J. Borucki, and J. H. Wofe. 1976. "Effect of Nearby Supernova Explosions on Atmospheric Ozone." *Nature* 203, no. 5576: 398–400.

Widhopf, G. F., L. Glatt, and R. F. Kramer. 1977. "Potential Ozone Column Increase Resulting from Subsonic and Supersonic Aircraft NOx Emissions." *AIAA Journal* 15: 1322–30.

Wilkniss, P. E., J. W. Swinnerton, R. A. LaMontagne, and D. J. Bressan. 1975. "Trichlorofluoromethane in the Troposphere, Distribution and Increase, 1971 to 1974." *Science* 187, no. 4179 (Mar. 7): 832–34.

Wine, P. H., et al. 1981. "Rate of Reaction of OH with HNO_3." *Journal of Geophysical Research* 86: 1105–10.

Winsemius, Peter. 1989. *Protecting the Global Atmosphere: Funding Mechanisms*. Interim Report to Steering Committee for Ministerial Conference on Atmospheric Pollution and Climate Change. Amsterdam: McKinsey.

Wirth, George F., Perry W. Brunner, and Ferial S. Bishop. 1982. "Regulatory Actions." In *Stratospheric Ozone and Man*. Ed. Frank A. Bower and Richard B. Ward. Vol. 2. Boca Raton, FL: CRC.

World Meteorological Organization. (WMO). 1979. *Report of World Climate Conference*. Geneva: WMO, Feb.

WMO. 1981. "Third WMO Statement on Modification of the Ozone Layer Due to Human Activities and Some Possible Geophysical Consequences." Nov. 25. Handwritten.

Wofsy, S. C., and M. B. McElroy. 1974. "HOx, NOx, and ClOx: Their Role in Atmospheric Photochemistry." *Canadian Journal of Chemistry* 52: 1582–91.

Wofsy, S. C., M. B. McElroy, and N. D. Sze. 1975. "Freon Consumption: Implications for Atmospheric Ozone." *Science* 187, no. 4176 (Feb. 14): 535–37.

Wofsy, Steven C., and Jennifer A. Logan. 1982. "Recent Developments in Stratospheric Photochemistry." In *Causes and Effects of Stratospheric Ozone Depletion: An Update*. Washington, DC: National Academy Press.

Wolf, Kathleen A. 1980. "Regulating Chlorofluorocarbon Emissions: Effects on Chemical Production." Prepared for U.S. EPA, N-1483-EPA. Santa Monica, CA: Rand, Aug.

Wood, Fred B. 1997. "Lessons in Technology Assessment: Methodology and Management at OTA." *Technological Forecasting & Social Change* 54, no. 2–3: 145–62.

Worrest, R. C., H. Van Dyke, and B. E. Thomson. 1976. "Impact of Enhanced Simulated Solar Ultraviolet Radiation upon a Marine Community." *Photochemistry and Photobiology* 27: 471–78.

Wuebbles, D. J. 1981. "The Relative Efficiency of a Number of Halocarbons for Destroying Stratospheric Ozone." LLNL UCID-18924. Livermore, CA: Lawrence Livermore National Laboratory.

Wuebbles, D. J., and P. S. Connell. 1984. "Interpreting the 1D Model Calculated Nonlinearities from Chlorocarbon Perturbations." Paper presented at International Workshop on Current Issues in Our Understanding of the Stratosphere and the Future of the Ozone Layer. Feldafing, Germany, June 11–16.

Wuebbles, Donald J. 1983. "Chlorocarbon Emission Scenarios: Potential Impact on Stratospheric Ozone." *Journal of Geophysical Research* 88, no. C2 (Feb. 20): 1433–43.

Wuebbles, Donald J., F. M. Luther, and J. E. Penner. 1983. "Effect of Coupled Anthropogenic Perturbations in Stratospheric Ozone." *Journal of Geophysical Research* 88, no. C2 (Feb.): 1444–56.

Wynne, Brian, et al. 2001. "Institutional Cultures and the Management of Global Environmental Risks in the United Kingdom." In *Learning to Manage Global Environmental Risks*. Ed. William C. Clark, Jill Jäger, Josee van Eijndhoven, and Nancy M. Dickson. Vol. 1. Cambridge, MA: MIT Press.

Young, Oran. 1989. "The Politics of International Regime Formation: Managing Natural Resources and the Environment." *International Organization* 43, no. 3 (Summer): 349–76.

Young, Oran R. 1994. *International Governance: Protecting the Environment in a Stateless Society*. Cornell Studies in Political Economy. Ithaca, NY: Cornell University Press.

Young, Oran R. 1999. *The Effectiveness of International Environmental Regimes*. Cambridge, MA: MIT Press.

Yung, Y. L., P. J. Pinto, R. T. Watson, and S. P. Sander. 1980. "Atmospheric Bromine and Ozone Perturbations in the Lower Stratosphere." *Journal of Atmospheric Sciences* 37: 339–53.

Zahniser, M. S., and Carlton J. Howard. 1978. "A Direct Measurement of the Temperature Dependence of the Rate Constant for the Reaction $HO_2 + O_3 \rightarrow 2O_2$." Paper presented to the WMO Symposium on the Geophysical Aspects and Consequences of Changes in the Composition of the Stratosphere. Toronto, June 26–30.

Zander, R. 1975. "Présence de HF dans la Stratosphère Supérieure." *C.R. Acad. Sc. Paris* B 281: 213–14.

Zellner, R. 1989. "Atmospheric Degradation Mechanisms of Hydrogen Containing Chlorofluorocarbons (HCFCs) and Fluorocarbons (HFCs)." In *Scientific Assessment of Stratospheric Ozone: 1989. Appendix, AFEAS Report*. Report of the Scientific Assessment Panel of the Montreal Protocol, WMO Global Research and Monitoring Project, rept. 20. Geneva: WMO, Jan.

Index